Notes on Numerical Fluid Mechanics and Multidisciplinary Design

Volume 127

About this Series

Notes on Numerical Fluid Mechanics and Multidisciplinary Design publishes state-of-art methods (including high performance methods) for numerical fluid mechanics, numerical simulation and multidisciplinary design optimization. The series includes proceedings of specialized conferences and workshops, as well as relevant project reports and monographs.

More information about this series at http://www.springer.com/series/4629

Rudibert King
Editor

Active Flow and Combustion Control 2014

Springer

Editor
Rudibert King
School of Process Sciences & Engineering
Institute of Process Engineering
Technische Universität Berlin
Berlin
Germany

ISSN 1612-2909 ISSN 1860-0824 (electronic)
ISBN 978-3-319-11966-3 ISBN 978-3-319-11967-0 (eBook)
DOI 10.1007/978-3-319-11967-0

Library of Congress Control Number: 2014949630

Springer Cham Heidelberg New York Dordrecht London

Printed on acid-free paper

Springer is part of Springer Science+Business Media (www.springer.com)

Preface

The field of active flow control (AFC) is of steadily increasing relevance in modern applications. As the demands on mobility and the efficiency of transport systems increase, new key technologies are necessary to meet these demands. It is believed that AFC offers possibilities for non-incremental increases in performance of many systems, thus forming an important part of future-oriented research.

Advanced courses for AFC are offered by many institutes, e.g., the American Institute of Aeronautics and Astronautics (AIAA), the European Research Community on Flow, Turbulence and Combustion (ERCOFTAC), the International Centre for Mechanical Sciences (CISM) or the von Karman Institute for Fluid Dynamics (VKI).

Books and even a dedicated journal, the INTERNATIONAL JOURNAL OF FLOW CONTROL, are published constantly.

The conference series ACTIVE FLOW CONTROL in Berlin is another part of this landscape of active flow control. After very successful conferences in 2006 and 2010, a continuation of the series was deemed appropriate. As with the start of the new Collaborative Research Center 1029 the focus of the Berlin AFC community has shifted towards active flow and combustion control in gas turbines, an addition of "Combustion" to the conference name was decided. Just as combustion control was part of the former AFC conferences, control of non-combustion processes is still part of the AFCC 2014.

As with the previous conferences in the series, the conference ACTIVE FLOW AND COMBUSTION CONTROL 2014 consisted only of invited lectures. To enable multidisciplinary discussion and collaboration between experimental, theoretical and numerical fluid dynamics, aerodynamics, turbomachinery, mathematics, control engineering, combustion, and computer science, parallel sessions were excluded.

The conference was organized by the aforementioned new Collaborative Research Center SUBSTANTIAL EFFICIENCY INCREASE IN GAS TURBINES THROUGH DIRECT USE OF COUPLED UNSTEADY COMBUSTION AND FLOW DYNAMICS, CRC 1029. This CRC 1029 inherits much of the experience and know-how of the organizers of the previous conferences AFC I and AFC II. At that time, work in Berlin concentrated around the CRC 557 CONTROL OF COMPLEX TURBULENT SHEAR FLOWS. Both CRCs were and are funded to a great part by the German Research Foundation (Deutsche Forschungsgemeinschaft, DFG). This financial support is gratefully acknowledged.

The focus of CRC 1029's research is the increase of the efficiency of a gas turbine by more than 10% by the exploitation and control of innovative combustion concepts and unsteady characteristics of a machine. The major contribution to an efficiency increase is expected from a thermodynamically motivated move from a constant-pressure to a constant-volume combustion. This will be done with the more classical pulsed detonation as well as with a new shockless explosion concept. As the pulsed combustion will give rise to severe consequences with respect to a stable operation of the compressor and turbine or a reliable cooling of the first stages of the turbine, new challenges for AFC arise. Additionally, the flow control methods developed in the CRC to handle these challenges should also offer an additional increase in efficiency even in a classical gas turbine.

While the increase in efficiency of a gas turbine is in itself a very promising application, it is far from beingthe only area of application for active flow control. Increased lift would allow for smaller, lighter engines. Also, steeper climbing and landing trajectories are enabled. The reduction of drag in flight conditions would directly reduce fuel consumption. Of course, the control of combustion is not only advantageous in aircraft engines. Gas turbines for energy generation, and of course also ground vehicle combustion engines may profit greatly from advanced control.

In this volume, besides international contributions, some of the first results of the newly founded CRC 1029 are documented. While some of this work is still in a relatively early stage, the obtained results show great promise for the future of the CRC.

All papers in this volume have been subjected to an international review process. We would like to express our sincere gratitude to all involved reviewers:

H. Pitsch, R. Klein, J. Moeck, J. Sesterhenn, V. Mehrmann, R. Meyer, P. Thamsen, W. Nitsche, O. Paschereit, D. Peitsch, K. Janscheck, S. Gugercin, D. Williams, S. Raghu, P. Benner, D. Thevenin, S. Henchandra, S. Leonov, S. Will, R. Joslin, D. Lignell, W. Schröder, E. Rogers, F. Haucke, D. Greenblatt, D. Henningson, A. Cavaliere, S. Grundmann, S. Heister, B. Lohmann, H. Schmidt, M. Juniper, R. Kielb, O. Semerao, J. Pfeiffer, K. Morris.

Finally, the members of CRC 1029 are indebted to their respective hosting organizations TU Berlin, FU Berlin, RWTH Aachen and DLR for the continuous support, and to Springer and the editor of the series NUMERICAL FLUID MECHANICS AND MULTIDISCIPLINARY DESIGN, W. Schröder, for handling this volume.

Last but not least we are indebted to N. Goldin and S. Stehr for the significant help in organizing and administrating CRC 1029, organizing the conference and compiling this volume.

Berlin
May 2014

Rudibert King
(Chairman of AFCC 2014 and CRC 1029)

Contents

Part I : Flow Control and Turbomachines

Active Boundary Layer Control with Fluidic Oscillators on Highly-Loaded Turbine Airfoils ... 3
Reinhard Niehuis, Marion Mack

Active Flow Control on a Highly Loaded Compressor Cascade with Non-steady Boundary Conditions ... 23
Marcel Staats, Wolfgang Nitsche, Inken Peltzer

Comparison of Iterative Learning and Repetitive Control Applied to a Compressor Stator Cascade ... 39
Simon J. Steinberg, Marcel Staats, Wolfgang Nitsche, Rudibert King

Feedforward Control of Lift Hysteresis during Periodic and Random Pitching Maneuvers ... 55
Xuanhong An, Lou Grimaud, David R. Williams

Dielectric Barrier Discharge Plasma Flow Control on a Vertical Axis Wind Turbine ... 71
David Greenblatt, Hanns Müller-Vahl, Ronen Lautman, Amos Ben-Harav, Ben Eshel

Mach-, Reynolds- and Sweep Effects on Active Flow Separation Control Effectivity on a 2-Element Airfoil Wing ... 87
Jochen Wild

Fundamentals in Coanda Flap Design ... 101
Rolf Radespiel, Marco Burnazzi

Heuristics for Effective Actuator and Sensor Placement in Feedback Flow Control ... 115
Kevin K. Chen, Clarence W. Rowley

An Interconnected Systems Approach to Flow Transition Control 131
Saulat S. Chughtai, Herbert Werner

Direct Numerical Simulation of Heat Transfer of a Round Subsonic Impinging Jet 147
Robert Wilke, Jörn Sesterhenn

Part II: Combustion

Low Temperature Gasoline Combustion – Potential, Challenges, Process Modeling and Control 163
Bastian Lehrheuer, Bastian Morcinkowski, Stefan Pischinger, Martin Nijs

Potential and Challenges of MILD Combustion Control for Gas Turbine Applications 181
Thivaharan Albin, Aline Aguiar da Franca, Emilien Varea, Stephan Kruse, Heinz Pitsch, Dirk Abel

Towards a Compressible Reactive Multiscale Approach Based on One-Dimensional Turbulence 197
Zoltan Jozefik, Alan R. Kerstein, Heiko Schmidt

Investigation of Equivalence Ratio Fluctuations on the Dynamics of Turbulent Lean Premixed Methane/Air Flames with a Linear-Eddy Model 213
Michael Oevermann, Christina Schrödinger, Christian O. Paschereit

Model-Based Detection of Misfirings in an Annular Burner Mockup 229
Sascha Wolff, Rudibert King

Part III: Constant Volume Combustion

Transient Mixing Enhancement of a Transverse Jet in Supersonic Cross Flow Using Pulse Detonation 247
Timothy Ombrello, Chung-Jen Tam, Willie Haw, Campbell Carter

An Experimental Study of Different Obstacle Types for Flame Acceleration and DDT 265
Joshua A.T. Gray, Christian O. Paschereit, Jonas P. Moeck

Investigation of Fluidic Devices for Mixing Enhancement for the Shockless Explosion Combustion Process 281
Bernhard C. Bobusch, Phillip Berndt, Christian O. Paschereit, Rupert Klein

Tailoring Fuels for a Shockless Explosion Combustor 299
Liming Cai, Heinz Pitsch

Unsteady Combustion Processes Controlled by Detailed Chemical Kinetics 317
Michael A. Liberman

Numerical Investigation of Reactive and Non-reactive Richtmyer-Meshkov Instabilities 343
Sergio Bengoechea, Lewin Stein, Julius Reiss, Jörn Sesterhenn

Part IV: Reduced Order Modeling

LQG-Balanced Truncation Low-Order Controller for Stabilization of Laminar Flows 365
Peter Benner, Jan Heiland

Model Reduction for DAEs with an Application to Flow Control 381
Jeffrey T. Borggaard, Serkan Gugercin

Model Reduction of Reactive Processes 397
Mathias Lemke, Agnieszka Międlar, Julius Reiss, Volker Mehrmann, Jörn Sesterhenn

Author Index 415

Part I
Flow Control and Turbomachines

Active Boundary Layer Control with Fluidic Oscillators on Highly-Loaded Turbine Airfoils

Reinhard Niehuis and Marion Mack

Universität der Bundeswehr München,
Institut für Strahlantriebe,
Werner-Heisenberg-Weg 39, 85577 Neubiberg, Germany
reinhard.niehuis@unibw.de

Abstract. The trend in jet engine development to reduce airfoil count is still ongoing in order to reduce weight and cost without sacrificing performance. In this context highly loaded airfoils have successfully been developed in the last decade. A better understanding of boundary layer transition and separation phenomena on turbine blades in combination with periodic incoming wakes was the key for this success. The Institute of Jet Propulsion of the Universität der Bundeswehr in Munich, Germany was heavily involved in the research effort from the very beginning. Since the flow gets prone to massive flow separation, associated with drastic increase of total pressure losses, further increased aerodynamic loading can only be achieved when applying passive or active methods to control the boundary layer on the blade surface. Reviewing shortly activities with passive devices this paper gives an overview of the research on active boundary layer control with fluidic oscillators on turbine blades.

Keywords: low pressure turbine, turbine aerodynamics, aerodynamic loading, loss characteristics, flow separation, boundary layer transition, boundary layer control, fluidic oscillator.

1 Introduction

Since the beginning of the jet engine age in general aviation continuous effort was spent in order to increase performance and to reduce specific fuel consumption for economic reasons. Due to the high demand air traffic is expected to continue to grow significantly in future and at the same rate as in the last decades (see e.g. Airbus [1]). This also means that carbon dioxide emission and the emission of other pollutants will increase by the same extent if more efficient equipment in aviation does not become available. Therefore, it is the challenge for industry to provide less pollutant and highly efficient ("greener") jet engines for ecological reasons which must fulfill current and future governmental regulations. It is the engine performance which is of particular importance in order to reduce fuel consumption and emissions but also in terms of the marketability of the entire aircraft. The development of more efficient jet engines is therefore characterized by continuously increasing bypass ratio and component efficiencies. In modern high bypass ratio engines, culminating in the upcoming geared

R. King (ed.), *Active Flow and Combustion Control 2014*, Notes on Numerical Fluid Mechanics and Multidisciplinary Design 127, DOI: 10.1007/978-3-319-11967-0_1

turbofan (see e.g. Rieger et al. [2]) the low pressure turbine (LPT) becomes a key component not only by its power and efficiency requirements but also decisive in order to meet the high efficiency demand of the entire propulsion systems. The increase in efficiency of the entire propulsion system can also be achieved by weight reduction of the engine, wherein the LPT is one of the major contributors. It is therefore an ongoing trend in LPT development to reduce the airfoil and stage count by increasing the aerodynamic loading of each airfoil row for weight and cost reasons. Pioneering work of Schulte and Hodson [3], Howell et al. [4], and Brunner et al. [5] was the starting point for the development of so-called high-lift bladings. Utilizing effects of unsteady wakes in multistage components the amount of blades needed for the duty of the airfoil rows could be reduced by 20% and even more while keeping the high efficiency level of conventional bladings. Those high-lift bladings are state-of-the-art in modern jet engines, as reported by Haselbach et al. [6] and Gier and Ardey [7].

By increasing the aerodynamic loading level beyond that of typical high-lift bladings the unsteady wake impingement on airfoils located downstream will not be able to prevent massive flow separation on the suction surface of turbine blades in the diffusing part of the airfoil. Open separation bubbles will occur resulting in much higher total pressure loss, which means that the efficiency of the LPT will drop to unacceptable low levels. Additionally, the flow turning cannot be maintained anymore such that the LPT fails to provide the required power to drive the low pressure compressor and the turbofan of the engine. This becomes particularly crucial at low Reynolds number applications, where flow separation is more likely to occur. In order to avoid massive flow separation, which is linked to dramatic loss in efficiency and performance, passive and/or active boundary layer control becomes necessary for blades exceeding the level of latest high-lift bladings.

After summarizing main objectives in the design of high performance turbine blades and reviewing shortly recent findings on passive boundary layer control in turbines, an overview of research activities performed at the Institute of Jet Propulsion of the Universität der Bundeswehr in Munich is given related to active boundary layer control. The development of suitable devices is outlined. Fluidic oscillators turned out to be particularly promising devices and it is demonstrated by testing of a turbine blade with very high aerodynamic loading at engine relevant flow conditions that high performance can be recovered by the developed fluidic oscillators.

2 Aerodynamic Performance of Low Pressure Turbine Airfoils

As outlined by Hourmouziadis [8] typical bladings of low pressure turbines in jet engines have to cover a wide Reynolds number range while maintaining highest efficiency. For take-off conditions the Reynolds number is highest and reaches its minimum Reynolds number at cruise conditions as depicted in fig. 1. It is especially the lowest Reynolds number which is the most critical case since best performance is required, but the flow gets particularly prone to flow separation on the suction surface of the airfoil. In order to achieve lowest losses it is beneficial to design airfoils which keep the boundary layer in the front part of the airfoil laminar by continuous acceleration and by moving the location of peak velocity as far as possible to the rear part of the airfoil (called rear loading). As was shown e.g. by Wakelam et al. [9] losses, however, will increase significantly due to severe flow separation (open separation without reattachment), if the peak velocity is located too far rearwards. The optimum location is such that the flow is able to withstand the deceleration in the rear part of the airfoil without massive flow separation.

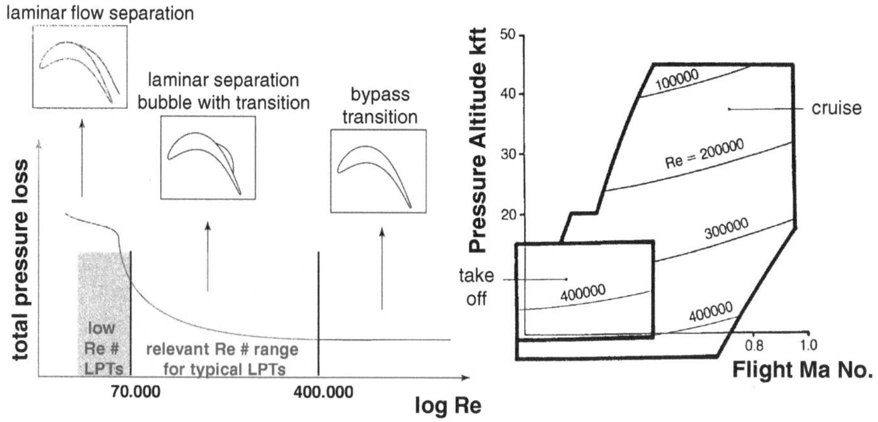

Fig. 1. Influence of Reynolds number on turbine aerodynamics (Hourmouziadis [8])

For typical applications in the Reynolds number range between 70,000 and 400,000, a transition from a laminar to a turbulent boundary layer state is present in the diffusing part, either by a bypass transition or a transition over a laminar separation bubble. It is known that a design accepting a moderate flow separation in this region in combination with transition from laminar to turbulent state results in lower total pressure losses, because this keeps the surface covered with turbulent boundary layers to a minimum. Increasing the Reynolds number for such a design leads to an earlier transition, which reduces the flow separation until no separation is present for a particular Reynolds number. For Reynolds numbers exceeding this level bypass transition is present, see fig. 1. In cases with very high aerodynamic loading and very high Reynolds numbers, a turbulent boundary layer separation is likely to occur. Also the very low Reynolds number range below 70,000, relevant for small jet engines and for cruise at very high altitudes, is especially challenging. Since no transition is likely to occur in the diffusing part open laminar flow separation is present if the aerodynamic loading becomes too high. This has to be avoided absolutely by a well-balanced airfoil design in order to maintain good LPT performance. Before this is elaborated further, first the experimental set-up and the instrumentation for the experimental studies are presented as well as relevant definitions will be given in the next chapter.

3 Experimental Set-Up and Instrumentation

3.1 High Speed Cascade Wind Tunnel

The high speed cascade wind tunnel (HGK) of the Institute of Jet Propulsion at the Universität der Bundeswehr München is a well-known test facility among experts in the turbomachinery world. It was first installed at DLR in Brunswick before it moved to Neubiberg in 1985, where it was heavily refurbished, rebuilt, and extended. Since then it is intensively used for cascade testing of compressor and turbine airfoils. The HGK is an open return wind tunnel that is enclosed within a large pressure tank as depicted in fig. 2. The wind tunnel itself consists of a six stage axial compressor, heat exchanger, settling chamber, nozzle, and the test section. The compressor is driven via gearbox and hydraulic coupling by a 1.3 MW electric motor, all located outside of the tunnel. More details of the wind tunnel can be found in Sturm and Fottner [10]. The air flow is continuously driven by the compressor through a cooler, which allows setting a constant temperature level, and a settling chamber before it is accelerated in the nozzle to the desired inflow condition in front of the cascade. The maximum cross section at the inlet of the test section is 0.5m x 0.3m. The air passes through the cascade, exits the test section, and recirculates back to the compressor inlet. The turbulence level of the flow can be adjusted by different turbulence grids. For the experimental results presented here the turbulence level was set to approximately 4%, which is a realistic level in typical LPTs.

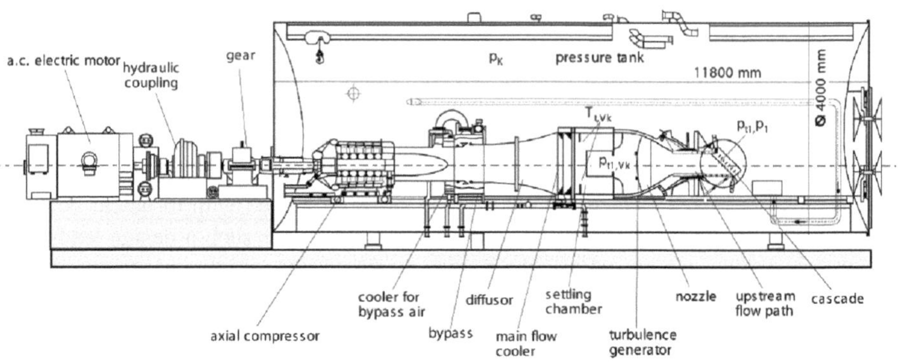

Fig. 2. Schematic of the High Speed Cascade Wind Tunnel

A particular feature of this test facility is that the pressure inside the tank can be adjusted between very low pressure of 3,000 Pa and ambient pressure by vacuum pumps. This allows to vary two important dimensionless similarity parameters independently, namely Mach and Reynolds number, both at engine relevant levels. The effect of periodic unsteadiness occurring within multistage turbomachines has been shown to be of significant importance to the overall aerodynamic performance, particularly in LPTs. It is therefore crucial to simulate and capture the physical effect of the wakes, also called rotor stator interaction, in order to obtain realistic loss levels.

For this purpose a wake generator can additionally be installed into the entire test set-up. The wake generator moves cylindrical bars of 2mm in diameter parallel to the inlet and exit plane in an endless way by two toothed belts. The pitch of the bars can be varied, typically between 40 and 80mm. Since the maximum bar velocity is limited to a maximum value of 40 m/s due to mechanical constraints the frequency of unsteady wake impingement is lower compared to real turbines. But as proven during the development of high lift airfoils it turned out that the dominant flow phenomena can be simulated with this device; see Acton and Fottner [11].

3.2 Turbine Cascades and Instrumentation

For the experimental investigations presented here two different turbine profiles are used, called T160 and T161. Both airfoils were designed by MTU Aero Engines, Munich. The profile T160 is derived from a low pressure turbine profile of a civil high bypass jet engine and is featuring a state-of-the-art airfoil with high flow turning of about 110 °. It is designed for an exit Mach number of 0.6 and a Reynolds number based on the true chord length of 200,000. The T161 is designed for the same operating point and the same velocity triangle. Compared to the T160 profile the aerodynamic loading of the profile T161 has been increased very significantly (Zweifel coefficient +25% compared to T160) by increasing the pitch to chord ratio, see Gier et al. [12]. A typical rear loading design was used in both cases. The T160 and T161 cascades consist of 7 blades in linear arrangements. With a chord length of about 0.07m and a blade height of about 0.2m at the outlet the aspect ratio is high enough to generate a two dimensional flow at midspan of the blades. The T160 as well as the T161 cascade have divergent side walls with a span ratio of 1 : 1.15 from inlet to outlet.

Both cascades are equipped with static pressure taps (inner diameter 0.6mm) along the suction and pressure surface at midspan on the blades adjacent to the central blade in the cascade arrangement. Since the suction side is of particular importance twenty pressure taps are put on this side, while the pressure side contains only six pressure taps. The pressure level is measured against the pressure inside the tank with a PSI 98RK system with 25 hPa or 345 hPa pressure transducers depending on the operating condition. The pressure distribution on the blade is presented here in terms of dimensionless pressure coefficient, defined as follows:

$$c_p = \frac{p_x - p_k}{p_{t1} - p_k} \,. \tag{1}$$

With an uncertainty of 0.05% of the full scale range, the pressure transducers allow to measure the pressure coefficient c_p with a maximum relative uncertainty of 0.2%. The loss characteristics of the cascades is determined by wake traverses with a calibrated five hole probe (head diameter 2.6 mm) in a plane 40% axial chord length downstream of the trailing edge of the central blade. In order to avoid any disturbances and to allow reliable loss measurements the central blade does not exhibit any pressure taps or other instrumentation. Losses are presented here by the dimensionless loss coefficient

$$\zeta = \frac{p_{t1}-p_{t2}}{p_{t1}-p_k} \quad . \tag{2}$$

Integral mixed out loss coefficients $\bar{\zeta}$ were determined applying the method of Amecke [13]. Using linear error propagating analysis the maximum uncertainty of the integral loss coefficient is calculated between 1% and 5%, where the highest uncertainty is present for the lowest Reynolds number. In the following the loss coefficients will be related to an arbitrarily chosen common reference value $\bar{\zeta}_{ref}$ in order to respect the proprietary nature of the data. In the studies presented here, also cases with boundary layer control by air blowing from holes in the blade surface are considered. For fair assessment and comparison with cases without air injection the addition of energy to the flow has to be accounted for when computing the loss coefficient of the cascade. Here, the method used by Ardey [14] is applied, which uses a corrected inlet total pressure and replaces p_{t1} in eq. (2), if blowing is activated

$$p_{t1}^{**} = \left(\frac{\dot{m}_1\, c_{p1}\, T_{t1}\, (p_{t1})^{\frac{1-\gamma}{\gamma}} + \dot{m}_i\, c_{pi}\, T_{ti}\, (p_{ti})^{\frac{1-\gamma}{\gamma}}}{\dot{m}_1\, c_{p1}\, T_{t1} + \dot{m}_i\, c_{pi}\, T_{ti}}\right)^{\frac{\gamma}{1-\gamma}} . \tag{3}$$

The injected mass flow, its total pressure and total temperature need to be measured in order to evaluate eq. (3). The total pressure of the secondary flow p_{ti} is measured in the plenum inside the blade. Therefore, the loss coefficient built with p_{t1}^{**} includes losses inside the blowing holes. In the context of blowing another quantity is introduced, named pressure ratio

$$R = \frac{p_{tpl}}{p_x} , \tag{4}$$

which relates the total pressure in the plenum to the static pressure at the location of the blowing hole. It is the driving force of the injected mass flow. The operating point of the cascade is defined by the inlet total pressure and inlet total temperature of the cascade and the tank pressure p_k. These quantities allow to determine the theoretical exit Mach and Reynolds number, assuming an isentropic adiabatic flow through the cascade, see e.g. Wakelam et al. [9]. The temperature is measured by four PT100 resistance temperature detectors and the pressures by a PSI 9116 scanner from pressure systems. In the studies presented here, the Mach number is kept constant, while the Reynolds number is varied over a large range between 50,000 and 400,000.

4 Discussion of Results

4.1 Influence of Blade Loading on LPT Performance

From the discussion in chap. 2, it becomes obvious that the flow in the diffusing part of an LPT airfoil is strongly affected by the aerodynamic loading, often expressed by the Zweifel coefficient, see e.g. Gier et al. [12]. The loading increases for instance by reducing the airfoil count while keeping the chord length unchanged in one blade row. Although the reduction of blade count is beneficial and attractive in terms of weight and cost reduction it can reduce component efficiency drastically due to higher total pressure losses generated especially in the diffusing part of the airfoils, see e.g. Gier et al. [12] and Praisner et al. [15]. The current state-of-the-art in turbine design are high lift blades, mentioned before, which utilize unsteady flow phenomena generated by the wakes of upstream blade rows in order to control boundary layer development on the airfoils. With proper designs the aerodynamic loading can be increased very significantly while keeping the efficiency unchanged at high levels. A reduction of airfoil count by 20% and more can be obtained compared to conventional airfoil designs. With this achievement, a further increase in aerodynamic loading and airfoil count reduction respectively can only be realized using passive or active boundary layer control, as is illustrated in the next figure.

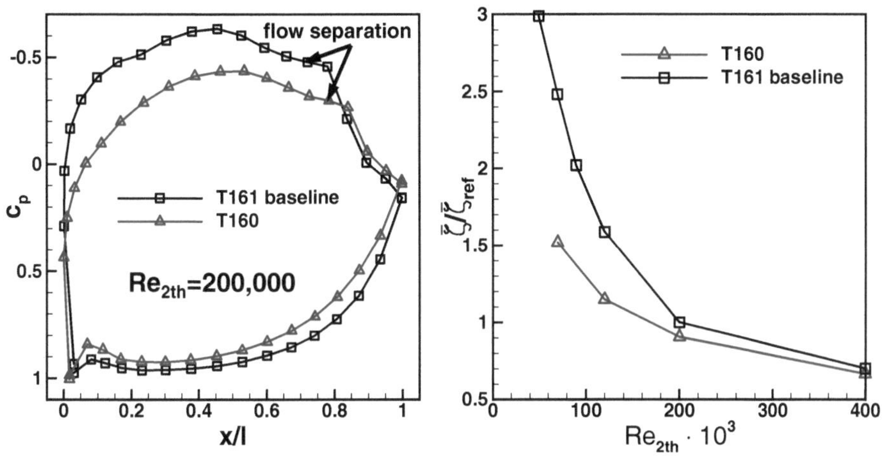

Fig. 3. left: c_p distribution of profiles T160 and T161 over dimensionless chord length x/l; right: total pressure loss over Reynolds number (Ma_{2th}=0.60, Tu_1=4%, Martinstetter et al. [16])

In fig. 3 (left) it is presented a comparison of the measured dimensionless pressure distribution at midspan of the two different turbine airfoils at design conditions and with steady inflow at Re_{2th}=200,000. In both cases a flow separation on the suction surface becomes visible as indicated. For the $Re_{2th} \leq$ 70,000 even an open flow separation was observed for T161 which is not present for T160, see Martinstetter et al. [16]. Looking at the total pressure losses expressed by the integral loss coefficient $\bar{\zeta}$ in fig. 3 (right), it can clearly be seen that the losses of T161 are significantly higher compared to T160, with an increasing difference with decreasing Reynolds number. For the lowest Reynolds number of 70,000 the loss is about 65% higher for T161 compared to T160, which is a dramatic increase and unacceptable for real turbine applications. It also becomes obvious that T161 becomes a very useful test candidate for efficiency recovery applying boundary layer control (blc) measures.

4.2 Passive and Active Boundary Layer Control on LPT Airfoils

From open literature it is known that passive blc methods can considerably reduce flow separation on LPT airfoils, if designed properly, see e.g. Martinstetter et al. [16] and Himmel et al. [17]. In fig. 4 the successful turbulator design T.ISA is shown, which has been developed at the authors' institute. For more details it is referred to [16]. The challenge in turbulator design is always to generate disturbances which are large enough to trigger flow transition from laminar to turbulent on the suction surface of the blade in order to prevent massive flow separation and loss increase in the low Reynolds number range. The disturbances, however, should be as small as possible to prevent substantial increase of total pressure losses at the high Reynolds number range, where the boundary layers are stable enough to withstand the deceleration in the rear part of the airfoil typically without any flow separation.

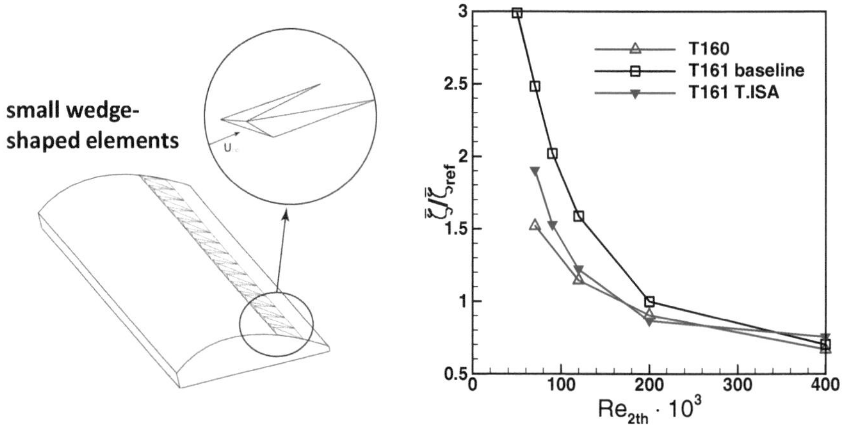

Fig. 4. Influence of 3d turbulators (T.ISA) on loss behavior of T161 in comparison to T160

Well-balanced designs are required to obtain best performance. As can be seen in fig. 4, with the T.ISA design the loss of T161 gets even lower at the design Reynolds number compared to the conventional design of T160 and is substantially lower for Re_{2th}<200,000 compared to the T161 without blc. However, the low loss level of T160 is not achieved. It also becomes obvious that the loss performance at higher Reynolds numbers is worse compared to T160 and is even higher compared to T161 without any turbulators. The loss increase in this Reynolds number regime is due to the fact that the flow in the rear part of the smooth T161 does not show any significant flow separation anymore and the flow disturbances introduced by the turbulator trigger an earlier transition of the boundary layer such that the amount of blade surface covered with turbulent flow increases. From this it can be concluded that it would be highly desirable to develop an active device which triggers transition in the low Reynolds number regime and can be switched off when not needed anymore.

Therefore, research at the authors' institute was performed looking at active blc by continuous blowing. For this purpose the T161 was equipped with a plenum inside the blade, fed by pressurized air. An array of equidistant holes along the blade height was integrated on the suction surface, each of them connected to the internal plenum, see fig. 5 (left). Extensive parameter studies have been performed considering the location of the holes on the suction surface, hole diameter, hole distance as well as different pressure ratios, as reported by Ludewig et al. [18]. Two configurations under consideration are depicted in fig. 5.

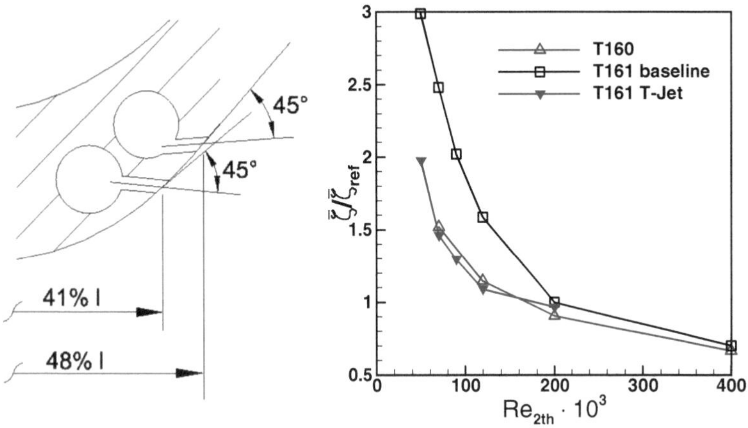

Fig. 5. Influence of continuous blowing on loss behavior of T161 in comparison to T160

In the experimental studies the blowing rate was tuned to provide the lowest possible total pressure loss at midspan. It turned out that minimum loss does not necessarily coincide with the minimum separation length on the suction surface. In fig. 5 (right) it is shown the loss behavior of T161 equipped with the blowing configuration and optimum blowing ratios in comparison to the uncontrolled T161 and T160. As can be seen the optimized blowing configuration is very successful and able to fully recover the loss deficit of the higher loaded T161. With activated blowing the total pressure loss (computed with eqs. (2) and (3)) is even slightly lower compared to the conventional design of T160 for $Re_{2th} \leq 120{,}000$. For $Re_{2th} = 200{,}000$ it becomes less effective and should be switched off for the higher Reynolds number range. Although the continuous blowing is successful in terms of loss reduction, fairly high mass flow rates are required to obtain best performance. Since the mass flow has to be provided by the compressor in a jet engine this means also negative impact on performance/efficiency of the entire system as well as a more complex system and a weight penalty. This is the reason why it is strived for blc methods which require no or less mass flow.

In an ongoing research effort it was therefore looked for innovative approaches to actively control the boundary layer on the suction surface of highly loaded turbine blades. From the development of boundary layers on flat plates it is known that after a certain surface length the originally laminar flow becomes turbulent. The transition process from laminar to turbulent state is initialized by so-called Tollmien-Schlichting waves; see e.g. Schlichting and Gersten [19]. Tollmien-Schlichting waves are first order instabilities which can be predicted by linear stability theory and which are receptive to particular frequencies. According to Wazzan et al. [20] three characteristic frequencies can be identified for each velocity profile. For incompressible flow they provide spatial stability charts which allow estimating these frequencies. Assuming that similar phenomena are also present in the boundary layer on turbine blades the excitation of Tollmien-Schlichting wave appears to be a promising approach to trigger transition. As outlined by Mack et al. [21] the required excitation frequencies have been estimated for the T161 profile at realistic operating conditions using next to stability charts [20] a Navier-Stokes flow solver and a boundary layer code. Depending on the Reynolds number it turned out that very high frequencies are required to promote Tollmien-Schlichting instabilities. For Re_{2th}=200,000 frequencies of about 9 to 15 kHz are requested. For lower Reynolds numbers the requested frequencies are much higher and add up to 25 to 37 kHz for Re_{2th}=50,000. This means that it becomes very challenging to develop blc devices providing such high frequencies and which are small enough to fit into a turbine blade and to operate at different pressure levels reliably in a turbomachinery environment.

Different potential candidates have been considered and evaluated by the authors. From the beginning it was clear that speakers are too big to be installed into a turbine blade. Also fast opening valves are too slow for the requirements mentioned above. Piezo actuators are known to provide the frequencies stated above, however, it turned out that the energy consumption is too high and the actuators installed as they would be in a turbine blade failed due to overheat already after a few seconds. Plasma actuators are expected to deliver very high frequencies, but were not considered so far, since they require high voltages, a risky and unfavorable issue for applications in turbines or jet engines. The most promising approach turned out to be the application of fluidic oscillators.

4.3 Development of Fluidic Oscillators for Turbine Applications

In open literature it is reported on various kinds of fluidic oscillators in different applications, see e.g. [22, 23, 24, 25, 26]. The underlying working principle is similar in all cases. Fluidic feedback oscillators generate an oscillating mass flow. As shown in fig. 6 fluid, accelerated in a nozzle, is entering a chamber with divergent walls. Due to instabilities and the Coanda effect the jet attaches arbitrarily to one of the walls and is entering primarily one of the feedback channels separated by a splitter. As a consequence, the pressure is rising inside this feedback channel, which pushes the jet flow in the next moment in time towards the other wall of the chamber. Now it is this feedback channel where the pressure increases pushing back the jet to its original direction and the same cycle starts again. If designed properly a periodic flow oscillation can be created with such a device.

Actually, it was the work published by Ries et al. [26] which gave first ideas of a feasible fluidic oscillator design for the authors and which had the potential providing high frequencies and fulfilling requirements in terms of installation space and durability. Like in all other designs known from literature the flow is leaving the device in the plane of the oscillator. In order to ease the embedding of fluidic oscillators in a turbine blade, the layout was changed such that the flow is leaving the fluidic oscillator through cylindrical holes perpendicular to the oscillator plane. In fact it turned out to be the first design of such a kind and extensive testing and development of new oscillator designs was required providing frequencies needed for the excitation of Tollmien-Schlichting waves on turbine blades. Many geometric parameters and also the location of the outlet holes were systematically varied with changing pressure ratio, but initially without any cross flow at the outlets of the oscillator. Some of these results were published by the authors in [21].

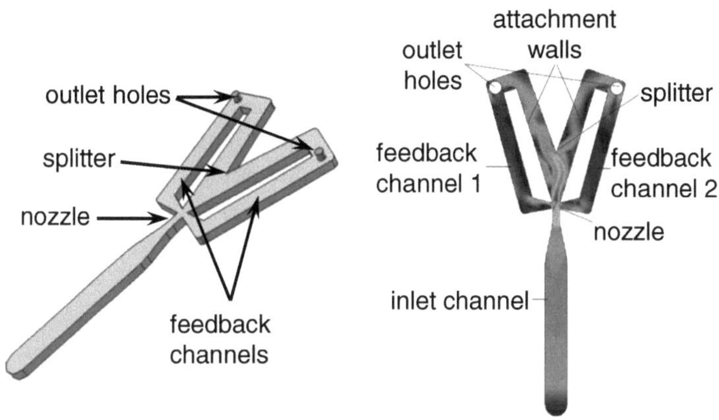

Fig. 6. Working principle and inner structure of a fluidic oscillator (Mack et al. [21])

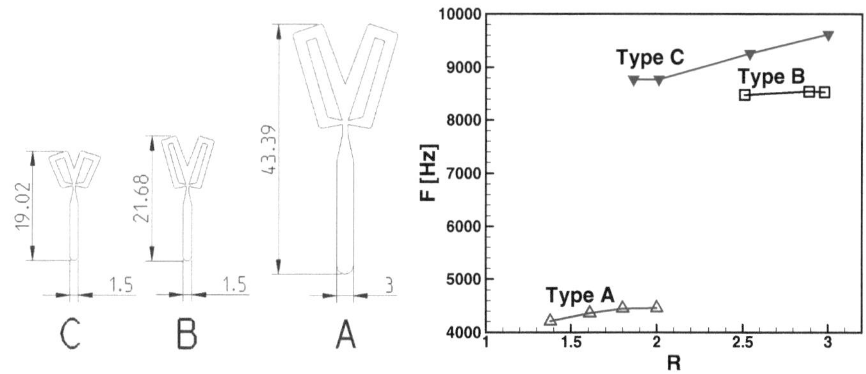

Fig. 7. Oscillator frequency F vs. pressure ratio R for 3 different geometries (Mack et al. [21]) (dimensions in mm)

As can be seen in fig. 7 the achieved oscillator frequency (measured with a single hot wire probe) is strongly dependent on the size of the fluidic oscillator. As expected the smallest size provides highest frequencies because it exhibits the shortest nozzle to splitter distance and the shortest feedback channels. In all cases a certain pressure ratio R is needed to initiate the oscillation. For the cases presented in fig. 8 the frequency is increasing with increasing pressure ratio. However, configurations have also been observed where frequency reaches a maximum at a certain pressure ratio and is decreasing again with further increasing pressure ratio, an effect which appears to be indicated for type A and B in fig. 7. A main finding is that the length over diameter ratio of the outlet hole has a strong influence on the obtained frequencies. As reported in Mack et al. [21] highest frequencies were obtained for a length over diameter ratio of about 5 for type C. The location of the hole along the flow channels inside the fluidic oscillator turned out to be of minor importance. Depending on its location different pressure ratios are required to initialize the oscillation. It needs to be mentioned that recent CFD solvers are really challenged with such devices and especially URANS solvers fail to predict the measured frequencies. More sophisticated turbulence models or LES methods might be future options to better understand the working principles of such small fluidic oscillators and to design better configurations. So far, the design approach has to rely on extensive experimental test results. Although not all flow phenomena inside the fluidic oscillator are fully understood, however, at least one promising configuration was found by the extensive experimental parameter study which provides frequencies in the order of magnitude of 10 kHz. This is still at the lower end of frequencies required to trigger transition on turbine blades as outlined above. Nevertheless it was decided to equip the T161 profile with an array of the most promising actuator found so far.

4.4 Active Boundary Layer Control with Fluid Oscillators on an LPT Airfoil

The fluidic oscillator of type C as described above was installed inside the three center blades of the T161 cascade. As illustrated in fig. 8 a linear array of nine equidistantly distributed oscillators was manufactured in the center portion of the blades. Each of the oscillators is connected to a common plenum which is fed with air from both sides of the blade. The pressure level is measured inside the plenum and it can be adjusted with a valve. Since each oscillator is equipped with two blowing holes, one on either side of the feedback channels, the blade contains 18 blowing holes on the suction surface covering a major part of the blade in spanwise direction as shown in fig. 8.

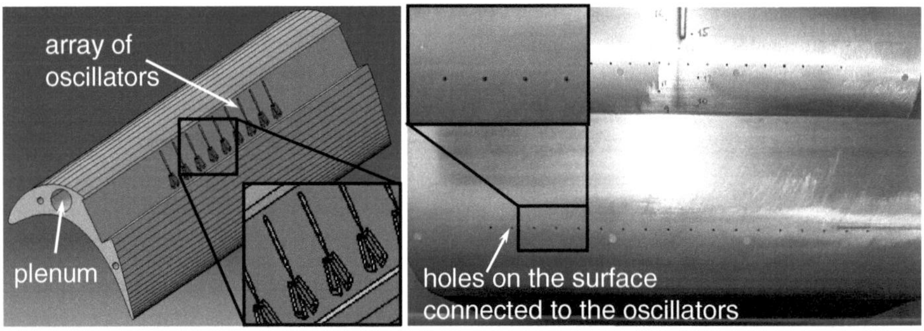

Fig. 8. Overview of turbine blade T161 equipped with fluidic oscillators (left: cut-away, right: view on the suction surface of two neighboring blades)

Keeping the fluidic oscillator shape unchanged three different blowing hole arrangements have been investigated so far. Layout I and II use cylindrical blowing holes perpendicular to the blade surface with a diameter of 1 mm and a length over diameter ratio of 5. In layout I the blowing holes are located close to the separation line at 60% chord length (cf. fig. 3) with a pitch to diameter ratio of 7.5. Layout II is placed further upstream at 51% chord length. Since the same oscillators are used and only the position of the outlet holes relative to the feedback channels is changed the distance between the holes changes too. With layout II the pitch to diameter ratio alters between 5 and 10. Layout III is at the same location as layout I but uses holes which are inclined to the blade tangential surface with an angle of 45°. In order to keep the length to diameter ratio the same as for layout I and II the diameter of the holes was increased by 40%.

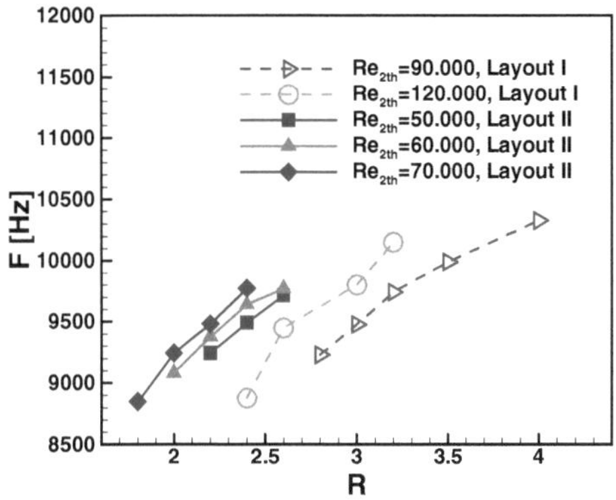

Fig. 9. Oscillation frequency of layout I and II vs. pressure ratio R

First tests with layout I and II have revealed that the oscillating frequency with cascade flow is in the same order of magnitude as expected from the basic tests without cross flow as described above. As is shown in fig. 9 the obtained oscillation frequency depends significantly on the pressure ratio. A certain minimum pressure ratio is needed to start the oscillation and the frequency is rising with increasing pressure ratio. Since the two layouts differ in their location with respect to the fluidic oscillator and also slightly in length to diameter ratio layout I requires higher pressure ratios to achieve the same frequency level as layout II. Comparing the characteristics of both layouts it appears that lower Reynolds numbers require higher pressure ratios. But this has to be substantiated further since a general trend cannot be derived from only two configurations.

By detailed experimental investigations in the high speed cascade wind tunnel at engine relevant Mach and Reynolds number levels it could be shown that the suction side boundary layer can successfully be controlled by fluidic oscillators, which will be further discussed with the following figures. In fig. 10 it is shown that the flow separation in the rear part of the suction surface is significantly reduced by activated blowing of layout I at Re_{2th}=50,000 resulting in lower total pressure losses as can clearly be recognized by the lower loss peak and the smaller wake width in the loss distribution over one blade pitch. Note that here and in the following figures the loss distribution is shifted in pitchwise direction such that the loss peak of the wake is always located at midpitch for ease of comparison and assessment. Increasing the pressure ratio, meaning higher frequencies according to fig. 9, reduces flow separation and losses even further until an optimum pressure ratio is reached. Higher pressure ratios then lead to higher losses again.

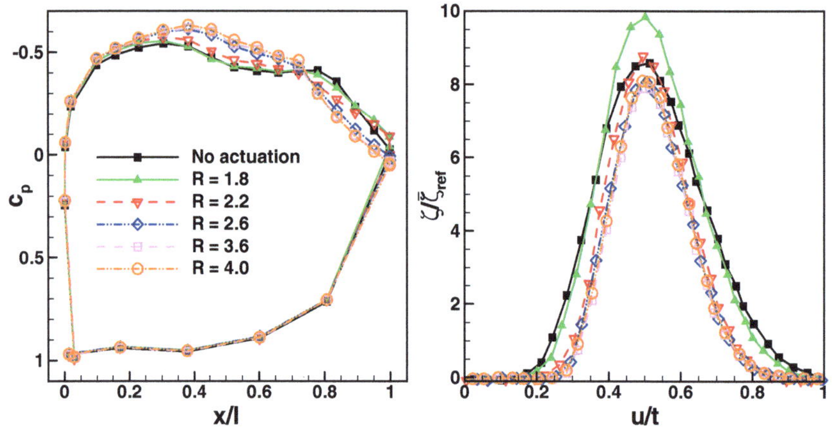

Fig. 10. c_p and loss distribution of T161 at various pressure ratios R (Re_{2th}=50,000, layout I)

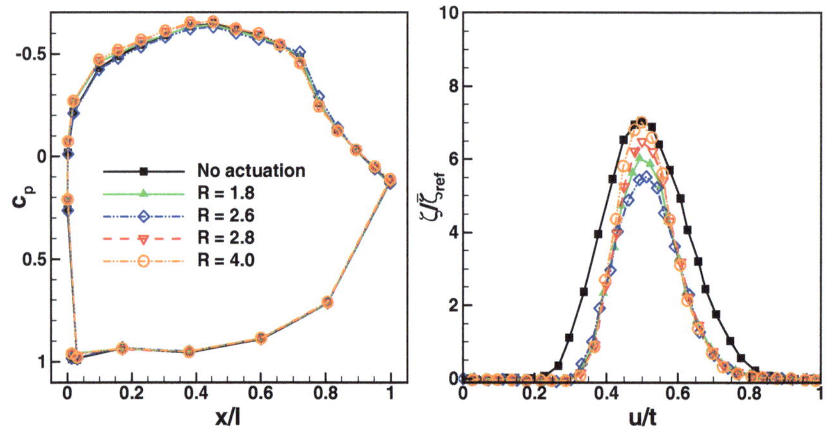

Fig. 11. c_p and loss distribution of T161 at various pressure ratios R (Re_{2th}=90,000, layout I)

As can be seen in fig. 11 the influence on the pressure distribution diminishes at Re_{2th}=90,000, but the influence on the loss distribution is still very significant. In fig. 12 the integral total pressure loss coefficients are plotted as function of pressure ratio R for different Reynolds numbers. Note that for each Reynolds number the integral loss is related to its respective value when blowing is deactivated. It can be seen that a significant loss reduction is perceived up to Re_{2th}=120,000, although the differences in static pressure distribution (not shown here) diminish even further compared to Re_{2th}=90,000. It can clearly be seen that after reaching a certain threshold pressure ratio a substantial loss reduction of more than 30% and even up to 45% can be obtained depending on the Reynolds number level. For $Re_{2th}\geq$200,000 no benefit is obtained anymore with layout I. The disturbances introduced by the activated fluidic oscillator obviously result here in a larger portion of the blade covered with turbulent boundary layers. As already discussed above an optimum pressure ratio can be recognized for each Reynolds number resulting in minimum losses. The losses increase again moderately when the pressure ratio or oscillation frequency respectively are higher than the optimum value. The highest relative loss reduction at all (-45%) is achieved for Re_{2th}=70,000, cf. fig. 12. For the lowest Reynolds number (Re_{2th}=50,000) the loss reduction is not as much, suggesting that the oscillation frequency might be off the optimum. According to theory, lower Reynolds numbers require very high oscillation frequencies which may not be achieved with layout I.

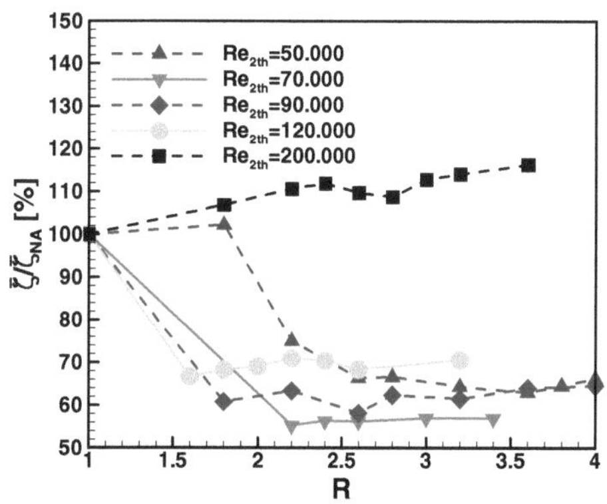

Fig. 12. Relative loss reduction by layout I vs. pressure ratio R at different Re number levels

Another interesting aspect is worth noting. As explained above, according to theory the required oscillation frequencies for triggering Tollmien-Schlichting waves were estimated to be as high as 25 kHz up to 37 kHz at Re_{2th}=50,000 for the T161 cascade. The measurements with layout I, however, were performed with considerably lower frequencies (about 10 kHz). Nevertheless, a very significant reduction in loss was found. From this it may be concluded that either the 2d theory is not fully applicable or more probably the transition process on the turbine blade can be initialized also at lower frequencies, meaning that the excitation of each and every Tollmien-Schlichting wave is not necessary. The excitation of even multiples of the waves is likely to be sufficient in order to force and enhance the transition process. But this has to be substantiated by further experiments. The next figures allow an assessment of the three different layouts.

While in fig. 13 and 14 it is depicted the pressure distribution on the blade and the loss distribution in the wake at Re_{2th}=50,000 and Re_{2th}=90,000 respectively, the achieved reduction of integral losses is presented in fig. 15 (left). In all cases the results of optimum pressure ratio are depicted. As can be seen it is layout II which results in the highest loss reduction at Re_{2th}=50,000. This is due to the strongest reduction of flow separation on the suction surface and the lowest loss peak in the wake obtained with layout II as revealed in figs. 13 and 14.

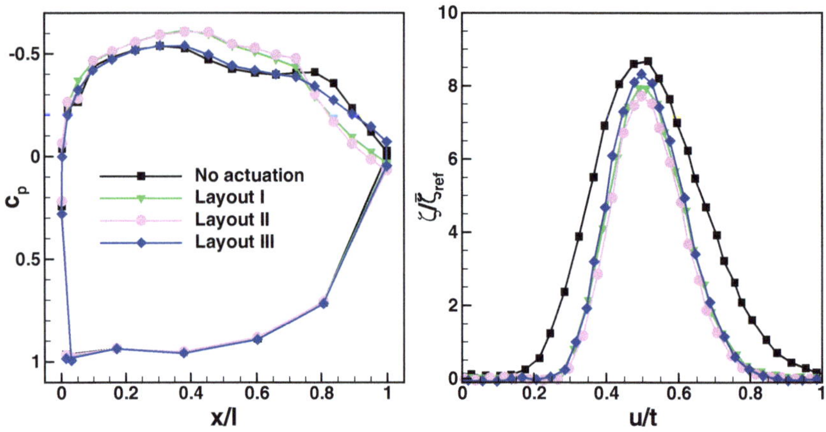

Fig. 13. c_p and loss distribution of T161, comparison of three layouts (Re_{2th}=50,000)

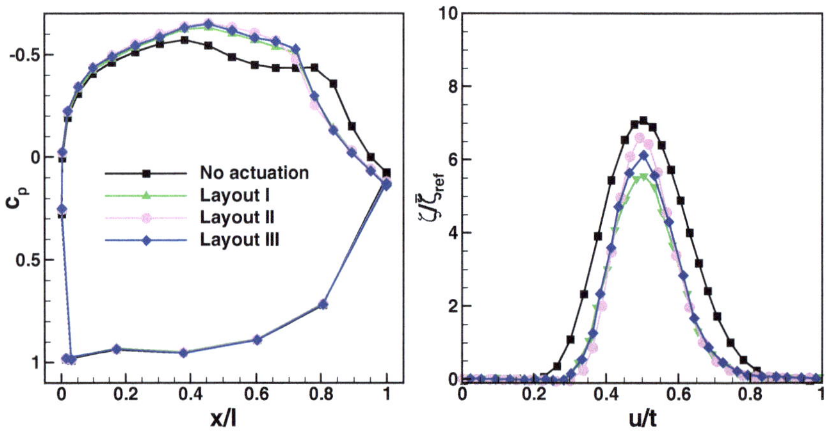

Fig. 14. c_p and loss distribution of T161, comparison of three layouts (Re_{2th}=90,000)

At the higher Reynolds number of 90,000 (and higher Reynolds numbers) it is layout I which provides lowest integral losses. From this it can be concluded that for the lower Reynolds number an earlier triggering of transition is necessary in order to control the boundary layer in the diffusing part of the blade appropriately. The more downstream location of layout I is obviously not able to provide early enough sufficient disturbances to the flow such that the extension of flow separation remains larger compared to layout II (fig. 13 left), associated with somewhat higher integral losses. The boundary layer control with both layouts turns out to be very successful as it reduces the total pressure losses of the T161 profile even below the level of the conventionally loaded profile T160 in the low Reynolds number regime. Another interesting result is that the overall performance of layout III is slightly worse compared to the other two layouts featuring blowing perpendicular to the blade surface. For the two investigated cases the inclined blowing is obviously not able to trigger transition as much as the other two designs. This is an indication that not only a specific oscillation frequency is needed to trigger transition but also a certain strength of the disturbance such that blowing normal to the surface is more effective compared to the inclined blowing. This is an observation which is also worth to be investigated further.

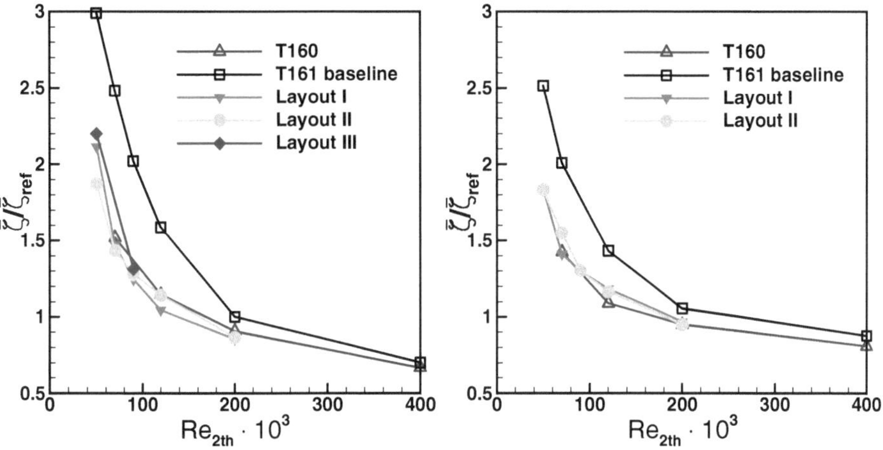

Fig. 15. Loss reduction potential by pulsed blowing (left: steady inflow, right: unsteady inflow)

The results presented above were obtained indeed at realistic engine relevant Mach and Reynolds numbers, but at steady inflow conditions (without wakes). As explained above the flow in multistage turbomachinery is inherently unsteady due to the wake generation from the upstream rows of airfoils. These effects can be simulated with the wake generator in the high speed cascade wind tunnel, as described before. Due to the fact that the periodically incoming wakes raise temporally the turbulence level inside the blade passage it is well-known that earlier transition is promoted at unsteady inflow conditions, which reduces the size of separated flow region on the blade. In case of T161 it was found that the open flow separation at Re_{2th}=50,000 and steady inflow conditions is not present anymore and reattaches well before the trailing edge at unsteady inflow conditions. From this it must be concluded that the positive effect of pulsed blowing by the fluidic oscillators will be less pronounced. As it is shown in fig. 15 (right) this is evidenced by the integral loss coefficient. Still, a significant loss reduction is achieved with activated blowing by layout I and II (layout III not tested so far), but the amount of loss reduction is less compared to the steady inflow condition. For reasons of space, not all aspects can be discussed here and the interested reader is referred to recent publications of the authors [27, 28].

5 Conclusions and Outlook

In this paper it is discussed that the increase of aerodynamic loading of turbine blades beyond the level of existing turbine designs is beneficial in terms of weight and cost reduction. However, this results in dramatic increase of total pressure loss due to massive flow separation, particularly in the low Reynolds number range relevant at cruise conditions of jet engines. Passive blc methods have shown to be successful in the low Reynolds number range, but they generate higher losses at higher Reynolds numbers. This deficit can only be avoided by applying active blc methods, which can be switched-off when not needed anymore at higher Reynolds numbers. Here a successful new approach is presented using fluidic oscillators, which allow pulsed blowing at very high frequencies. The intention of high frequency pulsed blowing is to force a boundary layer transition to turbulent state by excitating Tollmien-Schlichting waves on the suction surface in order to avoid massive flow separation in the diffusing part of the blade at very high aerodynamic blade loading.

With carefully designed fluidic oscillators pulsed blowing frequencies in the order of magnitude of 10 kHz were achieved resulting in an impressive loss reduction for a very high loaded turbine airfoil at engine relevant operating conditions in terms of Mach and Reynolds number as well as realistic unsteady flow conditions. For both steady inflow (no wakes) and unsteady inflow (with wakes) a full performance recovery was achieved by the actuators pushing the total pressure loss down to the level of conventionally loaded turbine airfoils. The devices are small enough to be fully integrated into the blade and offer the benefit, that no mechanical actuators or moving parts are needed, which makes them very reliable and particularly attractive for application in turbines. Also, in comparison to a continuous blowing much less air mass flow (reduction greater than 50%) is needed for blowing to achieve a greater reduction of total pressure losses, particularly at low Reynolds numbers.

An interesting result of the experimental investigations is that the transition could be triggered even at frequencies lower than the frequency estimated by known spatial stability charts. Inclined blowing was found to be less effective compared to blowing normal to the surface. More research into this direction is needed in order to further improve the understanding of the underlying physical phenomena and to optimize the pulsed blowing devices accordingly. Also the understanding of the complex unsteady flow phenomena inside the fluidic oscillators has to be improved in order to develop design capabilities for different applications in terms of optimum frequencies and amplitudes.

Acknowledgements. Part of the research results presented here received funding from the European Union 7th Framework Programme (FP7/2007-2013) under grant agreement no. 211861, DREAM and by the German Federal Ministry of Economics and Technology within the framework of "Lufo III" national research program, which is acknowledged here. The authors thank MTU Aero Engines for the permission to use the turbine cascade T161 for research purposes. The authors also thank Mrs. Julia Kurz for her extraordinary support in editing the graphics for this paper.

Nomenclature

c_p	dimensionless pressure coefficient, eq. (1) and specific heat at constant pressure
F	frequency
p	(static) pressure
l	chord length
$\dot{m}$	mass flow
Ma	Mach number
R	pressure ratio, eq. (4)
Re	Reynolds number
t	pitch
T	total temperature
Tu	turbulence intensity
u	pitchwise direction
x	coordinate in axial direction
γ	specific heat ratio
ζ	loss coefficient, eq. (2)
$\bar{\zeta}$	integral loss coefficient

Subcripts

1	plane upstream of cascade
2	plane downstream of cascade
k	pressure tank value
NA	no activation
pl	actuator plenum value
ref	reference condition
t	total or stagnation condition
th	theoretical value, based on no loss in cascade

Abbreviations

blc	boundary layer control
LPT	Low Pressure Turbine
HGK	high speed cascade wind tunnel

References

1. Airbus, Global Market Forecast "Future Journeys 2013-2014" (2013), `http://www.airbus.com/company/market/gmf2013`
2. Riegler, C., Bichlmaier, C.: The Geared Turbofan Technology - Opportunities, Challenges and Readiness Status. In: 1st CEAS European Air and Space Conference CEAS-2007-054 (2007)
3. Schulte, V., Hodson, H.P.: Unsteady Wake-Induced Boundary Layer Transition in High Lift LP Turbines. ASME J. Turbomach. 120, 28–35 (1998)
4. Howell, R.J., Hodson, H.P., Schulte, V., Stieger, R.D., Schiffer, H.-P., Haselbach, F., Harvey, N.W.: Boundary Layer Development in the BR710 and BR715 LP Turbines - The Implementation of High-Lift and Ultra-High-Lift Concepts. ASME J. Turbomach. 124, 385–392 (2002)
5. Brunner, S., Fottner, L., Schiffer, H.-P.: Comparison of Two Highly Loaded Low Pressure Turbine Cascades under the Influence of Wake-Induced Transition. ASME Paper 2000-GT-268 (2000)
6. Haselbach, F., Schiffer, H.P., Horsmann, M., Dressen, S., Harvey, N., Read, S.: The Application of Ultra High Lift Blading in the BR 715 LP Turbine. ASME J. Turbomach. 124, 45–51 (2002)
7. Gier, J., Ardey, S.: On the Impact of Blade Count Reduction on Aerodynamic Performance and Loss Generation in a Three-Stage LP Turbine. ASME Paper 2001-GT-0197 (2001)
8. Hourmouziadis, J.: Aerodynamic Design of Low-Pressure Turbines, AGARD Lecture Series No. 167 (1989)
9. Wakelam, C.T., Hoeger, M., Niehuis, R.: A Comparison of Three Low Pressure Turbine Designs. ASME J. Turbomach. 135, 051026 (2013)

10. Sturm, W., Fottner, L.: The High Speed Cascade Wind Tunnel of the German Armed Forces University Munich. In: 8th Symposium on Measuring Techniques for Transonic and Supersonic Flows in Cascades and Turbomachines (1985)
11. Acton, P., Fottner, L.: The Generation of Instationary Flow Conditions in the High Speed Cascade Wind Tunnel. In: 13th Symposium on Measurement Technique for Transonic and Supersonic Flows (1996)
12. Gier, J., Franke, M., Hübner, N., Schröder, T.: Designing Low Pressure Turbines for Optimized Airfoil Lift. ASME J. Turbomach. 132(3), 031008 (2010)
13. Amecke, J.: Auswertung von Nachlaufmessungen an ebenen Schaufelgittern. Report 67A49, AVA Göttingen, Germany (1967)
14. Ardey, S.: Untersuchung der aerodynamischen Effekte von Vorderkanten-Kühlluftausblasung an einem hochbelasteten Turbinengitter. Ph.D. thesis, Universität der Bundeswehr München, Neubiberg, Germany (1998)
15. Praisner, T.J., Grover, E.A., Knezevici, D.C., Popovic, I., Sjolander, S.A., Clarke, J.P., Sondergaard, R.: Towards the Expansion of Low-Pressure-Turbine Airfoil Design Space. ASME J. Turbomach. 135, 061007 (2013)
16. Martinstetter, M., Niehuis, R., Franke, M.: Passive Boundary Layer Control on a Highly Loaded Low Pressure Turbine Cascade. ASME Paper GT2010-22739 (2010)
17. Himmel, C.G., Thomas, R.L., Hodson, H.P.: Effective Passive Flow Control for Ultra-High Lift Low Pressure Turbines. In: ETC 8, Graz, Austria, March 23-27, pp. 17–27 (2009)
18. Ludewig, T., Mack, M., Niehuis, R., Franke, M.: Optimization of the Blowing Ratio for a Low Pressure Turbine Cascade with Active Flow Control. ETC2011 Paper No. 131 (2011)
19. Schlichting, H., Gersten, K.: Grenzschicht-Theorie. Auflage, vol. 10. Springer, Heidelberg (2006)
20. Wazzan, A.R., Okamura, T.T., Smith, A.M.O.: Spatial and Temporal Stability Charts for the Falkner–Skan Boundary Layer Profiles. Report No. DAC-67086, Douglas Aircraft Company (1968)
21. Mack, M., Niehuis, R., Fiala, A.: Parametric Study of Fluidic Oscillators for Use in Active Boundary Layer Control. ASME Paper No. GT2011-45073 (2011)
22. Gebhard, U., Hein, H., Schmidt, U.: Numerical Investigation of Fluidic Micro-Oscillators. Journal of Micromechanics and Microengineering 6, 115–117 (1996)
23. Chen, C.-K., Wang, L., Yang, J.-T., Chen, L.-T.: Experimental and Computational Analysis of Periodic Flow Structure in Oscillator Gas Flow Meters. Journal of Mechanics 22, 137–144 (2006)
24. Culley, D.E.: Active Flow Separation Control of a Stator Vane Using Surface Injection in a Multistage Compressor Experiment. ASME Paper GT2003-38863 (2003)
25. Cerreteli, C., Kirtley, K.: Boundary Layer Control with Fluidic Oscillators. J. Turbomach. 131(4), 41001 (2009)
26. Ries, T., Baumann, J., Rist, U., Raab, I., Staudacher, S.: LP Turbine Laminar Separation with Actuated Transition; DNS, Experiment and Fluidic Oscillator CFD. ASME Paper GT2009-59600 (2009)
27. Mack, M., Niehuis, R., Fiala, A., Guendogdu, Y.: Boundary Layer Control on a Low Pressure Turbine Blade by Means of Pulsed Blowing. ASME J. Turbomach. 135, 051023 (2013)
28. Mack, M., Brachmanski, R., Niehuis, R.: The Effect of Pulsed Blowing in the Boundary layer of a Highly Loaded Low Pressure Turbine Blade. ASME Paper No. GT2013-94566 (2013)

Active Flow Control on a Highly Loaded Compressor Cascade with Non-steady Boundary Conditions

Marcel Staats, Wolfgang Nitsche, and Inken Peltzer

Department of Aeronautics and Astronautics, TU Berlin,
Marchstraße 12-14, 10587 Berlin, Germany
marcel.staats@ilr.tu-berlin.de

Abstract. This paper discusses the effect of a periodical disturbance in the wake of a highly loaded axial compressor on the pressure distribution and the secondary flow characteristics of the turbo machinery stator blades. A large scale axial compressor cascade, consisting of six two dimensional passages, has been used for this investigation. The test rig is equipped with an active flow control system to enhance the operating range of the compressor. Results that have been achieved by means of side wall actuation are also presented.

1 Introduction

In modern aero engines more than 15 compressor stages are used to achieve the required total pressure rise [1]. One single compressor stage is capable of reaching a certain static pressure ratio of approximately $\frac{p_2}{p_1} = 1.2 - 1.4$, so that the fluid streams against a positive pressure gradient within the compressor, which can lead to flow separation and thus limit the operating range of the aero engine. In turbo machinery design, a satisfactory margin to the surge line of the compressor needs to be guaranteed and the de Haller criterion ($DH > 0.72$) has to be met. Compressors with lower deceleration ratios are highly loaded under design conditions and operate at the risk of stalling. The stator flow, especially of highly loaded stages, is characterized by very three-dimensional separation phenomena [2], [3], [4]. The corner stall leads to high flow losses in the region of the blade wall junction and limits the operating range and hence the efficiency of the compressor. Nowadays, combustion in turbo machinery takes place under almost isobaric conditions. It is a steady state process, where only a 4 to 6 % loss in pressure is caused in order to maintain film cooling on the first turbine stage. Applying isochoric combustion can lead to an overall increase in the efficiency of turbo machinery. This can be realized by the use of pulsed instead of continuous combustion. The former is a non-steady-state, which is carried out periodically. This causes periodic non-steady flow to the compressor. By the use of can-annular combustion chambers the disturbance, of the pulsed combustion, to the compressor can be minimized and kept locally, if they are closed one at a time, when combustion is running. The resulting flow separations are manageable by

R. King (ed.), *Active Flow and Combustion Control 2014*, Notes on Numerical Fluid Mechanics and Multidisciplinary Design 127, DOI: 10.1007/978-3-319-11967-0_2

means of active flow control (AFC). It has been shown, that flow separation can be avoided using passive flow control or AFC [5]. Different kinds of actuators can be used to achieve AFC. A good overview of different methods is presented in [6], [7]. In general two different kinds of active flow control approaches are used to ensure the stability of a highly loaded stage. Using side wall actuation mainly influences the secondary flow structures. It has been demonstrated that vortex generators on the side wall of a two-dimensional high-speed cascade can result in a reduced total pressure loss [8]. Blade actuators avoid flow separation on the blade itself. In earlier works the positive achievements using both side wall and blade actuators were investigated by [9],[10],[11] using pulsed air jets, where a reduction in loss of 10% was realized. The major advantage of pulsed air jets instead of steady ones, as used in [2], is the reduction of the necessary air mass flow rate. The pulsed jet uses the shear layer instabilities to induce vortices that help to re-energize the detached flow by cross-stream mixing.

2 Experimental Setup

For the experimental investigations a two-dimensional low speed compressor stator cascade is used, as shown in Figure 1 and Table 1. The test rig consists of seven blades to guarantee symmetric flow conditions in the measurement passage. The blades are mounted on a rotatable disk to adjust the inflow angle between $\alpha = 55° - 65°$. The design inflow angle is $\alpha = 60°$. Moveable tail boards are used to adjust the contour. A boundary layer suction is installed at the inlet of the cascade. Using the static pressure tabs upstream of the cascade makes it possible to measure in-flow conditions. The dynamic pressure is measured at a point in front of the nozzle in reference to the static pressure tab upstream of the measurement blade. 21 choking blades are mounted on top of the test

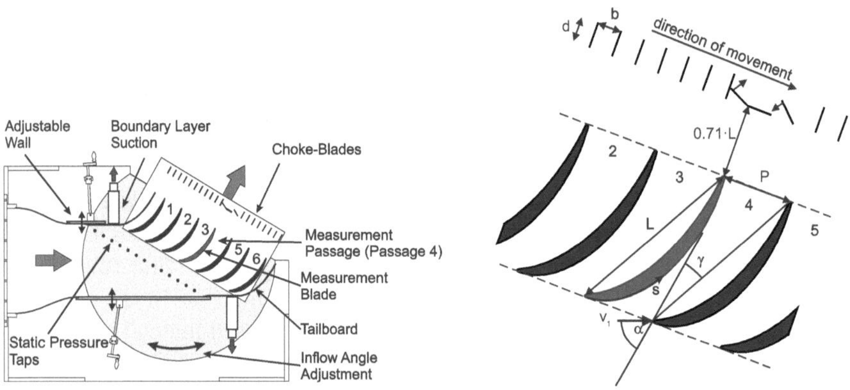

(a) cascade (b) measurement blade

Fig. 1. Experimental setup

rig, thus causing a maximum periodical wake blocking of 90% for any one passage, which results in a 13% all time blocking of the whole cascade wake area. The blade movements are described by a sinus motion and result in a passage blocking, which is shown in Figure 2a as a function of the phase angle for three passages. Passage three represents the passage on the pressure side of the measurement blade and passage five the one next to the measurement passage. The zero degree angle represents the case, where the choking is furthest away from the measurement blade and the $\varphi = 180°$ angle belongs to the case illustrated in Figure 1, where the choking blade in the wake of the measurement blade is fully closed. There are three choking blades per passage. All of them are driven by one tappet each, mounted on the side of the apparatus and rotated by a geared motor. Each tappet is installed with a phase angle shift of $\Delta\phi = 17.143°$. The choke apparatus is capable of delivering choking frequencies of up to $f_c = 2.5\,\text{Hz}$. Using the definition of the Strouhal number

$$Sr = \frac{f_c \cdot L}{v_1}, \tag{1}$$

where f_c represents the choking frequency, L the chord length of the stator blade and v_1 stands for the average flow speed, Strouhal numbers of up to $Sr = 0.04$ can be reached.

The following investigations are carried out at a Reynolds number of Re = 600000. The average speed at the inflow is about $v_1 = 25\,\text{m/s}$. Due to the choking of the cascade inflow conditions vary by $q/q_{mean} = \pm 2\%$, as can be seen in Figure 2b. For the following calculations the average dynamic pressure was used. The investigation of the effect of choking in respect of the pressure distribution,

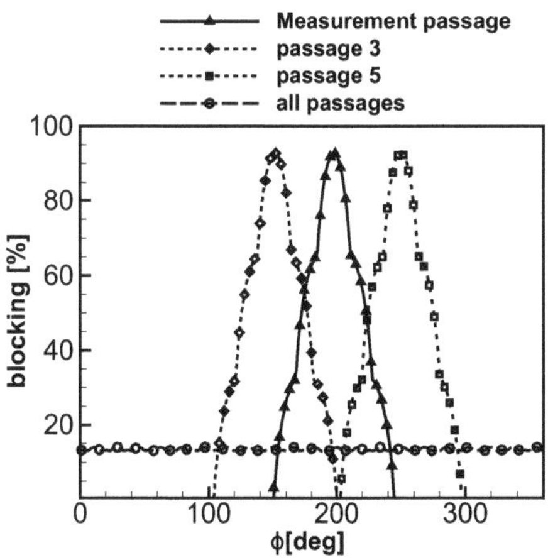

(a) choke apparatus blocking

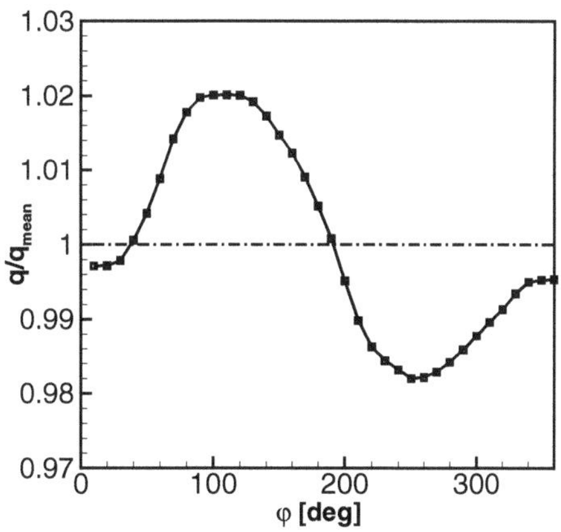

(b) impact on dynamic pressure

Fig. 2. Choke apparatus

Table 1. Cascade parameters

Parameter	Value
chord length	$L = 375\text{mm}$
blade pitch	$P = 150\text{mm}$
blade height	$H = 300\text{mm}$
inflow angle	$\alpha = 60^\circ$
stagger angle	$\gamma = 20^\circ$
Mach number	$\text{Ma} = 0.07$
Reynolds number	$\text{Re} = 600000$
choke-blade height	$d = 50\text{mm}$
choke-blade pitch	$b = 50\text{mm}$

as well as the secondary flow structures was carried out using different choking frequencies.

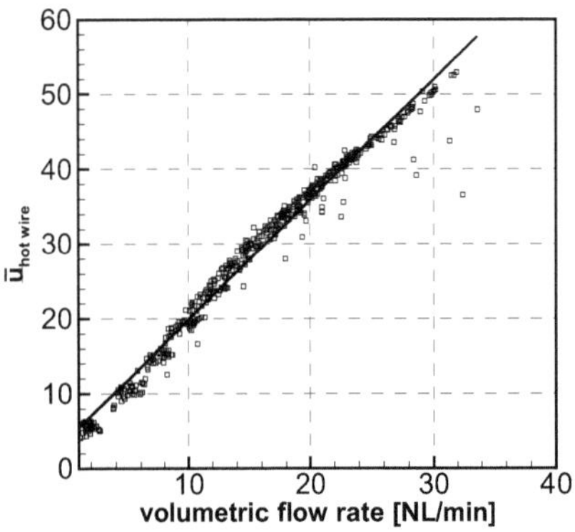

Fig. 3. SWA calibration

Every passage is equipped with two side wall actuators (SWA), that are mounted on each wall at $s/S = 14.5\%$. These actuators have rectangular slots with a length of $h = 20\,\text{mm}$ and a depth of $w = 0.4\,\text{mm}$. The blowing angle is $\varphi_{blow} = 15^\circ$. Both, variations in frequency and in amplitude were carried out. Before measurement, an outlet speed calibration in respect of the actuator volumetric flow was made using hot wire measurements on the outlet of the actuator at different frequencies. The linearized calibration function is shown in Figure 3. The actuators are calibrated for an outlet speed of up to $u_{jet} = 100\,\text{m/s}$ and $u_{jet}/v_1 = 4$ in relation to the average inflow conditions and a duty cycle of $DC = 50\%$. The specific characteristics of the actuators used are presented in [9]. The measurement blade is equipped with 44 pressure taps, 27 of which are installed on the suction side and 17 on the pressure side. Using a stepper motor, which the measurement blade is mounted to, the measurement cross section is moved through the passage with an increment of $\Delta y = 5\,\text{mm}$ to produce surface pressure data.

3 Time- and Phase-Averaged Baseflow

At first, the time-averaged results of the baseflow with periodical choking and no relation to any phase angle, is compared to the baseflow without choking. The periodical choking of the cascade does not only influence the pressure distribution, but also the secondary flow structures. In order to analyze the development of the secondary flow structures the standard deviation of the pressure coefficient was used. To distinguish the pressure fluctuations caused by the secondary flow from the comparably heavy fluctuations, caused by periodical choking, a high-pass filter at $f_{pass} = 4\,\text{Hz}$ was applied. The quality of the pressure transducer signals are increased by wavelet de-noising. In Figure 4a and Figure 4b the time averaged flow structures are represented by $RMS(cp')$ and cp contour plots. All negative x/H values represent the baseflow and all positive ones the flow with the $f_c = 2\,\text{Hz}$ choking, which corresponds to a Strouhal number of $Sr = 0.03$. The baseflow shows a distinctive suction peak in the cp distribution at $s/S = 10\% - 15\%$. This suction peak is less distinctive in the $f_c = 2\,\text{Hz}$ case, and lower pressure coefficients are found all over the profile's suction side. The negative pressure gradient from the leading edge to the suction peak stabilizes the laminar flow in both cases. Downstream this peak the pressure rises along the passage, which leads to the laminar separation bubble (LSB) and a turbulent reattachment afterwards, as illustrated in Figure 4a. The visualization of the baseflow without choking shows the LSB to be at about $s/S = 25\%$. Due to the turning of the profile the fluid is forced to stream against a positive pressure gradient, which in turn leads to a destabilization of the flow and causes the corner vortex breakdown at $s/S = 30\%$. Further downstream, this causes a blocking of the passage and results in flow separation at $s/S = 65\%$.

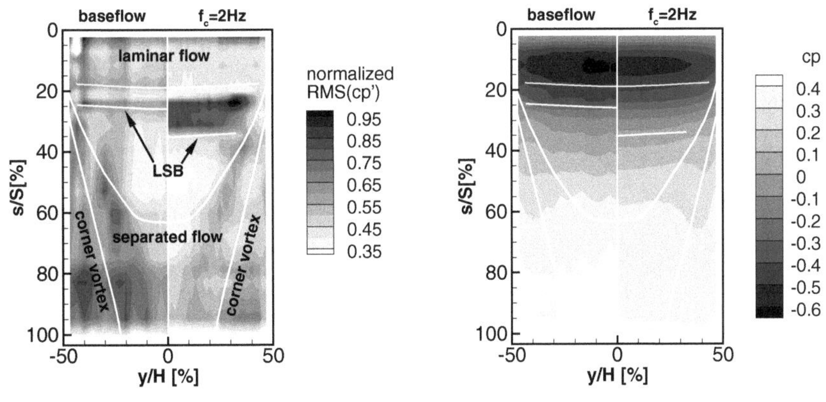

(a) $RMS(cp')$ (b) cp

Fig. 4. Time-averaged flow structure on suction side

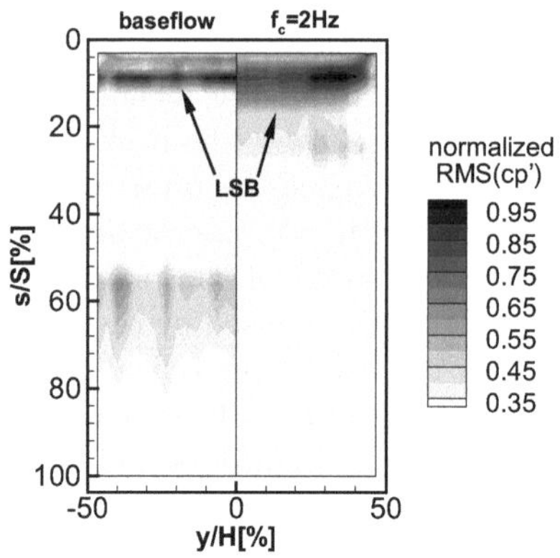

Fig. 5. Time-averaged $RMS(c_p')$ on pressure side

The baseflow is well documented in [12]. Comparing the fundamental flow structures of the choked case with the baseflow leads to the conclusion that the size of the LSB becomes larger in the choked case. The periodical choking influences the inflow condition in a way that the Reynolds number changes periodically. This affects the position of the LSB and causes it to oscillate in respect of the phase angle of the choking apparatus. In addition, the incidence angle of the profile also changes with regard to the phase angle, as will be presented later. The $RMS(cp')$ distribution indicates an upstream shift of the LSB at a certain phase angle to $s/S = 23\%$. It is not only the position of the LSB which is influenced by the periodical choking, but along with it also the corner vortex breakdown. The position of the corner vortex breakdown is still $s/S = 30\%$ and in time-averaged data higher blocking of the passage leads to an earlier flow separation resulting in a lower static pressure rise of the choked passage. The analysis of the blades pressure side is shown in Figure 5. The baseflow shows a LSB at $s/S = 10\%$. Due to the choking this LSB oscillates in respect of the phase angle as well. This movement causes a greater coverage in the time-averaged plot. Figure 6a displays the trailing edge pressure coefficient with a view to the blade height coordinate for two different

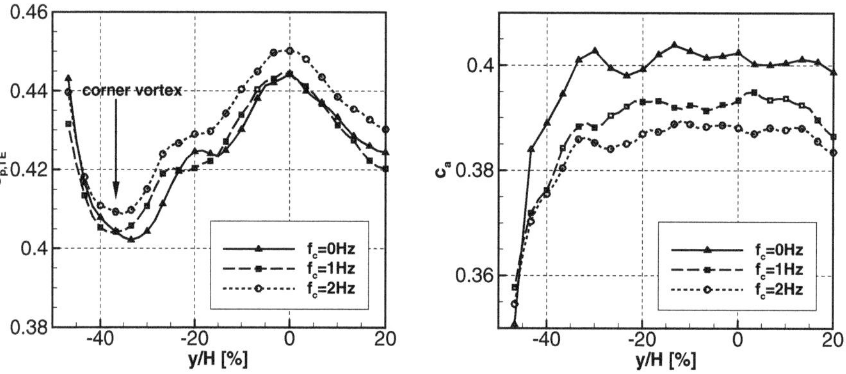

(a) time-averaged $c_{p,TE}$ (b) time-averaged c_a

Fig. 6. trailing edge pressure coefficients $c_{p,TE}$ and lift coefficient c_a

choking frequencies compared with the baseflow. Because of symmetric conditions only half of the blade height is depicted here. Taking the $f_c = 1\,\text{Hz}$ case into account, reveals that the trailing edge pressure coefficient is not shifted to overall higher values to a great extend. In the centre of the blade the pressure coefficient of the $f_c = 1\,\text{Hz}$ case shows a good agreement with the baseflow. For the coordinates of $y/H = -20\%$ and less the corner vortex causes lower pressure rises and it becomes apparent that the corner vortex structure is similarly developed in both investigated cases, when choking is applied. Considering the $f_c = 2\,\text{Hz}$ situation, an overall shift of the trailing edge pressure towards higher pressure coefficients can be recognized. The section-wise integration of the blades pressure distribution is plotted in Figure 6b. It follows that periodical choking leads to decreasing time-averaged lift coefficients all over the blade height. In the following considerations the pressure data are phase-averaged using 36 discrete phase angles. By integrating the pressure distribution over the measurement blade's midsection, the lift coefficient is calculated and displayed in respect to the phase angle in Figure 7d regarding four different cases. In order to compare these values to the baseflow, that curve is also depicted. It can be found that the lift coefficient for that case is approximately $c_a = 0.4$. In all other cases illustrated, this figure oscillates around a mean value. This behavior indicates a change in incidence angle in relation to the phase angle. Taking the $f_c = 1\,\text{Hz}$ case into account, the largest influence of the choking can be found in the lift coefficient behavior. For all choking frequencies considered a rising edge ranging from $\varphi = 0°$ to $\varphi = 100°$ can be observed. A relatively sharp falling edge is found between $\varphi = 170°$ and $\varphi = 240°$. One phase can be subdivided into two cases. Regarding the phase angles in between $\varphi = 180°$ and $\varphi = 360°$ only passages facing the suction side (ss) of the blade are generally affected by the choking apparatus, whilst for all other phase angles choking takes place in passages facing the blade's pressure side (ps). It can be said that the choking between $\varphi = 50°$ and $\varphi = 200°$ causes effects, that are comparable to a rise in incidence angles, resulting in a higher lift coefficient, whereas decreasing lift coefficients or incidence angles are observed for other phase angles. The direct influence of the choking apparatus on the measurement blade takes place at phase angles between $\varphi = 120°$ to approximately $\varphi = 220°$, where a transition to lower lift coefficients is illustrated in Figure 7d, as choking takes place in the direct wake of the measurement blade. The rest of the plots in Figure 7 present the distribution of the pressure coefficient at the mid-section for different phase angles. Each figure contains $f_c = 1\,\text{Hz}$, $f_c = 1.5\,\text{Hz}$, $f_c = 2\,\text{Hz}$ and $f_c = 2.5\,\text{Hz}$ choking frequency line plots. By means of comparing these pressure distributions to the $f_c = 0\,\text{Hz}$ (baseflow) case, this line is also plotted in each figure.

Taking Figure 7a into account a far off pressure distribution of all cases shown in respect of the baseflow can be detected. Comparing the four different cases with each other leads to a good agreement, which compares to Figure 7d, as all choking frequencies discussed reach similar c_a values at this certain phase angle. Inspecting the suction surface the low pressure peak of the baseflow is found

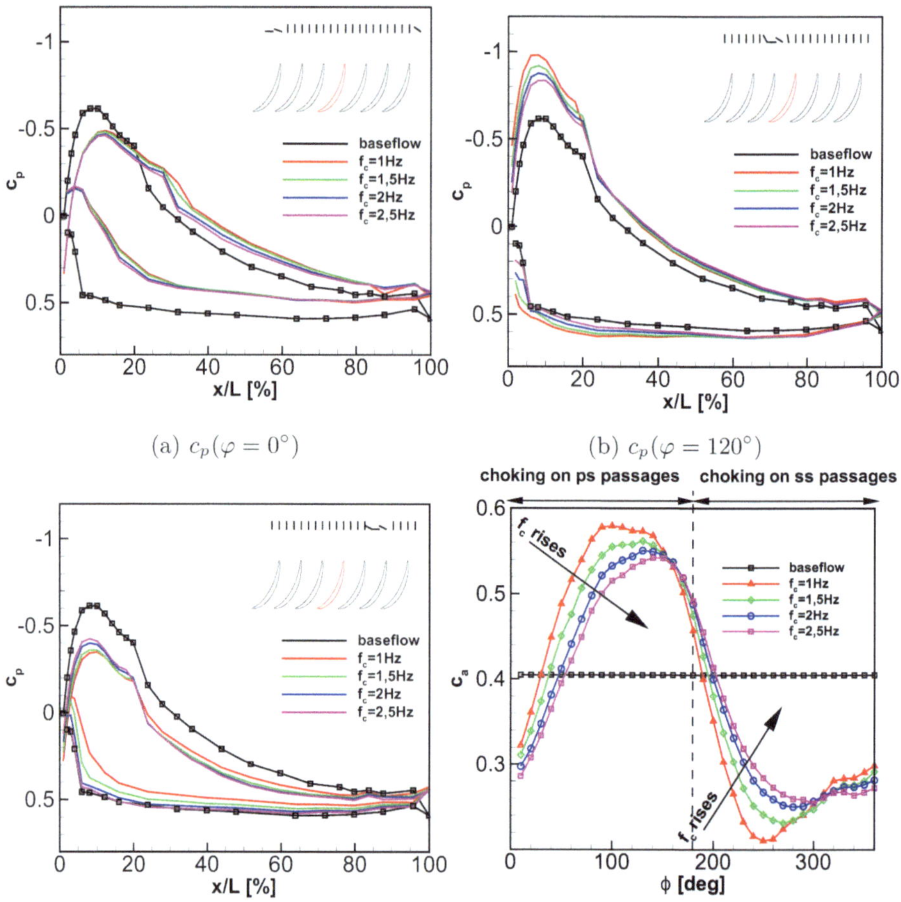

(a) $c_p(\varphi = 0°)$

(b) $c_p(\varphi = 120°)$

(c) $c_p(\varphi = 260°)$

(d) $c_a(\varphi)$

Fig. 7. Phase-averaged pressure distributions at midsection for different choking frequencies and phase angles

at $x/L = 8\%$, whereas a clear shift of the peak away from the leading edge to about $x/L = 12\%$ results at this phase angle. The positive pressure gradient downstream of this point is lower in contrast to the baseflow and a shift in the LSB is detected. Taking the $f_c = 1\,\text{Hz}$ case into account, the position of the LSB is found to be at $x/L = 30\%$. It can be observed, that the LSB is located slightly upstream for higher choking frequencies. Basically, there is a phase shift due to the fluid's inertia causing the reaction of the low choking frequencies to appear faster and more distinctly. The lines for the different choking frequencies obviously follow the $f_c = 1\,\text{Hz}$ case. This performance can also be observed in Figure 7d. At this phase angle the LSB is shifted downstream. After the turbulent reattachment the pressure distribution moves towards lower pressure coefficients until it reaches the trailing edge. Taking into account that the static pressure rise of the passage corresponds to the static pressure on the trailing edge of the blade and is a measure of the blade loading, it can be said, that for $\varphi = 0°$ the blade loading is lower compared to the baseflow. Looking at the pressure sides c_p-distribution an appropriate shift of the pressure towards lower pressure coefficients compared to the baseflow can be observed. Due to the effect of choking a distinct LSB can be found for all cases shown near the profile's leading edge at $x/L = 5\%$. Compared to the baseflow, the LSB is shifted downstream as well.

Figure 7b shows the midsection's pressure distribution at the phase angle of $\varphi = 120°$. This case refers to a phase angle with a rather big lift coefficient, as can be seen in Figure 7d. At the suction side the pressure peak reaches $c_p = -1$ regarding the $f_c = 1\,\text{Hz}$ case. Again the phase shift causes the pressure peaks of all cases discussed to be arranged depending on their choking frequency. The higher the choking frequency gets, the less distinct the pressure peaks are. For the choking frequencies greater then $f_c = 1\,\text{Hz}$ the LSB is at the same coordinate as in the baseflow. The high pressure peak of the $f_c = 1\,\text{Hz}$ case causes the laminar separation to be shifted upstream, which results in an earlier turbulent reattachment. After this x/L coordinate the pressure coefficients do not differ much from each other until they reach the trailing edge, where all cases discussed involving choking end up roughly at the same value. Having said that they are still lower compared to the baseflow pressure coefficient. At the pressure side the LSB is situated just after the blades leading edge, followed by a rapid static pressure rise, which ends in a pressure plateau. This topology is similar to the baseflow characteristic.

In Figure 7c the phase angle is $\varphi = 260°$ and the choking still moves away from the measurement blade. Lower c_p-values can be observed all over the suction side's pressure distribution in comparison to the baseflow. The suction peak is roughly at the same position, as is the LSB. The pressure coefficients at the trailing edge are higher compared to the baseflow. On the pressure surface the pressure distribution for the $f_c = 2\,\text{Hz}$ and $f_c = 2.5\,\text{Hz}$ case is very much related to the baseflow distribution. Taking the $f_c = 1.5\,\text{Hz}$ and $f_c = 1\,\text{Hz}$ case into consideration it can be found that they are situated slightly below the baseflow. Comparing the effect of choking from Figure 7c with the impact shown in Figure 7a a shift to lower pressure coefficients in all cases and x/L positions can be found. Figure 7a shows a less distinct pressure gradient after the suction peak compared to Figure 7c, which results in a downstream shift of the LSB, which can not be observed in Figure 7c. The choking has multiple impacts on the secondary flow structure regarding different phase angles. These effects are best investigated by taking the standard deviation of the pressure coefficient into account. The LSB, just like the corner vortex is characterized by an increasing standard deviation. The time-averaged secondary flow structures have been shown in Figure 8 for the $f_c = 2\,\text{Hz}$ choking frequency. All plots are subdivided and show the baseflow on the left hand side. On top c_p isolines are plotted as well. Taking the phase angles of $\varphi = 20°$, $\varphi = 70°$ and $\varphi = 320°$ into account these phase angles are close to the beginning or the ending of one phase,

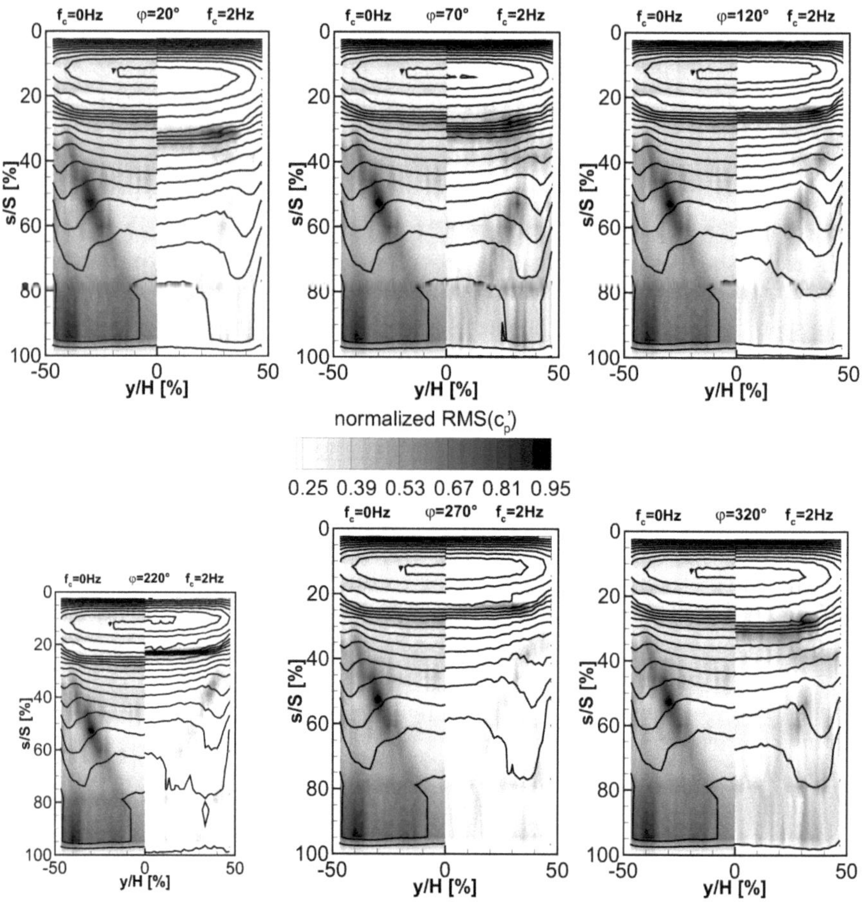

Fig. 8. Comparison of $RMS(c_p')$ regarding the baseflow and $f_c = 2\text{Hz}$ choking frequency

which means the choking indirectly influences the measurement blade's flow and causes a reduced pressure gradient on the suction side as described above. Again a downstream shift of the LSB is found in these cases. The pressure fluctuations caused by the corner vortex are found to be less distinct for the $\varphi = 20°$ case compared with the baseflow and the vortex breakdown takes place further downstream. $\varphi = 70°$ represents a case, where the choking is moving towards the measurement blade and the lift coefficient increases again (Figure 7d). This causes a stabilization of the secondary flow structures and results in a more definite corner vortex structure. The breakdown of the corner vortex is now shifted upstream and takes place at about $s/S = 35\%$. At the phase angle of $\varphi = 320°$ the corner vortex structure is almost imperceptible. This leads to the idea that the solid structure of the corner vortex disappears from this phase angle onwards and builds up again, until it stabilizes at $\varphi = 70°$. An even more definite flow structure regarding the corner vortex and the LSB can be found in the $\varphi = 120°$ case. Here, the breakdown appears at $x/L = 30\%$. The overall flow structure is comparable to the one in the baseflow. The LSB is situated roughly at the same spot ($x/L = 27\%$) and the corner vortex appearance is almost identical. Later on, at $\varphi = 220°$ the vortex breakdown position does not change. The LSB is shifted clearly upstream to $x/L = 22\%$. Looking at $\varphi = 270°$ the LSB is shifted slightly downstream compared with the $\varphi = 220°$ case. The corner vortex structure is beginning to destabilize, which results in a slightly bent structure and finally leads to its disappearance, as discussed above.

4 Active Flow Control Results

Taking the cascade's baseflow into account a relatively large flow separation can be observed at $s/S = 65\%$. In earlier works this separation was successfully avoided using multiple actuators simultaneously. The secondary flow structures can be well influenced using the side wall actuators, but with only minor influence on the trailing edge separation, as can be seen in Figure 9. The standalone use of the side wall actuators clearly shows, that only half of the pressure rise is achieved in comparison with the use of the entire actuator system, consisting of side wall and blade actuators. Using the entire actuator system separation is fully avoided and the pressure rises up to the trailing edge. The following investigations are based on side wall actuators only and introducing c_μ the momentum of the actuators is in relation with the momentum of the passage flow:

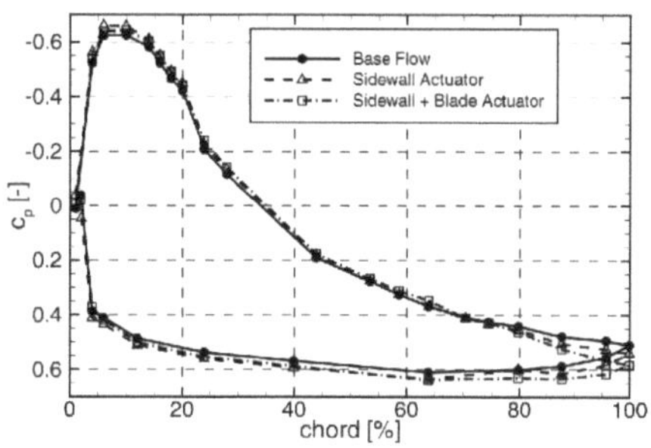

Fig. 9. Influence of actuation on c_p distribution [10]

$$c_\mu = \frac{\dot{m} \cdot \bar{u}}{q_1 \cdot A_1}, \tag{2}$$

were $\dot{m}$ describes the mass flow rate through the actuators, $\bar{u}$ the mean actuator jet velocity, q_1 the dynamic pressure at the inflow and A_1 the inlet plane of one passage. Figure 10 illustrates the change in the trailing edge pressure coefficient due to the actuation for different momentum coefficients. The choking frequency was varied and two particular actuation frequencies were investigated.

It follows, that even for $c_\mu = 0.1\%$ an increase of about 4% in time-averaged trailing edge pressure coefficient can be observed for all choking frequencies investigated. When a higher momentum is applied a linear rise in the trailing edge pressure occurs with a very low spread until reaching $c_\mu = 0.2\%$.

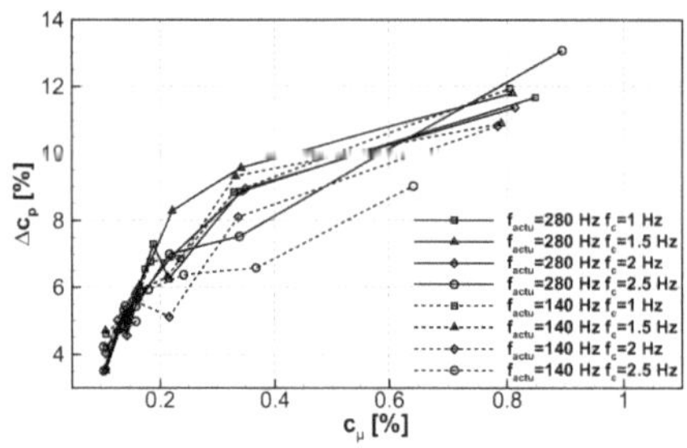

Fig. 10. Achievement in trailing edge pressure coefficient

From there on, it is possible to achieve a further increase in the trailing edge pressure with growing momentum coefficients but in a non-linear way and with a bigger spread in actuation efficiency for the cases discussed. The time-averaged plot shows that improvements of $\Delta c_p = 10\%$ by means of moderate momentum coefficients are feasible. Due to the pulsed combustion the flow is highly non-steady, which causes a different actuation success with regard to each phase angle. Figure 11 examines the phase-averaged c_p improvements taking the Strouhal number into account, discussing three different actuating amplitudes. As already mentioned the time-averaged $c_{p,TE}$ improvement is feasible for very low actuation amplitudes. All cases shown in Figure 11 demonstrate different impacts of the actuation on the trailing edge pressure rise regarding the phase angle. In the cases discussed the Strouhal number is calculated as

$$F^+ = \frac{f \cdot l}{v_1}, \tag{3}$$

where l describes the length from the actuator slot to the trailing edge of the blade.

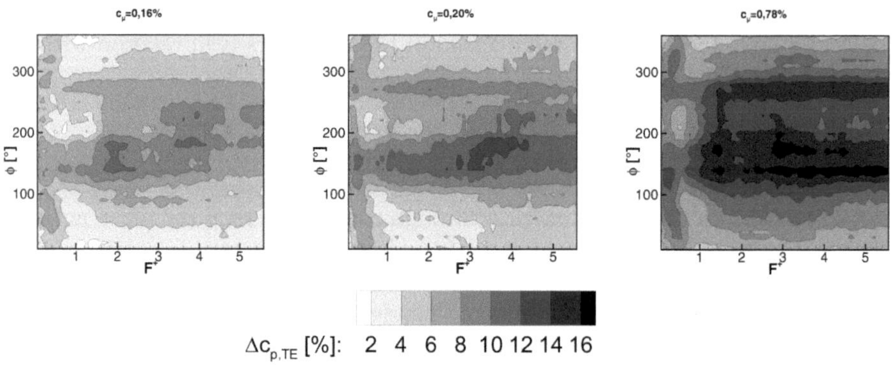

Fig. 11. Development of phase averaged actuation efficiency using different Strouhal numbers and momentum coefficients and $f_c = 1.5$Hz

Going from low to high amplitudes, it can be found that the overall success of the actuation rises rapidly. The best AFC achievements can be observed at phase angles ranging from $\varphi = 100°$ to $\varphi = 300°$. Actuation at high amplitudes can lead to local improvements of up to $\Delta c_{p,TE} = 17\%$ at $\varphi = 150°$, as is evident from the right-hand side illustration in Figure 11. At low amplitudes different phase angles require distinct Strouhal numbers for the best increase in the trailing edge pressure. This would result in a phase angle depending change in actuation frequency. Taking the case of $c_\mu = 0.16\%$ into account, it follows, that actuation works best between $\varphi = 120°$ and $\varphi = 180°$ corresponding to $F^+ = 2$, whilst for phase angles between $\varphi = 180°$ to $\varphi = 250°$ actuation corresponding to $F^+ = 4$ works best for a change in the trailing edge pressure. At very high, as well as at very low phase angles the least increase is observed in the trailing edge pressure coefficient, as the choking takes place furthest away from the measurement blade. Taking the phase angle dependant lift coefficient in Figure 7d into account, it is remarkable that the phase angles of the transition from the rising- and falling edges to the local extremes of the c_a-curve correspond to lower achievements in AFC results in Figure 11. This is best observed in cases covering $c_\mu = 0.2\%$ and $c_\mu = 0.78\%$. Figure 12a shows the effect of the side wall actuation on the non-steady flow to be similar in respect of the flow without choking, as presented in Figure 9.

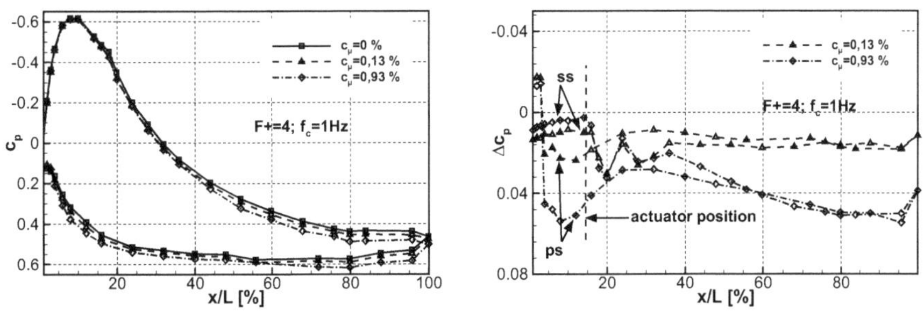

(a) c_p distribution (b) change of c_p due to x/L position

Fig. 12. Time-averaged c_p distribution in midsection due to actuation

Two different momentum coefficients are depicted and compared with the baseflow. Figure 12b discusses the local changes in pressure distribution caused by the AFC. It stands out that the suction peak is not greatly influenced by the side wall actuation. For both cases observed a parallel shift towards slightly higher c_p values can be seen up to the actuators position of $x/L = 14.5\%$. From $x/L = 40\%$ onwards the benefit for the pressure coefficient stays constant for the $c_\mu = 0.13\%$ case. The parallel shift of the c_p distribution can also be found in Figure 12a. The $c_\mu = 0.93\%$ case shows a much more increasing pressure coefficient until reaching $x/L = 80\%$, from where it stays at a constant level. It follows that the AFC has an impact on the pressure side as well. Using AFC along the whole blade a pressure rise on the pressure side can be observed. The positive area between the suction- and the pressure surface Δc_p distribution indicates the benefit of the operating range of the stator due to the actuation.

5 Conclusion and Outlook

Within the Collaborative Research Centre 1029 a highly loaded linear compressor cascade is used, with major three-dimensional flow structures and a trailing edge separation under design conditions causing high losses without AFC. This stator cascade was now used under periodical non-steady conditions, as they are dictated by pulsed combustion. The flow in the six stator passages was influenced using slotted side wall actuators, as they were also used earlier within the Collaborative Research Centre 557. The investigations on the baseflow showed major effects of the choking regarding every phase angle. Investigating the blade's pressure distribution showed an oscillating incidence angle and Reynolds effects for all phase angles. It was found that higher choking frequencies have lower interference of the stators flow characteristics than extremely low ones and are likely to be preferred for pulsed combustion in real. The excitation by the choking apparatus causes major impacts on the secondary flow structures. The laminar separation bubble, as well as the corner vortices are subject to change. Applying the AFC has shown that the effect of the choking can be handled with moderate momentum. It is possible to raise the passage's static pressure ratio, thereby increasing the operating range of the stator. Future investigations will need to undertake wake measurements, for an explicit statement regarding total pressure loss and static pressure rise. Particle Image Velocimetry will help to define the assumed changes of incidence angle and furthermore show the phase dependent secondary flow structures. In future, the actuator system will be extended by blade actuator components to further increase the systems efficiency and finally fully avoid the trailing edge flow separation. Within the Collaborative Research Centre 1029 a three dimensional test rig will be set up, to investigate the effects of the periodical non-steady conditions on the three dimensional flow. The AFC concept will be adapted to that test rig later.

Acknowledgements. The authors gratefully acknowledge support by the Deutsche Forschungsgemeinschaft (DFG) as part of collaborative research center SFB 1029 "Substantial efficiency increase in gas turbines through direct use of coupled unsteady combustion and flow dynamics".

References

1. Lord, W.K., MacMartin, D.G., Tillman, T.G.: Flow Control and Opportunities in Gas Turbine Engines. In: AIAA 2000-2234 (2000)
2. Nerger, D.: Aktive Strömungskontrolle in ebenen Statorgittern hoher aerodynamischer Belastung durch Ausblasen. PhD thesis. Technische Universität Braunschweig (2009)

3. Ma, W., et al.: Experimental Study of Corner Stall in a Linear Compressor Cascade. Chinese Journal of Aeronautics 24, 235–242 (2011)
4. Gbadebo, S.A., Cumpsty, N.A., Hynes, T.P.: Three-Dimensional Separations in Axial Compressors. Journal of Turbomachinery 127, 331–339 (2005)
5. Gad el Hak, M.: Passive, Active and Reactive Flow Management. In: Flow Control (1998)
6. Cattafesta, L., Sheplak, M.: Actuators for Active and Flow Control. Annu. Rev. Fluid Mech. 43, 247–272 (2011)
7. Greenblatt, D., Wygnanski, I.J.: The Control of Flow Separation by Periodic Excitation. Progress in Aerospace Sciences 36, 487–545 (2000)
8. Hergt, A., et al.: Loss Reduction in Compressor Cascades by Means of Passive Flow Control. In: ASME Turbo Expo. (2008)
9. Hecklau, M., et al.: Experimental and numerical results of active flow control on a highly loaded stator cascade. Proceedings of the Institution of Mechanical Engineers, Part A: Journal of Power and Energy 225, 907–918 (2011)
10. Hecklau, M., Zander, V., Peltzer, I., Nitsche, W., Huppertz, A., Swoboda, M.: Experimental AFC Approaches on a Highly Loaded Compressor Cascade. In: King, R. (ed.) Active Flow Control II 2010. NNFM, vol. 108, pp. 171–186. Springer, Heidelberg (2010)
11. Zander, V., et al.: Active Flow Control by Means of Synthetic Jets on a Highly Loaded Compressor Cascade. Proceeedings of the Institution of Mechanical Engineers, Part A: Journal of Power and Energy 225, 897–906 (2011)
12. Zander, V., Nitsche, W.: Control of secondary flow structures on a highly loaded compressor cascade. Proceedings of the Institution of Mechanical Engineers, Part A: Journal of Power and Energy 227, 674–682 (2013)

Comparison of Iterative Learning and Repetitive Control Applied to a Compressor Stator Cascade

Simon J. Steinberg[1], Marcel Staats[2], Wolfgang Nitsche[2], and Rudibert King[1]

[1] Technische Universität Berlin, Fachgebiet Mess- und Regelungstechnik, Hardenbergstraße 36a, 10623 Berlin, Germany
{simon.steinberg,rudibert.king}@tu-berlin.de
[2] Technische Universität Berlin, Fachgebiet Aerodynamik, Marchstraße 12, 10587 Berlin, Germany
{marcel.staats,wolfgang.nitsche}@ilr.tu-berlin.de

Abstract. This contribution deals with learning control algorithms applied as closed-loop flow controllers. The system to be controlled is an experimental linear compressor stator cascade with sidewall actuation. A damper flap device located downstream of the trailing edges of the blades generates periodic disturbances of the individual passage flows. By learning from period to period, the controllers are able to successfully decrease the effect of the disturbance on the pressure coefficient distribution of the blades' suction surface.

Keywords: Iterative Learning Control, Repetitive Control, Active Flow Control.

1 Introduction

Novel combustion processes such as the Pulsed Detonation Propulsion (PDP) [25], [18], [5] have the potential to increase the efficiency of gas turbines and jet engines significantly. In contrast to conventional combustion processes, the PDP leads to considerable pressure peaks inside the combustion chamber of the Pulsed Detonation Engine (PDE). These pressure fluctuations are transferred into other parts of the machine, e.g. the turbine and the compressor. The operation of the compressor, however, tends to become unstable when experiencing downstream pressure oscillations. This is due to flow separations induced by periodically enhanced positive pressure gradients along the flow direction. Increasing the compressor's safety margins by blade design measures is not an option. It often leads to a loss of compressor efficiency which may be greater than the gain in efficiency achieved by the novel combustion process. Alternatively, stable operation of the compressor can be ensured by means of closed-loop active flow control [19], [20]. In this contribution, the main objective of the feedback flow controller is the rejection of the disturbances seen in a cascade and induced by downstream pressure fluctuations.

A linear compressor stator cascade [12], [31] is considered. To imitate the pressure fluctuation induced by a PDE the individual flow passages are gradually blocked

R. King (ed.), *Active Flow and Combustion Control 2014*, Notes on Numerical Fluid Mechanics and Multidisciplinary Design 127, DOI: 10.1007/978-3-319-11967-0_3

by damper flaps located downstream of the stator blades. For closed-loop control, the state of the passage flow is monitored by a real-time accessible control variable. The control variable is based on static pressure measurements taken directly from the suction surface of the middle blade of the cascade. Furthermore, a sidewall actuation device is used to realise appropriate actions when disturbances are detected. Specifically, the cascade flow is manipulated actively by injecting pressurised air into the middle passage via slots in the cascade's sidewalls. A closed-loop flow controller adjusts the actuation amplitude, where the control command is calculated based on the control variable mentioned previously.

Due to the working principle of PDE combustion chambers, the disturbance in the compressor flow will appear in a periodic manner with a frequency assumed to be known. Control algorithms such as Iterative Learning Control (ILC) [4] and Repetitive Control (RC) [29], [23] are especially suited to deal with such periodic processes. This paper analyses and compares both concepts when implemented as flow controllers for the linear compressor stator cascade.

The remainder of the paper is organized as follows. Section 2 briefly reviews ILC and RC. Section 3 describes the compressor stator cascade in detail and Sec. 4 introduces the surrogate control variable and the control architecture. Experimental closed-loop results are presented in Sec. 5 and in Sec. 6 a conclusion is drawn.

2 Methods

Two optimal synthesis methods for learning control systems, i.e. optimisation based ILC [9] and $\mathcal{H}_\infty$-optimal RC [6], are revisited in this section. Whereas a time-domain description will be given for ILC, frequency-domain approaches will be better suited for RC.

2.1 Iterative Learning Control

System Description. Consider a linear, time-invariant, single-input single-output system in its discrete-time state space representation

$$\boldsymbol{x}(k+1) = \boldsymbol{A}\boldsymbol{x}(k) + \boldsymbol{b}\,u(k)\ ,\ \ \boldsymbol{x}(0) = \boldsymbol{x}_0 \tag{1}$$

$$y(k) = \boldsymbol{c}^T\boldsymbol{x}(k) + d(k)\ , \tag{2}$$

where $u(k) \in \mathbb{R}$ is the input, $\boldsymbol{x}(k) \in \mathbb{R}^{n_x}$ the state, $y(k) \in \mathbb{R}$ the output, and $d(k) \in \mathbb{R}$ the disturbance at time step $k = t/T_s \in \mathbb{N}_0$. The actual time is denoted by $t \in \mathbb{R}_+$ and $T_s \in \mathbb{R}_+$ is the sampling time. The initial condition is $\boldsymbol{x}_0 \in \mathbb{R}^{n_x}$. $\boldsymbol{A} \in \mathbb{R}^{n_x \times n_x}$ is the discrete-time state matrix, $\boldsymbol{b} \in \mathbb{R}^{n_x}$ the discrete-time input vector, and $\boldsymbol{c} \in \mathbb{R}^{n_x}$ the output vector of the state space model. The system (1), (2) is assumed to be asymptotically stable and to possess the relative degree $\varrho = 1$. Introducing the lifted input, disturbance, and output vector for $n_p \in \mathbb{N}$ time steps of a period

$$\boldsymbol{u} = \Big(u(0),\ u(1), \dots,\ u(n_p - 1)\Big)^T \in \mathbb{R}^{n_p}\ , \tag{3}$$

$$\boldsymbol{d} = \Big(d(1),\ d(2), \dots,\ d(n_p)\Big)^T \in \mathbb{R}^{n_p}\ , \tag{4}$$

$$\boldsymbol{y} = \Big(y(1),\ y(2), \dots,\ y(n_p)\Big)^T \in \mathbb{R}^{n_p}\ , \tag{5}$$

the dynamic system (1), (2) can be written in its lifted representation [4], [29]

$$\boldsymbol{y} = \boldsymbol{\Phi}\boldsymbol{u} + \boldsymbol{F}\boldsymbol{x}_0 + \boldsymbol{d}\ . \tag{6}$$

$\boldsymbol{\Phi} \in \mathbb{R}^{n_p \times n_p}$ and $\boldsymbol{F} \in \mathbb{R}^{n_p \times n_x}$ describe the impact of the input trajectory $\boldsymbol{u}$ and the initial state $\boldsymbol{x}_0$ on the output trajectory $\boldsymbol{y}$, respectively, with

$$\boldsymbol{\Phi} = \begin{pmatrix} \boldsymbol{c}^T\boldsymbol{b} & 0 & \dots & 0 \\ \boldsymbol{c}^T\boldsymbol{A}\boldsymbol{b} & \boldsymbol{c}^T\boldsymbol{b} & \dots & 0 \\ \vdots & \vdots & \ddots & \vdots \\ \boldsymbol{c}^T\boldsymbol{A}^{n_p-1}\boldsymbol{b} & \boldsymbol{c}^T\boldsymbol{A}^{n_p-2}\boldsymbol{b} & \dots & \boldsymbol{c}^T\boldsymbol{b} \end{pmatrix} \tag{7}$$

$$\boldsymbol{F} = \Big((\boldsymbol{c}^T\boldsymbol{A})^T, (\boldsymbol{c}^T\boldsymbol{A}^2)^T, \dots, (\boldsymbol{c}^T\boldsymbol{A}^{n_p})^T\Big)^T\ . \tag{8}$$

Introducing the reference $r(k) \in \mathbb{R}$ and the control error $e(k) = r(k) - y(k)$ at time step k, the corresponding trajectories are defined as

$$\boldsymbol{r} = \Big(r(1),\ r(2), \dots,\ r(n_p)\Big)^T \in \mathbb{R}^{n_p}\ , \tag{9}$$

$$\boldsymbol{e} = \Big(e(1),\ e(2), \dots,\ e(n_p)\Big)^T \in \mathbb{R}^{n_p}\ , \tag{10}$$

respectively.

Optimal Iterative Learning Control with Quadratic Cost Function. The optimisation based ILC algorithm [1], [9], [24], [4] minimises a quadratic cost function iteratively. We assume the reference $\boldsymbol{r}$, the initial state $\boldsymbol{x}_0$, and the disturbance $\boldsymbol{d}$ to be iteration-invariant, i.e. they are identical for all iterations. In contrast, the input $\boldsymbol{u}_i$, the output $\boldsymbol{y}_i$, and the control error $\boldsymbol{e}_i$ vary along the iterations and are therefore supplied with an iteration index $i \in \mathbb{N}_0$.

As an iteration-variant cost function we introduce

$$J_i(\boldsymbol{u}_i) = J_e(\boldsymbol{u}_i) + J_u(\boldsymbol{u}_i) + J_{\Delta,i}(\boldsymbol{u}_i) \tag{11}$$

consisting of three parts. $J_e(\boldsymbol{u}_i)$ takes into account the control error, $J_u(\boldsymbol{u}_i)$ considers the control effort, and $J_{\Delta,i}(\boldsymbol{u}_i)$ includes the *change* of the control command in the *iteration-domain*. We use

$$J_e(\boldsymbol{u}_i) = \frac{1}{2}\left(\boldsymbol{r} - \boldsymbol{\Phi}\boldsymbol{u}_i - \boldsymbol{F}\boldsymbol{x}_0 - \boldsymbol{d}\right)^T \boldsymbol{W}_e \left(\boldsymbol{r} - \boldsymbol{\Phi}\boldsymbol{u}_i - \boldsymbol{F}\boldsymbol{x}_0 - \boldsymbol{d}\right) , \tag{12}$$

$$J_u(\boldsymbol{u}_i) = \frac{1}{2}\boldsymbol{u}_i^T \boldsymbol{W}_u \boldsymbol{u}_i , \tag{13}$$

$$J_{\Delta,i}(\boldsymbol{u}_i) = \frac{1}{2}\left(\boldsymbol{u}_i - \boldsymbol{u}_{i-1}\right)^T \boldsymbol{W}_\Delta \left(\boldsymbol{u}_i - \boldsymbol{u}_{i-1}\right) . \tag{14}$$

The weighting matrices $\boldsymbol{W}_e \in \mathbb{R}^{n_p \times n_p}$, $\boldsymbol{W}_u \in \mathbb{R}^{n_p \times n_p}$, and $\boldsymbol{W}_\Delta \in \mathbb{R}^{n_p \times n_p}$ are assumed to be symmetric and to be positive definite. To obtain a learning controller that minimises (11), the weighting matrices must be chosen such that the Hessian matrix

$$\boldsymbol{H} = \left(\boldsymbol{\Phi}^T \boldsymbol{W}_e \boldsymbol{\Phi} + \boldsymbol{W}_u + \boldsymbol{W}_\Delta\right) \tag{15}$$

is positive definite. If all weighting matrices $\boldsymbol{W}_e$, $\boldsymbol{W}_u$, $\boldsymbol{W}_\Delta$ are positive definite, this condition is satisfied automatically.

By setting the derivative of $J_i(\boldsymbol{u}_i)$ with respect to $\boldsymbol{u}_i$ equal to $\boldsymbol{0}$, the optimal update law

$$\boldsymbol{u}_i = \boldsymbol{Q}\boldsymbol{u}_{i-1} + \boldsymbol{L}\boldsymbol{e}_{i-1} \tag{16}$$

is obtained. The so–called Q–Filter $\boldsymbol{Q}$ and learning operator $\boldsymbol{L}$ read as follows

$$\boldsymbol{Q} = \boldsymbol{H}^{-1}\left(\boldsymbol{\Phi}^T \boldsymbol{W}_e \boldsymbol{\Phi} + \boldsymbol{W}_\Delta\right) , \tag{17}$$

$$\boldsymbol{L} = \boldsymbol{H}^{-1}\boldsymbol{\Phi}^T \boldsymbol{W}_e . \tag{18}$$

Hence, knowing the state space model of the system to be controlled and specifying the weighting matrices $\boldsymbol{W}_e$, $\boldsymbol{W}_u$, and $\boldsymbol{W}_\Delta$ fully determines the update law (16). Note that $\boldsymbol{W}_u = \mathcal{O}$ leads to $\boldsymbol{Q} = \boldsymbol{I}$.

2.2 Repetitive Control

System Description. As previously mentioned, a frequency-domain approach is chosen here. Consider a linear, time-invariant, single-input single-output, discrete-time system in the i–th iteration

$$Y_i(z) = G(z)U_i(z) + D(z) \tag{19}$$

where $Y_i(z) \in \mathbb{C}$, $U_i(z) \in \mathbb{C}$, and $D(z) \in \mathbb{C}$ are the z–transforms of the output $y_i(k)$, the input $u_i(k)$, and the disturbance $d(k)$, respectively. The plant transfer function $G(z) \in \mathbb{C}$ is assumed to be asymptotically stable and to have a relative degree of $\varrho = 1$. As before, the disturbance is assumed to be iteration-invariant.

$\mathcal{H}_\infty$ – Optimal Repetitive Control. The $\mathcal{H}_\infty$–synthesis framework [26] has been successfully used for the synthesis of (robust) repetitive controllers in the past [6], [2], [21], [30]. Consider the following frequency domain update law

$$U_i(z) = Q(z)U_{i-1}(z) + z\,L(z)E_{i-1}(z)\ , \tag{20}$$

where the Q-filter transfer function $Q(z) \in \mathbb{C}$ with $\|Q(z)\|_\infty \le 1$ and the learning transfer function $L(z) \in \mathbb{C}$ are assumed to be asymptotically stable, i.e. all poles of $Q(z)$ and $L(z)$ lie inside the unit circle around the origin of the complex plane. Accordingly, $E_i(z) \in \mathbb{C}$ is the z-transform of the control error of the i–th iteration which is calculated from the iteration-invariant z-transform $R(z)$ of the reference $r(k)$, i.e. $E_i(z) = R(z) - Y_i(z)$. Combining (19) and (20) leads to the error transition equation

$$\underbrace{\Big(E_\infty(z)-E_i(z)\Big)}_{\Delta E_i(z)\in\mathbb{C}} = \underbrace{\Big(Q(z) - G(z)\,z\,L(z)\Big)}_{M(z)\in\mathbb{C}}\underbrace{\Big(E_\infty(z)-E_{i-1}(z)\Big)}_{\Delta E_{i-1}(z)\in\mathbb{C}}\ , \tag{21}$$

where $E_\infty(z) = \lim\limits_{i\to\infty} E_i(z) = \lim\limits_{i\to\infty} E_{i-1}(z)$ is the converged control error with

$$E_\infty(z) = \underbrace{\big(1 - M(z)\big)^{-1}\big(1 - Q(z)\big)}_{S_\infty(z)\in\mathbb{C}}\big(R(z) - D(z)\big)\ . \tag{22}$$

The $\mathcal{L}_2$–norms $\|R(z)\|_2$, $\|D(z)\|_2$ are assumed to be finite. A sufficient convergence condition for the learning system (19), (20) is [6]

$$\|M(z)\|_\infty < 1\ . \tag{23}$$

If (23) is satisfied, $\|\Delta E_i(z)\|_2 \le \|\Delta E_{i-1}(z)\|_2$ holds and $\|S(z)\|_\infty$ is finite which implies that the $\mathcal{L}_2$–norm of the converged control error $\|E_\infty(z)\|_2$ is finite as well. For a given Q-filter $Q(z)$, the $\mathcal{H}_\infty$–optimal learning operator $L(z)$ can be calculated by

$$L(z) = \underset{L(z)\in\mathcal{H}^\infty}{\arg\min}\Big\{Q(z) - G(z)\,z\,L(z)\Big\}\ . \tag{24}$$

This optimisation problem can be solved by means of well-known algorithms [7], [8], [16]. To this end, the error transition equation (21) is rewritten as

$$\begin{pmatrix}\Delta E_i(z)\\ Y_L(z)\end{pmatrix} = \underbrace{\left(\begin{array}{c|c} Q(z) & -G(z)\,z\\ 1 & 0\end{array}\right)}_{P(z)\in\mathbb{C}^{2\times2}}\begin{pmatrix}\Delta E_{i-1}(z)\\ U_L(z)\end{pmatrix}, \tag{25}$$

where $P(z)$ is a so-called generalised plant model with $U_L(z)$ and $Y_L(z)$ as shown in Fig. 1. When $Q(z)$ is specified, $P(z)$ can be used as input to corresponding software packages, e.g. the one described in [3], to determine $L(z)$.

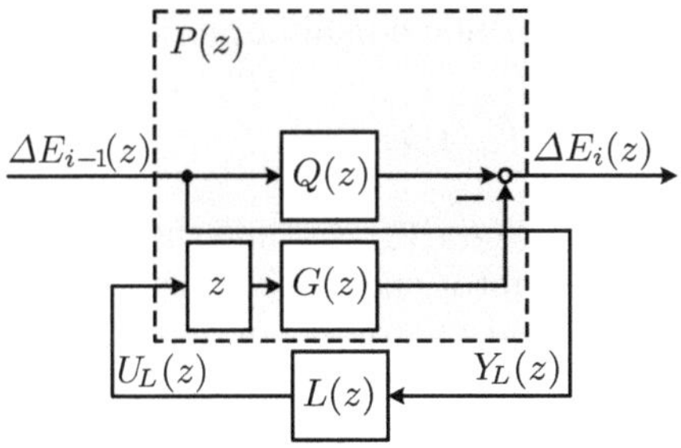

Fig. 1. Generalised plant model $P(z)$ for nominal $\mathcal{H}_\infty$– synthesis of an RC

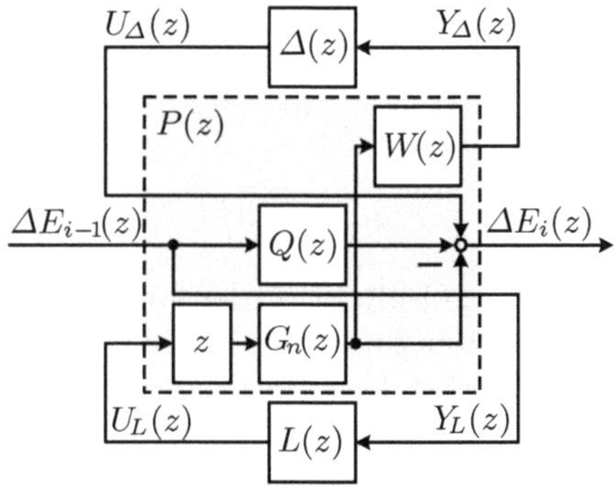

Fig. 2. Generalised plant model $P(z)$ for robust $\mathcal{H}_\infty$– synthesis of an RC

Robust $\mathcal{H}_\infty$ – Optimal Repetitive Control. The plant is now assumed to contain uncertainties. It is described by a nominal model $G_n(z) \in \mathbb{C}$ and an uncertainty description. $G_n(z)$ is assumed to possess a relative degree of $\varrho = 1$. Considering a multiplicative output uncertainty description we have

$$G_p(z) = \Big(1 + \Delta(z)W(z)\Big)G_n(z) \;, \tag{26}$$

where $G_p(z) \in \mathbb{C}$ is the so-called perturbed plant model, $\Delta(z) \in \mathbb{C}$ is a dynamic normalised uncertainty with $\|\Delta(z)\|_\infty \leq 1$, and $W(z) \in \mathbb{C}$ is a weighting function describing the frequency dependent amplitude of the uncertainty. Now, the generalised plant model becomes

$$\begin{pmatrix} Y_\Delta(z) \\ \Delta E_i(z) \\ Y_L(z) \end{pmatrix} = \underbrace{\left(\begin{array}{c|c|c} 0 & 0 & W(z)G_n(z)z \\ 1 & Q(z) & -G_n(z)z \\ 0 & 1 & 0 \end{array}\right)}_{P(z)\in\mathbb{C}^{3\times 3}} \begin{pmatrix} U_\Delta(z) \\ \Delta E_{i-1}(z) \\ U_L(z) \end{pmatrix} . \tag{27}$$

The generalized plant model for the synthesis of a robust $\mathcal{H}_\infty$ – optimal RC is illustrated in Fig. 2.

3 Experimental Setup

Both control algorithms are applied to a linear compressor stator cascade with periodic disturbances. This section describes the cascade itself, the disturbance generator device and the actuation system. To avoid any confusion concerning the typography, it should be mentioned that all compressor related variables are presented in the normal font, e.g. x, whereas all control theory and signal processing related variables are presented in italic type, e.g. x.

3.1 Compressor Stator Cascade

The compressor stator cascade is described in detail in [13], [31]. The cascade consists of seven blades (controlled diffusion airfoils) bounded by two tailboards. The cascade is integrated into an open wind tunnel. The inflow conditions are monitored by 24 static pressure taps (*Honeywell, MPXV7002DP*) located at $x_a/L = 0.5$ upstream of the blade tips, where L is the chord length of the blades. These sensors are integrated into the sidewalls of the wind-tunnel. Uniform inflow conditions are provided by adjustable tailboards, two boundary layer suction devices, and variable wind-tunnel walls, see Fig. 3. The geometric features and the operating point of the cascade are listed in Tab. 1.

3.2 Disturbance Generator

The disturbance generator device is described in detail in [27]. The device consists of 21 damper flaps, positioned at $x_b/L = 0.71$ downstream of the trailing edges of the blades, see Fig. 3. The damper flaps can be sequentially opened and closed with a frequency up to 2.5 Hz. Thereby, a travelling area corresponding to approximately 90% of the cross-section of one cascade passage is closed at all times. This is schematically illustrated in Fig. 3 as well. By reducing the outlet flow cross-section of the cascade the device induces periodic pressure fluctuations in the cascade flow passages. The damper flaps are driven such that neighbouring passages are blocked one after the other. Hence, the sequence of blocked passages is 1–2–3–4–5–6–1–2–etc., where the numbers correspond to the passage labels given in Fig. 3. When referring to the *middle passage*, passage number 4 is addressed.

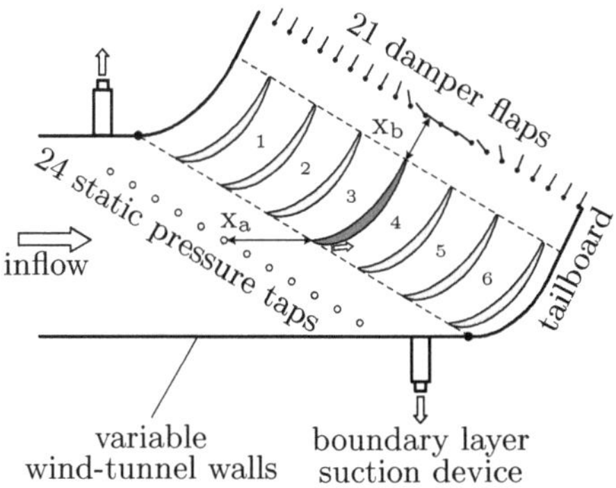

Fig. 3. Stator cascade with damper flaps

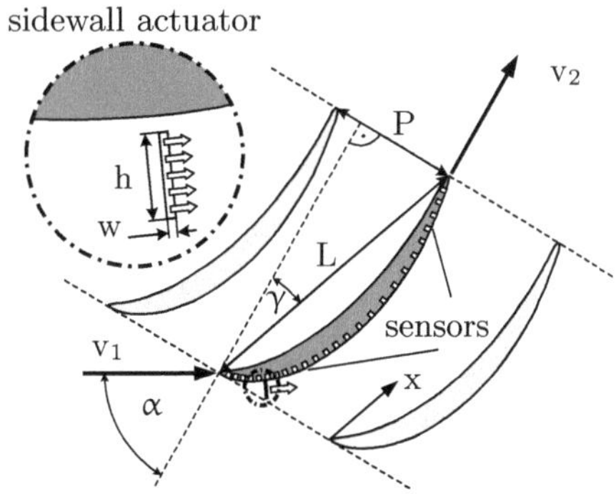

Fig. 4. Compressor blade geometry and sidewall actuator slot. Geometry variables see Table 1.

3.3 Sidewall Actuation

The passage flow is actively manipulated by sidewall actuation. Pressurised air is injected into the middle passage via sidewall slots ($\mathrm{h} \times \mathrm{w} = 0.0200\,\mathrm{m} \times 0.0004\,\mathrm{m}$, injection angle 15°) located at a chord length of $\mathrm{x/L} = 0.1$, see Fig. 4. Each sidewall slot is connected to an individual pressure vessel with volume $V_d = 1.26 \cdot 10^{-4}\,\mathrm{m}^3$ as illustrated in Fig. 5. The corresponding pressure levels $\mathrm{p_{jet,L}}$, $\mathrm{p_{jet,R}}$ (*Festo, SPTW–B6R*) and the injected mass flows $\dot{\mathrm{m}}_{\mathrm{jet,L}}$, $\dot{\mathrm{m}}_{\mathrm{jet,R}}$ (*Festo, SFAB–200U*) are measured. To achieve a symmetric actuation, this sensor information is processed by an subordinate feedback-linearising controller [15] which adjusts the pressure level in the pressure vessels by commanding the corresponding proportional directional valves (*Festo, MPYE–5*).

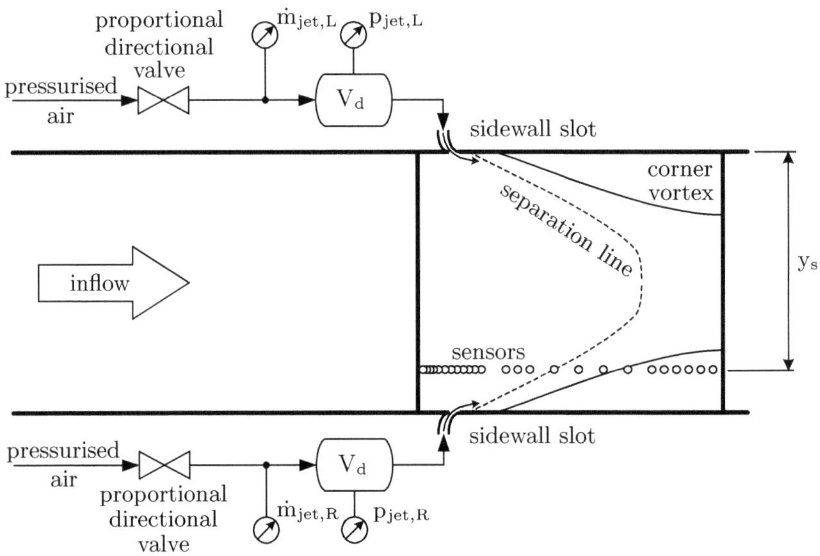

Fig. 5. Sidewall actuation and sensor positions on the middle blade's suction surface

The impact of sidewall actuation on the passage flow has been extensively investigated in the past [12], [11], [10], [31]. The two main effects are the manipulation of secondary flow phenomena, i.e. detrimental corner vortices are pushed towards the sidewalls when the actuation mass flow is increased, and the increase of the overall cascade mass flow at high actuation amplitudes which results in a more distinctive $\mathrm{c_p}$-suction peak on the blades' suction surfaces.

4 Closed-Loop Active Flow Controller

In this contribution, the plant to be controlled is the middle passage of the compressor stator cascade including the sidewall actuation system. As the sidewall actuators are controlled in an subordinate control loop, the control input

$u(k)$ determined by the learning controllers equals the reference signals for these subordinate loops. The main objective of this outer-loop controller is to reject disturbances $d(k)$ which are induced by the damper flaps.

4.1 Surrogate Control Variable

For the purpose of implementing a closed-loop active flow controller, a real-time accessible control variable $y(k)$ describing the impact of the sidewall actuation and the disturbance generator device onto the passage flow needs to be determined. To this end, the $\mathrm{c_p}$-distribution on the middle blade's suction surface was measured for different actuation and disturbance amplitudes, where $\mathrm{c_p}$ is the pressure coefficient. The pressure coefficient at position (x, y) is given by

$$\mathrm{c_p}(\mathrm{x}, \mathrm{y}) = \frac{\mathrm{p_s}(\mathrm{x}, \mathrm{y}) - \mathrm{p_1}}{\frac{1}{2}\rho \mathrm{v}_1^2}, \tag{28}$$

where $\mathrm{p_s}(\mathrm{x}, \mathrm{y})$ is the static pressure at the corresponding position on the suction surface of the blade, ρ is the density of air, and $\mathrm{p_1}$ is the static pressure of the passage inflow. Furthermore, $\mathrm{v_1}$ is the inlet flow speed.

Measurements were taken using a sensor (*SensorTechnics, RMOH*) grid of 57×25 locations on the suction surface. This data was analysed by means of a Principal Component Analysis (PCA) [17]. The PCA has proven to be a useful tool to detect meaningful sensor positions for compressor stator cascade in a former study, see [28]. Fig. 6 shows the impact of the actuation by presenting the first Principal Component (PC) of the $\mathrm{c_p}$-distribution when the actuation amplitude is varied. Black to grey shaded areas indicate a large variance of the local $\mathrm{c_p}$-value when the actuation amplitude is varied in a range from 1.05 bar to u_{max}, for different values of u_{max}. The locations of the sensors that provide an informative real-time accessible surrogate control variable are schematically illustrated in Fig. 5. This sensor line is located at $\mathrm{y_s/H} = 0.85$. The surrogate control variable $y(k)$ itself is calculated by weighting this real–time accessible sensor data with the corresponding first PC describing the impact of the actuation on the $\mathrm{c_p}$-distribution at $u_{\mathrm{max}} = 4.05$ bar, see Fig. 6 (bottom, right).

For sake of completeness, Fig. 7 shows the first (left) and second PC (right) of the impact of the running disturbance generator on the $\mathrm{c_p}$-distribution.

4.2 Dynamic Plant Model

The dynamic input-output behaviour of the plant was determined by means of identification experiments. To this end, the input was chosen as a Pseudo-Random-Binary-Signal (PRBS) containing frequencies of up to 50 Hz and the plant's response, i.e. the progression of the surrogate control variable, was recorded. Both signals were used within the Prediction Error Method (PEM) [22] to determine a dynamic plant model. Since the cascade represents a non-linear system, it cannot be described well by a single linear model. To address this, the amplitude and the mean value of the input PRBS was varied to scan

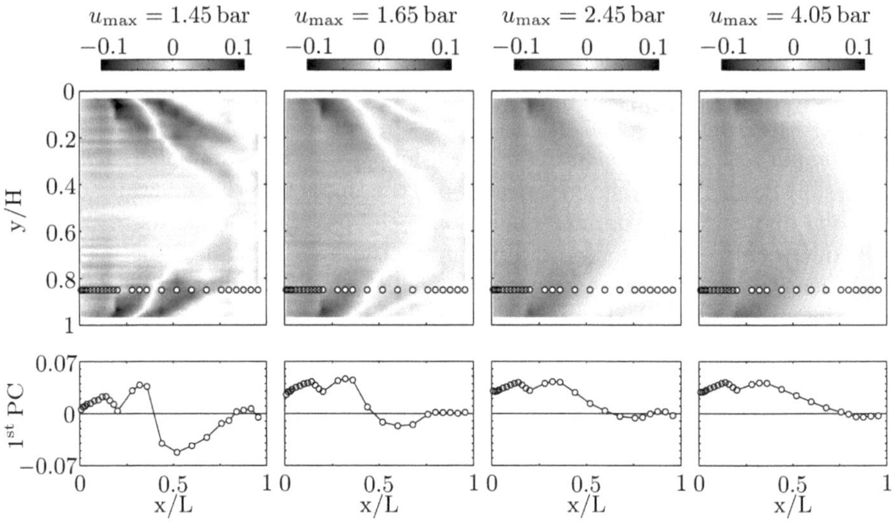

Fig. 6. Top: First PC that describes the impact of the actuation on the c_p-distribution of the suction surface of the middle blade for different maximum actuation amplitudes u_{max}. Bottom: PC profiles at $y/H = 0.85$.

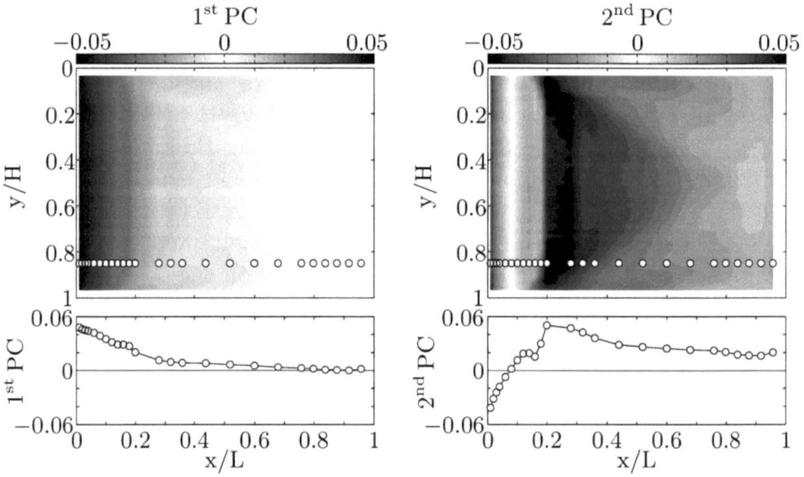

Fig. 7. Top: First (left) and second (right) PC describing the impact of the running disturbance generator on the c_p-distribution of the suction surface of the middle blade. Bottom: PC profiles at $y/H = 0.85$.

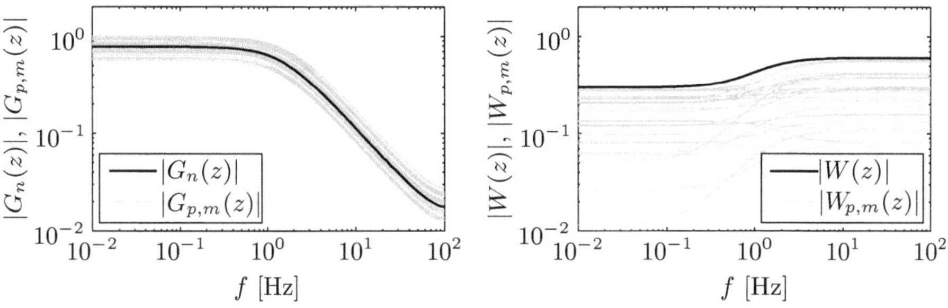

Fig. 8. Left: Gain of the frequency responses of the nominal plant model $G_n(z)$ and the perturbed plant models $G_{p,m}(z)$. Right: Absolute value of the uncertainty weighting function $W(z)$ and the individual relative uncertainties $W_{p,m}(z) = \big(G_{p,m}(z) - G_n(z)\big)G_n^{-1}(z)$.

the nonlinear system for locally linearised models. By doing so, 26 plant models $G_{p,m}(z)$, $m \in \{1, 2, ..., 26\}$ were determined, each describing the input-output behaviour of the cascade locally around a corresponding operating point. Fig. 8 (left) shows the frequency response of the identified plant models $G_{p,m}(z)$. Furthermore, a nominal model $G_n(z)$ was chosen, such that the uncertainty weight $W(z)$ in (26) remains smaller than 1 for all frequencies, see Fig. 8 (right).

4.3 Repetitive Controller

The digital signal processor (*dSPACE, DS1005*) performing the online calculations runs with a sampling time of $T_s = 0.005$ s. All following discrete-time models refer to this sampling time.

The outcome of the model identification is the nominal plant model

$$G_n(z) = \frac{-0.0346}{z - 0.9556}\,, \tag{29}$$

and the uncertainty weight

$$W(z) = \frac{0.6z - 0.5870}{z - 0.9565}\,. \tag{30}$$

$G_n(z)$ and $W(z)$ are used to build a generalised plant model (27). To fully specify $P(z)$, a Q-filter has to be chosen. Here, a low pass filter

$$Q(z) = \frac{0.0801z^2 + 0.1193z + 0.0115}{z^3 - 0.8419z^2 + 0.0759z - 0.0231} \tag{31}$$

is used. The generalised plant model is used for an $\mathcal{H}_\infty$-synthesis of a robust RC. The synthesis result is shown in Fig. 9. Based on the RC sensitivity $S_\infty(z)$, see (22), the bandwidth of the learning system is approximately 3.9 Hz. The worst-case learning rate is $M_{\mathrm{wc}} = 0.664$ at 2.33 Hz.

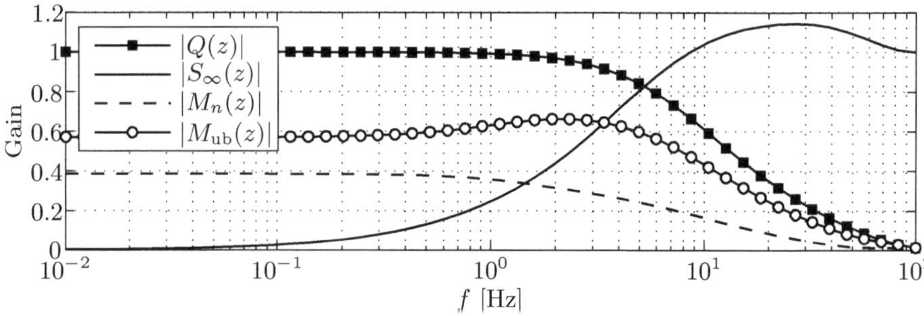

Fig. 9. Frequency response of the Q-Filter $Q(z)$, the RC sensitivity $S_\infty(z)$, and the nominal error transition transfer function $M_n(z) = \Big(Q(z) - G_n(z)\, z\, L(z)\Big)$ as well as its upper bound $M_{\text{ub}}(e^{j\omega T_s}) = \max_{\Delta\in\mathcal{H}^\infty,\, |\Delta|\leq 1} \Big(Q(e^{j\omega T_s}) - G_p(e^{j\omega T_s})\, e^{j\omega T_s}\, L(e^{j\omega T_s})\Big)$.

4.4 Iterative Learning Controller

The ILC synthesis is based on the state space representation of the nominal plant model. The weighting matrices in (12), (13), and (14) are chosen as $\boldsymbol{W}_e = 10^2 \cdot \boldsymbol{I}$, $\boldsymbol{W}_u = 10^0 \cdot \boldsymbol{I}$, and $\boldsymbol{W}_\Delta = 5 \cdot 10^1 \cdot \boldsymbol{I}$, respectively.

5 Closed-Loop Experiments

This section presents results obtained from closed-loop experiments. The experiments were designed to demonstrate and to compare the capability of an optimal ILC against a robust $\mathcal{H}_\infty$– optimal RC for the purpose of rejecting periodic disturbances. To this end, the damper flap device is driven with a frequency of 2.0 Hz, which corresponds to $n_p = 100$ at $T_s = 0.005$ s. This leads to a periodic disturbance in the passage flow. The objective of the learning controllers, i.e. the ILC and the RC, is to reduce the impact of the disturbances on the surrogate control variable as much as possible. The results of the closed-loop experiments are shown in Fig. 10.

Concerning the 2-norm of the error of one iteration $\sqrt{\mathbf{e}_i^T \mathbf{e}_i}$, the ILC performs better than the RC, see Fig. 10 (left). Regarding the progress of the control error during one iteration, see Fig. 10 (right, top), it can be seen that none of the controllers is capable of rejecting the disturbance in the first part of the period, i.e. for $t \in [0\,\text{s},\ 0.21\,\text{s}]$. This is due to the fact that the sidewall actuation device can only blow but not suck, see Fig. 10 (right, bottom). In the second part of the iteration, i.e. for $t \in [0.21\,\text{s}, 0.5\,\text{s}]$, both controllers perform well. However, the RC reacts more aggressively to non-periodic disturbances. This is not beneficial, since it often results in less favourable initial conditions for the next iteration.

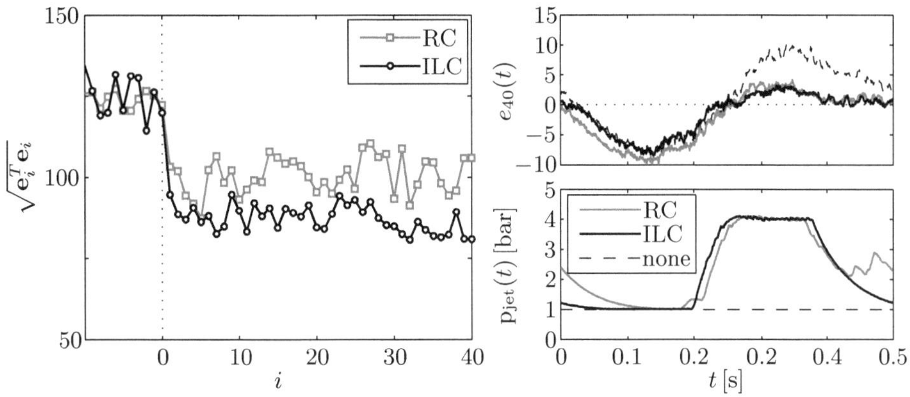

Fig. 10. Left: Comparison between RC (grey, square) and ILC (black, circle) with respect to the development of $\sqrt{\mathbf{e}_i^T \mathbf{e}_i}$ over the iterations. Right top: Trajectory of the control error in the 40–th iteration, RC (grey, solid), ILC (black, solid), no control (black, dashed). Right bottom: Trajectory of the actuation amplitude in the 40-th iteration that has been realised by the subordinate closed-loop.

Table 1. Cascade geometric data and operating point

Parameter	Symbol	Value
Chord length	L	0.375 m
Blade height	H	0.300 m
Blade pitch	P	0.150 m
Stagger angle	γ	20°
Inflow angle	α	60°
Turning	$\Delta\alpha$	60°
De Haller number (v_2/v_1)	DH	0.5
Inlet flow speed	v_1	24.5 m/s
Mach number	Ma	0.07
Reynolds number based on the chord length	Re	600000
Turbulence intensity	Tu	0.02

6 Conclusion

ILC and RC were utilised successfully within an active flow control scenario. Both controllers are capable of decreasing the impact of periodic disturbances. We found the ILC algorithm to obtain better closed-loop results.

The achieved error transition transfer function of the RC learning system is mainly a result of the uncertainty description, e.g. a small uncertainty description $\|W(z)\|_\infty$ leads to a controller which achieves a small value of $\|M(z)\|_\infty$. Hence, for a given uncertain plant there is not much scope left to influence the closed-loop behaviour of the RC learning system. This includes the way the controller handles (unavoidable) non-periodic disturbances as they appear within the presented experiment. Extended generalised plant models that result from considerations presented in [14] may lead to better results.

The ILC approach, however, allows for more degrees of freedom within the controller synthesis. This is due to the three weighting matrices $\boldsymbol{W}_e$, $\boldsymbol{W}_u$, and $\boldsymbol{W}_\Delta$. This greater flexibility allows designing an ILC with a desired learning rate. Especially when it comes to non-periodic disturbances, a less aggressive controller, i.e. one using a $\boldsymbol{W}_\Delta$ with larger elements, turned out to be more beneficial.

Acknowledgements. The authors gratefully acknowledge support by the Deutsche Forschungsgemeinschaft (DFG) as part of collaborative research center SFB 1029 "Substantial efficiency increase in gas turbines through direct use of coupled unsteady combustion and flow dynamics".

References

1. Amann, N., Owens, D.H., Rogers, E.: Iterative learning control for discrete-time systems with exponential rate of convergence. IEE Proceedings - Control Theory and Applications 143(2), 217–224 (1996)
2. Amann, N., Owens, D.H., Rogers, E., Wahl, A.: An $\mathcal{H}_\infty$ approach to linear iterative learning control design. International Journal of Adaptive Control and Signal Processing 10(6), 767–781 (1996)
3. Balas, G., Chiang, R.Y., Packard, A., Safonov, M.G.: Robust Control Toolbox^{TM} User's Guide R2012b. The MathWorks, Inc. (2012)
4. Bristow, D.A., Tharayil, M., Alleyne, A.G.: A survey of iterative learning control. IEEE Control Systems 26(3), 96–114 (2006)
5. Coleman, M.L.: Overview of pulse detonation propulsion technology. Tech. rep., Chemical Propulsion Information Agency (2001)
6. de Roover, D.: Synthesis of a robust iterative learning controller using an $\mathcal{H}_\infty$ approach. In: Proceedings of the 35th Conference on Decision and Control, Kobe, Japan, pp. 3044–3049 (December 1996)
7. Doyle, J., Glover, K., Khargonekar, P., Francis, B.: State-space solutions to standard $\mathcal{H}_\infty$ and $\mathcal{H}_2$ control problems. IEEE Transactions on Automatic Control 34(8), 831–847 (1989)
8. Gahinet, P., Apkarian, P.: A linear matrix inequality approach to $\mathcal{H}_\infty$– control. International Journal of Robust and Nonlinear Control 4(4), 421–448 (1994)
9. Gunnarsson, S., Norrlöf, M.: On the design of ILC algorithms using optimization. Automatica 37(12), 2011–2016 (2001)
10. Hecklau, M., Gmelin, C., Nitsche, W., Thiele, F., Huppertz, A., Swoboda, M.: Experimental and numerical results of active flow control on a highly loaded stator cascade. Proceedings of the Institution of Mechanical Engineers, Part A: Journal of Power and Energy 225(7), 907–918 (2011)
11. Hecklau, M., van Rennings, R., Zander, V., Nitsche, W., Huppertz, A., Swoboda, M.: Particle image velocimetry of active fow control on a compressor cascade. Experiments in Fluids 50(4), 799–811 (2011)
12. Hecklau, M., Wiederhold, O., Zander, V., King, R., Nitsche, W., Huppertz, A., Swoboda, M.: Active separation control with pulsed jets in a critically loaded compressor cascade. AIAA Journal 49(8), 1729–1740 (2011)

13. Hecklau, M., Zander, V., Peltzer, I., Nitsche, W., Huppertz, A., Swoboda, M.: Experimental AFC approaches on a highly loaded compressor cascade. In: King, R. (ed.) Active Flow Control II 2010. NNFM, vol. 108, pp. 171–186. Springer, Heidelberg (2010)
14. Helfrich, B.E., Lee, C., Bristow, D.A., Xiao, X.H., Dong, J., Alleyne, A.G., Salapaka, S.M., Ferreira, P.M.: Combined $\mathcal{H}_\infty$-feedback control and iterative learning control design with application to nanopositioning systems. IEEE Transactions on Control Systems Technology 18(2), 336–351 (2010)
15. Isidori, A.: Nonlinear Control Systems, 3rd edn. Springer, Berlin (2001)
16. Iwasaki, T., Skelton, R.: All controllers for the general $\mathcal{H}_\infty$ control problem: LMI existence conditions and state space formulas. Automatica 30(8), 1307–1317 (1994)
17. Jackson, J.E.: A User's Guide to Principal Components. John Wiley & Sons, New York (1991)
18. Kailasanath, K.: Recent developments in the research on pulse detonation engines. American Institute of Aeronautics and Astronautics Journal 41(2), 145–159 (2003)
19. King, R. (ed.): Active Flow Control. NNFM, vol. 95. Springer, Heidelberg (2006)
20. King, R. (ed.): Active Flow Control II. NNFM, vol. 108. Springer, Heidelberg (2010)
21. Li, J., Tsao, T.C.: Robust performance repetitive control systems. Transactions of the ASME Journal of Dynamic Systems, Measurement, and Control 123(3), 330–337 (2001)
22. Ljung, L.: System Identification: Theory for the User, 2nd edn. Prentice Hall, Englewood Cliffs (1999)
23. Longman, R.W.: Iterative learning control and repetitive control for engineering practice. International Journal of Control 73(10), 930–954 (2000)
24. Norrlöf, M.: Iterative Learning Control, Analysis, Design, and Experiments. Ph.D. thesis, Linköping Universitet, Sverige, Sweden (2000)
25. Roy, G., Frolov, S., Borisov, A., Netzer, D.: Pulse detonation propulsion: challenges, current status, and future perspective. Progress in Energy and Combustion Science 30(6), 545–672 (2004)
26. Skogestad, S., Postlethwaite, I.: Multivariable Feedback Control, Analysis and Design, 2nd edn. John Wiley & Sons, New York (2005)
27. Staats, M., Nitsche, W., Peltzer, I.: Active flow control on a highly loaded compressor cascade with non-steady boundary conditions. In: King, R. (ed.) Active Flow and Combustion Control III. NNFM, vol. 127, pp. 23–38. Springer, Heidelberg (2015)
28. Steinberg, S.J., Tiedemann, C., King, R., Peitsch, D.: Identification of surrogate control variables for a robust active flow controller of an experimental high speed stator cascade. In: Proceedings of the ASME Turbo Expo 2013, San Antonio, Texas, USA, vol. 4 (2013), Paper No. GT2013-94179
29. Wang, Y., Gao, F., Doyle III, F.J.: Survey on iterative learning control, repetitive control, and run-to-run control. Journal of Process Control 19(10), 1589–1600 (2009)
30. Weiss, G., Häfele, M.: Repetitive control of MIMO systems using $\mathcal{H}_\infty$ design. Automatica 35(7), 1185–1199 (1999)
31. Zander, V., Hecklau, M., Nitsche, W., Huppertz, A., Swoboda, M.: Active flow control by means of synthetic jets on a highly loaded compressor cascade. Proceedings of the Institution of Mechanical Engineers, Part A: Journal of Power and Energy 225(7), 897–906 (2011)

Feedforward Control of Lift Hysteresis during Periodic and Random Pitching Maneuvers

Xuanhong An, Lou Grimaud, and David R. Williams

Fluid Dynamics Research Center, Illinois Institute of Technology,
Chicago IL 60616, USA
david.williams@iit.edu
http://fdrc.iit.edu

Abstract. A feedforward controller based on the nonlinear Goman-Khrabrov state space model is designed to compensate for the adverse lift and moment effects resulting from an airfoil pitching in a post-stalled state. The lift hysteresis is shown to be strongly dependent on the reduced frequency, so controllers based on simple synchronization with angle of attack are not likely to work at off-design conditions. The Goman-Khrabrov state space model is shown to accurately predict the lift over a wide range of frequencies and under quasi-random pitching conditions. The model is implemented in a feedforward controller that adjusts the amplitude of a short-duty cycle piezo-electric actuator located near the leading edge of the airfoil. Both periodic and quasi-random pitching motions within the stalled region are used to demonstrate the effectiveness of the control approach. The relatively long time delay believed to be associated with the recovery of the separated flow state limits the controllable bandwidth to $k < 0.1$.

Keywords: flow control, separation, hysteresis, feedforward.

1 Introduction

It is well known that pitching wings near post-stall conditions can produce dynamic stall vortices (DSV) that lead to strong unsteady loads and negative pitching moments. The instantaneous lift value can be 20 percent larger than the static value during pitch-up, and can drop to values 30 percent lower than static during pitch-down. The adverse effects of DSV's appear on the blades of vertical axis wind turbines at low tip-speed-ratios [1], and they appear on rotorcraft during high-speed forward flight. The power available from wind turbines and the flight envelope for rotorcraft blades could be extended, if a method of controlling the lift and pitching moment is developed that is effective over a wide range of operating conditions.

Several active flow control techniques have demonstrated the ability to mitigate the effects of the dynamic stall vortex. Active suction applied near the leading edge of an airfoil undergoing single pitch maneuvers at different pitch rates was used by Karim & Acharya [2] and Alrefai & Acharya [3] to suppress

R. King (ed.), *Active Flow and Combustion Control 2014*, Notes on Numerical Fluid Mechanics and Multidisciplinary Design 127, DOI: 10.1007/978-3-319-11967-0_4

DSV formation, to delay the detachment of the vortex, and to maximize unsteady lift at different pitch rates and angles of attack. Greenblatt & Wygnanski [4,5,6] studied the effects of continuous and intermittent pulsed-blowing excitation on the dynamic stall process during single-frequency periodic pitching motion. Post & Corke [7,8] used plasma actuators synchronized with the angle of attack to control leading-edge flow separation and the DSV formation at a dimensionless frequency $k = \pi fc/U = 0.08$. Transitory control with short bursts from pulsed combustion actuators was shown by Woo et al. [9,10,11] to significantly increase the lift during periodic pitching motion at conditions both in and out of stall. The pulsed combustion actuators were synchronized with the pitching motion. The number of bursts and the phase of the start of control relative to the pitching phase angle were shown to be important to the control effectiveness. Transitory control of stall was achieved for single frequency oscillation of the airfoil at frequency, $k = 0.072$. LePape, et al. [12] used mechanically deployable vortex generators at the leading edge of an airfoil to mitigate the adverse effects of dynamic stall. Their system was also synchronized with blade angle of attack, and the performance evaluated at a frequency $k = 0.1$. Based on these investigations it is clear that the transitory lift and moment history can be strongly influenced by leading edge actuation. Furthermore, the timing of actuation relative to the motion of the airfoil also has a significant effect on the controlled lift history.

The time-periodic actuation techniques described above were studied at single frequency conditions corresponding to $k \sim 0.1$, which is typical of rotorcraft flying at high speeds [13]. The dynamic stall vortex formation time changes with changing pitch frequency, which causes the unsteady aerodynamic lift and drag amplitudes and phases (relative to the motion) of separated flows to be strongly dependent on the frequency. Therefore, one cannot assume that optimized control at one frequency will apply to a different frequency. For example, Granlund, et al. [14] showed for the angle of attack, $\alpha = 15^o$, that the fluctuating lift amplitude in a surging flow is 2.5 times larger than the quasi-steady flow value when $k = 0.6$ and only half as large as the quasi-steady value when $k = 1.2$. Therefore, if flight conditions are different from those at which the control was derived, then the control cannot be expected to be effective. In fact, it is even possible for adverse DSV effects to be enhanced by a mistuned control system. Changing flight speed, executing maneuvers, or flight in gusting flow conditions are situations where we expect single-frequency, synchronized control to fail.

A model-based closed loop controller is expected to be more robust to changing flight conditions. Kerstens, et al. [15] demonstrated the effectiveness of using a dynamic model for external disturbances in a 'randomized-gust suppression' controller for a wing at 20 degrees angle of attack. A transfer function for the lift response to randomized surging flow disturbances was obtained from experimental measurements, and was used to design a feedforward controller that was responsible for most of the lift suppression. The measured lift was used as a feedback signal to a robust controller that compensated for inaccuracies in the feedforward model.

A similar approach is used in this paper, where we investigate the ability of the nonlinear Goman-Khrabrov [16] state-space model (G-K model) to act as a model for generalized pitching disturbances when the airfoil is in a separated flow condition. Pitching motion of the wing produces DSV's that create a lift hysteresis loop, and the G-K model is shown to accurately predict the resulting lift histories. The G-K model is combined with an inverted plant model to form a feedforward controller that is tested at different frequencies and with quasi-random pitching maneuvers around a mean angle of attack. The objective is to demonstrate control of non-periodic pitching motion of the airfoil.

The performance limitations of the feedforward controller and the reasons behind the limitations are identified. The longer term goal is to include a robust feedback controller that will compensate for the inaccuracies of the G-K feedforward model.

2 Experimental Setup

The experiments were conducted in the Andrew Fejer Unsteady Flow Wind Tunnel at the Illinois Institute of Technology. The cross sectional dimensions of the test section are 600 mm x 600 mm, and it has a length of 3.1m. A cut-away diagram of the test article is shown in Fig. 1a. The wing is nominally two-dimensional with an NACA 0009 profile (c = 245 mm, S = 600 mm), which spans the width of the wind tunnel test section. There is a 2 mm gap between the wing and the sidewalls of the wind tunnel to allow the wing to pitch without interference from the side walls. The freestream speed was U = 3 m/s, which gave a chord based Reynolds number Re_c = 57,000. At 3 m/s the turbulence level of the freestream velocity is 0.6 percent over a band from 0.1 Hz to 30 Hz.

The leading edge of the wing consisted of four actuator segments as shown in Fig. 1a and Fig. 1b. Each leading-edge segment contained two actuator housings, which were constructed from Duraform^{TM}nylon in a 3D selective laser sintering rapid proto-typing machine. The actuator slots are located at x/c = 0.077, and the jet exits in the downstream direction at a 30° angle relative to the surface of the wing. The piezo-electric actuators were powered by a PiezoDrive PDL200 amplifier. The resonant frequency of the actuators was 400 Hz, and the maximum voltage to the piezo-disc was 80 V.

The convective time t_{conv} = c/U = 0.082 s is used to normalize time throughout this investigation, so that dimensionless time is t^+ = t/t_{conv}.

Direct measurements of the forces and moments were obtained with an ATI Inc., Nano-17 transducer located inside the wing at x/c = 0.33, which can be seen in Fig. 1a. All moment measurements are converted to a reference point at the quarter-chord location.

The wing was mounted to the force/moment transducer, which was connected to the pitch-plunge mechanism consisting of two computer controlled Copley servo tubes. The two servo tubes enable the pivot point for the pitching motion to be changed. For the results presented in this paper, the pivot point was at the force balance location, x/c = 0.33. Pitch rates were restricted to 2 Hz or less ($k < 0.55$) to avoid over-stressing the force transducer.

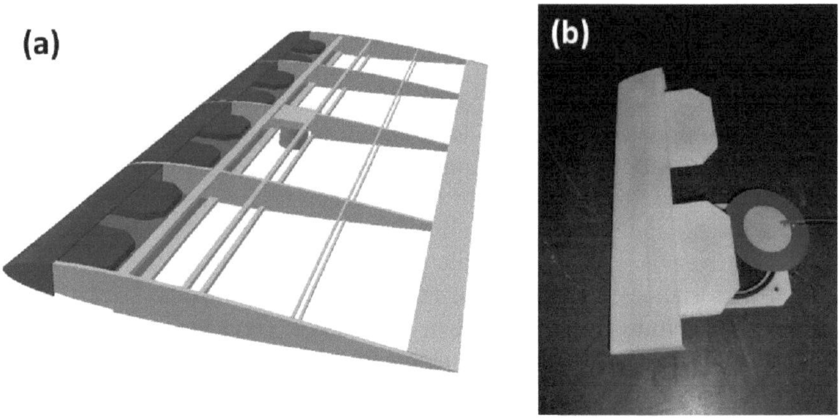

Fig. 1. Wing and actuator segment. a) Drawing of NACA 0009 wing internal structure with four actuator segments and force transducer. b) one leading-edge actuator segment showing exit slots, actuator housing, and piezoelectric components.

The blockage ratio of the wing varied from 7 percent to 10 percent as the wing pitched between 10 degrees and 14 degrees angle of attack. No corrections were applied to the data for blockage. The feedforward control experiments were done using Matlab-Simulink models implemented on a dSPACE 1104 system running at 10,000 cycles/sec.

2.1 Baseline Lift and Moment Dependence on Angle of Attack

The steady state lift and pitching moment coefficients were measured by slowly pitching the wing from -5 degrees to 25 degrees angle of attack at a pitch rate of 0.1 degrees per second. The conditions for actuation are discussed in the next section. The results are shown in Fig. 2 where the solid line is the case without flow control, and the dashed line shows the effect of continuous actuation with $C\mu = \frac{\rho u^2 A_{actuator}}{0.5\rho U^2 A_{wing}} = 0.01$. The force and moment data were low-pass filtered with a cutoff frequency of 100 Hz, which leaves some randomness in the figures. Error bars showing 90% confidence intervals were computed at $\alpha = 0^o, 10^o$, and 20°, and are comparable to the randomness in the C_L and C_m values. Although the wing is symmetric, $C_L \geq 0$ when $\alpha = 0^o$ because the supporting sting introduces an asymmetry in the flow. At $\alpha = 12^o$ the active flow control increases the static lift coefficient from $C_L = 0.95$ to 1.32. Active flow control delays stall and extends the zero pitching moment from 9 degrees to 12 degrees angle of attack.

2.2 Actuation and the Plant Model

Eight piezo-electric disc actuators ("synthetic jets") were used as the flow control actuators. Margalit, et al. [17] showed that short duty cycle (burst mode) actuation is more effective than sinusoidal, while Woo & Glezer [10] demonstrated

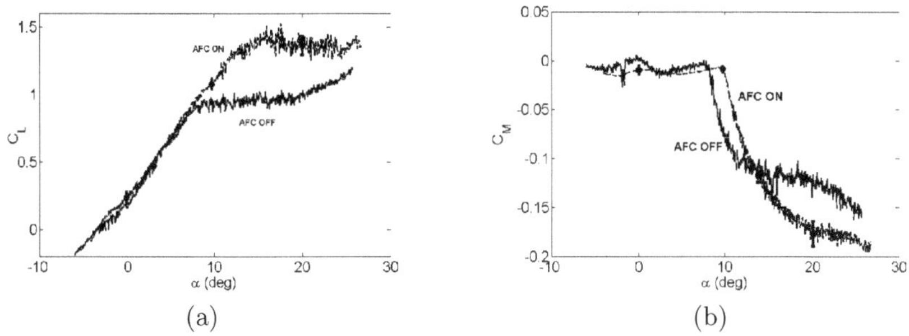

Fig. 2. Static maps of lift coefficient and pitching moment variation with angle of attack. Solid lines are the baseline cases without flow control. The dashed lines show the effect of continuous excitation and the delay of static stall from $\alpha = 9^o$ to 12°.

that actuators operated in a "burst mode" can be very effective in controlling separated flows under transitory conditions. For our experiment an extensive parameter study was conducted, in which the pulse width t_p and the time between pulses t_b were varied to find the optimum operating conditions. The maximum lift increment was obtained when the pulse width, $t_p = 0.12t_{conv}$ and the time between pulses was $t_b = 3.38t_{conv}$, however, it was necessary to shorten the time between pulses to $t_b = 1.56t_{conv}$ to achieve higher actuator bandwidth for the pitching control experiments. The shorter t_b value reduced the maximum achievable lift increment from ΔC_L= 0.3 to 0.2. The 7.7 percent duty cycle was fixed, and the actuator operated as a continuous train of pulses. For flow control applications only the pulse amplitude, A, was varied.

A hot wire anemometer measured the jet velocity at the exit of the actuator jet, which is shown by the dashed line in Fig. 3. The voltage signal to the piezoelectric actuator is indicated by the solid line. The 400 Hz pulse rate is clearly visible in the hot wire signal, and spectrum analysis shows the bandwidth of the actuator to be 80 Hz ($k \approx 20$). However, the effective bandwidth is determined by the time interval between bursts, which occur at a frequency $\frac{1}{t_b} = 7.8$ Hz, or $k = 2.0$. It will be shown later that the effective bandwidth of actuation is approximately 20 times larger than the response of the controlled system, so the actuator does not limit the bandwidth of hysteresis control.

Bench top hot-wire anemometer measurements of the flow velocity at the exit of the actuator with 80 volts excitation determined that the peak velocity was 6.2 m/s. Defining the momentum coefficient with the peak velocity gives C_μ=0.021 at 80 V peak voltage.

The dynamic response of the separate flow over the airfoil to actuator input is modeled by a first order transfer function with time delay $G_p(s)$, which is obtained from experimental data. An example of 1.5 cycles of the dynamic lift increment change that occurs when the wing is fixed at $\alpha = 12^o$ and the actuator voltage is varied with a triangle waveform with a two second period is shown in Fig.4. The lift increment drops below the mean C_L value in the trough of

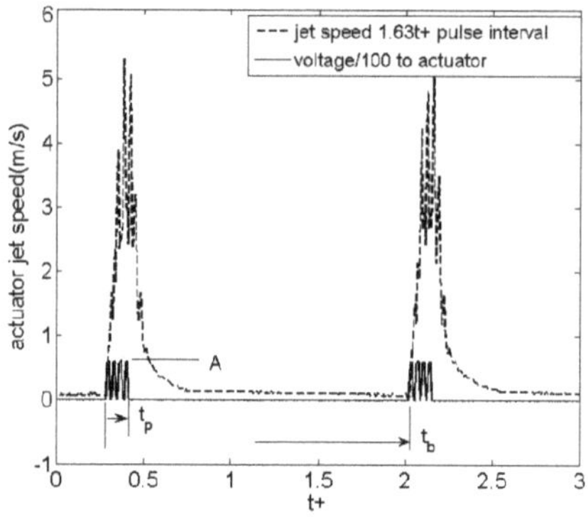

Fig. 3. Actuator pulse duration time $t_p = 0.12t_{conv}$ and time between pulses $t_b = 1.63t_{conv}$. The actuator duty cycle is fixed and only the amplitude of the voltage signal is varied with time.

the wave, which suggests non-minimum phase behavior. However, the first order transfer function with time delay does not capture this feature. The delay of the lift relative to the actuator input signal is typical of separated flow response, and is connected with the formation and convection of a leading edge vortex over the wing [18]. The high frequency oscillations that can be seen in the C_L signal in Fig. 4 result from the individual actuator pulses with period t_b. Leading edge vortices are formed with each actuator pulse which produce a lift increment fluctuation of approximately $\Delta C_L = 0.1$. We did not attempt to capture the higher frequency lift oscillations, because a high-order transfer function would be required for the plant model.

The prediction error method [19](Matlab system identification toolbox) was used with the data shown in Fig. 4 to obtain a first-order transfer function with a time delay. A similar form for the plant model was used by Kerstens, et al. [15] for the gust suppression controller. The identified model is $G_p(s) = \frac{0.713}{(0.2974s+1)} e^{(-0.086s)}$, and the predicted response is shown by the line labelled $G_p(s)$ in Fig.4. Clearly the low order of the transfer function is not capable of tracking the higher frequency oscillations caused by the actuator's short duty cycle, but it does reproduce the lower frequency lift response reasonably well. The time constant for the plant is approximately 0.30 seconds, which corresponds to a dimensionless frequency of $k = 0.86$.

3 Results

3.1 Hysteresis and Synchronized Open-Loop Control

Lift hysteresis was produced by pitching the airfoil about a mean angle of attack of 12 degrees with amplitude of 2 degrees, $\alpha(t) = 12^o + 2^o sin(2\pi f t)$. The hysteresis without active flow control is shown in Figs. 5a and 5b as the gray dashed

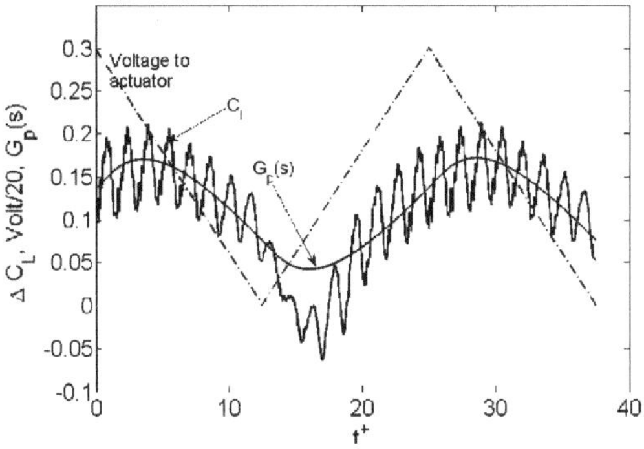

Fig. 4. Measured lift coefficient increment (solid line) at $\alpha = 12^o$ in response to a triangle wave input voltage (dash-dot line) to the actuator is compared to the predicted response from the transfer function, $G_p(s)$. The period of the triangle wave is 2 s, corresponding to $k = 0.13$.

line, and the lift at static conditions is shown by the light gray line. Open-loop forcing experiments were conducted to determine if the fixed duty cycle actuator operated with a variable amplitude was capable of modifying the lift hysteresis loop. A triangle wave signal varying between 0 V and 60 V was applied to the amplitude of the actuator, which was synchronized with the pitching motion of the wing. The two second period ($24t_{conv}$) of the triangle wave matched the period of the pitching motion. A time delay was added to the voltage signal to shift the phase of the periodic actuation relative to the pitching motion.

In Fig. 5a the relative phase between the wing and actuator voltage level is zero, such that the voltage amplitude is zero when the angle of attack is at $\alpha = 12^o$. The controlled case (black solid line) in Fig. 5a shows a large change in the hysteresis loop occurs when flow control is applied. The hysteresis loop is significantly reduced in Fig. 5a, and the range of lift coefficient values is increased. The large amplitude, high frequency oscillations in the lift seen in the controlled case are caused by the short duty cycle actuation. When the actuation is delayed by $\phi = 144^o$ relative to the pitching motion, then the hysteresis loop is significantly increased as shown in Fig. 5b. The dependence of the shape of the hysteresis loop on the phase of actuation indicates that the timing of the actuation is important, which confirms the results observed by Woo, et al. [8-10] As long as the motion is periodic, then some control of the hysteresis is possible by synchronizing the actuator with the pitch angle of the wing. Synchronized control at a single frequency of pitching was used by Post & Corke [8], Woo & Glezer [10]. In both cases the actuator amplitude is fixed and the onset of actuation relative to the pitch angle is optimized for a single frequency of motion. If the frequency changes or the motion of the wing is not periodic, then synchronized control may not work or may even increase the adverse effects of hysteresis. A better approach is to include a model for the lift

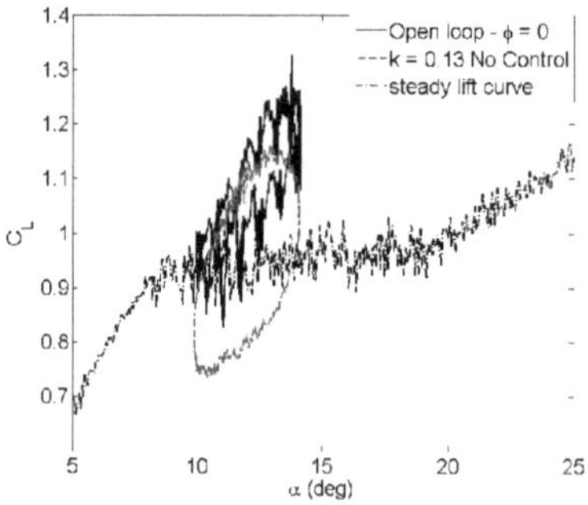

(a)

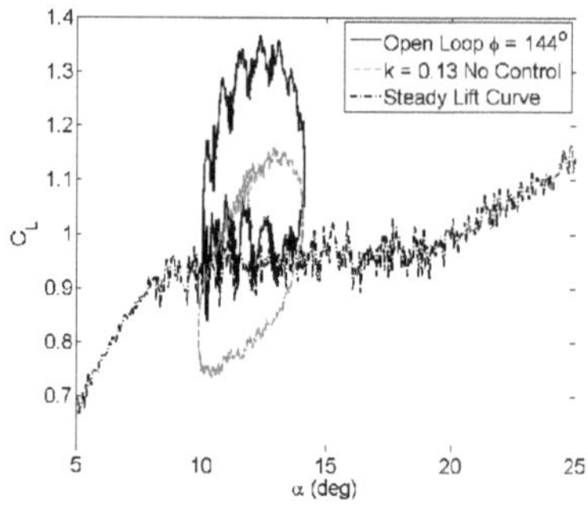

(b)

Fig. 5. Lift hysteresis produced by the wing pitching about $\alpha = 12^o$. The uncontrolled case is shown by the dashed line. Open-loop forcing with a triangle wave excitation (0 - 80 V) that is synchronized with the pitching motion is shown by the solid line. a)$\phi = 0^o$ between actuation and pitching motion; b)$\phi = 144^o$ between actuation and pitching motion.

response to external disturbances that can compensate for changing frequencies and amplitudes of pitching motion. A disturbance model for the lift hysteresis resulting from the pitching motion is needed. In the following section we apply the Goman Khrabrov model, and test its ability to predict lift under different periodic and quasi-random pitching conditions.

3.2 G-K Model Prediction of Lift

The Goman- Khrabrov [16] (G-K) state-space model is used to predict the lift during arbitrary pitching motions of the wing. The G-K model uses two time constants (τ_1 and τ_2), and an internal dynamic variable X that nominally represents the degree of flow attachment over a wing. Fully attached flow corresponds to $X = 1$, and fully separated flow is $X = 0$. The function X_o vs α shown in Fig. 6 was obtained by fitting equation (2) to a steady state lift curve. The time constants τ_1 and τ_2 were obtained by fitting the $C_L(\alpha)$ hysteresis maps for the condition k = 0.13. In reality, X_o is not correlated with the actual separation line location, but is a function that is adjusted to fit the static lift (C_L vs. α) obtained from the experimental data. The quasi-steady position of the separation point is given by the function $X_o(\alpha)$, which is shown as the dark line in Fig.6. The time lag associated with dynamic stall vortex formation and its convection over the wing is represented by $\tau_2 = 0.35s, (4.3t_{conv})$, which is approximately four convective times. The relaxation time constant is $\tau_1 = 0.30s, (3.6t_{conv})$. The Euler method of integration was used to compute $X(\alpha(t))$ from equation (1) during real-time control experiments. The gray line in Fig. 6 shows the predicted hysteresis in X when the pitch rate is $k = 0.13$. The instantaneous lift coefficient was found from equation (2).

$$\tau_1 \frac{dX}{dt} + X = X_o(\alpha - \tau_2\dot{\alpha}) \tag{1}$$

$$C_L(\alpha, X) = 2\pi\alpha(0.4 + 0.6X) + 0.1 \tag{2}$$

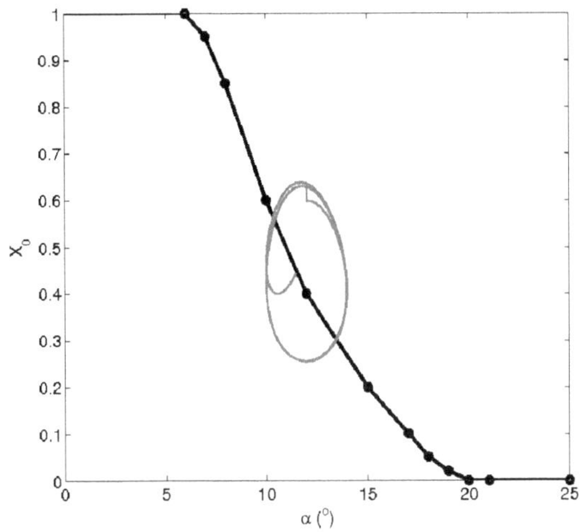

Fig. 6. Quasi-steady and unsteady position of the separation point used as $X_0(\alpha)$ in the G-K model. The hysteresis loop is produced by the time delay, $\tau_2\dot{\alpha}$, for periodic pitching motion $k = 0.13$.

Periodic Motion. The ability of the G-K model to predict the lift hysteresis that occurs with periodic pitching motion is shown four different pitching rates, $k = 0.064, k = 0.13, k = 0.26$, and $k = 0.51$ in Figs.7a-d, respectively. The range of pitch angle is fixed by $\alpha(t) = 12^o + 2^o sin(2kt^+)$. The measured static lift curve and the G-K model fit to the static lift curves are shown by the dotted and solid gray lines. The measured lift hysteresis loop is indicated by the black dashed line, and the G-K model prediction is shown by the solid black line.

The lift hysteresis has a strong dependence on frequency. At low frequency ($k = 0.064$) the loop is almost circular in shape. It becomes more elongated and elliptic as the frequency increases. At $k = 0.51$ the hysteresis is very narrow and the lift coefficient amplitude reaches a peak value of 1.3.

The G-K model tracks the changes in lift during periodic motions quite well. The initial value of C_L for the integration of equations 1 and 2 is set equal to

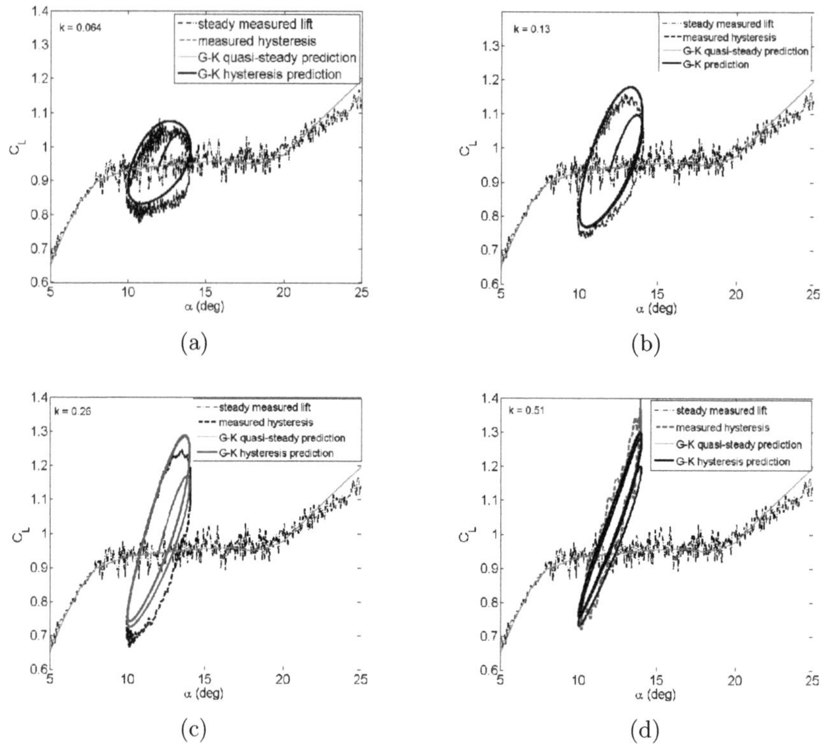

Fig. 7. Measured and G-K model predictions of lift hysteresis for quasi-steady and pitching conditions. a) pitching rate $k = 0.064$; b) pitching rate $k = 0.13$; c) pitching rate $k = 0.26$; d) pitching rate $k = 0.51$.

the static lift value, which is responsible for the short transient observed in the G-K model prediction lines. The model converges to the hysteresis loop within one cycle of motion.

Quasi-random Motion. As discussed in the Introduction, a controller is needed that is capable of operating at different frequencies and with arbitrary (random) maneuvers. The ability of the G-K model to predict the lift produced by random pitching motion is shown in Fig. 8. The quasi-steady pitch motion was derived by superposing 10 sinusoidal signals with random initial phases relative to each other. The highest frequency in the signal corresponded to $k = 0.51$. The G-K model prediction for the random pitching is shown by the black solid line, and is compared to the experimental measurement and a simple attached flow estimate of $C_L = 2\pi\alpha(t)$, which is offset to overlay the experimental data. The experimental lift values have steeper gradients and reach larger amplitudes than predicted by the attached flow model. The G-K model prediction is closer to the experiment than the attached flow model.

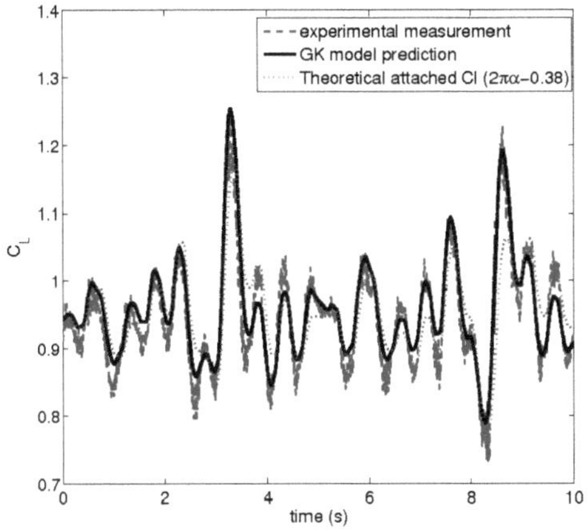

Fig. 8. Comparison of the G-K model prediction of the quasi-random lift signal, the measured lift, and the theoretical attached flow ($C_L = 2\pi\alpha - 0.38$) produced by a quasi-random pitching motion.

3.3 Feedforward Control Architecture

The information provided by the G-K model was used in the feedforward control architecture shown in Fig. 9. A similar control architecture was used successfully by Kerstens, et al. [15] to suppress lift oscillations on a wing experiencing a quasi-random gusting flow. The disturbance model, G_d is the nonlinear Goman-Khrabrov model in state space form. The pitch oscillations are treated as disturbances, d, which become input to the feedforward controller $K_d(s)$. The feedforward controller was constructed from $K_d = G_f\tilde{G_n}^{-1}G_d$. The transfer function $\tilde{G_n}^{-1}$ is the inverse of the first order plant model without the time delay. The term $G_f(s) = \frac{0.713}{0.2974s+0.713}$ is a first order low-pass filter required to make K_d causal. A proportional/derivative transfer function (P-D) following the feedforward controller was added to include an offset phase in the control. For the data presented P = 1 and D = 0.2. The dashed line indicates a lift coefficient feedback signal, which has not been implemented yet in the controller. Similarly, the feedback controller K was not used in the data presented, consequently, the reference lift coefficient C_{Lref} is input directly to the plant $G_p(s)$.

3.4 Feedforward Control Results

The control objective is to eliminate the lift hysteresis by maintaining a constant value of $C_{Lref} = 1.05$ while the wing is pitching with either a periodic or random maneuver. Under ideal circumstances the feedforward controller will provide an actuator input signal that exactly compensates for the lift deviations caused by the pitching maneuver or “disturbance”.

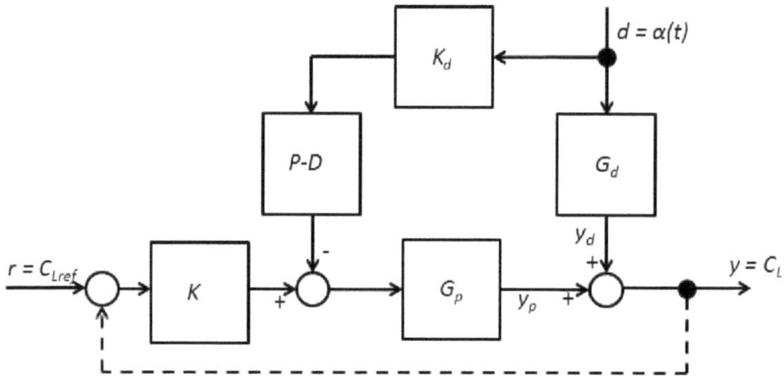

Fig. 9. Control loop architecture used to mitigate the lift hysteresis loop. The design follows the controller used by Kerstens, et al. [15] to alleviate random gust effects on a three dimensional wing.The dashed line shows a feedback path, although it was not used in the feedforward control experiments.

It is clear from the results in Fig. 10 that the feedforward controller is capable of reducing the size of the hysteresis loop, and the controller did make the necessary adjustments to the phase and amplitude of actuation when the frequency was changed. The high frequency fluctuations in the lift associated with the actuator bursts are visible.

When the pitch rate was larger than $k = 0.1$ the controller was not effective in reducing the size of the hysteresis loops. We attribute this limitation to the relatively long recovery time of the separated flow following a burst from the actuator. The lift coefficient increases rapidly ($t^+ \approx 2-3$) to its maximum value following a burst from the actuator, but the flow requires $t^+ \approx 10 - 15$ for C_L to reduce its value to its original state. The actuator cannot create a negative lift increment, it can only stop operating and wait for the flow to recover, which limits the bandwidth of the available control.

Quasi-random Motion. A quasi-random signal for angle of attack was constructed by superposing 10 waves with a random phase between each component. The experimental and G-K model predictions of lift without flow control are shown as the gray dashed and gray solid lines, respectively. The black solid line shows the lift response with the feed-forward controller activated. As before, the feedforward control attempts to maintain $C_L = 1.1$. The controller is clearly adjusting to the randomized lift as desired, but fluctuations are still large. The implementation of the feedback control is expected to help correct for these inaccuracies.

It is our opinion that better control can be achieved with an improved plant model that more accurately represents the response of the separated flow to actuator input. A robust feedback controller should be able to better compensate for the time delays inherent in the recovery phase of the flow separation. While there is clearly room for improvement in the controller design, the feedforward

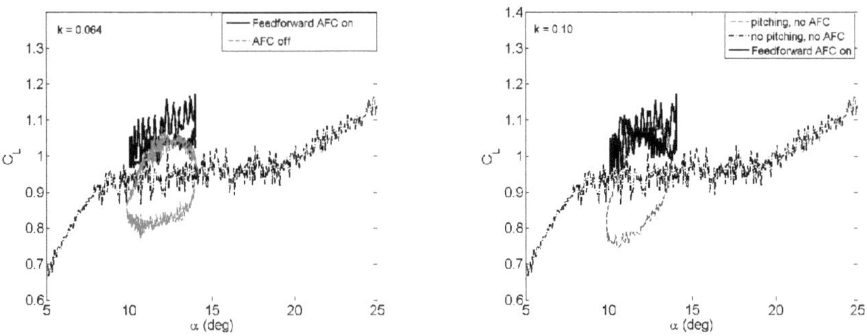

(a) (b)

Fig. 10. Feedforward controller designed to maintain a constant lift coefficient ($C_L = 1.05$) reduces the size of the lift hysteresis loop during periodic pitching maneuver. a) $k = 0.064$; b)$k = 0.10$

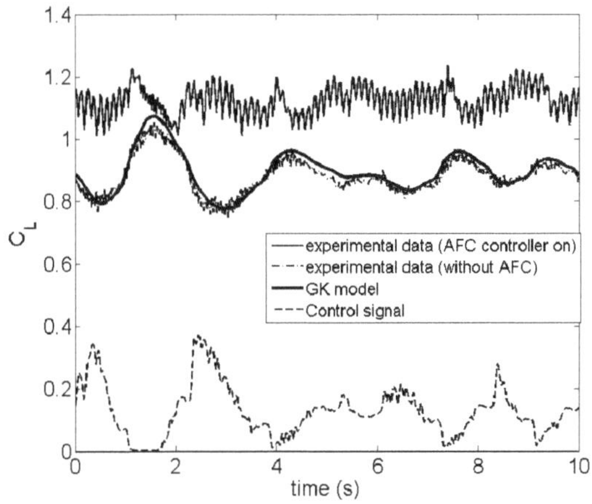

Fig. 11. Feedforward control of a quasi-random lift signal pitching about $\alpha_0 = 12^o$. The controller attempts to maintain constant $C_L = 1.1$. The G-K model ability to predict the lift coefficient variation without AFC is compared with direct measurements of C_L.

controller based on the G-K model has demonstrated a significant advancement compared to synchronization based controllers.

4 Conclusions

The ability of the Goman-Khrabrov state space model to predict lift oscillations and hysteresis was tested in wind tunnel experiments with a two-dimensional wing. Both periodic and quasi-random angle of attack profiles were used to pitch the model around a post stall mean angle of attack of 12 degrees with dimensionless frequencies ranging from $k = 0.064$ to 0.51. The G-K model was able to accurately reproduce the lift hysteresis curves that occur when the pitching motion is periodic, and the G-K model accurately tracked the measured lift time series when the pitching motion is quasi-random.

The G-K model is then used as the model for external disturbances in a feedforward control architecture whose objective is to eliminate lift hysteresis. The feedforward controller was designed as a combination of the G-K model, a low-pass filter, and an inverted form of the plant transfer function. The latter model was the least accurate model. It was shown that lift hysteresis can be substantially suppressed with feedforward control alone, provided $k < 0.1$. The feedforward control also compensates for randomized lift, when a quasi-random pitch input is applied. Without the feedback control component, the controller was not able to compensate for all of the time delays in the plant, which caused deviations in the lift from the set point value of the controller. While there are still improvements to be made, in this paper we have demonstrated the basic feedforward control architecture that can compensate for changing flight conditions.

Acknowledgments. This work was supported by the U.S. Air Force Office of Scientific Research (FA9550-12-1-0075) with program manager Dr. Douglas Smith. The design and construction of the actuated NACA-0009 wind tunnel model by Mr. Simeon Iliev is gratefully acknowledged.

References

1. Sasson, B., Greenblatt, D.: Effect of Leading-Edge Slot Blowing on a Vertical Axis Wind Turbine. AIAA J. 49, 1932–1942 (2011)
2. Karim, M., Acharya, M.: Suppression of Dynamic Stall Vortices over Pitching Airfoils by Leading-Edge Suction. AIAA J. 32, 1647–1655 (1994)
3. Alrefai, M., Acharya, M.: Controlled Leading-Edge Suction for Management of Unsteady Separation over Pitching Airfoils. AIAA J. 32, 2327–2336 (1996)
4. Greenblatt, D., Wygnanski, I.: Dynamic stall control by oscillatory excitation. AIAA Paper 99-3121 (1999)
5. Greenblatt, D., Wygnanski, I.: The Control of Flow Separation by Periodic Excitation. Progress in Aerospace Sciences 36, 487–545 (2000)
6. Greenblatt, D., Wygnanski, I.: Effect of Leading-Edge Curvature and Slot Geometry on Dynamic Stall Control. In: AIAA 1st Flow Control Conference. AIAA (June 2002)
7. Post, M., Corke, T.: Separation Control Using Plasma Actuators: Dynamic Stall Control on an Oscillating Airfoil. AAA Paper 2004-2517 (2004)
8. Post, M., Corke, T.: Separation Control Using Plasma Actuators: Dynamic Stall Vortex Control on Oscillating Airfoil. AIAA J. 44, 3125–3135 (2006)
9. Woo, G., Crittenden, T., Glezer, A.: Transitory Control of a Pitching Airfoil Using Pulse Combustion Actuation. In: 4th Flow Control Conference, AIAA Paper 2008-4324 (2008)

10. Woo, G.T.K., Glezer, A.: Transitory Control of Dynamic Stall on a Pitching Airfoil. In: King, R. (ed.) Active Flow Control II 2010. NNFM, vol. 108, pp. 3–18. Springer, Heidelberg (2010)
11. Woo, G., Crittendon, T., Glezer, A.: Transitory Separation Control over a Stalled Airfoil. AIAA Paper 2010–861 (2010)
12. LePape, A., Costes, M., Richez, F., Joubert, G., David, F., Deluc, J.-M.: Dynamic Stall Control Using Deployable Leading-Edge Vortex Generators. AIAA J. 50, 2135–2145 (2012)
13. McCroskey, W.J.: Dynamic Stall of Airfoils and helicopter Rotors. AGARD Rept. 595, Neuilly sur Seine, France (1972)
14. Granlund, K., Ol, M., Monnier, B., Williams, D.: Airfoil longitudinal gust response in separated vs. attached flows. Phys. Fluids 26 (2014)
15. Kerstens, W., Pfeiffer, J., Williams, D., King, R., Colonius, T.: Closed-Loop Control of lift for Longitudinal Gust Suppression at Low Reynolds Numbers. AIAA J. 49, 1721–1728 (2011)
16. Goman, M., Khrabrov, A.: State-Space Representation of Aerodynamic Characteristics of an Aircraft at High Angles of Attack. J. Aircraft 31, 1109–1115 (1994)
17. Margalit, S., Greenblatt, D., Seifert, A., Wygnanski, I.: Delta Wing Stall and Roll Control Using Segmented Piezoelectric Fluidic Actuators. J. Aircraft 42, 698–709 (2004)
18. Albrecht, T., Weier, T., Gerbeth, G., Monnier, B., Williams, D.: Separated Flow Response to Single Pulse Actuation. In: 51st Aerospace Sciences Meeting, AIAA Paper 2013-849 (2013)
19. Skogestad, S., Postlethwaite, I.: Multivariable feedback control - Analysis and design. John Wiley & Sons, Chichester (1996)
20. Williams, D., Kerstens, W., Pfeiffer, J., King, R., Colonius, T.: Unsteady Lift Suppression with a Robust Closed Loop Controller. In: King, R. (ed.) Active Flow Control II 2010. NNFM, vol. 108, pp. 19–30. Springer, Heidelberg (2010)

Dielectric Barrier Discharge Plasma Flow Control on a Vertical Axis Wind Turbine

David Greenblatt, Hanns Müller-Vahl, Ronen Lautman, Amos Ben-Harav, and Ben Eshel

Faculty of Mechanical Engineering,
Technion, Israel Institute of Technology, Technion City, Haifa 32000, Israel

Abstract. Unsteady flow separation was partially controlled on a double-bladed H-Rotor vertical axis wind turbine model using pulsed dielectric barrier discharge plasma actuators at wind speeds between 4.4m/s and 7.1m/s. With pulsations applied in an open-loop manner on the inboard side of the blades, flowfield measurements showed that the development and shedding of the dynamic stall vortex could be modified. Pulsations were then applied in a feed-forward manner by predetermining the plasma-pulsation initiation and termination azimuth angles. These angles were selected on the basis of wind speed and turbine rotational speed, with the objective of maximizing the net turbine output power. Remarkably, a net turbine power increase of more than 10% was measured. For the purposes of power regulation, a hysteresis controller was applied to the turbine subjected to a fluctuating wind profile. Control produced a 7% increase in net power and a reduction from ±6.5% to ±1.3% in power fluctuations.

1 Introduction

Wind energy is the fastest-growing renewable energy resource, with an average increase of 28% in new installations each year. In addition to the development of multi-megawatt wind farms, significant resources are being channeled to off-grid or feed-in small wind turbines (SWTs; ≤100 kilowatts). Recent demand has driven renewed interest in vertical axis wind turbines (VAWTs) [1-8] due to their insensitivity to wind direction, proximity of the generator to the ground and relatively low noise levels. Furthermore, blade profile uniformity along the span can significantly reduce manufacturing costs. A drawback of VAWTs is the dynamic stall [9] their blades experience as they are pitched beyond their static stall angle [10,11]. This has several undesirable effects: there is a sharp drop in blade lift, producing large power fluctuations and large unsteady loads are imposed on the generator and drive train [12]. In many cases the problem is compounded in built-up areas or on buildings where VAWTs are often required to operate in highly unsteady winds and at low rotational speeds. This forces the turbine to operate in a partially stalled state that is exacerbated at high wind-speed when the potential for power generation is maximized.

R. King (ed.), *Active Flow and Combustion Control 2014*, Notes on Numerical Fluid Mechanics and Multidisciplinary Design 127, DOI: 10.1007/978-3-319-11967-0_5

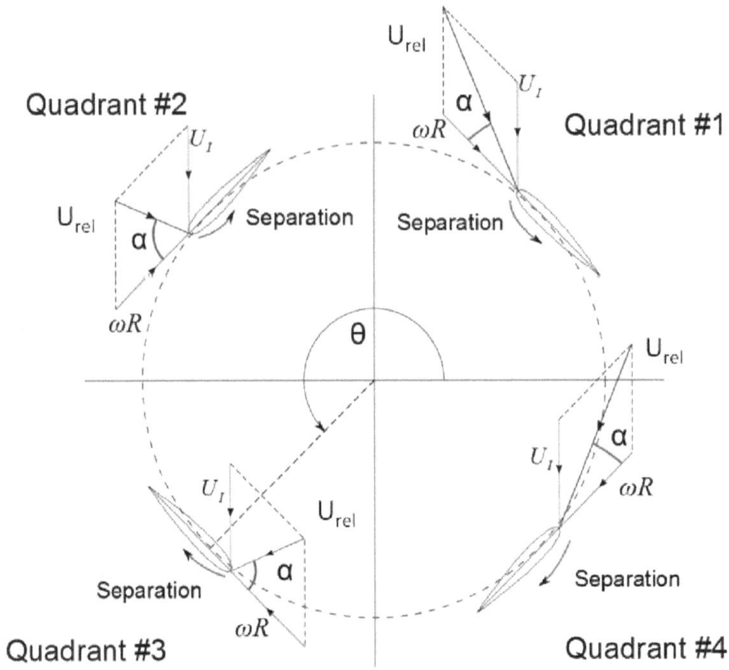

Fig. 1. Top view schematic of a hypothetical four-bladed VAWT showing the velocity components relative to the blade depending upon the position of the blade

The control of unsteady separation, also known as dynamic stall, on airfoils has been achieved by means of introducing periodic perturbations such as zero mass-flux blowing [13-16] and dielectric barrier discharge (DBD) plasma actuation [17]. Recently, our group manipulated dynamic stall on the blades of a small high-solidity vertical axis wind turbine by means of DBD plasma actuators installed on the blade leading-edges [18,19]. A parametric study was conducted to assess the effects of non-dimensional frequency and duty cycle, with the objective of increasing the net power output resulting from control in an open-loop manner. Actuators were configured to control separation on the upwind half of the turbine azimuth and wind speeds in the range 4.4m/s to 7.1m/s were considered. Dynamic stall on VAWTs is unique in that it occurs alternately on the inboard or the outboard surfaces of the blades depending on whether they are upwind or downwind. A schematic showing an idealized VAWT plan view in fig. 1 illustrates the differences between inboard and outboard dynamic stall. Here U_I is the wind velocity immediately upwind of the turbine, U_{rel} is the velocity relative to the blade and α is the local instantaneous angle of attack. The curved arrows indicate dynamic stall occurring on the inboard surface of the blade when the blade is in the upwind half of the azimuth (quadrants #1 and #2). Likewise, the curved arrows indicate dynamic stall occurring on the outboard surface when the blade is in the downwind half of the azimuth (quadrants #3 and #4).

The objective of this paper is to demonstrate a closed-loop control scheme applied to a VAWT. Initially Particle Image Velocimetry (PIV) data were acquired to determine the physical mechanism of control. Following this, a feed-forward controller was developed to initiate and terminate plasma actuation at predetermined initiation and termination angles, θ_i and θ_t. With this controller, plasma pulsation frequency and pulsation duty cycle were varied systematically to maximize turbine performance. Finally, a simple feedback hysteresis control scheme (on–off controller) was employed to increase the net power output and regulate power fluctuations.

2 Experimental Setup

The turbine employed in this investigation (see fig. 2) was an H-rotor (diameter D=0.5m and height H=0.6m) with two composite NACA 0015 blades (chord length c=0.15m) connected to a vertical shaft by means of horizontal adapters. The setup, described in detail in [18], was inverted such that the upper blade tip was closely aligned with the upper adapter. This formed a 25cm (1.67c) free end below the lower adapter (fig. 2a) to facilitate PIV measurements midway between the adapter and the blade-tip. This location was considered to be sufficiently far from both the adapter and the blade-tip so as to be representative of the turbine blade flowfield (see section 3 below). The turbine was installed in a low-speed wind tunnel (dimensions 1.0m× 0.61m×2.0m) with a maximum speed (U_∞) of 55m/s and maximum turbulence level

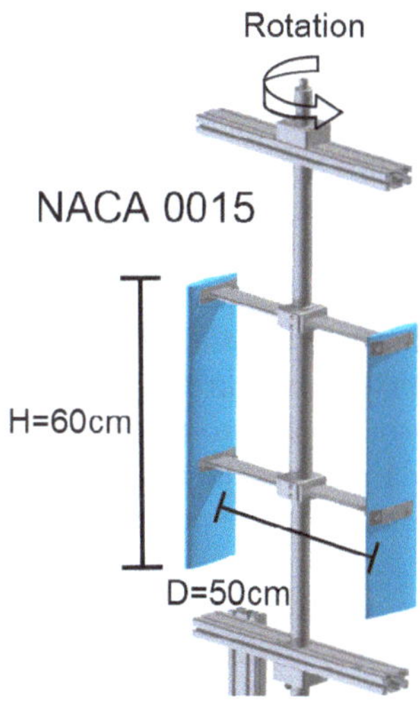

(a)

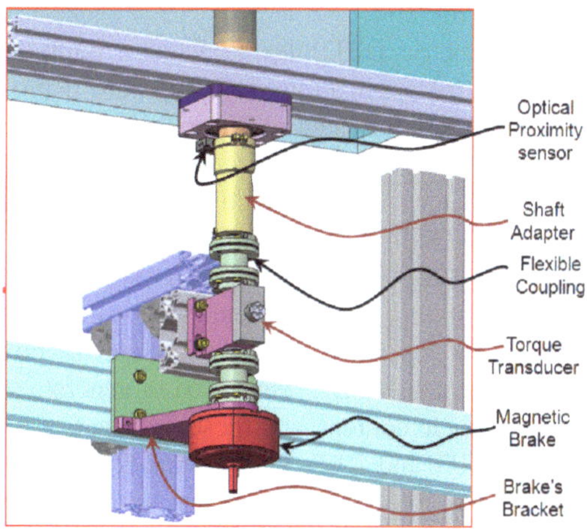

(b)

Fig. 2. (a) Schematic of the turbine mounted in the wind tunnel test section, showing the main dimensions; (b) Schematic of the custom-designed dynamometer with shaft adapter for the slip-ring (not shown)

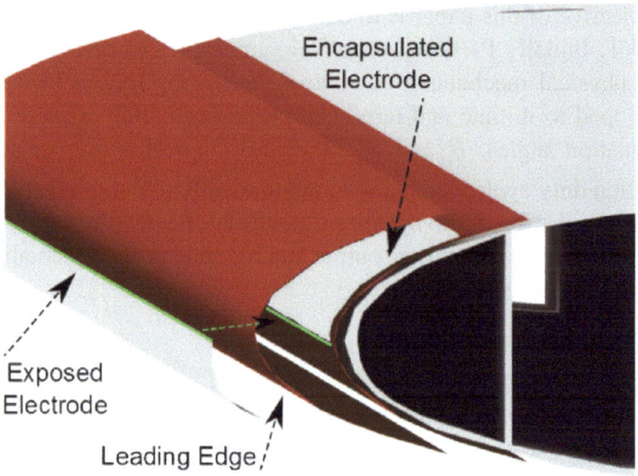

Fig. 3. Schematic of the DBD plasma actuators considered in this study showing the exposed and encapsulated electrode alignment at the leading-edge

of 0.1% relative to the free-stream velocity. The tunnel velocity was monitored by means of a Pitot-static probe, mounted immediately upstream of the turbine. Turbine power P was measured by means of a dynamometer, comprising a Magtrol Magnetic Brake (1.4Nm maximum; Coulomb damping) and a Kistler torque/rotational speed transducer (maximum nominal torque T of 1.56Nm ± 0.2%FSO) and flexible couplings (see fig. 2b).

Asymmetric DBD plasma actuator configurations, which are commonly-used for flow control applications [17,20], were deployed at the leading-edges of the blades (fig. 3). They consisted of 70μm thick exposed and encapsulated electrodes that were separated by 125μm of Kapton© polyimide tape and aligned with the leading-edge. Each actuator was configured to act inboard on the blades and was thus only active on the upwind half of the turbine. Plasma was generated using two Minipuls high-voltage inverters (GBS Elektronik) that require a dc voltage (V_{in}) and current (I_{in}) input, thereby facilitating simple calculation of the gross power supplied to the actuation system ($P_{in}=V_{in}I_{in}$). Note that this was not the power dissipated in the actuators. Signals of 8kVpp at f_{ion}=8kHz were used to generate the plasma, as these produced the smallest harmonic content, and these were pulsed at $f_p = 500$Hz with duty cycles (DC) of between 1% and 20%. This f_{ion} and f_p combination thus produced truncated pulsation cycles; the consequences of this are discussed in section 4. Slip-rings were used at both ends of the shaft to transfer the high-voltage stationary ac signal to the rotating system and these are described in detail in [17].

The PIV data were acquired with a 200 mJ double-pulsed Nd-YAG laser and a 4 megapixel CCD camera. The laser light sheet was oriented horizontally and perpendicular to the blade plane at a spanwise distance of 0.83c from both the adapter and the blade free edge (see fig. 1a). A seeding generator, producing particles 1μm diameter, was placed upstream of the wind tunnel inlet. Sufficient seeding density to produce high-quality PIV measurements was achieved by filling the entire laboratory space with seeding particles. A mirror was placed beneath the test section at a 45° angle and the PIV camera was placed in front of the mirror. The use of the mirror was necessary to orient the camera axis horizontally and thereby avoid accumulation of seeding particles on the lens. 3M red reflective tape was placed on all aluminum components. In combination with a filter mounted in front of the camera lens that permits only light of the Nd-YAG laser (532nm) to be captured, this significantly reduced reflections and thereby improved the data quality. Additional details of the PIV setup can be found in [19].

3 Dynamic Stall Vortex Control

For all PIV data sets, 400 image pairs were acquired with a pulse separation of $\Delta t = 25\mu s$. The data were processed with a multi-pass algorithm using 64×64 pixel interrogation windows initially and 16×16 pixel windows for the final results. These measurements were made with the pulsed plasma operating continuously, thus we refer to this as open-loop control. The velocity component ωR resulting from the rotation of the blade was subsequently added to the velocity fields so as to present the data in the coordinates of a stationary blade. The vorticity distributions were calculated from the mean velocity fields and non-dimensionalized with respect to the chord length and the wind tunnel speed:

$$\omega_z = \left[\frac{\partial v}{\partial x} - \frac{\partial u}{\partial y}\right] \cdot \frac{c}{U_\infty} \tag{1}$$

Example PIV data sets for baseline and open-loop control cases recorded at $\theta = 120°$ (quadrant #2) are shown in figs. 4a and 4b respectively. Under baseline conditions, the initial tip speed ratio of $\lambda_0 \equiv \omega R / U_\infty = 1.37$ corresponds to an angle-of-attack of approximately 47° at this azimuthal position; hence this can be considered as very deep stall. Not surprisingly, the baseline PIV data shows a large dynamic stall vortex (DSV) in the process of being shed from the leading-edge on the inboard side of the blade (fig. 4a). In addition, a large region of vortical flow is present above the aft region of the blade. A similar phenomenon was observed by [21] on a pitching NACA 0018 airfoil and is believed to originate from trailing-edge separation on thick airfoils, particularly at low Reynolds numbers. A feature generally not observed on oscillating airfoils is the strong shear layer that forms at the trailing-edge. This is attributed to the small radius of the turbine relative to the chord length, c/R=0.6. After the baseline data had been recorded, the plasma pulsations were initiated with the brake torque left unchanged. As a result of control the turbine speed increased and PIV data was recorded once it had settled to new final steady state at $\lambda_F = 1.48$ (fig. 4b). With control the vortex is smaller and closer to the surface of the blade. Aft on the blade, the vortical region is stronger and the magnitude of the flow towards the surface is greater. These observations are consistent with the higher tip speed ratio and represent an

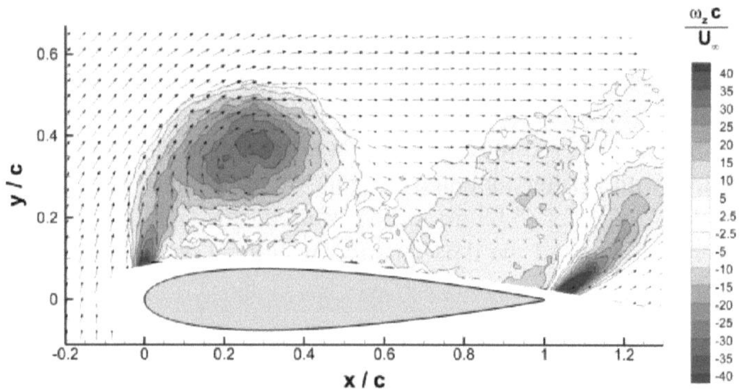

(a)

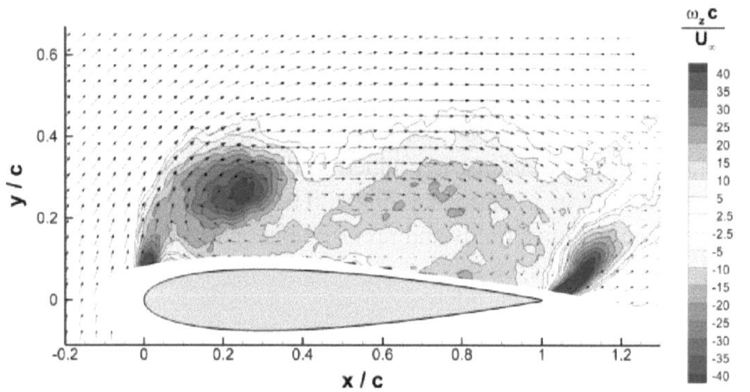

(b)

Fig. 4. Phase averaged velocity and vorticity fields recorded at $\theta = 120°$ with U_∞=7m/s showing (a) the baseline at $\lambda_0 = 1.37$ and (b) pulsed plasma at $\lambda = 1.48$. Pulsed plasma perturbations are at 500Hz and 20% duty cycle.

increase in measured power of 8%. It is evident from these data that the measured increases in turbine power result from a modification of the DSV and the aft flowfield.

Increasing the brake torque, and thereby returning the turbine to $\lambda = 1.37$ with plasma pulsations, resulted in a smaller visible effect on the DSV even though the power increased by 16.4%. The small effect was attributed to the exceedingly large maximum angle-of-attack (approximately 47°). A similar experiment, carried out at a lower brake setting with $\lambda_0 = 1.74$ and hence maximum angle-of-attack of 35°, produced a greater effect on the DSV [19] although elimination of dynamic stall was certainly not achieved. This is mainly due to low thrust produced by the actuators; indeed the maximum torque measured with the actuators alone is at most two orders of magnitude less than the measured changes. Nevertheless, in future investigations significantly greater control authority could be exerted over the vortex with larger amplitude pulsed-plasma perturbations [23].

4 Feed-Forward Control

The data presented in the previous section was acquired in an open-loop manner with the actuators operating continuously and, as such, the plasma pulsations and azimuthal angle were independent of one another. The purpose of the feed-forward control system developed here was to initiate and terminate plasma actuation at predetermined azimuthal angles θ. The system consisted of a wheel, two metallic clips and a proximity transducer (see fig. 5). The wheel was printed using Object Fullcure720® and incorporated an inner and outer circular ring of holes producing a discretization of 1°. The metallic clips could be located independently on the holes and were always located with a spacing of 180° relative to each other; the two-clip arrangement ensured a trigger for each blade, namely at θ_i and at $\theta_i + 180°$. The wheel was bolted to

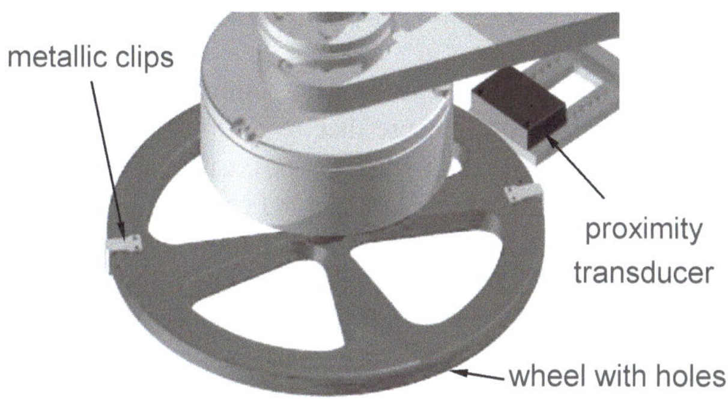

Fig. 5. Schematic of the wheel attached to the brake shaft with attached metallic clips. Sensing of clips by the proximity transducer triggers the pulsed-plasma initiation

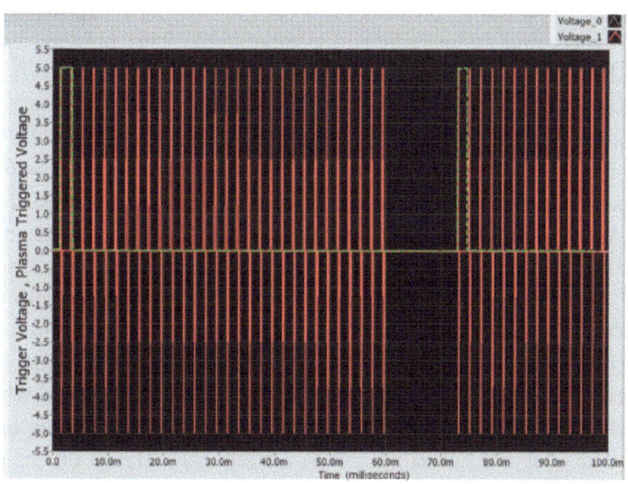

Fig. 6. LabVIEW screenshot showing synchronization of the trigger signal and pulsed plasma driving signal. (Green line represents the trigger; red lines represent the plasma pulses.)

the bottom brake shaft as shown in fig. 5. A bracket was mounted to the frame of the wind tunnel upon which a proximity transducer (TL-18 Series KFPS) was located. With this arrangement, arbitrary angles could be selected to trigger, or initiate, the plasma pulsations. For each trigger signal, a user defined number of plasma pulses N was generated where the frequency and duty cycle could be independently varied (see fig. 6). For the present experiments, these were maintained constant during each turbine transient. With the plasma initiation angle and the number of pulses predetermined, the termination angle was calculated according to:

$$\theta_t = \theta_i + \omega N / f_p \quad (2)$$

where ω is the measured turbine rotational speed. This facilitated the definition of the turbine duty cycle, namely:

$$DC_T = (\theta_t - \theta_i)/180° \quad (3)$$

where the angles are measured in degrees. The parameters on the right hand side of equation 2 were always chosen to ensure that $\theta_t < 180°$. Simultaneous measurements were made of the tunnel speed (Pitot-static probe), trigger signal, turbine torque and rotational speed, and plasma dc supply voltage and current (see section 2).

To determine the effect of initiation and termination angles, the following plasma excitation conditions were selected: 8kVpp, $f_p = 500\text{Hz}$ and 20% duty cycle. Experiments were performed at tip speed ratios in the range $1.38 \leq \lambda \leq 1.53$. For each combination of θ_i and θ_t, after a steady state had been reached, the change in turbine rotational speed ΔRPM relative to the baseline was recorded. Initially, the metallic clips were located on the wheel to produce a plasma initiation angle of $\theta_i = 0°$. The number of plasma pulses was set to N=35 to give a termination angle θ_t slightly less than $180°$. N was then systematically reduced until N=5 and following this, the metallic clips were relocated to produce $\theta_i = 10°$. A starting N<35 was selected to ensure that $\theta_t < 180°$ and again reduced systematically. This process was repeated up to, and including, $\theta_i = 80°$, where performance increases were significantly smaller. A final data set was acquired with the clips at $\theta_i = 75°$.

A selection of data generated in this manner is summarized on the plot shown in fig. 7 where the standard deviation for a 95% confidence interval is displayed for a sample data set. The ΔRPM data are directly proportional to changes in turbine power $P_F - P_0 = T(\Delta\omega)$, where the subscripts refer to the "final" actuated and "initial (0)" baseline states, because the brake torque T was maintained constant. The graph shows a number of important results of feed-forward control. Starting with $0° \leq \theta_i \leq 30°$, when $\theta_t < 60°$ there is a small negative effect, namely the turbine speed reduces slightly ($\Delta\omega < 0$) when plasma is operating in the $0° < \theta < 60°$ range. The reason for this is not clear, but perhaps the perturbations cause mild separation when the flow is fully attached. This effect has been observed in flow control studies on airfoils with attached boundary layers, but using steady air jets [22]. Based on the data shown in [26], it is speculated that applying plasma actuation at 100% duty cycle may have a similar effect and could possibly be used for braking the turbine, but this was not investigated here.

More importantly for purposes of increasing performance, when perturbations are introduced at $40° \le \theta_i \le 70°$, there is always a positive result. Note, however, that the effect saturates at $\theta_t \approx 110°$. This is an important observation because it means that the additional pulsations at $\theta_t > 110°$, and therefore additional power supplied to the actuators, do not produce an additional increase in the turbine power. Increasing θ_i to 75° resulted in a smaller RPM increment than that measured with $\theta_i \approx 70°$ and this effect was even more pronounced for $\theta_i = 80°$. Overall, these results indicate optimum values of $\theta_i \approx 70°$ and $\theta_t \approx 110°$ since this combination produced the largest performance enhancement but only required the plasma to operate throughout a small azimuthal range. The corresponding range of angles of attack can be calculated as $\alpha = \tan^{-1}\left[\sin\theta / (\lambda + \cos\theta)\right]$ when the induced velocity upstream of the turbine is assumed to be equal to the wind tunnel velocity [19]. This approximation yields plasma initiation angles of attack significantly higher than the static stall angle, namely $23° \le \alpha \le 35°$.

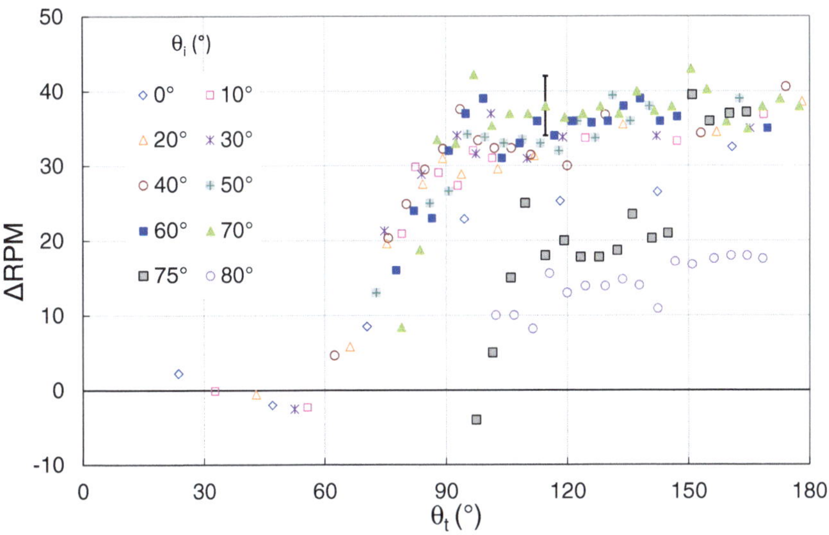

Fig. 7. Effect of azimuthal angle initiation and termination of pulsed plasma actuation: $0° \le \theta_i \le 80°$, $f_p = 500\text{Hz}$, 20% Duty Cycle, $U_\infty = 7\text{m/s}$, initial $\lambda_0 = 1.38$. Sample standard deviation bar shown for 95% confidence interval.

With the turbine duty cycle minimized, a systematic set of experiments was performed with a view to further reducing DC and P_{in}. In general, the measurement of power dissipated in the actuator requires additional electronics. Fortunately, the inverters employed here are supplied with a low-voltage dc source. Hence the power consumed by the actuators, including all losses, was calculated by the dc $P_{in}=V_{\text{in}}I_{in}$ as described in section 2.

This facilitated the calculation of maximum gross and net power changes produced by actuation which were calculated as percentages according to:

$$\Delta P_{\text{gross}} = [(P_F - P_0)/P_0]\times 100 \tag{4}$$

$$\Delta P_{\text{net}} = [(P_F - P_0 - P_{\text{in}})/P_0]\times 100 \tag{5}$$

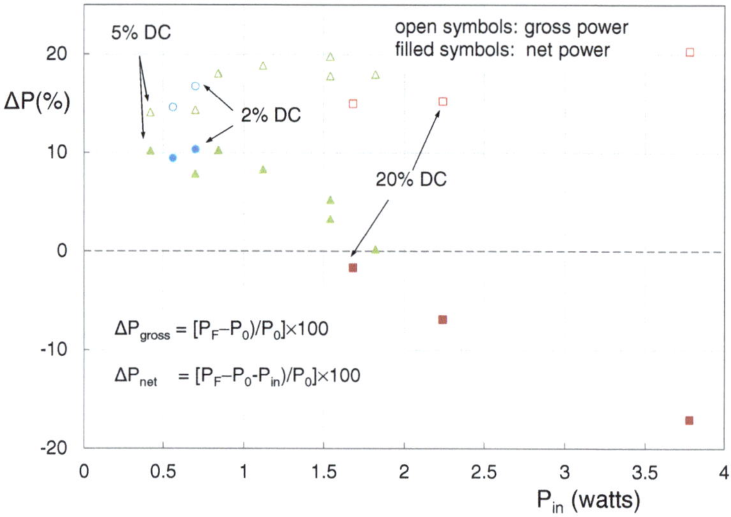

Fig. 8. Gross and net turbine power increase in percentage as a function actuator power. Actuation conditions: $\theta_i = 70°$; $U_\infty = 7$m/s ; initial $\lambda_0 = 1.36$; $5.3 \le F^+ \le 7.3$

Results calculated according to equations 4 and 5 are presented in fig. 8 and are shown as a function of the power input to the actuators. The power varies mainly as result of the reduction the turbine duty cycle (corresponding to a reduction of N) while the different symbols represent different pulsation duty cycles. Open and filled symbols represent gross and net power output respectively. The gross change in power varies from 15% to 20% for these experiments. Note, however, that DC=20% does not produce a net turbine power increase but almost breaks even. For DC=2% and 5%, net positive turbine power is produced with comparable results of approximately 10% in both cases. Despite this positive result, the experiments were hampered by the combination of f_{ion}=8kHz (which produced the least harmonic content) and $f_p = 500$Hz (which was the lowest viable pulsation frequency; see discussion below). This resulted in truncated pulsation cycles for all duty cycles used here, namely those that were not integer multiples of $f_p / f_{\text{ion}} \times 100 = 6.25$. An oscilloscope signal indicated that truncated signals, or a fraction of a cycle for DC<6.25%, resulted in the excitation of the inverter resonance that, for these experiments, was approximately 17kHz. Future experiments on this turbine will maintain full, non-truncated, cycles in order to establish an accurate representation of power consumption.

As alluded to above, and consistent with a previous open-loop investigation [18], the turbine performance was reduced with successive reductions in f_p. This was counterintuitive because the largest performance gains were recorded with reduced frequencies in the range of $5.3 \le F^+ \equiv f_p c / U_{\mathrm{rel}} \le 7.3$ ($f_p = 500\mathrm{Hz}$) even though the optimum value for separation control with periodic excitation is typically on the order of $F^+ \approx 2$ [14, 24]. It was speculated that by lowering f_p, too few plasma pulses were produced in the region where dynamic stall was occurring, namely in quadrants #1 and #2 identified in fig. 1. The number of pulses generated within the effective azimuth range can be estimated by substituting $\theta_i = 70°$ and $\theta_t = 110°$ in equation (2) and rearranging, to produce: $N = \mathrm{Z}(2\pi f_p / 9\omega)$. Pulsing the plasma actuators at $f_p = 500\mathrm{Hz}$ gives $N = 9$, reducing the frequency to $F^+ = 2$ results in N values of 2 or 3 which appear to be insufficient to produce a meaningful effect on dynamic stall. However, with the turbine stationary, the largest changes to torque were observed at $F^+ \approx 2$; these changes were approximately twice as large as those observed with $5.3 \le F^+ \le 7.3$ [18]. This is in line with established findings on separation control, indicating that a larger performance enhancement could potentially be achieved if a sufficient number of plasma pulses could be generated at a lower reduced frequency within the effective azimuth range. The simplest way to overcome this problem is to increase the turbine radius. For example, increasing the radius by a factor of 4, namely to R=1m, would facilitate $N = 13$ at $F^+ \approx 2$ for a turbine operating at the same λ. Furthermore, with the present inverter, lower plasma duty cycles (≤1.5%) could be employed.

Without accounting for the power supplied to the actuation system, a comparison of feed-forward versus open-loop control showed the former to be superior by 7.5% to 12.0% depending on the control parameters. The reason for this relatively small difference was the large disparity between the pulsation frequency and turbine rotational frequency, namely $2\pi f_p / \omega = \mathrm{O}(10^2)$. Therefore, due to the random nature of open-loop control, a similar but smaller number of pulsations were produced in the azimuth demarcated by $70° \le \theta \le 110°$.

5 Feedback Hysteresis Control

With power-efficient feed-forward control demonstrated, attention was focused on rudimentary feedback control, namely hysteresis (or on-off) control. The objective was to regulate the turbine's power when it was subjected to unsteady wind conditions. In contrast to feed-forward control that was applied within the azimuth of each rotation, feedback control was based on the integral rotational speed of the turbine. Thus both feed-forward and feedback control were applied simultaneously, with the former nested within the latter. Simulated unsteady wind conditions were generated by using a system of louver vanes mounted downstream of the turbine at the test section exit. The system consisted of 13 counter-rotating louvers and was driven by a 0.75 kW servo-motor with a 5:1 gear (see fig. 9a). The motor drove the vanes via the gear mechanism shown in fig. 9b thereby varying the exit area and consequently the head-loss, for details see [25].

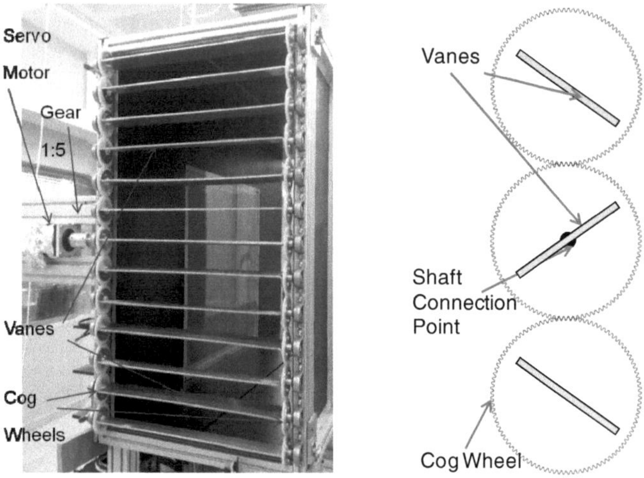

(a) (b)

Fig. 9. (a) Photograph of the louver mechanism attached to the wind tunnel exit, shown with the louvers fully open; (b) cross-section schematic of three vanes illustrating the gear mechanism

For all experiments performed, the tunnel wind speed was varied between approximately 6.05m/s and 6.60 m/s. Wind fluctuation periods between 20 seconds and 80 seconds were considered. For all experiments, the feed-forward parameters were employed as in section 4 with DC=20% and the feedback logic was as follows:

$$P_{out} - P_{set} < 0: \quad \text{actuation on}$$
$$P_{out} - P_{set} \geq 0: \quad \text{actuation off}$$

As in the previous section, the brake setting is maintained constant and thus changes to turbine RPM are directly proportional to the turbine power. Therefore, selecting P_{set} is equivalent to selecting RPM_{set}. This logic is shown in the block diagram of fig. 10. It also includes the nested feed-forward scheme which adapts θ_i and N to the unsteady operating conditions to maintain the optimum values derived in section 4. For purposes of illustration we show here wind fluctuation periods of 40 seconds together with measured turbine RPM in figs. 11a and 11b for baseline and controlled cases respectively. These data are also broadly consistent with those at lower and higher wind fluctuation frequencies. In the uncontrolled case, the wind turbine can be seen to "track" the wind speed fluctuations. This was expected because the turbine behaves like a non-linear first-order system with a time constant of approximately 10 seconds. It can be shown from fig. 11a that the relative changes to turbine power are larger than the relative changes in wind speed: ±4.3% fluctuations in wind speed result in ±6.5% fluctuations in power output.

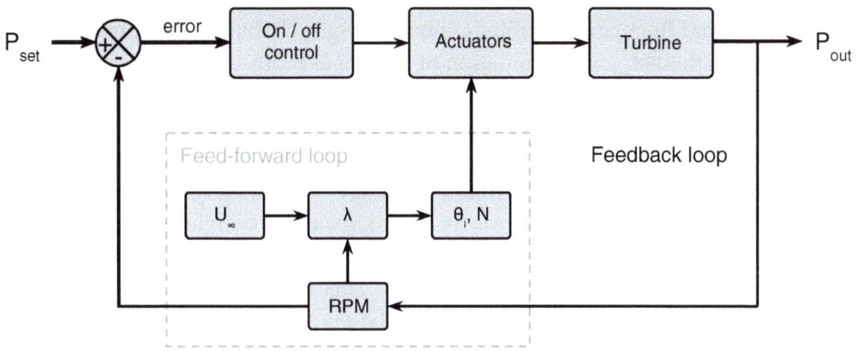

Fig. 10. Block diagram showing the feedback control scheme applied once per second. The nested feed-forward control was applied within the turbine azimuth or rotation.

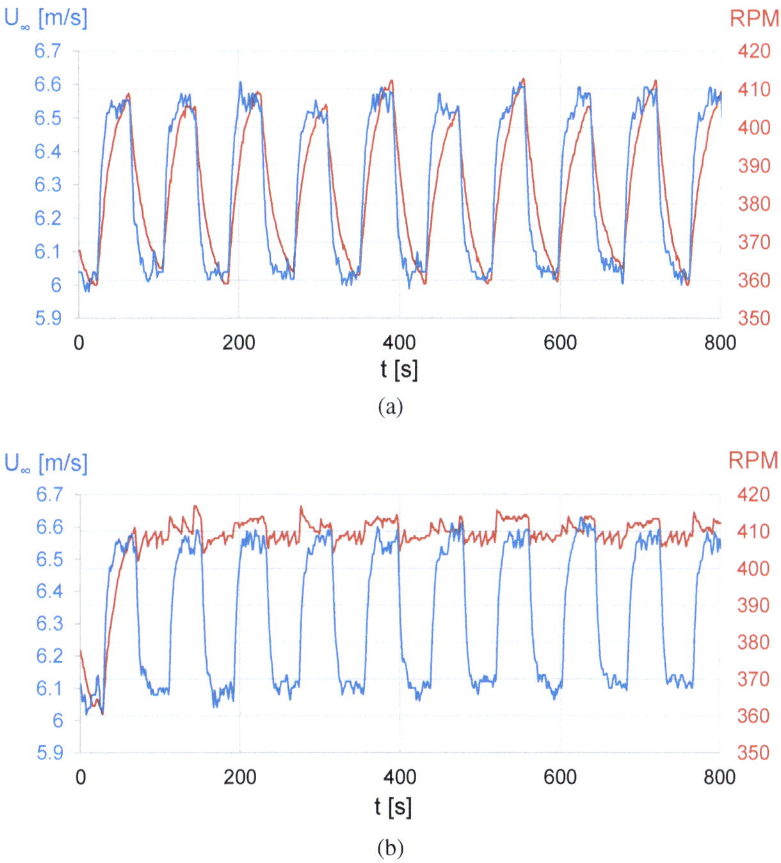

Fig. 11. (a) Wind speed fluctuations shown for a 40 second period together with turbine RPM, proportional to output power: for (a) the uncontrolled, no plasma actuation case, and (b) the feedback control scheme illustrated in fig. 10

The same wind fluctuation profile is shown in fig. 11b, but here feedback control is implemented with RPM_{set} =410. Initiation of control at t=70s produces a response similar to a step function input to the first-order system. Indeed, a linearized first-order model can easily be implement into the feedback control scheme but was not done during these preliminary experiments. After a steady state has been reached, the turbine RPM and, therefore, power fluctuations are dramatically reduced to ±1.3%. Moreover, the mean net turbine power is approximately 7% higher than in the uncontrolled case. A potential improvement to the present setup would be to replace the feed forward control with a feedback scheme. This could be achieved, for example, by placing sensors on the blade surface with the purpose of sensing unsteady separation or incipient dynamic stall. An alternative would be to use adaptive control based on instantaneous measurements of turbine power, thus eliminating the need for wind speed (and direction) measurements.

6 Conclusions

This paper summarized our recent results relating to the use of DBD plasma actuators as a means for controlling the flow, and hence the power and loads, on a 2-bladed vertical axis wind turbine. Open-loop control, feed-forward control and combined feed-forward/feedback control schemes were considered.

Baseline PIV data showed a large dynamic stall vortex in the process of being shed from the blade. With actuation applied in an open-loop manner, the vortex was smaller and closer to the surface. At lower maximum angles-of-attack the effect of pulsations on the dynamic stall was greater, but separation was not fully controlled.

Feed-forward perturbations were effective at increasing turbine power when applied at angles significantly larger than the static stall angle, in the approximate range $23° \leq \alpha \leq 35°$. Furthermore, successively reducing the active azimuthal angle range and reducing the pulsations duty cycles produced a net positive turbine power increase of 10%. Thus feed-forward control was consistently superior to open-loop control due to the reduced turbine duty cycle and elimination of the adverse effect of actuation under pre-stall conditions.

Feedback (hysteresis) control resulted in a reduction in turbine power fluctuations from ±6.5% to ±1.3% and a mean power increase of 7%. Potential improvements to the present control scheme include a dynamical model of the turbine and replacement of the feed-forward control by feedback control based on sensors on the turbine blade and/or adaptive control based on instantaneous measurements of turbine power.

References

1. MacPhee, D., Beyene, A.: Recent Advances in Rotor Design of Vertical Axis Wind Turbines. Wind Engineering 36(6), 647–666 (2012)
2. Islam, M., Ting, D.S.-K., Fartaj, A.: Aerodynamic models for Darrieus-type straight-bladed vertical axis wind turbines. Renewable and Sustainable Energy Reviews 12, 1087–1109 (2008)

3. Howell, R., Qin, N., Edwards, J., Durrani, N.: Wind tunnel and numerical study of a small vertical axis wind turbine. Renewable Energy 35, 412–422 (2010)
4. McLaren, K., Tullis, S., Ziada, S.: Computational fluid dynamics simulation of the aerodynamics of a high solidity, small-scale vertical axis wind turbine. Wind Energy 15, 349–361 (2012)
5. Scheurich, F., Fletcher, T.M., Brown, R.E.: Simulating the aerodynamic performance and wake dynamics of a vertical-axis wind turbine. Wind Energy 14, 159–177 (2011)
6. Marsh, G.: Offshore reliability. Renewable Energy Focus, 62–65 (May/June 2012)
7. Shires, A.: Design optimisation of an offshore vertical axis wind turbine. Energy 166(1), 7–18, Proceedings of the Institution of Civil Engineers, doi.org/10.1680/ener.12.00007.
8. Akimoto, H., Tanaka, K., Uzawa, K.: Floating axis wind turbines for offshore power generation – a conceptual study. Environmental Research Letters 6(4), 044017 (2011), doi:10.1088/1748-9326/6/4/044017
9. Carr, L.W.: Progress in the analysis and prediction of dynamic stall. AIAA Journal of Aircraft 25(1), 6–17 (1988)
10. Simão Ferreira, C.J., van Zuijlen, A., Bijl, H., van Bussel, G.J.W., van Kuik, G.A.M.: Simulating dynamic stall in a two-dimensional vertical-axis wind turbine: Verification and validation with particle image velocimetry data. Wind Energy 13(1), 1–17 (2009)
11. Simão Ferreira, C.J., van Kuik, G.A.M., van Bussel, G.J.W., Scarano, F.: Visualization by PIV of dynamic stall on a vertical axis wind turbine. Experiments in Fluids 46(1), 97–108 (2009)
12. Paraschivoiu, I.: Double-multiple streamtube model for Darrieus wind turbines. In: Second DOE/NASA Wind Turbines Dynamics Workshop, NASA CP-2186, Cleveland, OH, pp. 19–25 (February 1981)
13. Greenblatt, D., Neuburger, D., Wygnanski, I.: Dynamic Stall Control by Intermittent Periodic Excitation. AIAA Journal of Aircraft 38(1), 188–190 (2001)
14. Greenblatt, D., Wygnanski, I.: Dynamic stall control by periodic excitation. Part 1: NACA 0015 Parametric Study. AIAA Journal of Aircraft 38(3), 430–438 (2001)
15. Greenblatt, D.: Active Control of Leading-Edge Dynamic Stall. International Journal of Flow Control 2(1), 21–38 (2010)
16. Greenblatt, D., Wygnanski, I.: The control of separation by periodic excitation. Progress in Aerospace Sciences 36(7), 487–545 (2000)
17. Post, M.L., Corke, T.C.: Separation control using plasma actuators: dynamic stall vortex control on oscillating airfoil. AIAA Journal 44, 3125–3135 (2006)
18. Greenblatt, D., Schulman, M., Ben-Harav, A.: Vertical axis wind turbine performance enhancement using plasma actuators. Renewable Energy 37(1), 345–354 (2012)
19. Greenblatt, D., Ben-Harav, A., Mueller-Vahl, H.: Dynamic Stall Control on a Vertical-Axis Wind Turbine Using Plasma Actuators. AIAA Journal 52(2), 456–462 (2014), doi:10.2514/1.J052776.
20. Corke, T.C., Post, M.L., Orlov, D.M.: SDBD Plasma Enhanced Aerodynamics: Concepts, Optimization and Applications. Progress in Aerospace Sciences 43, 193–217 (2007)
21. Müller-Vahl, H., Strangfeld, C., Nayeri, C.N., Paschereit, C.O., Greenblatt, D.: Thick Airfoil Dynamic Stall. In: Euromech Colloquium 528 "Wind Energy and the Impact of Turbulence on the Conversion Process. Springer (2013) (in press)
22. Attinello, J.S.: Design and engineering features of flap blowing installations. In: Lachmann, G.V. (ed.) Boundary Layer and Flow Control. Its Principles and Application, vol. 1, pp. 463–515. Pergamon Press, New York (1961)

23. Thomas, F.O., Corke, T.C., Iqbal, M., Kozlov, A., Schatzman, D.: Optimization of Dielectric Barrier Discharge Plasma Actuators for Active Aerodynamic Flow Control. AIAA Journal 47(9), 2169–2178 (2009)
24. Greenblatt, D., Göksel, B., Rechenberg, I., Schüle, C., Romann, D., Paschereit, C.O.: Dielectric Barrier Discharge Flow Control at Very Low Flight Reynolds Numbers. AIAA Journal 46(6), 1528–1541 (2008)
25. Furman, Y., Müller-Vahl, H., Greenblatt, D.: Development of a Low-Speed Oscillatory Flow Wind Tunnel. In: 51st AIAA Aerospace Sciences Meeting, Grapevine, Texas, AIAA Paper 2013-0505 (2013)
26. Göksel, B., Greenblatt, D., Rechenberg, I., Nayeri, C.N., Paschereit, C.O.: Steady and Unsteady Plasma Wall Jets for Separation and Circulation Control. In: 3rd AIAA Flow Control Conference, San Francisco, California, USA, June 5-8, AIAA Paper 2006-3686 (2006)

Mach-, Reynolds- and Sweep Effects on Active Flow Separation Control Effectivity on a 2-Element Airfoil Wing

Jochen Wild

DLR German Aerospace Center, 38108 Braunschweig, Germany
Jochen.Wild@dlr.de
http://www.dlr.de/en/as

Abstract. Over the last decade numerous experimental and numerical studies have been conducted by DLR together with partners regarding the maturation of active flow separation control technologies at low speed high-lift conditions. The DLR-F15 high-lift airfoil has been established as a common test bed for such kind of investigations within the German national framework and in the scope of European Research. The present paper highlights major results of this research regarding flow separation suppression by means of pulsed blowing.

1 Introduction

Over the past two decades at least, active flow control (AFC) has been investigated to suppress flow separation. Most investigations have been carried out at Universities in proprietary facilities. Despite the fact that the basic principles are understood, the technology lacks maturity for implementation on real aircraft, especially at transport aircraft category. A major shortcoming is that most of the investigations have been carried out at 2D airfoils and low Reynolds numbers. For the application at multi-element wings of transport aircraft two main effects need to be addressed for further maturation of AFC: the wing sweep and the Reynolds number. Only Seifert [1] reported high Reynolds number investigations at a 2D airfoil with AFC applied to the leading edge and at the kink of a chamber flap. AFC at the leading edge of a swept wing by constant blowing was performed within the EC project EUROLIFT II [2].

In 2006 DLR, in close partnership with the Universities TU Berlin and TU Braunschweig, and Airbus Operations, started to tackle this gap and to study AFC for separation control towards real application. First steps were the setup of a common experimental database featured by a common model geometry, the DLR-F15 airfoil [3]. First tests were performed at medium Reynolds numbers to qualify the experimental setup and to define promising AFC concepts for the control of separation at a two-dimensional 2-element airfoil model [4]. In 2010 high Reynolds number test with this model were performed to qualify the AFC for Reynolds numbers relevant for transport aircraft applications. In 2011 the model setup was modified to incorporate a mounting as a swept cantilever wing.

R. King (ed.), *Active Flow and Combustion Control 2014*, Notes on Numerical Fluid Mechanics and Multidisciplinary Design 127, DOI: 10.1007/978-3-319-11967-0_6

This setup was used to qualify the AFC actuation at relevant sweep angles. This contribution summarizes these activities and gives on overview on relevant findings.

2 The DLR-F15 High-Lift Airfoil Model in AFC Setup

The DLR-F15 wind tunnel model shown in 1 is a 2D wall-to-wall high-lift model. The modular main wing allows exchanging leading and trailing edges. Therefore, different types of high-lift elements can be investigated and compared at the same baseline geometry. The clean wing section is derived from a generic high-lift wing investigated in the national funded project ProHMS [5] and represents a state of the art transonic turbulent airfoil for a modern civil transport aircraft. The high-lift devices are mounted on continuously adjustable brackets, allowing the free positioning of each element in all three degrees of freedom. The model is equipped with about 220 static pressure taps. One dense pressure distribution is located in the center section and is used for the integration of the aerodynamic force and moment coefficients. In addition, two less dense pressure distributions are located closed to the tunnel walls in order to assess the two-dimensionality of the flow. As described in [3], it has been discovered that the pressure distribution is not dense enough for an accurate integration of drag coefficients, leading to errors of up to 20%. The pure integration error for lift coefficients is about 1% and for the pitching moment coefficient an accuracy of about 3% is achieved. Nevertheless, since the deviation is systematical, the quantification of the effects of coinfiguration changes including the application of AFC is still possible.

2.1 Geometry

The setup of interest for flow control is a 2-element configuration that features a clean leading edge, and a single-slotted flap. The flap setup has been adjusted in numerous experiments to provide a specific flow separation pattern to simultaneously study active flow separation control at various positions of the high-lift airfoil. The flap setting denoted as '2eFC' and listed in Tab. 1 shows a separated flap flow over the complete range of angle of attack. Additionally, the stall is triggered by a turbulent leading edge separation as has been proven by examination of the pressure distributions around airfoil stall.

Table 1. geometric characteristics of the DLR-F15 model in AFC setup

variable	*symbol*	*unit*	*value*
clean chord	c	mm	600
flap chord	c_F	$\%c$	28
flap deflection	δ_F	$^\circ$	45.011
flap gap	g_F	$\%c$	2.653
flap overlap	ovl_F	$\%c$	0.505

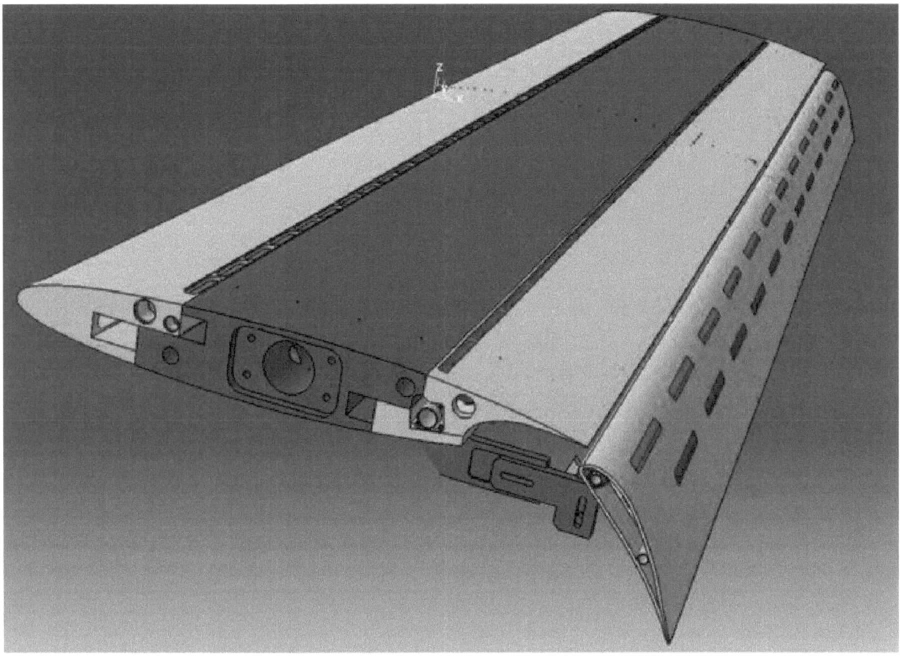

Fig. 1. DLR-F15 modular high-lift airfoil model

2.2 Active Flow Control Actuation

For the control of the flap separation pulsed blowing through slots has been chosen as AFC technology. The principle was introduced by Tinapp and Nitsche [6] and Petz and Nitsche [7] using a combination of two NACA airfoils as representation of a high-lift airfoil. Initially the blowing was obtained by loudspeakers providing a sinusoidal signal. Improvements of the technology led to a pulsed blowing from pressurized air with a close to rectangular signal of the exit velocity by the use of fast-switching solenoidal valves (see Haucke and Nitsche [8]). It has been found favorable to have two actuation positions, leading to the odd-even distributed slots at 20%c and 50%c as shown in Fig. 2.

At the leading edge fluidic vortex generator jets (VGJ) are used to generate longitudinal vortices that energize the boundary layer. Earlier experiments showed that the favorable position for those VGJs is at the lower side close to the leading edge (see Scholz et al. [9,10]). It has been shown that the vortices survive the strong adverse and favorable pressure gradients along the leading edge suction pressure peak, and the efficiency is better due to the higher velocity ratio at the actuator position. Fig. 3 shows the implemented position of the leading edge actuation. The blowing is generated similarly to the flap actuation by pressurized air and fast-switching solenoidal valves.

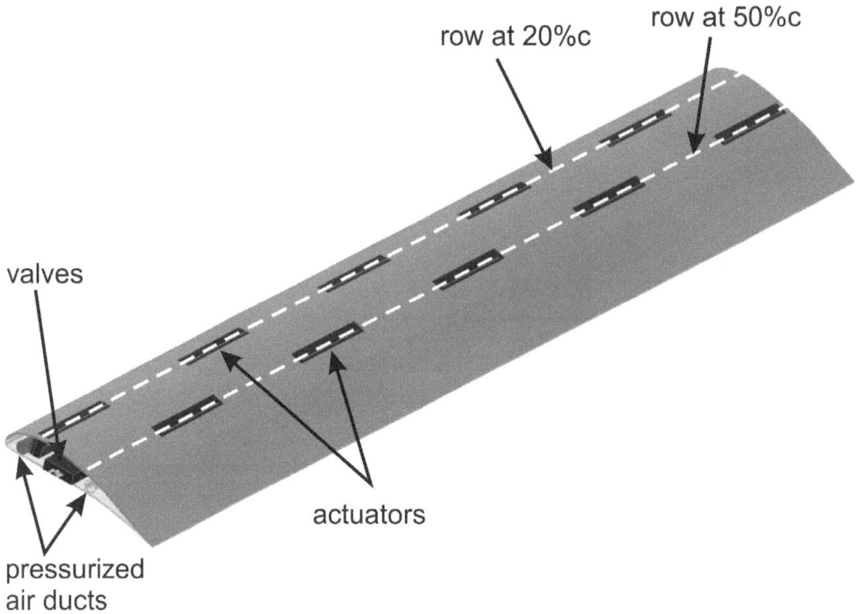

Fig. 2. Flap actuation by two staggered rows of blowing slots

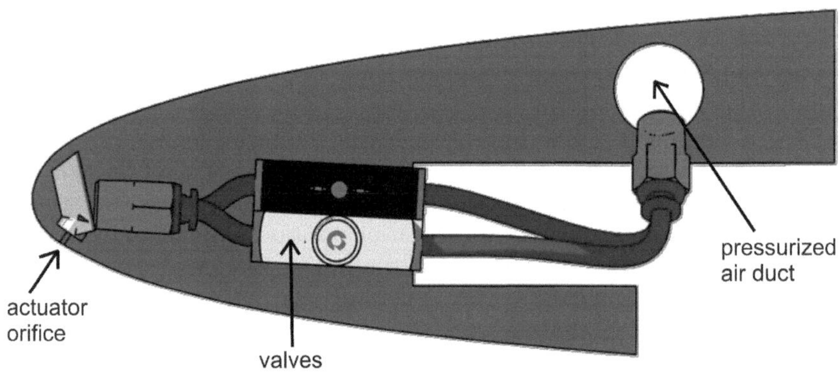

Fig. 3. Leading edge actuation by fluidic vortex generator jets at the lower leading edge

2.3 Wind Tunnel Setups

First studies of AFC at the DLR-F15 airfoil have been conducted with the 2D wing section model in the DNW-NWB atmospheric low-speed facility in Braunschweig, Germany (Fig. 4). The tunnel is a closed loop wind tunnel with a $3.25 \times 2.8m^2$ test section. The model is vertically mounted wall-to-wall between two turn tables. The maximum operating speed of the tunnel is at 90m/s, but the tests were performed up to a flow velocity of 65m/s resulting in a Mach number of $M_\infty = 0.2$ and a corresponding Reynolds number of $Re_\infty = 3 \times 10^6$ based on the airfoil model clean chord.

Fig. 4. DLR-F15 model mounted wall-to-wall in DNW-NWB atmospheric low speed tunnel

In order to study the Mach and Reynolds number effects separately and to achieve Reynolds numbers relevant to flight conditions, tests have also been performed at the DNW-KKK cryogenic wind tunnel in Cologne, Germany (Fig. 5). It is of the same type as the DNW-NWB, but with a slightly smaller test section of $2.4 \times 2.4m^2$. Due to the cryogenic environment at temperatures down to $T = 100K$ the KKK test envelope can cover a Reynolds number range of $Re_\infty = 1.4 \times 10^6 \ldots 15.7 \times 10^6$ at Mach numbers between $M_\infty = 0.1 \ldots 0.25$. The test matrix for AFC was arranged that distinct Mach- and Reynolds sensitivities can be derived. Fig. 6 shows the test conditions providing two Reynolds number sensitivities at $Re_\infty = 4.5 \times 10^6$ and $Re_\infty = 7 \times 10^6$, and two Mach number sensitivity rows at $M_\infty = 0.15$ and $M_\infty = 0.2$. Transition fixation has not been applied to observe the real Reynolds number trend.

For studying the sweep effect the 2D model was mounted as a 30°-swept cantilever wing in DNW-NWB (Fig. 7). The wing tip was covered by a tiptank-like device to prevent flow around the flap side edge and the feed pressurized air from the top side back into the flap. Within these tests transition fixation has been applied at the wing and flap leading edges. Additional tests with the DLR-F15 model have been performed recently at the University of Southampton and at ONERA in Lille, France.

Fig. 5. DLR-F15 model mounted wall-to-wall in DNW-KKK cryogenic low speed tunnel

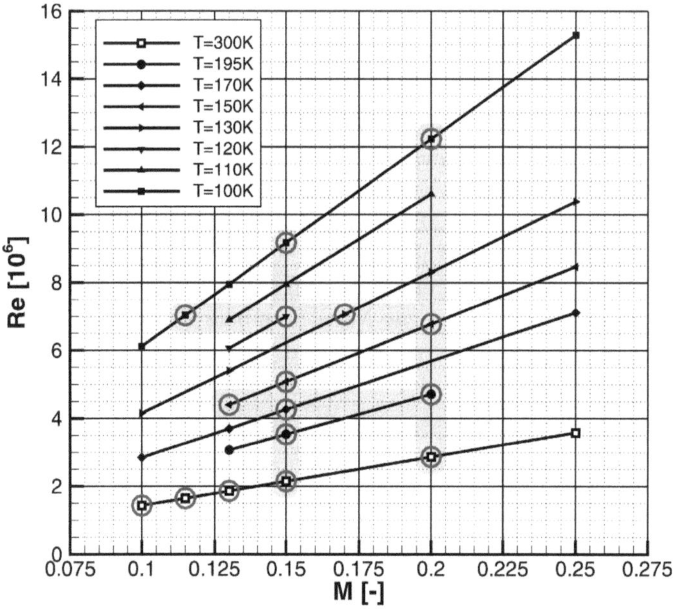

Fig. 6. Tested envelope for Mach and Reynolds numbers sensitivities of the DLR-F15 model in the DNW-KKK cryogenic low speed tunnel

Fig. 7. DLR-F15 model mounted as 30°-swept cantilever wing in DNW-NWB atmospheric low speed tunnel

Beside DLR's own original model, a series of models with the same shape is available and has been tested by partners within a series of collaborative projects. Own models with AFC technology exist at the TU Berlin, TU Braunschweig, and TU Dresden for research in their own facilities. A derivative has also been built by INCA, Romania, for research conducted in the scope of the AVERT project and was measured at INCAS and Airbus Filton.

3 Active Flow Separation Control

Before examining the active flow control effectivity, a clarification has to be given for the specification of the blowing. The most expressive value for the active flow control input is the momentum coefficient c_μ that relates the momentum of the blowing jets to the free stream momentum. Nevertheless, this value is hard to be controlled in an experimental setup. The experiments were conducted at defined mass flows. This implies that the corresponding momentum coefficient varies with the Mach number. During the cryogenic test the pressurized flow had to be heated to prevent freezing of the valves. This also implies also a varying momentum coefficient depending on the temperature of the cryogenic flow. For the sake of brevity of this paper only the general trends are reported in the following. For detailed information, the reader is referred to [11,12,13,14]. In all cases reported here regarding the Mach and Reynolds number sensitivity for each actuation a mass flow rate has been selected that corresponds to a momentum coefficient in the order of $c_\mu \approx 0.5\%$, respecting that the coefficient depends on the Reynolds number due to the severe density change of the fluid. For the sweep variation the mass flow rate has been normalized to the effective span of the wing to be able to compare the 2D and 2.5D setups.

3.1 Reynolds Number Effects

Fig. 8 shows the maximum lift coefficient variation with Reynolds number for both types of actuations compared to the non-actuated flow at an inflow Mach number of $M_\infty = 0.15$. First, the strong sensitivity of the non-actuated case at the low Reynolds numbers is to be highlighted, corresponding to a change of the separation mechanism shifting from a turbulent leading edge separation towards the main wing trailing edge. With the leading edge actuation the benefit at higher Reynolds numbers gets limited, and the gain in maximum lift coefficient remains quite constant for $Re > 3.5 \times 10^6$. For the flap actuation, where the flap stays attached for all Reynolds numbers, the gain by AFC is similar at all Reynolds numbers, only slightly reduced toward the highest Reynolds number due to a reduction of the baseline flap separation.

For the leading edge actuation, the second indicator is the incidence where the maximum lift coefficient is obtained (Fig. 9 left). The variation over Reynolds number directly follows the trend seen for the maximum lift coefficient itself. For the flap actuation, the lift coefficient at zero incidence is of more importance The right subfigure of Fig. 9 compares the dependency of this on the Reynolds number for the actuated and non-actuated case. The drop at higher Reynolds number gets more pronounced and it has to be reminded that the momentum coefficient drops with increasing Reynolds number in this cryogenic setup.

3.2 Mach Number Effects

Fig. 10 shows the Mach number sensitivity of the maximum lift coefficient at an inflow Reynolds number of $Re_\infty = 4.5 \times 10^6$ for the actuated and non-actuated flow. The leading edge actuation preserves the trend of the non-actuated flow of a slight increase of the maximum lift coefficient with Mach number. For the flap actuation the benefit is reducing with Mach number. This indicates that for the flap actuation the effectivity is more directly related to the momentum coefficient than the leading edge actuation.

For the incidence of maximum lift coefficient (Fig. 11 left) the trends for the leading edge actuation are similar as for the Reynolds number sensitivity. The non-actuated case shows a first increase at low Mach number with a constant drop towards high Mach number. The leading edge actuation shows a continuous drop with the Mach number resulting in the highest benefit at the lowest inflow velocity.

The zero lift coefficient for the non-actuated case (Fig. 11 right) shows an expected trend due to the raise of compressibility of the flow. The flap actuation shows a maximum at $M\_infty = 0.15$. The drop towards higher Mach number is again closely related to the drop in momentum difference between the actuator jet flow and the surrounding flow.

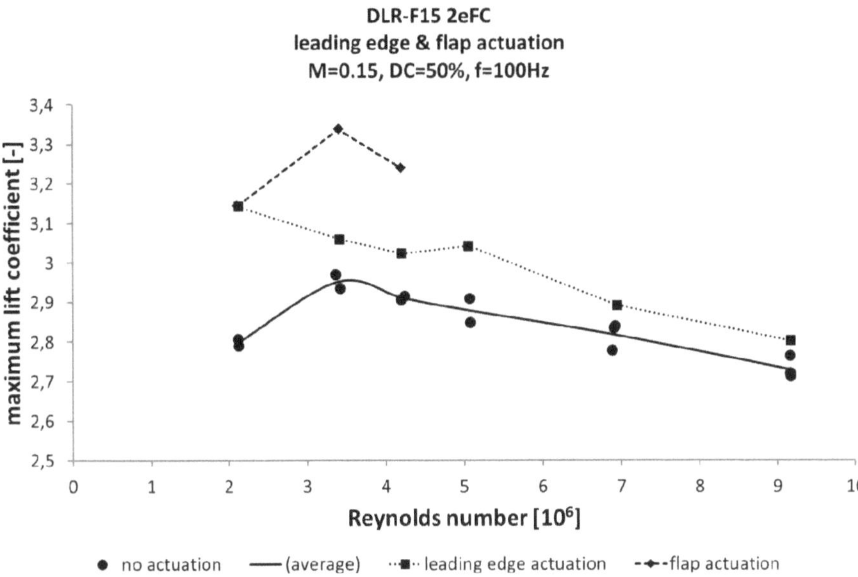

Fig. 8. Dependency of AFC effectivity for maximum lift coefficient on Reynolds number coefficient

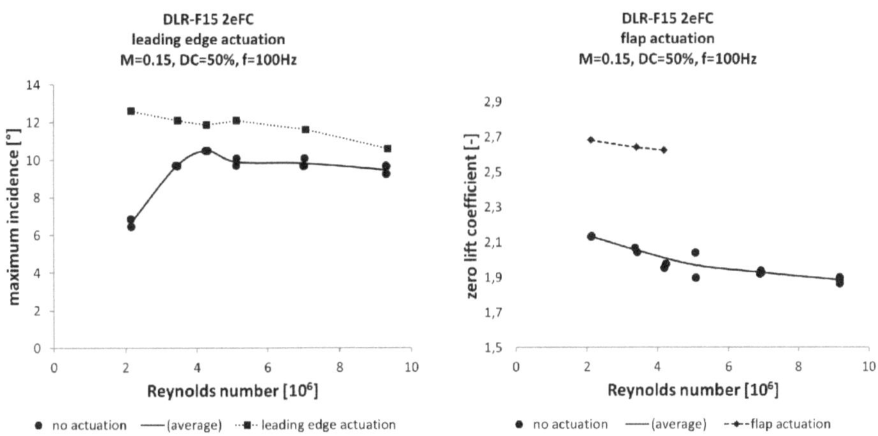

Fig. 9. Dependency of AFC effectivity on Reynolds number: (left) incidence of maximum lift coefficient; (right) lift coefficient at zero incidence

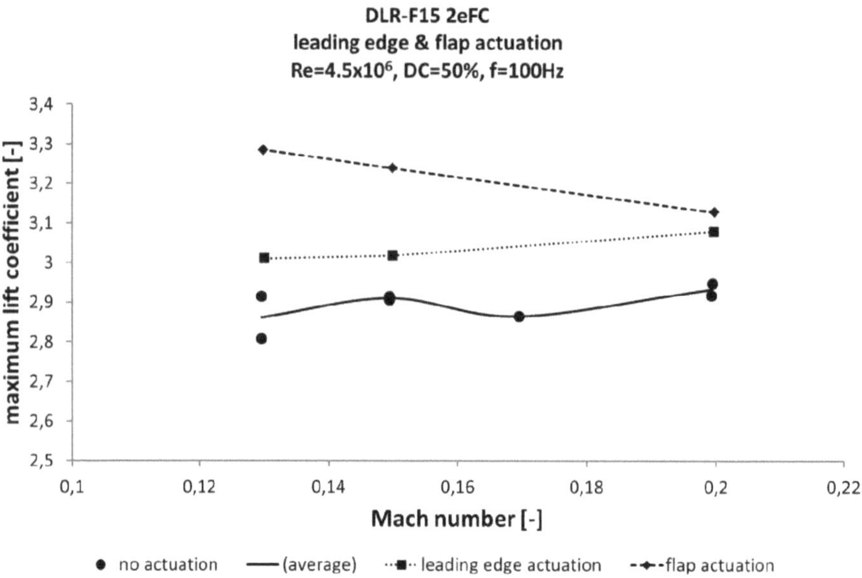

Fig. 10. Dependency of AFC effectivity for maximum lift coefficient on Reynolds number coefficient

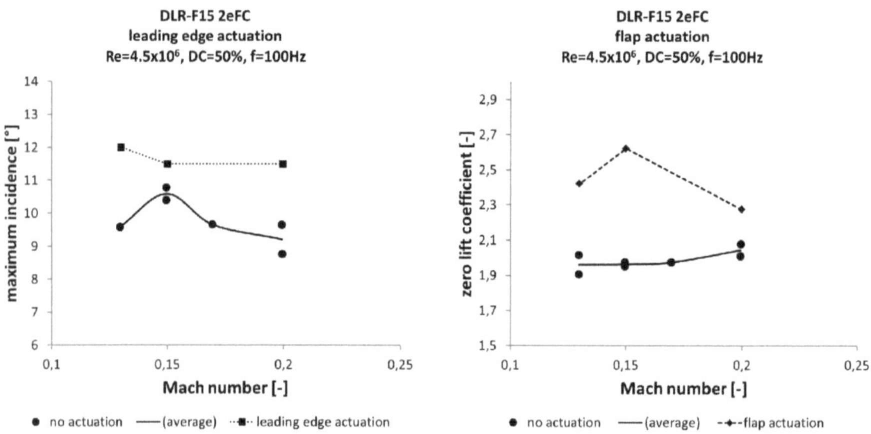

Fig. 11. Dependency of AFC effectivity on Mach number: (left) incidence of maximum lift coefficient; (right) lift coefficient at zero incidence

3.3 Sweep Effects

The analysis of the sweep effect has been performed at equivalent flow conditions. For this, the Mach number of the inflow for the swept wing has been increased to provide an equal Mach number normal to the leading edge. Since the tests with the swept wing were all run under ambient conditions in DNW-NWB, all tests were performed at the lowest Reynolds number.

Fig. 12 shows the dependency of the obtained maximum lift coefficient on the normalized mass flow per span in comparison of the straight 2D wing and the swept cantilever wing for an equivalent 2D Mach number $M_{2D} = 0.15$. The maximum lift coefficient is also scaled to the 2D equivalent leading edge normalized values. First it can be seen that for the non-actuated case both wings show a similar equivalent maximum lift, which underlines the comparability of the swept wing with respect to the ideal 2D flow. As a consequence, the application of AFC at the flap shows comparable effectivity and efficiency. The application of AFC at the leading edge in contrast shows a reduced effectivity and an unstable behavior, but without degradation below the non-actuated case. This change of the behavior compared to the ideal 2D case is currently attributed to the switch of the leading edge VGJ configuration from counter-rotating in 2D to co-rotating at the swept wing to account for the additional cross-flow component.

With regards to the incidence of maximum lift coefficient, shown left in Fig. 13, it is seen that the leading edge actuation is still able to delay the stall onset by

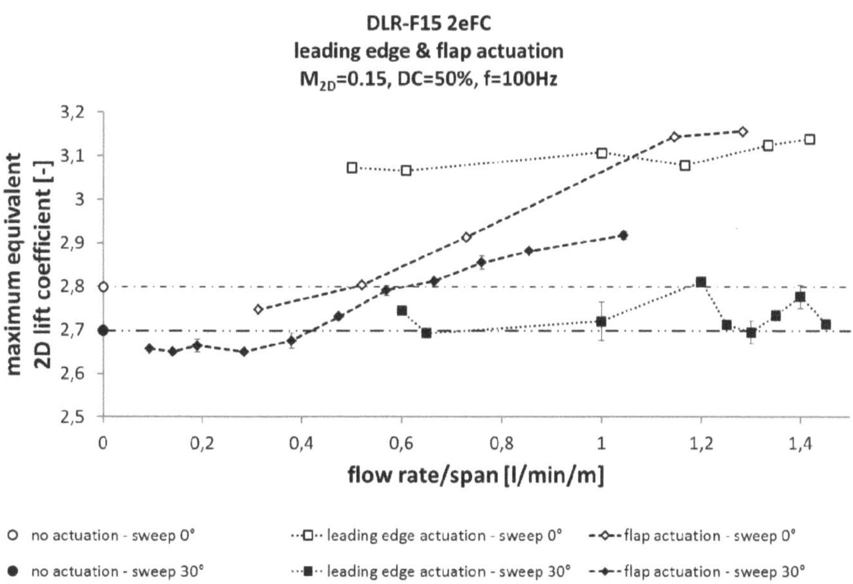

Fig. 12. Dependency of AFC efficiency for maximum lift coefficient on wing sweep

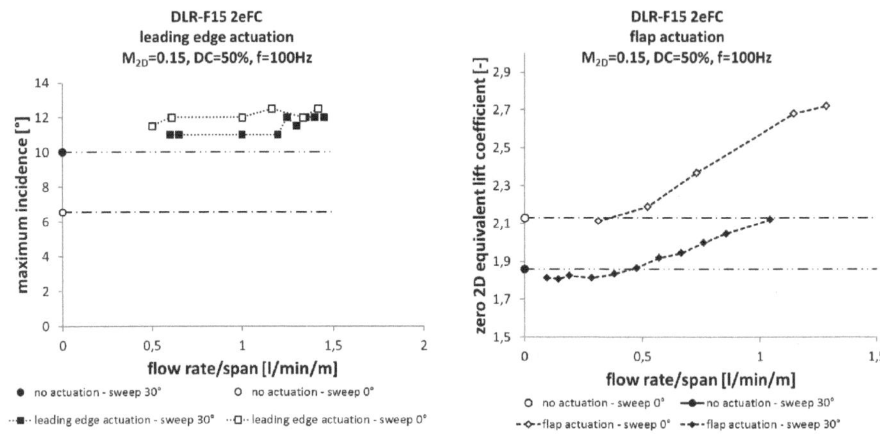

Fig. 13. Dependency of AFC efficiency on wing sweep: (left) incidence of maximum lift coefficient; (right) lift coefficient at zero incidence

some few degree, but this does not imply an additional benefit in maximum lift coefficient, which is most likely due to the onset of separation by additional mechanisms not influenced by the leading edge actuation. It has to be mentioned, that the incidence is not corrected here to account for the 2D equivalent flow conditions, since beside the pure geometrical correspondence an additional influence is given by the limited span of the cantilever wing.

The AFC at the flap shows similar behavior for the zero lift coefficient as for the maximum lift coefficient (Fig. 13 right). The level of the non-actuated case has changed for similar reasons as above, since the geometrical zero incidence does not necessarily correspond to the equivalent flow condition. To this regard the efficiency indicated by the slope of the curves is slightly reduced for the swept wing case, but it has to be respected that the real onflow velocity is higher at the swept wing and therefore the absolute momentum coefficient is lower.

4 Summary

This contributions gives an overview on experimental investigations with the purpose to identify the sensitivity of active flow separation control on three major parameters that play a role with regard to real aircraft application, which are the Mach number, the Reynolds number and the wing sweep. A series of wind tunnel tests in ambient and cryogenic conditions have been performed to identify those parameters that are urgently needed to mature AFC technology towards aircraft application.

Concluding, two major trends are observed throughout the experiments. First, the AFC application at the flap to prevent its local separation at high deflection angles shows to be effective with only minor sensitivity on the investigated parameters. Decreasing effectivity at constant mass flow rates can at first be attributed to the corresponding decrease in effective momentum introduced into the flow.

The application of AFC at the leading edge shows decreasing effectivity when compared to the idealized 2D flow at low Reynolds number. This change is mainly attributed to the change of the flow separation mechanism at the main wing shifting from the leading edge towards the trailing edge. For the variation of the Mach number the effectivity is maintained regardless of the reduction of momentum coefficient as explained above. Especially at the swept wing, where the effectivity is partly unstable, the AFC application at the leading edge is in all cases able to delay the incidence of the wing stall by some degrees.

Acknowledgments. The two-dimensional measurements in the DNW-NWB facility have been partially funded by the German Ministry of Economics via the German " Luftfahrtforschungsprogramm III" within the compound project " IHK". The high-Reynolds-number tests have been also supported by the German Ministry of Economics via the German " Luftfahrtforschungsprogramm IV" within the compound project " M-Fly/Aeronext". The tests with the swept model have been supported by the Joint Undertaking " CleanSky" within the frame of the Joint Technology Initiative " Smart Fixed Wing Aircraft (SFWA)". All experimental work has been performed in close partnership with the Universities TU Berlin and TU Braunschweig, and Airbus Operations GmbH, Bremen. Here to be mentioned explicitly are the contributions of F. Haucke and W. Nitsche (TU Berlin), P. Scholz, M. Casper, C.P. Hühne and R. Radespiel (TU Braunschweig), B. Gölling and H. Bieler (Airbus), V. Ciobaca, M. Sitzmann and M. Gülzow (DLR).

References

1. Seifert, A., Pack, L.G.: Oscillatory Control of Separation at High Reynolds Number. In: 36th Aerospace Sciences Meeting & Exhibit, AIAA-Paper 98-0214 (1998)
2. Rudnik, R.: European High Lift Programme II - Final Publishable Report (2007), Accessible by `http://transport-research.info`
3. Wild, J.: Mach and Reynolds Number Dependencies of the Stall Behavior of High-Lift Wing-Sections. Journal of Aircraft 50(4), 1202–1216 (2013), doi:10.2514/1.C032138, ISSN 0021-8669
4. Wild, J., Wichmann, G., Haucke, F., Peltzer, I.: Large-Scale Separation Flow Control Experiments Within the German Flow Control Network. In: 47th AIAA Aerospace Sciences Meeting including The New Horizons Forum and Aerospace Exposition, AIAA-Paper 2009-0530 (2009), doi:10.2514/6.2009-530
5. Dargel, G., Hansen, H., Wild, J., Streit, T., Rosemann, H., Richter, K.: Aerodynamische Flügelauslegung mit multifunktionalen Steuerflächen. In: DGLR Jahrbuch 2002. DGLR, vol. 1, p. 1605 (2002) ISSN 0070-4083
6. Tinapp, F., Nitsche, W.: Separation Control on a High-Lift Configuration by Periodic Excitation. In: Wagner, S., Rist, U., Heinemann, H.-J., Hilbig, R. (eds.) New Results in Numerical and Experimental Fluid Mechanics III. NNFM, vol. 77, pp. 42–49. Springer, Heidelberg (2002)

7. Petz, R., Nitsche, W.: Active Separation Control on the Flap of a Two-Dimensional Generic High-Lift Configuration. Journal of Aircraft 44(3), 865–874 (2007), doi:10.2514/1.25425, ISSN 0021-8669
8. Haucke, F., Nitsche, W.: Active Flow Control on the Flap of a 2D High-Lift Wing Section at Re=10e6. In: 29th AIAA Applied Aerodynamics Conference, AIAA-Paper 2011–3359 (2011), doi:10.2514/6.2011-3359
9. Scholz, P., Ortmanns, J., Kähler, C.J., Radespiel, R.: Leading Edge Separation Control by Means of Pulsed Jet Actuators. AIAA Journal 46(4), 837–846 (2008), doi:10.2514/1.26176, ISSN 0001-1452
10. Scholz, P., Kähler, C., Radespiel, R., Wild, J.: Georg Wichmann Active Control of Leading Edge Separation Within the German Flow Control Network. In: 47th AIAA Aerospace Sciences Meeting Including The New Horizons Forum and Aerospace Exposition, AIAA-Paper 2009-0529 (2009), doi:10.2514/6.2009-529
11. Casper, M., Scholz, P., Radespiel, R., Ciobaca, V., Wild, J.: Separation Control on a High-Lift Airfoil using Vortex Generator Jets at High Reynolds numbers. In: 41st AIAA Fluid Dynamics Conference and Exhibit, AIAA-Paper 2011–3442 (2011), doi:10.2514/6.2011-3442
12. Hühne, C.-P., Mahmood, S., Scholz, P., Radespiel, R.: Active flow control at a swept wing: A comparison of numerical and experimental results. In: 31st AIAA Applied Aerodynamics Conference, AIAA-Paper 2013-2798 (2013), doi:10.2514/6.2013-2798
13. Haucke, F., Nitsche, W.: Active Separation Control on a 2D High-Lift Wing Section Towards High Reynolds Number Application. In: 31st AIAA Applied Aerodynamics Conference, AIAA-Paper 2013-2514 (2013), doi:10.2514/6.2013-2514
14. Wild, J.: Experimental Determination of Wing Sweep Effect on the Efficiency of Active Flow Separation Control. In: Greener Aviation 2014 Conference, Brussels, March 12-14 (2014)

Fundamentals in Coanda Flap Design

Rolf Radespiel and Marco Burnazzi

Institute of Fluid Dynamics, Technische Universität Braunschweig,
Hermann-Blenk-Str. 37, 38108 Braunschweig, Germany
r.radespiel@tu-bs.de

Abstract. Using the Coanda-effect in active high lift flaps opens the way to achieve large lift coefficients, needed for the design of aircraft with short take-off and landing capabilities. In transport aircraft applications overall aircraft design requirements also call for low power consumption of the active high-lift system. Moreover, these aircraft need a reasonable stall angle for their operations. The present contribution describes recent research that aims at high-lift augmentation to be achieved with low rates of active blowing and with suited angle-of-attack ranges. The characteristics of well designed internally blown flaps are addressed. Of similar importance is the design of the leading edge where droop noses and slats may be introduced to avoid locally large boundary layer losses. Moreover, blowing of a Coanda wall jet over a flap can be combined with boundary layer suction in order to yield even higher efficiencies of the overall high-lift system.

Keywords: Active high lift, Coanda jet, stall angle, boundary layer suction.

1 Introduction

The wings of transport aircraft must have systems to generate high lift coefficients for take-off and landing so that the wing size can be chosen for efficient cruise operation. These systems usually exhibit a Fowler-type high lift flap at the trailing edge and a slat at the leading edge. Maximum sectional lift coefficients obtained with these slotted devices mounted on a wing section are in the range of 4.0-4.5, according to Wild[1]. This provides a typical commercial aircraft of the A320-class with the ability to operate from runways with around 1600m in length and above[2]. Future requirements for commercial aviation call for reducing the airport community noise, for making point-to-point connections available to more customers, and for reducing aircraft fuel consumption and operating cost[3]. These requirements will strengthen the role that the segment of smaller, regional aircraft will play in future commercial aviation. These aircraft will have to combine cruise efficiency with short take off and landing capabilities[4], in order to allow operation at regional airports with noise sources from engine and airframe kept at the lowest possible level. Overall design of cruise efficient aircraft states that substantial reductions of runway length are only possible by increasing the maximum lift coefficient by significant factors accompanied by a moderate increase of the installed engine thrust[5,6]. This can be accomplished by using active

R. King (ed.), *Active Flow and Combustion Control 2014*, Notes on Numerical Fluid Mechanics and Multidisciplinary Design 127, DOI: 10.1007/978-3-319-11967-0_7

high-lift systems, where the lift is augmented by a well designed blowing system. Active circulation control can provide the required lift coefficients without using flow through gaps, which are identified as major source of airframe noise during approach and landing[7]. This high-lift approach is adopted by the Collaborative Research Center SFB 880 of the Technische Universität Braunschweig[6].

Using blowing as a means for circulation control is not new and reviews of this technology appeared continuously over the last decades[8-12]. The blowing concepts shown in Figure 1 use the turbulent flow entrainment of a tangentially blown wall jet to keep the flow attached along the wall geometry with a high rate of turning. The blunt trailing edge concept allows a direct control of the flow direction leaving the airfoil trailing edge and hence, circulation control. Unfortunately, one needs continuous blowing during cruise with a blunt trailing edge, in order to keep cruise drag low[13]. Therefore, the design of cruise-efficient aircraft would rather prefer a movable at the trailing edge[14] to create the suited Coanda surface at take-off and landing conditions, hence the term Coanda flap. This leads to the employment of an internally blown flap. This flap may therefore be also viewed as special case of circulation control.

Fig. 1. Internal blowing schemes to obtain high lift coefficients

An important requirement for transport aircraft applications calls for low blowing power of the active control system, as otherwise a significant engine growth will occur compared to using conventional high-lift systems. Early experiments revealed the importance of defining a suited slot height and Poisson-Quinton[15] defined the momentum coefficient, in order to characterize the Coanda jet:

$$C_\mu = \frac{v_j \dot{m}_j}{\frac{1}{2}\rho_\infty v_\infty^2 S_{ref}},$$

where v_j and $\dot{m}_j$ are the velocity and the mass flow of the jet through the exit section of the plenum, $1/2\rho_\infty v_\infty^2$ is the dynamic pressure of the freestream, and S_{ref} is the reference surface. Using this normalization, the effort to provide blowing may be understood as an additional drag coefficient. This interpretation of the momentum coefficient is obvious for applications where high-pressure air for blowing is provided by bleed from the aero engine compressor and hence, a loss of engine thrust in the order of wall-jet momentum flux is expected.

Until the late 1990s mostly experimental approaches were used to explore the potentials of improving the blowing efficiency in circulation control. A good account of these works can be found in the collective volume edited by Joslin and Jones[10].

A special challenge associated with these experimental approaches is the large overall effort needed to achieve trustworthy experimental results. It turns out that the requirements in model accuracy around the blowing slot are very high, and the engineering work needed to obtain a homogeneous and well defined wall jet along the span is significant[13,16]. Therefore, the currently existing experimental data base for efficient blowing configurations in circulation control represents rather low Reynolds numbers of around $1 \cdot 10^6$.

Numerical flow simulation capability offers new opportunities to the development of efficient lift augmentation systems since it allows ways of detailed flow analyses not feasible with experimental techniques. Moreover, numerical flow simulations allow access to flight Reynolds numbers. However, numerical predictions of circulation control by wall jets depend on the ability to simulate complex turbulent shear layers in adverse pressure gradients that exist downstream of the blowing slot. The task is particularly demanding because circulation control using blown Coanda surfaces involves the prediction of turbulent shear flows with significant streamwise curvature. The set up of suited numerical simulations for aerodynamic analysis was studied by a number of researchers[17-22]. It was found that the high flow gradients in wall-normal and tangential directions are only captured using significantly denser meshes, compared to usual airfoil simulations. Typically, 10[5] grid points are needed for an accurate two-dimensional RANS simulation[22]. Furthermore, 2D computations for airfoils at high lift coefficients obtained by circulation control need significant corrections if a comparison with wind tunnel data is sought. The corrections have their origin in the separation and roll-up of the wind tunnel sidewall boundary layer, which is assumed to generate an induced angle of attack in the model plane of symmetry[22]. Moreover, standard turbulence models based on the eddy viscosity concept failed to predict accurate levels of Reynolds stresses along curved Coanda surfaces. This problem could be overcome by using turbulence models with curvature corrections[18,21,22].

The present contribution builds on the described status of validation in numerical flow predictions of circulation control. It presents recent results in the aerodynamic development of Coanda flaps with high blowing efficiency using numerical simulations for flow analysis. Section 2 of the paper describes the numerical method used. Section 3 explains the design of blowing slots and Coanda flap geometries, whereas Section 4 displays the advantages of using leading edge devices for protection against leading edge stall. Section 5 shows the potentials offered by providing high-pressure air by distributed compressors in the wing, as boundary layer suction and wall jet blowing can be combined to yield higher overall efficiencies.

2 Numerical Simulation Method

The flow solver employed to perform flow analysis is the DLR TAU Code[23]. The TAU Code solves the Reynolds-Averaged Navier-Stokes (RANS) equations by using a finite volume approach. The turbulence model is given by Spalart and Allmaras with a correction due to flow rotation and curvature[24]. This module improves the physical modeling of the one-equation turbulence model in regions where the streamlines have a high curvature. This is important for the simulation of the Coanda effect, which is based on the equilibrium between the inertia forces, pressure forces and the momentum transport in the direction normal to the convex surface. The numerical set up was assessed by using grid convergence studies and comparing with wind tunnel experiments. Grid convergence was investigated by computing flow solutions on grids with systematically reduced grid spacing in all coordinate directions. A sample grid is displayed in Fig. 2 which also displays the typical geometries of flap and blowing slots.

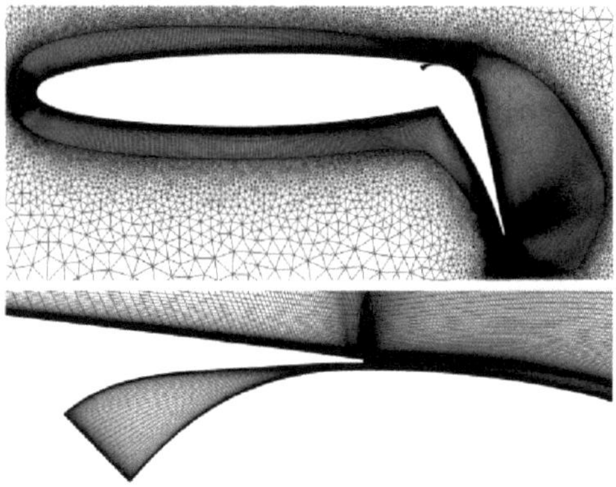

Fig. 2. Hybrid grid with around 230000 points employed for numerical flow analyses

The evaluation of grid convergence indicates that the numerical errors in aerodynamic coefficients of lift, drag and pitching moments are less than 1% for the standard grid shown in Fig. 2. Wind tunnel experiments with an airfoil model with a representative flap were recently performed in the low-speed wind tunnel of Technische Universität Braunschweig at a Reynolds number of 1•10[6]. These wind tunnel tests resulted in oil flow pictures, pressure distributions, aerodynamic coefficients and flow field data from PIV for different angles of attack and blowing rates[25,26]. The test data were compared to corresponding numerical simulations. The results of 2D flow solutions exhibited significant discrepancies to the measured data, as previously observed in other circulation control simulations[18,21,22]. As described above, these discrepancies were associated with the side-wall boundary layer interactions, resulting in wind tunnel blockage and induced angle of attack. The discrepancies became very small, if a 3D numerical rebuilding of the wind tunnel flow was employed[25,26]. The results of these validation studies lead to the conclusion, that the current numerical set up of flow simulations for Coanda flaps is valid.

3 Blowing Slot and Flap Geometry

This chapter describes numerical flow analyses aimed at improving the lift gain due to blowing by careful adjustments of the design parameters of the Coanda flap while the leading edge was geometrically fixed. The transonic airfoil DLR F15 is selected as a reference geometry. The internal flow towards the slot is defined by a converging slot as displayed in Fig. 2. The design studies assumed steady blowing to produce turbulent wall jets that exploit the Coanda effect for effective flow turning. The free stream flow conditions were Re=20•10^6, and M=0.15. The most important design parameters turned out to be momentum coefficient of blowing, flap deflection angle, η, and the blowing slot height. The optimal value of momentum coefficient was obtained by selecting just the value, for which incipient flow separation occurs right at the flap trailing edge. This represents the kink in the sketched graph of Fig. 3 (left), where the initially high lift gain with respect to blowing momentum reduces. Below this value, blowing functions as boundary layer control, while above blowing provides supercirculation. Using this optimal setting, flap angle and blowing momentum coefficient increase for increased lift targets, Fig. 3 (right), whereas optimal slot heights are found to be rather small, with values of around 0.0006 times the airfoil chord length[27]. Surprisingly, the optimum slot height is independent of the flap angle.

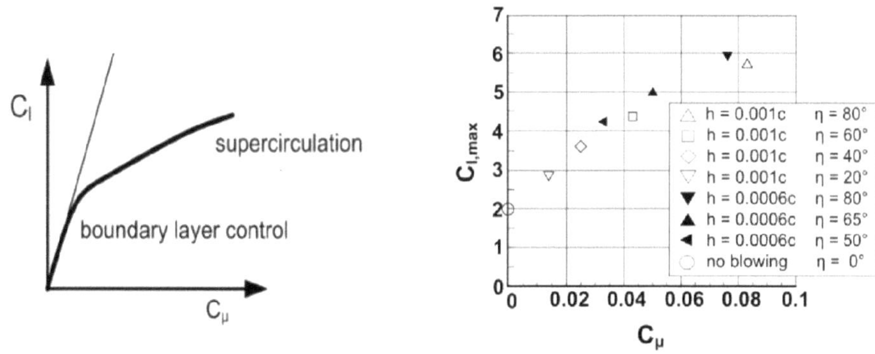

Fig. 3. Regimes of circulation control and most efficient blowing achieved DLR F15 airfoil, relative flap length 0.3c, Re=20•10^6

The detailed curvature distribution of the Coanda surface used as flap knuckle shape was found less important. Values of the radius of curvature of around 0.07 times the chord length are a reasonable design choice. Also the flap length suited to achieve high lift gains can be identified. The largest lift gains are obtained with flap lengths of 0.25-0.30 times the airfoil chord[28]. With these design choices typical lift gains over blowing momentum (lift gain factor) of 80 are obtained at a lift coefficient around 4 and with a flap deflection angle of 50°, whereas this value was reduced to 50 at a lift coefficient around 6 and flap deflection of 80°. It was generally observed though, that the angle of attack of maximum lift reduces significantly at higher blowing rates. The maximum lift coefficient of 5 with a flap angle of 65° in Fig. 3 is obtained at an angle α_{max}= 2°. This low stall angle makes it impossible to exploit the high-lift coefficients on the wing of a transport aircraft, since the resulting aircraft angle of attack at landing will be negative. Therefore, leading edge devices for stall delay are investigated in the next chapter.

4 Leading Edge Geometry

The use of leading edge devices for delaying stall to higher angles of incidence is generally well established. Here, recent experimental data of Wild[1] are presented in Table 1 as examples for the effectiveness of two possible solutions: The slat and the flexible droop nose.

Table 1. Enhancement of high-lift behaviors of DLR F15 airfoil using leading edge devices at M=0.2 and Re=$5 \cdot 10^6$, according to Wild[1]

	$c_{l,max}$	α_{stall}
 Fowler flap alone	3.2	8.5°
 Fowler flap and slat	4.6	31°
 Fowler flap and flexible droop nose	3.6	14°

The present research into suited means to delay stall on high-lift Coanda flap configurations compares three approaches: rigid droop nose, flexible droop nose and slat. The aerodynamic design of these devices assumed that 20% of the chord was available for housing the leading edge device. The Coanda flap had a relative length of 0.25 times the airfoil chord and the flap deflection was kept constant at 65°.

The rigid droop nose is obtained by deflecting the leading edge downward, without changing its shape. Therefore, this technique does not involve structural deformations, except for the hinge-cover sheet at the lower surface and can be realized by a simple rotation of the nose around a hinge. Thanks to its efficiency and mechanical simplicity, this device is currently employed by some of the current commercial transport aircraft. The rigid droop nose is characterized by two strong peaks of low pressure as the deflection of the nose creates a new region of high curvature on the suction side. Variation of the deflection angle showed that the best balance between the two peaks is obtained with δ=30°. For this case, the improvement of lift with respect to the baseline configuration is quite significant, as shown in Fig. 4.

The flexible droop nose employs a morphing structure. In our aerodynamic shape design, both, camber and thickness of the leading edge was systematically varied. The best solutions are obtained by smoothly increasing the slope of the camber line until the leading edge, where a local value of 90° is observed. The local airfoil thickness is smoothly increased to the leading edge, where a value of 1.6 times the original airfoil thickness is employed. More design details are given by Burnazzi and Radespiel[29]. The resulting contour is shown in Fig. 4 along with its lift curve. An impressive gain of maximum lift and stall angle is achieved which is much larger than the improvements previously observed for flexible droop noses in combination with a Fowler flap, see Table 1.

The slat configuration is the one shown in Table 1, named 3eOpt. The slat has a deflection angle of 28°. Slat angle and gap were defined by using numerical optimization of the 2D flow[1]. For the current application with the Coanda flap, the performance of the slat could likely be somewhat improved by adapting slat angle and its location to the high circulation generated by the active flap. However, the present configuration is considered as representative of flow features and performance trends that an optimized geometry would show, at least for moderate blowing momentum coefficients. The results shown Fig.4 for the relatively high blowing rate, C_μ=0.06, are rather unexpected, since the slat does not take advantage of its extended chord and the gap flow, relative to the rigid droop nose.

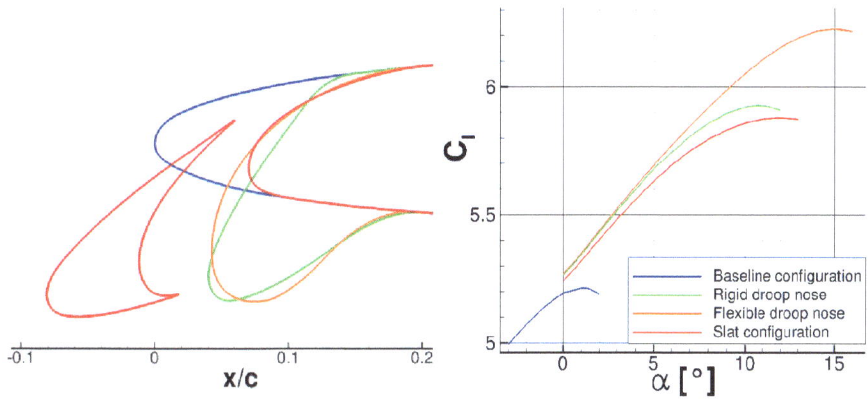

Fig. 4. Comparison of leading-edge configurations at C_μ=0.06, Re=12•10^6, M=0.15

A more detailed analysis of the aerodynamic performance is possible by comparing pressure distributions and the development of the boundary layer momentum thickness just upstream of the location where the Coanda wall jet is introduced. This integral quantity may be understood as the resistance of the incoming flow to the intended flow turning. Fig. 5 shows significant differences in the boundary layer behavior. Boundary layer losses rise rapidly with α and C_μ for the clean nose, due to the sensitivity of the nose suction peak. The flexible nose, on the other hand, shows rising losses with α but not with C_μ . This is the key to achieving high stall angles at high C_μ. The slat nose shows an adverse trend with respect to C_μ. However, inspection of the velocity profiles in Fig. 6 displays that significant losses come with the slat wake and these are not augmented by Coanda jet blowing. This is finally reflected in the resulting gains in lift and stall angle due to blowing as displayed in Fig. 7. We observe different behaviors of the droop nose variants compared to the slat. The slat is the preferred aerodynamic solution for blowing momentum coefficients below 0.045, whereas the flexible droop nose is better for larger blowing rates.

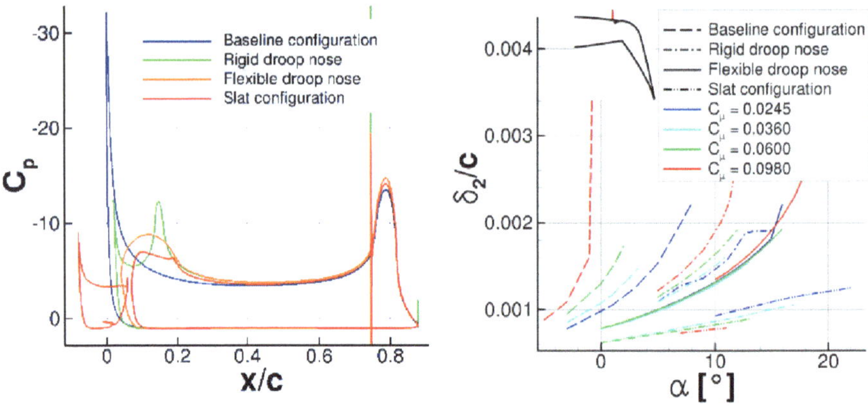

Fig. 5. Pressure distributions for C_μ=0.06, α=1° and development of boundary layer momentum thickness at x/c=0.73, Re=12·10^6, M=0.15

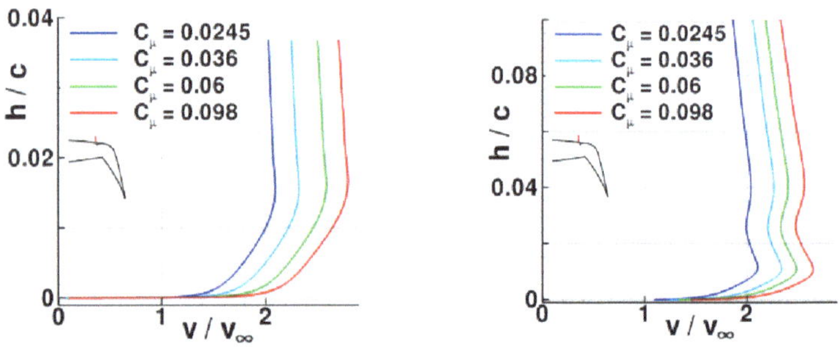

Fig. 6. Velocity profiles of flexible droop nose (left) and slat (right) at x/c=0.73, α=1°, Re=12·10^6, M=0.15

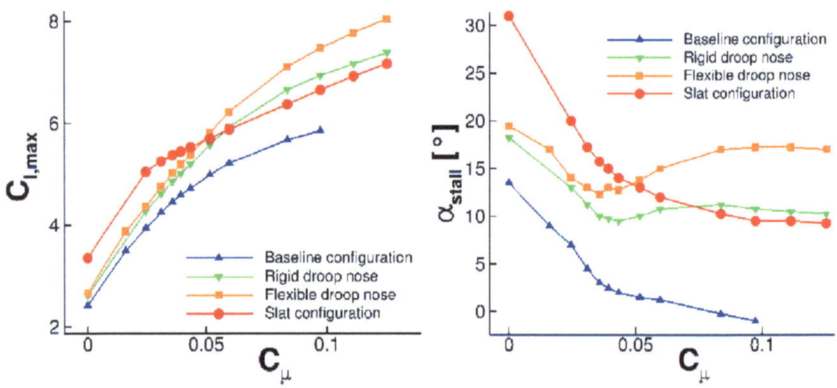

Fig. 7. Maximum lift coefficient and stall angle over wall jet momentum, Re=12·10^6, M=0.15

5 Synergies of Blowing and Suction

While most of the current active high lift approaches assume that the compressed air needed for blowing is provided by using bleed air from the aircraft engine, an alternative approach is to generate compressed air by a set of electrical compact compressors which are integrated into the wing near the flap, as sketched in Fig. 8. With this approach, aerodynamic design plays an important role, as it is responsible for the aerodynamic efficiency of the high-lift flap, as well as for providing air to the compressors by means of a suction slot located on the suction side of the airfoil at a suited position on the wing.

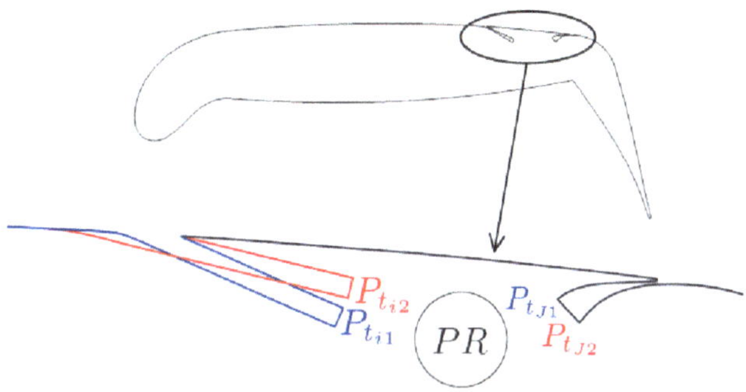

Fig. 8. Sketch of the interaction of suction and blowing through the pressure ratio, PR, of the compressor

The present analysis investigates the design sensitivities of using wall suction. The objectives of suction are to provide air with high pressure recovery to the compact compressor and to improve the lift coefficient of the airfoil. Unfortunately, these two objectives behave differently as the geometrical suction parameters are varied. Therefore a combined parameter is needed to compare the benefit of the tested geometries[30]. For this purpose the total pressure obtained at the end of the suction duct is used to balance the jet momentum coefficient, by assuming a constant pressure ratio of the compressor. This results in a balanced lift coefficient that is a suited quantity to compare and evaluate different geometries. In the present work we consider three geometrical suction parameters: Suction inlet angle β, slot diffuser angle γ, and slot location x along the airfoil, as sketched in Fig. 10. For comparing different suction geometries one first defines one of the configurations as a reference. For a given momentum coefficient of the wall jet, and ensuring the same mass flow rate of slot and jet, the numerical flow simulation yields the total pressures $P_{t_{i1}}$ and $P_{t_{J1}}$, according to Fig. 8. The pressure ratio of the compressor is defined as

$$PR = \frac{P_{t_{J1}}}{P_{t_{i1}}}.$$

For the second configuration the total pressure of the wall jet is then given by

$$P_{t_{J2}} = PR \cdot P_{t_{i2}} .$$

This balancing approach takes the effects of static pressure at the suction slot inlet and the aerodynamic efficiency of the inlet providing pressure recovery into account.

The reference geometry has the flexible droop nose of Chapter 3, and the suction slot is defined by β=10°, γ=0°, x/c=0.61. Note that the reference slot location is just downstream of the assumed wing box, upstream of the wing spoiler. The width of the inlet slot is always defined so that the bulk Mach number of the inlet exit face is 0.1. Further reference parameters are C_μ=0.0356, M=0.15, Re=12•10^6. The medium grid used before is now modified to include the suction slot. It contains about 250000 points, and a typical discretization is displayed in Fig.9.

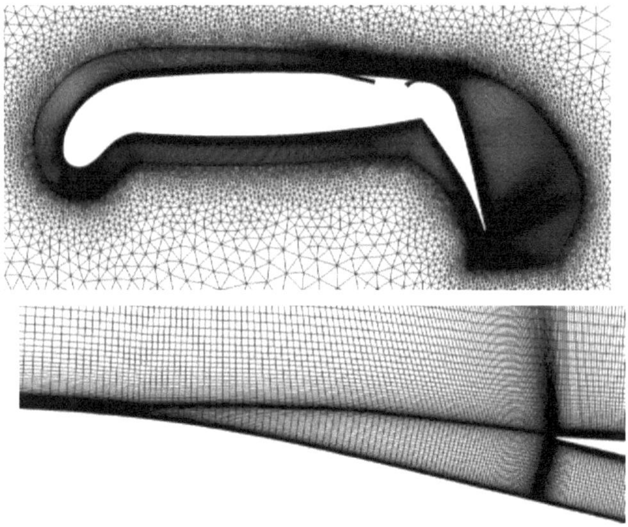

Fig. 9. Hybrid grid with 250000 points employed for suction simulation

As the maximum lift coefficient of the airfoil without suction is $c_{l,max}$=5.01, the results shown in Fig.10 indicate that significant lift gains are obtained with suction. The graphs with broken lines denote solutions for fixed jet momentum coefficient, C_μ=0.0356, whereas the full lines labeled "p_t balanced" include the interactions between inlet and blowing slot as described above. Good inlet geometries have inlet angles between 10 and 20 deg and a rather small diffuser angle. Diffuser angles of γ>5° lead to significant total pressure losses at the inlet exit face and hence, losses in the balanced airfoil lift. It also appears that forward inlet locations are aerodynamically more effective than locations behind the wing box. However, such inlet positions would present a great challenge to wing structure design. Suction on the high-lift flap does not result in any outer flow improvement as seen from the non-balanced result.

Nevertheless, the higher static pressure at the inlet location on the flap results in an increment of wall jet total pressure. This effect of the static pressure at the inlet location improves the balanced lift compared to the unbalanced value. Our overall assessment of design solutions for a suited suction slot concludes that the reference β=10°, γ=0°, x/c=0.61 is most likely the best solution.

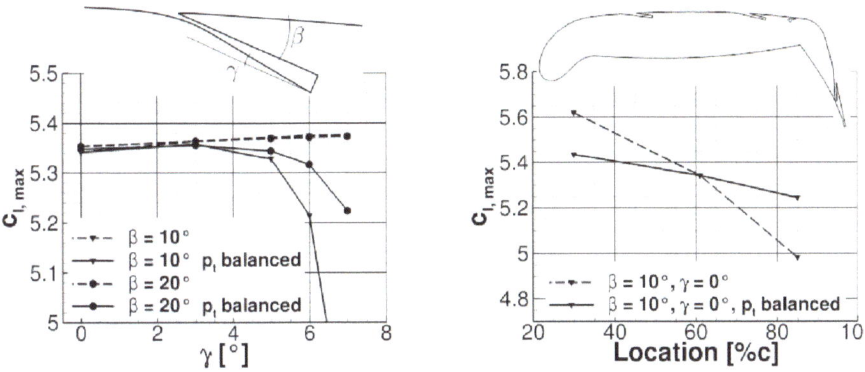

Fig. 10. Influence of suction slot geometry (left) and suction slot position (right)

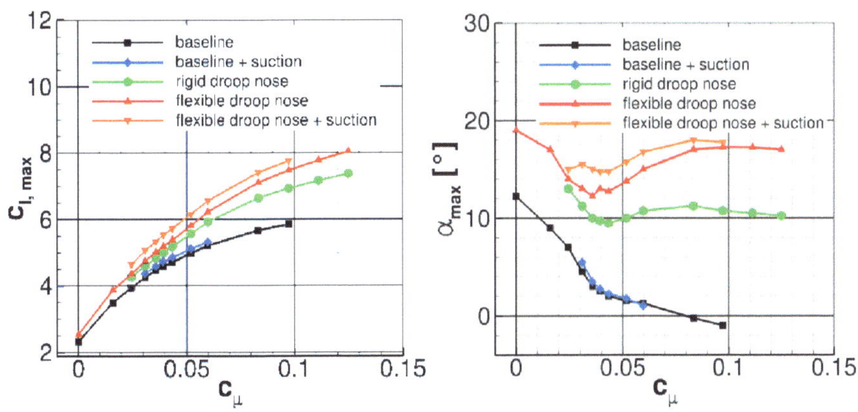

Fig. 11. Maximum lift coefficient and stall angle over wall jet momentum for Re=12·10^6, M=0.15, suction slot parameters β=10°, γ=0°, x/c=0.61

This reference suction design is finally used for comparison of airfoil performance with and without suction as shown in Fig. 11. It appears that applying suction in combination with the flexible droop nose is much more effective than the application of suction on the baseline airfoil. Suction results in significant reductions of the blowing rate, in order to achieve a give lift coefficient, and it extends the stall angle of attack by up to 5°. More detailed results of the flow simulations with suction are given by Burnazzi and Radespiel[30].

6 Conclusions

The present contribution describes recent aerodynamic research to improve the efficiency of active blowing in high-lift Coanda flaps. The importance of the geometrical parameters in Coanda flap design, and the need for applying leading edge devices for delaying airfoil stall are addressed. It is found that droop nose devices are more effective in combination with Coanda flaps as compared to conventional Fowler flaps. Furthermore, the exploitation of synergies between Coanda wall jet blowing and suction is investigated. It appears that significant gains in lift augmentation are possible, by applying suction with well designed suction slots.

The overall results of the present Coanda flap integration studies are promising. For example, a given maximum lift coefficient of five is obtained with 32% less momentum of the Coanda jet when a flexible droop nose is applied. Larger reductions, i.e. 43%, are possible when combining the flexible droop nose with the option to suck in the mass flow of the Coanda jet at an appropriate location on the airfoil. These results underline the importance of the "design for flow control" principle. Attaching an active Coanda Flap on a clean wing section in order to achieve high lift coefficients will result in a flow that is dominated by the viscous losses resulting from the strong suction peak at the leading edge. This stiffness of the aerodynamic flow characteristics is greatly relaxed by introducing a shape adaptive droop nose, thereby yielding room for further approaches to improve on the blowing efficiency.

The Collaborative Research Center SFB 880 plans further work into these approaches. Specific research projects address technologies for structural design of shape-adaptive droop noses on airfoils and 3D wings, compact electrical compressors, and integrated actuators and sensors for closed-loop flow control of Coanda jets on high-lift flaps.

References

1. Wild, J.: Mach and Reynolds Number Dependencies of the Stall Behavior of High-Lift Wing-Sections. Journal of Aircraft 50(4), 1202–1216 (2013)
2. Werner-Spatz, C., Heinze, W., Horst, P.: Improved representation of high-lift devices for a multidisciplinary conceptual aircraft design process. Journal of Aircraft 46, 1984–1994 (2009)
3. Flightpath2050 – Europe's Vision for Aviation, Report on the High Level Group on Aviation Research (2011), `http://www.acare4europe.org` ISBN 978-92-79-19724-6
4. Henke, R., Lammering, T., Anton, E.: Impact of an Innovative Quiet Regional Aircraft on the Air Transportation System. Journal of Aircraft 47(3), 875–886 (2010)
5. Werner-Spatz, C., Heinze, W., Horst, P., Radespiel, R.: Multidisciplinary conceptual design for aircraft with circulation high-lift systems. CEAS Aeronautical Journal 2, 145–163 (2012)
6. Radespiel, R., Heinze, W.: SFB 880 - fundamentals of high-lift for future commercial aircraft. CEAS Aeronautical Journal (2014), doi:10.1007/s13272-014-0103-6
7. Pott-Pollenske, M., Alvarez-Gonzalez, J., Dobrzynski, W.: Effect of Slat Gap on Farfield Radiated Noise and Correlation with Local Flow Characteristic. AIAA 2003-3228 (2013)

8. Nielsen, J.N., Biggers, J.C.: Recent Progress in Circulation Control Aerodynamics. AIAA Paper 87-001 (1987)
9. Yaros, S.F., et al.: Synergistic Airframe-Propulsion Interactions and Integrations. NASA/TM-1998-207644 (1998)
10. Joslin, R.D., Jones, G.S.: Applications of Circulation Control Technology. Progress in Astronautics and Aeronautics, vol. 214. AIAA (2006)
11. Radespiel, R., Pfingsten, K.-C., Jensch, C.: Flow Analysis of Augmented High-Lift Systems. In: Radespiel, R., Rossow, C.-C., Brinkmann, B.W. (eds.) Hermann Schlichting – 100 Years. NNFM, vol. 102, pp. 168–189. Springer, Heidelberg (2009)
12. David, D., Marshall, D.D., Jameson, K.K.: Overview of Recent Circulation Control Modeling Activities at Cal Poly. AIAA 2010-0348 (2010)
13. Jones, G.S.: Pneumatic Flap Performance for a Two-Dimensional Circulation Control Airfoil. In: Joslin, D., Jones, G.S. (eds.) Applications of Circulation Control Technology. Progress in Astronautics and Aeronautics, vol. 214. AIAA (2006)
14. Englar, R.J., Huson, G.G.: Development of Advanced Circulation Control Using High-Lift Airfoils. Journal of Aircraft 21(7), 476–483 (1984)
15. Poisson-Quinton, P., Lepage, L.: Survey of French research on the control of boundary layer and circulation. In: Lachmann, G.V. (ed.) Boundary Layer and Flow Control: Its Principles and Application, vol. 1, pp. 21–73. Pergamon Press, New York (1961)
16. Alexander, M.G., Anders, S.G., Johnson, S.K.: Trailing Edge Circulation Control of an Airfoil at Transonic Mach numbers. In: Joslin, D., Jones, G.S. (eds.) Applications of Circulation Control Technology. Progress in Astronautics and Aeronautics, vol. 214. AIAA (2006)
17. Slomski, J.F., Gorski, J.J., Miller, R.W., Marino, T.A.: Numerical Simulation of Circulation Control Airfoils as Affected by Different Turbulence Models. AIAA paper 2002-0851 (2002)
18. Swanson, R.C., Rumsey, C.L., Anders, S.G.: Progress towards Computational Method for Circulation Control Airfoils. AIAA Paper 2005-0089 (2005)
19. Chang, I.P.A., Slomski, J., Marino, T., Ebert, M.P., Abramson, J.: Full Reynolds-Stress Modeling of Circulation Control Airfoil. In: Joslin, R.D., Jones, G.S. (eds.) Applications of Circulation Control Technology. Progress in Astronautics and Aeronautics, vol. 214, ch. 17, pp. 445–466. American Institute of Aeronautics and Astronautics, Inc. (2006)
20. Baker, W.J., Paterson, E.G.: RANS CFD simulation of a circulation-control foil: validation of performance, flow field, and wall jet. AIAA Paper 2006–3010 (2006)
21. Pfingsten, K.C., Jensch, C., Körber, K.V., Radespiel, R.: Numerical simulation of the flow around circulation control airfoils. In: First CEAS European Air and Space Conference, Berlin (2007)
22. Swanson, R.C., Rumsey, C.L.: Computation of circulation control airfoil flows. Computers and Fluids 38, 1925–1942 (2009)
23. Schwamborn, D., Gerhold, T., Heinrich, R.: The DLR TAU-Code: Recent applications in Research and Industry. In: ECCOMAS CFD, Egmond aan Zee, The Netherlands (2006)
24. Shur, M.L., Strelets, M.K., Travin, A.K., Spalart, P.R.: Turbulence Modeling in Rotating and Curved Channels: Assessing the Spalart-Shur Correction. AIAA Journal 38, 784–792 (2000)
25. Pfingsten, K.C., Cecora, R.D., Radespiel, R.: An experimental investigation of a gapless high-lift system using circulation control. In: Proc. Katnet II Conference, Bremen (2009)
26. Pfingsten, K.C.: Experimentelle und numerische Untersuchung eines spaltlosen Hochauftriebssystems mit Zirkulationskontrolle. Campus Forschungsflughafen FB 2011-04 (2011) ISBN 978-3-8440-0099-3

27. Jensch, C., Pfingsten, K.C., Radespiel, R.: Numerical Investigation of Leading Edge Blowing and Optimization of the Slot and Flap Geometry for a Circulation Control Airfoil. In: Dillmann, A., Heller, G., Klaas, M., Kreplin, H.-P., Nitsche, W., Schröder, W. (eds.) New Results in Numerical and Experimental Fluid Mechanics VII. NNFM, vol. 112, pp. 183–190. Springer, Heidelberg (2010)
28. Jensch, C., Pfingsten, K.C., Radespiel, R., Schuermann, M., Haupt, M., Bauss, S.: Design Aspects of a Gapless High-Lift System with Active Blowing. In: DLRK 2009, Aachen (2009)
29. Burnazzi, M., Radespiel, R.: Design of a Droopnose Configuration for a Coanda Active Flap Application. AIAA 2013-487 (2013)
30. Burnazzi, M., Radespiel, R.: Numerical Simulations of an Airfoil Equipped with Suction and Blowing for Active High-Lift. In: Proc. Deutscher Luft- und Raumfahrtkongress, Stuttgart, Germany, September 10-12 (2013)

Heuristics for Effective Actuator and Sensor Placement in Feedback Flow Control

Kevin K. Chen and Clarence W. Rowley

Princeton University, Princeton, NJ 08544, USA
{kkchen,cwrowley}@princeton.edu

Abstract. Actuator and sensor placement can be just as consequential for the performance of localized feedback flow control as controller design. Yet, effective placement is not well understood, and the use of suboptimal placements is common. This manuscript reports descriptions and characteristics of effective actuator and sensor placements for optimal flow control. We review $\mathscr{H}_2$ optimal placements in the linearized Ginzburg–Landau and Orr–Sommerfeld/Squire models of fluid flow. We then analyze the feedback control of these models by relating physical observations with mathematical tools. Although these tools do not fully predict optimal placements, they do reveal patterns that most or all effective placements share. Most notably, effective actuator–sensor placements provide good authority over unstable modes and transient growth, and avoid large time lags between inputs and outputs.

1 Introduction

Every application of localized feedback flow control must address a basic question: where should the actuators and sensors be located in the flow? This placement can impact the control performance just as much as the controller dictating actuation signals from sensor signals. By and large, research in actuator and sensor placement has trailed behind research in controller design. To fill this gap, this manuscript attempts to characterize effective placements for flow control.

A full review of actuator and sensor placement would be too expansive to provide here. Instead, we discuss varying points of view from which different communities have tackled this problem. Control theorists often approach placement theory in ways that are broadly applicable to partial differential equation systems [1]. Much of this research focuses on algorithms for optimizing actuator and sensor locations, but many key aspects are still unsolved. For instance, many articles (e.g., [2,3]) discuss convergence toward optimal actuator placements, but do not simultaneously solve for optimal sensor placements. The flexible structures community has put forth a sizable body of research (e.g., [4,5]). Nevertheless, the structures' governing dynamics are typically described by normal linear operators, and are therefore better behaved than those of fluid flows, for which the linearized dynamics are often highly non-normal. In fluid mechanics, a number of researchers [6,7,8,9,10] have corroborated the hypothesis that control is especially effective when actuators and sensors reside in absolutely unstable,

R. King (ed.), *Active Flow and Combustion Control 2014*, Notes on Numerical Fluid Mechanics and Multidisciplinary Design 127, DOI: 10.1007/978-3-319-11967-0_8

or "wavemaker," regions. Nevertheless, few have described effective actuator and sensor placements in general frameworks, and guess-and-check (e.g., [6,7,11]) and other suboptimal placement methodologies (e.g., [12,13]) remain common.

The purpose of this manuscript is to provide qualitative characterizations of effective actuator and sensor placements for linear time-invariant (LTI) feedback flow control. We utilize an optimization technique described in our previous work [10,14] to solve for $\mathscr{H}_2$ optimal placements in the linearized Ginzburg–Landau and the Orr–Sommerfeld/Squire models of fluid flow. We then describe features common to these optimal control systems. This analysis is based on mathematical concepts—such as right-half-plane poles and zeros, time lags, eigenmodes, sensitivity to spatially localized perturbations, optimal transient growth and disturbances, and impulse responses—as well as physical intuition.

We organize the manuscript as follows. Sec. 2 reviews $\mathscr{H}_2$ optimal control and placement. Sec. 3 then reviews the linearized Ginzburg–Landau and Orr–Sommerfeld/Squire models of fluid flow, along with the optimal placements in those systems. Sec. 4 describes common features of optimal actuator and sensor placements, and Sec. 5 provides concluding remarks.

2 $\mathscr{H}_2$ Optimal Control and Placement

The chief goal of the theory in this section is the optimization of closed-loop control performance simultaneously over all actuator positions, sensor positions, and LTI controllers. This section is a concise summary of [10,14].

We assume a spatial variable x, time t, dynamical operator $\mathcal{L}$, and a linear model $\dot{q}(x,t) = \mathcal{L}q(x,t)$ of fluid flow for the state q (e.g., a velocity perturbation). If the dynamics are subject to the effects of disturbances $d(x,t)$, as well as r actuators with spatial authority $\{b_j(x)\}_{j=1}^r$ and input signals $\{u_j(t)\}_{j=1}^r$, then

$$\dot{q}(x,t) = \mathcal{L}q(x,t) + \sum_{j=1}^{r} b_j(x)u_j(t) + d(x,t) \ . \tag{1}$$

Furthermore, if we assume p sensors with spatial authority $\{c_k(x)\}_{k=1}^p$ and sensor noise $\{n_k(t)\}_{k=1}^p$, as well as the inner product $\langle q_1(x), q_2(x)\rangle = \int_\Omega \bar{q}_2(x)q_1(x)\,dx$ over the domain Ω (where $\overline{(\cdot)}$ is the complex conjugate), then the sensor outputs are

$$y_k(t) = \langle q(x,t), c_k(x)\rangle + n_k(t) \tag{2}$$

for $k = 1,\ldots,p$. In this setup, the objective of $\mathscr{H}_2$ optimal control is to minimize a cost function on q and $\{u_j\}_{j=1}^r$ by computing $\{u_j\}_{j=1}^r$ from $\{y_k\}_{k=1}^p$.

In discrete space, we let $\boldsymbol{q}(t)$ be a vector representation of $q(x,t)$ over discrete points in the domain. We then bundle the disturbances and sensor noise in the vector $\boldsymbol{w}(t)$, and we let $\boldsymbol{z}(t)$ be the vector representation of the $\mathscr{H}_2$ optimal control cost function. We further let $\boldsymbol{u}(t)$ and $\boldsymbol{y}(t)$ be vectors respectively stacking the signals $\{u_j(t)\}_{j=1}^r$ and $\{y_k(t)\}_{k=1}^p$. The matrices $\boldsymbol{A}$, $\boldsymbol{B}_1$, $\boldsymbol{B}_2$, $\boldsymbol{C}_1$, $\boldsymbol{C}_2$, $\boldsymbol{D}_{12}$,

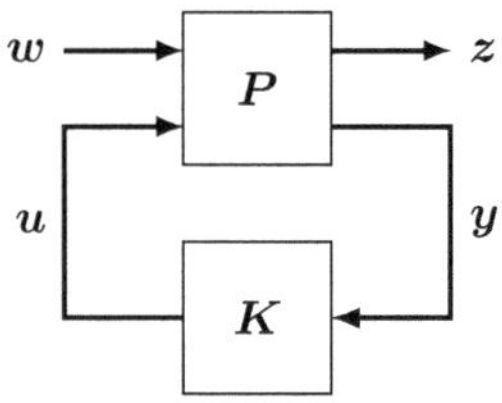

Fig. 1. The closed-loop system $\boldsymbol{G}$

and $\boldsymbol{D}_{21}$ then relate these variables via the "plant" $\boldsymbol{P}$, given by

$$\begin{bmatrix} \dot{q} \\ \hline z \\ y \end{bmatrix} = \left[\begin{array}{c|cc} A & B_1 & B_2 \\ C_1 & 0 & D_{12} \\ C_2 & D_{21} & 0 \end{array}\right] \begin{bmatrix} q \\ \hline w \\ u \end{bmatrix} . \tag{3}$$

We connect the continuous-space (1, 2) and discrete-space (3) formulations by describing the matrices in Sec. 3.

The $\mathscr{H}_2$ optimal controller is the LTI system $\boldsymbol{K}$, arranged as in Fig. 1, that minimizes the $\mathscr{H}_2$ norm of the closed-loop system $\boldsymbol{G}$. This $\mathscr{H}_2$ norm $\|\boldsymbol{G}\|_2$ is essentially an average-case measure of the gain provided by $\boldsymbol{G}$ from $\boldsymbol{w}$ to $\boldsymbol{z}$. Given a unit white noise input $\boldsymbol{w}(t)$ and the expected value operator $E(\cdot)$, the norm satisfies $\|\boldsymbol{G}\|_2^2 = E\left(\lim_{T\to\infty} \int_{-T}^{T} \|\boldsymbol{z}(t)\|_2^2\, dt/(2T)\right)$. The computation of $\boldsymbol{K}$ and $\|\boldsymbol{G}\|_2$ is a standard procedure (see [15]).

Following [10,14], we define the $\mathscr{H}_2$ optimal actuator and sensor locations to be the placements (i.e., the choice of $\boldsymbol{B}_2$ and $\boldsymbol{C}_2$, given appropriate constraints) such that the resulting $\|\boldsymbol{G}\|_2$ is globally minimized. In [10], we proposed an efficient method for computing the gradient of $\|\boldsymbol{G}\|_2$ with respect to actuator and sensor locations, allowing for gradient-based optimization. This technique is a more efficient variant of [4], but [14] uses adjoint operators to speed up the computation further. In this manuscript, we first sample the entire domain of actuator and sensor placements. Local minima in the sampling then serve as initial conditions for Broyden–Fletcher–Goldfarb–Shanno quasi-Newton minimization; we take the best of these results to be the globally optimal placement.

3 Fluid Flow Models

3.1 Overview

For computational tractability, we analyze the optimal actuator and sensor placements in the linearized Ginzburg–Landau and the Orr–Sommerfeld/Squire equations. These systems are complementary in the study of shear flows, as they respectively model streamwise and transverse variations in flow perturbations.

In our fluid models, we actuate and sense the native state variables directly. Although this is not particularly realistic, it simplifies the control problem and makes it generalizable. We model localized actuators in (1) with the Gaussian functions $b_j(x) = \exp\left((x - x_{\mathrm{a},j})^2/(2\sigma^2)\right) / \left(\sqrt{2\pi}\sigma\right)$, where $x_{\mathrm{a},j}$ is the jth actuator location, and σ is a small Gaussian width. Denoting the spatial discretization of $b_j(x)$ by $\boldsymbol{b}_j$, $\boldsymbol{B}_2 = \begin{bmatrix}\boldsymbol{b}_1 & \cdots & \boldsymbol{b}_r\end{bmatrix}$. Similarly, we model localized sensors in (2) with $c_k(x) = \exp\left((x - x_{\mathrm{s},k})^2/(2\sigma^2)\right) / \left(\sqrt{2\pi}\sigma\right)$, where $x_{\mathrm{s},k}$ is the kth sensor location. If $(\cdot)^{\mathrm{H}}$ denotes the conjugate transpose, $\boldsymbol{H}$ is a weighting matrix for trapezoidal integration such that the spatial inner product is equivalently $\langle \boldsymbol{q}_1, \boldsymbol{q}_2 \rangle = \boldsymbol{q}_2^{\mathrm{H}} \boldsymbol{H} \boldsymbol{q}_1$, and $\boldsymbol{c}_k$ is the discretization of $c_k(x)$, then $\boldsymbol{C}_2 = \begin{bmatrix}\boldsymbol{c}_1 \cdots \boldsymbol{c}_p\end{bmatrix}^{\mathrm{H}} \boldsymbol{H}$.

In our $\mathscr{H}_2$ optimal control setup, we assume a standard linear quadratic Gaussian (LQG) problem. Thus, the objective of the control is to minimize $E(\boldsymbol{q}^{\mathrm{H}}\boldsymbol{Q}\boldsymbol{q} + \boldsymbol{u}^{\mathrm{H}}\boldsymbol{R}\boldsymbol{u})$, given the disturbances and sensor noise in (1, 2) and appropriate choices of the state cost matrix $\boldsymbol{Q} \geq 0$ (i.e., positive semidefinite) and the input cost matrix $\boldsymbol{R} > 0$ (i.e., positive definite). We pick $\boldsymbol{Q}$ such that $\boldsymbol{q}^{\mathrm{H}}\boldsymbol{Q}\boldsymbol{q}$ is the kinetic energy of the state. Furthermore, if $\boldsymbol{d}(t)$ is the discretization of the disturbance $d(x,t)$ in (1), and the sensor noise $n_k(t)$ (see (2)) is stacked in the vector $\boldsymbol{n}(t)$, then we define the disturbance covariance $\boldsymbol{W} = E(\boldsymbol{d}\boldsymbol{d}^{\mathrm{H}}) \geq 0$ and sensor noise covariance $\boldsymbol{V} = E(\boldsymbol{n}\boldsymbol{n}^{\mathrm{H}}) > 0$. We pick $\boldsymbol{W}$ so that the disturbance is effectively a unit white noise uncorrelated in space and time. Finally, in the plant (3), we assign $\boldsymbol{w} = \begin{bmatrix}\boldsymbol{d}^{\mathrm{H}} & \boldsymbol{n}^{\mathrm{H}}\end{bmatrix}^{\mathrm{H}}$ and

$$\boldsymbol{B}_1 = \begin{bmatrix}\boldsymbol{W}^{1/2} & \boldsymbol{0}\end{bmatrix} , \quad \boldsymbol{C}_1 = \begin{bmatrix}\boldsymbol{Q}^{1/2} \\ \boldsymbol{0}\end{bmatrix} , \quad \boldsymbol{D}_{12} = \begin{bmatrix}\boldsymbol{0} \\ \boldsymbol{R}^{1/2}\end{bmatrix} , \quad \boldsymbol{D}_{21} = \begin{bmatrix}\boldsymbol{0} & \boldsymbol{V}^{1/2}\end{bmatrix} . \tag{4}$$

The following sections describe the choice of $\mathcal{L}$ in (1)—and consequently, $\boldsymbol{A}$ in (3)—as well as $\boldsymbol{Q}$, $\boldsymbol{R}$, $\boldsymbol{W}$, and $\boldsymbol{V}$ for the fluid flow models. They also report the optimal actuator and sensor locations in these systems. We employ pseudospectral differentiation with software based on [16].

3.2 Linearized Ginzburg–Landau Equation

This section is a brief summary of [10,14]. The linearized Ginzburg–Landau equation models small fluid perturbations near a Hopf bifurcation [17,18]. Letting $x \in \mathbb{R}$ be the primary direction of the fluid flow, and choosing some $\mu(x) : \mathbb{R} \to \mathbb{R}$, $\nu \in \mathbb{C}$, and $\gamma \in \mathbb{C}$, the linearized Ginzburg–Landau operator is

$$\mathcal{L} = \mu(x) - \nu\frac{\partial}{\partial x} + \gamma\frac{\partial^2}{\partial x^2} . \tag{5}$$

The state variable is $q(x,t) : \mathbb{R} \times \mathbb{R} \to \mathbb{C}$, and $\mathrm{Re}(q)$ represents a velocity perturbation in the flow. The boundary conditions are that $\lim_{x\to\pm\infty} q = 0$. For a comprehensive review of the equation and its control, see [13]. We choose $\mu(x) = \mu_0 - 0.04 - 5 \cdot 10^{-3}x^2$, $\mu_0 = 0.41$, $\nu = 2 + 0.4i$ (with i the imaginary unit), and $\gamma = 1 - i$, as in [10,13,14]. This system is unstable, strongly non-normal, and highly convective in the positive x direction. The amplification region—that is, where $\mu(x) > 0$—is $[-8.60, 8.60]$, and we choose $\sigma = 0.4$.

To compute $\boldsymbol{A}$ from $\mathcal{L}$, we employ pseudospectral differentiation with Hermite polynomials. Also, we choose $\boldsymbol{Q} = 49\boldsymbol{H}$, $\boldsymbol{R} = \boldsymbol{I}$, $\boldsymbol{W} = \boldsymbol{I}$, and $\boldsymbol{V} = 4 \cdot 10^{-8}\boldsymbol{I}$,

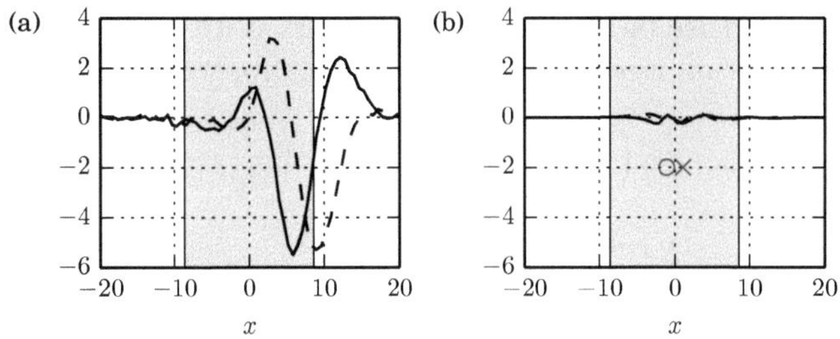

Fig. 2. Snapshots of Re(q) (solid) and Im(q) (dashed) in a neutrally stable linearized Ginzburg–Landau simulation ($\mu_0 = 0.398$) with white noise disturbances. In this and all following plots, the amplification region is shaded gray. (a) Without control. (b) With sensor noise and LQG control at the optimal actuator ($\circ$) and sensor placement ($\times$), shown below.

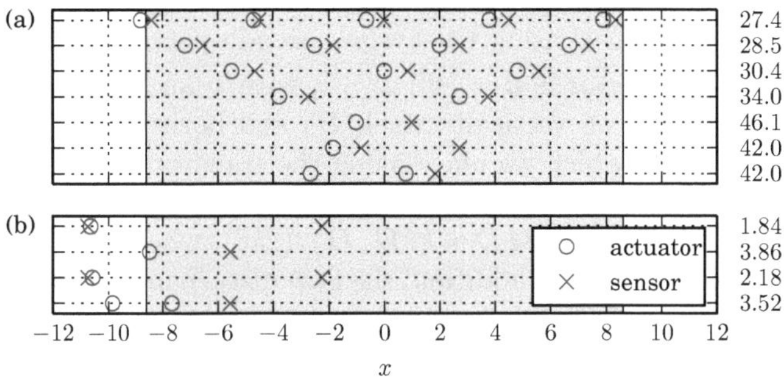

Fig. 3. Optimal actuator and sensor placements in the linearized Ginzburg–Landau system. Each row shows the solution for a different number of actuators and sensors, and values of $\|\boldsymbol{G}\|_2$ are shown on the right. (a) Flow disturbances everywhere. (b) Flow disturbances introduced only at $x_{\mathrm{d}} = -11.0$. The top row has two collocated actuators.

thereby emphasizing the state disturbance and state cost. With 100 grid points, the leading eigenvalues of $\boldsymbol{A}$ match the analytic eigenvalues of $\mathcal{L}$ presented in [13], and the optimal placement results are converged. (For a discussion of convergence to the continuous-space solution, see [2].)

We solve for the $\mathscr{H}_2$ optimal actuator and sensor positions (see [14]), with varying numbers of each; see Fig. 2 and Fig. 3(a). Fig. 3(b) is a special case where the flow disturbance $d(x,t)$ is a localized, single-channel white noise scaled by $\exp\left((x - x_{\mathrm{d}})^2/(2\sigma^2)\right) / \left(\sqrt{2\pi}\sigma\right)$, with the disturbance location $x_{\mathrm{d}} = -11$.

3.3 Orr–Sommerfeld/Squire Equations

In shear flows, the Orr–Sommerfeld and Squire equations model transverse variations in exponential mode amplitudes [19]. Let $y \in \mathbb{R}$ be the transverse spatial variable, replacing x in (1, 2). Also, let $\alpha, \beta \in \mathbb{R}$ be streamwise and spanwise wavenumbers, and $k = \sqrt{\alpha^2 + \beta^2}$. Next, let $u_0(y)$ be the streamwise velocity of a base flow, and Re be the Reynolds number. With the vertical velocity and vorticity perturbation mode amplitudes $v(y,t), \omega(y,t) : \mathbb{R} \times \mathbb{R} \to \mathbb{C}$, we define

$$\mathcal{L}_{\mathrm{OS}} = i\alpha \left(u_0'' - u_0 \left(\frac{\partial^2}{\partial y^2} - k^2 \right) \right) + Re^{-1} \left(\frac{\partial^4}{\partial y^4} - 2k^2 \frac{\partial^2}{\partial y^2} + k^4 \right) \tag{6a}$$

$$\mathcal{L}_{\mathrm{S}} = -i\alpha u_0 + Re^{-1} \left(\frac{\partial^2}{\partial y^2} - k^2 \right) , \tag{6b}$$

so that

$$\mathcal{L} : \begin{bmatrix} v \\ \omega \end{bmatrix} \mapsto \begin{bmatrix} (\frac{\partial^2}{\partial y^2} - k^2)^{-1} \mathcal{L}_{\mathrm{OS}} & 0 \\ -i\beta u_0' & \mathcal{L}_{\mathrm{S}} \end{bmatrix} \begin{bmatrix} v \\ \omega \end{bmatrix} . \tag{6c}$$

The Orr–Sommerfeld and Squire systems are respectively the top and bottom rows of (6c). At the endpoints of the domain, $v = v' = \omega = 0$. Our base flow is the plane Poiseuille flow through a channel bounded by $y = \pm 1$. Thus, $u_0(y) = y^2 - 1$, and we choose $\sigma = 0.047$.

We compute $\boldsymbol{A}$ from $\mathcal{L}$ using Chebyshev pseudospectral differentiation. The discrete-space state vector is $\boldsymbol{q} = \begin{bmatrix} \boldsymbol{v}^{\mathrm{H}} & \boldsymbol{\omega}^{\mathrm{H}} \end{bmatrix}^{\mathrm{H}}$, with $\boldsymbol{v}$ and $\boldsymbol{\omega}$ representing $v(y,t)$ and $\omega(y,t)$. Similarly, we vertically stack $\boldsymbol{B}_2$ and horizontally stack $\boldsymbol{C}_2$. Each actuator and sensor—which we place along the transverse direction y—acts on either $\boldsymbol{v}$ or $\boldsymbol{\omega}$. To penalize the kinetic energy of the flow perturbation in the LQG design, we set $\boldsymbol{Q} = \mathrm{diag}(\boldsymbol{D}^{\mathrm{H}} \boldsymbol{H} \boldsymbol{D} + k^2 \boldsymbol{H}, \boldsymbol{H})/(2k^2)$, where $\boldsymbol{D}$ is the Chebyshev differentiation matrix [19]. In addition, the flow disturbances are an uncorrelated unit white noise in all three components of velocity; thus, $\boldsymbol{W} = \mathrm{diag}(\boldsymbol{I}, k^2 \boldsymbol{I})$ [19]. Finally, $\boldsymbol{R} = \boldsymbol{I}$ and $\boldsymbol{V} = 0.01\boldsymbol{I}$. With 125 nodes, the leading eigenvalues of $\boldsymbol{A}$ match Fig. 3.1 of [19], and the optimal placements are converged.

We consider three sets of parameters in this section. For the spanwise waves given by $Re = 10^4$, $\alpha = 1$, and $\beta = 0$, (6c) decouples into the Orr–Sommerfeld and Squire systems; we assume single-input, single-output (SISO) control in each. We also consider the streamwise streaks given by $Re = 10^4$, $\alpha = 0$, and $\beta = 1$ (see Fig. 4), as well as the oblique waves given by $Re = 5 \cdot 10^3$ and $\alpha = \beta = 1$. In these two cases, we consider two-input, two-output (2I2O) control, where the locations of the v-actuator and sensor are y_{a}^v and y_{s}^v, and the locations of the ω-actuator and sensor are y_{a}^ω and y_{s}^ω. The off-diagonal term in (6c) causes small changes in v to have large effects on ω. Therefore, even when $|v| \ll |\omega|$ (e.g., Fig. 4), the v-actuator and sensor placement is still relevant.

Fig. 5 shows the globally optimal actuator and sensor placements, as well as placements that are locally, and nearly globally, optimal. We also remark that this Orr–Sommerfeld/Squire system is symmetric about $y = 0$; for each optimal placement, we do not show the equivalent solution reflected about $y = 0$.

4 Observations and Heuristics

In this section, we discuss various considerations, observations, and traits we observe in feedback flow control systems with effective actuator and sensor placements. These topics are not disjoint, but are rather related in various ways. Each section within briefly describes a particular feature of interest.

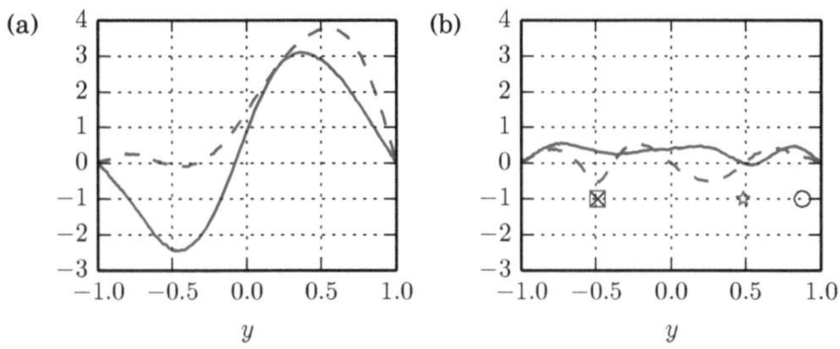

Fig. 4. Snapshots of Re(ω) (solid) and Im(ω) (dashed) in an Orr–Sommerfeld/Squire simulation with white noise disturbances, using $Re = 10^4$, $\alpha = 0$, and $\beta = 1$; v (not shown) is very small. (a) Without control. (b) With sensor noise and LQG control at the optimal placement (○: v actuator; ×: v sensor; □: ω actuator; ☆: ω sensor), shown below.

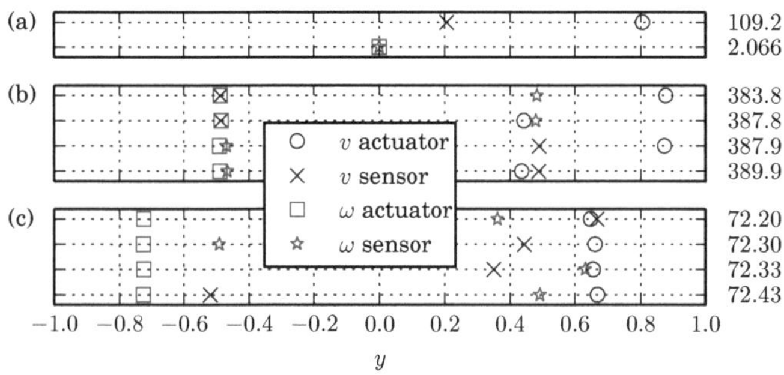

Fig. 5. As Fig. 3, but for the Orr–Sommerfeld/Squire system. (a) The decoupled SISO systems with $Re = 10^4$, $\alpha = 1$, and $\beta = 0$. (b) The coupled 2I2O system with $Re = 10^4$, $\alpha = 0$, and $\beta = 1$, showing both globally (top row) and locally optimal solutions. (c) As (b), but with $Re = 5 \cdot 10^3$ and $\alpha = \beta = 1$.

4.1 Fundamental Requirements

In rigorous control theory, it is possible to identify and characterize plants for which good control performance is never possible with any LTI controller. This section briefly summarizes ideas from [15].

Two basic characteristics that all plants must possess for closed-loop stability are *stabilizability* and *detectability*. That is, for closed loop stability, the actuators must be able to act on all unstable adjoint eigenmodes of $\mathcal{L}$, and the sensors must be able to observe all unstable direct eigenmodes. Comparing Fig. 6(a) and Fig. 7(a), for instance, we observe that the linearized Ginzburg–Landau plant is stabilizable and detectable at the optimal actuator and sensor placement. If the sensor is very far upstream or the actuator is very far downstream, however, they lose authority over the unstable mode, leading to high values of $\|\boldsymbol{G}\|_2$, and therefore, poor control performance. In the decoupled Orr–Sommerfeld and Squire systems, the optimal placement in Fig. 8 has good actuator and sensor

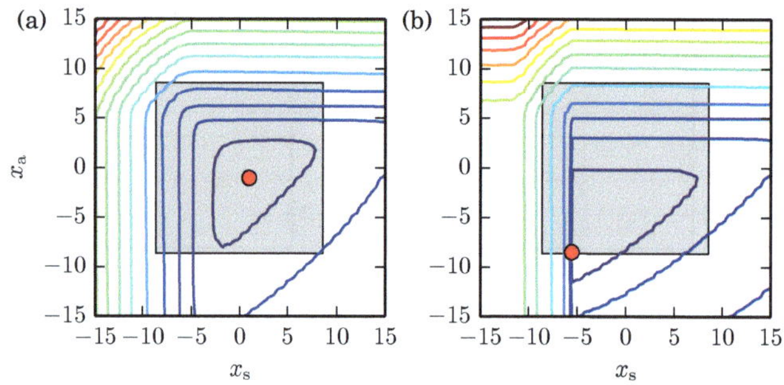

Fig. 6. Optimal placement (●) and $\log_{10} \|\boldsymbol{G}\|_2$ contours in the linearized Ginzburg–Landau system, plotted against the actuator placement x_a and the sensor placement x_s. (a) With spatially uncorrelated flow disturbances; the contour levels are 2, 2.5, ... (b) Disturbances only at $x_d = -11.0$; the contour levels are 1, 1.5, ...

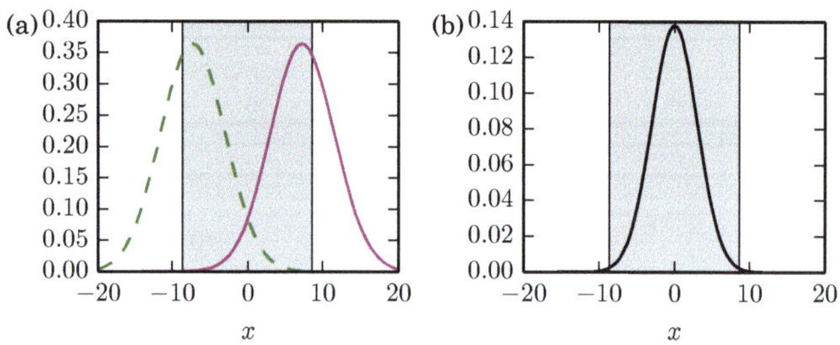

Fig. 7. (a) Magnitudes of leading direct (solid magenta) and adjoint (dashed green) modes of the linearized Ginzburg–Landau system; $\lambda = 0.012 - 0.648\jmath$. (b) The corresponding sensitivity magnitude $|\zeta(x)|$.

authority over the unstable (or least stable) modes in Fig. 9. Other placements lead to poorer modal controllability and observability.

Plant poles and zeros can also limit the efficacy of LTI control. Given the transfer function $\boldsymbol{P}(s) = \boldsymbol{C}_2(s\boldsymbol{I} - \boldsymbol{A})^{-1}\boldsymbol{B}_2$ of a plant (where we do not assume any direct feedthrough from $\boldsymbol{u}$ to $\boldsymbol{y}$), the poles and zeros are the values of $s \in \mathbb{C}$ for which $\boldsymbol{P}(s)$ diverges or is rank-deficient, respectively. For good disturbance and noise rejection, plant poles p and zeros z with positive real parts should roughly satisfy $|p| < |z|/4$ [15]. In Fig. 6(b), a small upstream shift in the optimal sensor placement causes $\|\boldsymbol{G}\|_2$ to rise extremely quickly. This may be because a plant zero crosses into the complex right half-plane, violating the pole–zero condition. Control designers should place actuators and sensors so as to avoid small right-half-plane zeros, but this discussion is beyond the scope of this manuscript.

Finally, if a system has right-half-plane poles p as well as time delays τ from actuators to sensors, then $\tau < 1/(2|p|)$ is generally required for good disturbance and noise rejection [15]. Although our fluid models do not strictly contain time

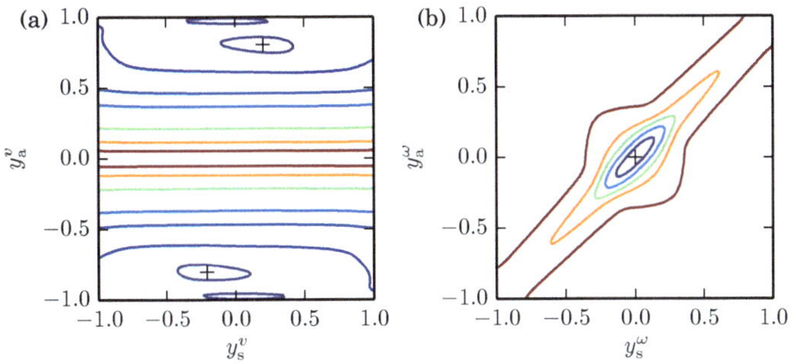

Fig. 8. (a) The optimal placements (+) and $\log_{10} \|\boldsymbol{G}\|_2$ contours in the decoupled Orr–Sommerfeld system, with $Re = 10^4$, $\alpha = 1$, and $\beta = 0$. The contour levels are 2.04, 2.1, 2.3, 2.5, 3.0, 3.5, 4.0. (b) As (a), but with contours of $\|\boldsymbol{G}\|_2$ in the decoupled Squire system. The contour levels are 2.10, 2.15, ..., 2.30.

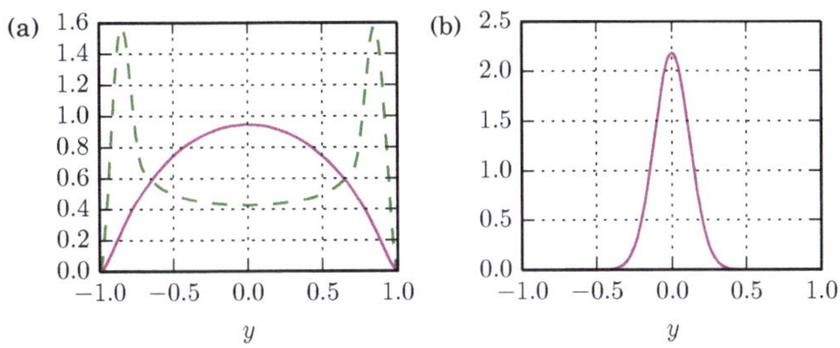

Fig. 9. Magnitudes of leading direct (solid magenta) and adjoint (dashed green) modes. (a) The decoupled Orr–Sommerfeld system; $\lambda = 0.004 - 0.238i$. (b) The decoupled Squire system; $\lambda = -0.007 - 0.993i$. The direct and adjoint eigenmodes overlap.

delays, the Ginzburg–Landau model does exhibit time lags, since convection predominantly drives the dynamics. Fig. 6 confirms that when the actuator is far upstream of the sensor—leading to large time lags—the control performance decreases. In this system, it is generally preferable for the two to be nearly collocated, as Fig. 3 shows. An advantage of placing actuators slightly upstream of sensors is that the actuation's transient growth provides stronger sensor outputs.

4.2 Eigenmode-Based Analyses

Direct and Adjoint Eigenmodes. As a direct extension of the stabilizability and detectability discussion in Sec. 4.1, it seems natural to place actuators and sensors where the most unstable adjoint and direct eigenmodes are respectively large. This approach is common [12,13], and it works well for the decoupled Orr–Sommerfeld and Squire equations at $Re = 10^4$, $\alpha = 1$, and $\beta = 0$. Comparing Fig. 8(b) and Fig. 9(b), we confirm that the modal analysis exactly predicts the optimal vorticity placement of $y_a^\omega = y_s^\omega = 0$. Observing the peaks Fig. 9(a), we also predict an optimal placement of $y_a^v = \pm 0.848$ and $y_s^v = 0$, yielding $\|\boldsymbol{G}\|_2 =$

109.6. This is not far from the true optimum[1] $y^v_{\rm a} = \pm 0.789$ and $y^v_{\rm s} = \pm 0.246$, with $\|\boldsymbol{G}\|_2 = 109.2$.

We must emphasize, however, that placement by eigenmode analysis is generally suboptimal in fluid systems. It is well established [19] that the eigenmodes of non-normal operators do not individually predict transient growth. Thus, even if a control system can prevent the long-time blow-up of unstable modes, it may be very poor at controlling transient growth. The linearized Ginzburg–Landau system's eigenmodes predict an optimal SISO placement of $x_{\rm a} = -7.28$ and $x_{\rm s} = 7.28$, with $\|\boldsymbol{G}\| = 259$ (Fig. 7(a)), but the true optimum is $x_{\rm a} = -1.03$ and $x_{\rm s} = 0.98$, with $\|\boldsymbol{G}\|_2 = 46.1$ (Fig. 6(a)). The analysis also fails in the coupled Orr–Sommerfeld/Squire equations, where the comparatively large off-diagonal term in $\mathcal{L}$ (6c) causes strong non-normality. The general failure of this approach necessitates alternate analyses, which we describe below.

Sensitivity Analysis. The sensitivity analysis [9,18] determines how much a certain eigenvalue of $\mathcal{L}$ will move in the complex plane if $\mathcal{L}$ experiences a point perturbation. This is related to the ideas of absolute instability and wavemakers in local flow stability theory. Given a perturbation strength ds and location $x = \xi$, and denoting the Dirac delta function by δ, we suppose that $d\mathcal{L} = \delta(x-\xi)\,ds$. Furthermore, we consider the eigendecomposition $\mathcal{L}\phi_j = \lambda_j\phi_j$, as well as the eigendecomposition $\mathcal{L}^*\psi_j = \bar{\lambda}_j\psi_j$ of the adjoint operator $\mathcal{L}^*$. The sensitivity function is then

$$\zeta_j(\xi) = \frac{d\lambda_j}{ds}(\xi) = \frac{\bar{\psi}_j(\xi)\phi_j(\xi)}{\langle \phi_j, \psi_j \rangle} . \tag{7}$$

The sensitivity analysis is particularly useful with collocated actuators and sensors, where the feedback control perturbs $\mathcal{L}$ in our prescribed way. Therefore, the regions where $\zeta_j(x)$ is large are the areas where feedback control can have a considerably favorable effect on the closed-loop eigenvalues. In the SISO linearized Ginzburg–Landau system, we can use the sensitivity in Fig. 7(b) to set $x_{\rm a} = x_{\rm s} = 0$, yielding $\|\boldsymbol{G}\|_2 = 54.2$. This is close to the true optimum $x_{\rm a} = -1.03$ and $x_{\rm s} = 0.98$, with $\|\boldsymbol{G}\|_2 = 46.1$ (see Fig. 3(a) and Fig. 6(a)).

The sensitivity analysis fails to predict the optimal placements in the Orr–Sommerfeld/Squire system, however. Fig. 5 reveals that these optimal placements generally do not involve nearly collocating actuators and sensors, and this sensitivity analysis is therefore not directly applicable. It is necessary to consider alternate means of evaluating effective placements in non-normal systems, which the next section discusses.

4.3 Transient Growth

Optimal Growth and Disturbances. The purpose of optimal growth theory (see [19]) is to analyze non-normal transient growth in ways that the eigenmode

[1] The computation reports that $y^v_{\rm a} = \pm 0.970$ and $y^v_{\rm s} = \mp 0.046$ is the global optimum, with the slightly lower $\|\boldsymbol{G}\|_2 = 109.1$ (see Fig. 8(a)). This placement appears to violate the boundary conditions considerably, however, so we reject this solution.

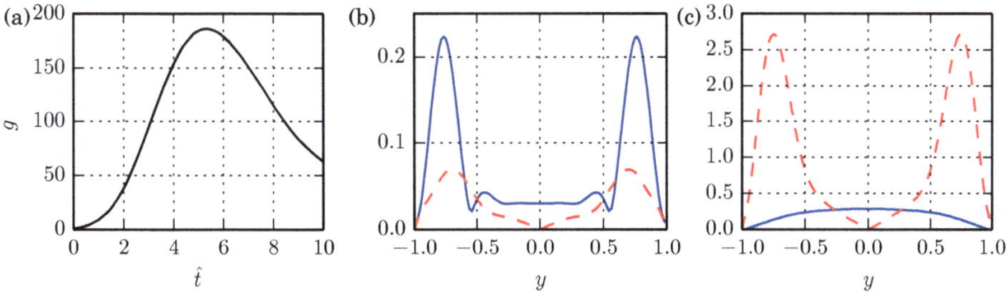

Fig. 10. The Orr–Sommerfeld/Squire system's optimal growth with $Re = 5 \cdot 10^3$ and $\alpha = \beta = 1$. (a) The optimal energy growth $g(\hat{t})$. (b) The optimal initial disturbance for the optimal $\hat{t} = 5.31$, shown as $|v|$ (solid blue) and $|\omega|$ (dashed red). (c) The disturbance at $\hat{t} = 5.31$.

analysis (Sec. 4.2) cannot. The optimal energy growth $g(t)$ dictates the largest energy amplification that a dynamical operator $\boldsymbol{A}$ (or equivalently, $\mathcal{L}$) can provide out of all possible initial conditions. If, given some $\boldsymbol{M}$, $\|\boldsymbol{q}\|_{\mathrm{E}} = \|\boldsymbol{M}^{1/2}\boldsymbol{q}\|_2$ is a norm whose square is the state's energy, then

$$g(t) = \max_{\boldsymbol{q}(0)\neq\boldsymbol{0}} \frac{\|\boldsymbol{q}(t)\|_{\mathrm{E}}^2}{\|\boldsymbol{q}(0)\|_{\mathrm{E}}^2} = \|\boldsymbol{M}^{1/2} e^{\boldsymbol{A}t} \boldsymbol{M}^{-1/2}\|_2^2 \ . \tag{8}$$

For the largest singular value of $\boldsymbol{M}^{1/2} e^{\boldsymbol{A}t} \boldsymbol{M}^{-1/2}$, the corresponding right singular vector is the optimal initial disturbance giving rise to the optimal growth, and the corresponding left singular vector is that disturbance at time t [19].

The optimal initial disturbances and the evolved disturbances hint at effective actuator and sensor placements, respectively. The reason for this is intuitive: the optimal initial disturbance indicates regions where actuation can have the largest short-term effect on the state, and the evolved disturbance represents the transient growth that the sensor can most easily detect. This is in contrast to the eigenmode analysis (Sec. 4.2), which only considers the infinite time horizon.

The choice of t, however, is not clear. In the stable Orr–Sommerfeld/Squire system in Fig. 10, it suffices to examine the time t that maximizes $g(t)$. Defining a characteristic time $t_{\mathrm{c}} = 1/\min_j |\lambda_j|$, we use the scaled time $\hat{t} = t/t_{\mathrm{c}}$ in Fig. 10(a) and find that the optimal growth is largest at $\hat{t} = 5.31$. Observing the peaks in the optimal initial disturbance (Fig. 10(b)), this predicts the optimal placement $y_{\mathrm{a}}^{v} = 0.77$ or -0.77, and $y_{\mathrm{a}}^{\omega} = 0.70$ or -0.70. Similarly, Fig. 10(c) predicts $y_{\mathrm{s}}^{v} = 0$ and $y_{\mathrm{s}}^{\omega} = 0.75$ or -0.75. The placement $y_{\mathrm{a}}^{v} = \pm 0.77$, $y_{\mathrm{a}}^{\omega} = \mp 0.70$, $y_{\mathrm{s}}^{v} = 0$, and $y_{\mathrm{s}}^{\omega} = \pm 0.75$ yields $\|\boldsymbol{G}\|_2 = 78.0$, which is close to the optimal $\|\boldsymbol{G}\|_2 = 72.2$ at $y_{\mathrm{a}}^{v} = \pm 0.648$, $y_{\mathrm{a}}^{\omega} = \mp 0.725$, $y_{\mathrm{s}}^{v} = \pm 0.667$, and $y_{\mathrm{s}}^{\omega} = \pm 0.361$ (Fig. 5(c)).

Our example linearized Ginzburg–Landau system is unstable, however, implying that $g(\hat{t})$ must diverge (Fig. 11(a)). Nonetheless, we can still examine the optimal initial and evolved disturbance for the arbitrarily chosen $\hat{t} = 1$. The optimal initial disturbance (Fig. 11(b)) predicts $x_{\mathrm{a}} = -1.55$, and the disturbance at $\hat{t} = 1$ (Fig. 11(c)) predicts $x_{\mathrm{s}} = 1.55$, yielding $\|\boldsymbol{G}\|_2 = 48.3$. This is very close to the true optimum $x_{\mathrm{a}} = -1.03$ and $x_{\mathrm{s}} = 0.98$, with $\|\boldsymbol{G}\|_2 = 46.1$.

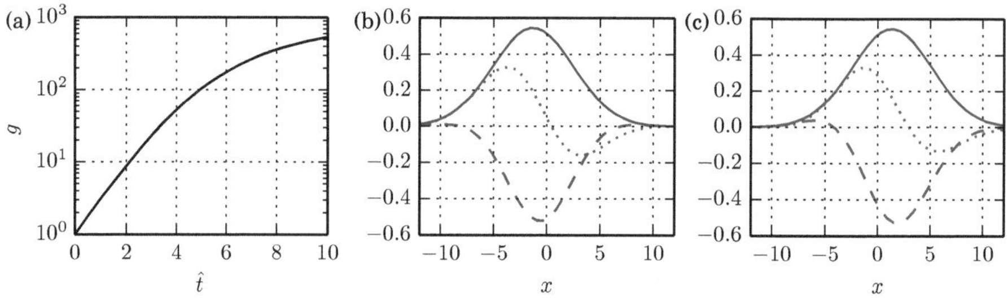

Fig. 11. The linearized Ginzburg–Landau system's optimal growth. (a) The optimal energy growth $g(\hat{t})$. (b) The optimal initial disturbance for $\hat{t} = 1$, shown as real (dashed) and imaginary (dotted) parts, and magnitude (solid). (c) The disturbance at $\hat{t} = 1$.

Impulse Responses. Since we assume a linear framework in (1, 2), impulse responses from actuators and adjoint impulse responses from sensors are effectively Green's functions. For non-normal systems, the impulse responses reveal each actuator's ability to affect the state via the transient growth of its impulse; the adjoint impulses also reveal each sensor's ability to detect transient dynamics. The impulse responses are given by (1, 2) with $u_j(t) = \delta(t)$ for one value of j, and $d(x,t)$, $n_k(t)$, and all other $u_j(t)$ equal to zero. To compute the adjoint impulse responses, we use the adjoint operator $\mathcal{L}^*$ instead of $\mathcal{L}$, and we swap $c_k(x)$ and $b_j(x)$, as well as $u_j(t)$ and $y_k(t)$, with the impulse now on $y_k(t)$.

The impulse responses do not directly predict optimal actuator and sensor placements, but they do reveal why certain placements may be effective or ineffective. In the Orr–Sommerfeld/Squire system shown in Fig. 12(a), we find that the impulse from the v-actuator at the optimal $y_{\text{a}}^v = 0.877$ creates an extremely large transient growth in ω, as a result of the off-diagonal term in (6c). This implies that the actuator has significant authority over ω, using very little effort. Furthermore, the sensor at the optimal $y_{\text{s}}^\omega = 0.483$ registers a very large output from this impulse (Fig. 13(a)), indicating strong feedback. Although the impulse from the ω-actuator at $y_{\text{a}}^\omega = -0.488$ (Fig. 12(b)) does not appear to have a considerable effect, it does act in the $y < 0$ domain, where the v-actuator does not have as much authority over ω as in $y > 0$. We remark that neither actuator has an appreciable effect on ω at $y \approx 0$, but the underlying dynamics naturally suppress ω there. Finally, we reach similar conclusions from the adjoint impulse responses at $y_{\text{s}}^v = -0.487$ and $y_{\text{s}}^\omega = 0.483$; see Fig. 12(c, d) and Fig. 13(b).

We also briefly mention that the input–output impulse responses in Fig. 13 exhibit a nearly nonexistent time lag. This is a favorable characteristic in light of the fundamental requirements that Sec. 4.1 described.

4.4 Remarks

Physical Intuition. Many of the conclusions we drew with mathematical tools are also possible using physical intuition. We highlight a few examples here.

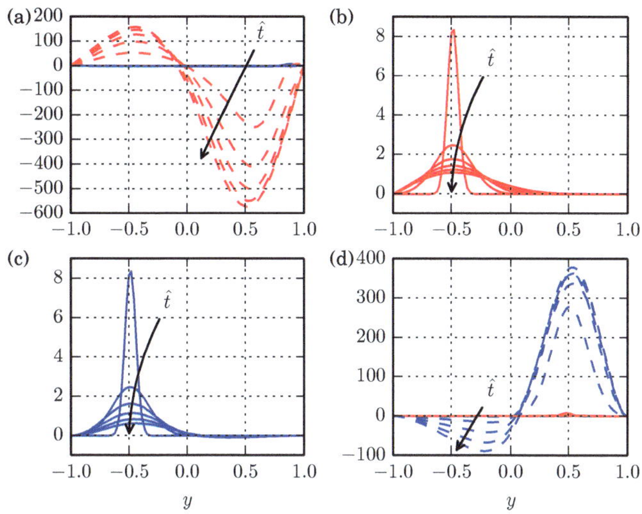

Fig. 12. Impulse responses of the Orr–Sommerfeld/Squire system with $Re = 10^4$, $\alpha = 0$, and $\beta = 1$ at the optimal placement, showing $\hat{t} = 0, 0.04, \ldots, 0.2$. (a) Impulse in v from $y_a^v = 0.877$ (solid blue: $\text{Re}(v)$; dashed red: $\text{Im}(\omega)$). (b) Impulse in ω from $y_a^\omega = -0.488$ ($\text{Re}(\omega)$ shown). (c) Adjoint impulse in v from $y_s^v = -0.487$ ($\text{Re}(v)$ shown). (d) Adjoint impulse in ω from $y_s^\omega = 0.483$ (dashed blue: $\text{Im}(v)$; solid red: $\text{Re}(\omega)$).

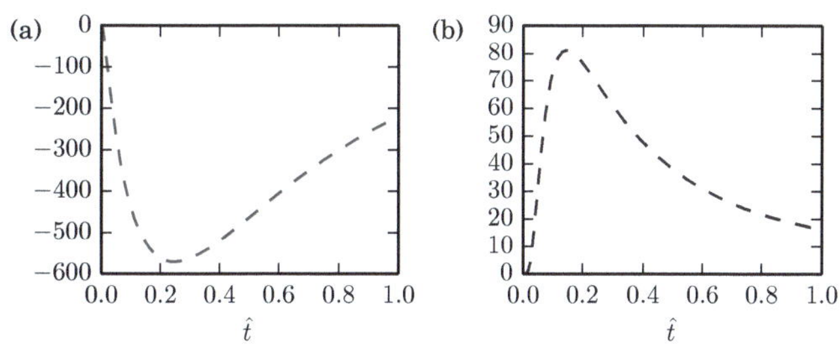

Fig. 13. Output traces in Fig. 12. (a) $\text{Im}(\omega)$ sensor output at $y_s^\omega = 0.483$ from the actuator impulse at $y_a^v = 0.877$ (see Fig. 12(a)). (b) $\text{Im}(v)$ adjoint actuator output at $y_a^v = 0.877$ from the adjoint sensor impulse at $y_s^\omega = 0.483$ (see Fig. 12(d)).

First, Sec. 4.1 explained effective and ineffective placements in the linearized Ginzburg–Landau system (Fig. 6) using stabilizability, detectability, and time lags. We intuit that in a convection-driven system, downstream actuators cannot influence much of the domain, and upstream sensors cannot detect the most relevant disturbances. Furthermore, sensors far downstream of actuators cannot detect actuator-induced growth until the growth has evolved and amplified for many convective time units. These are all recognizably unfavorable conditions.

Next, Sec. 4.2 justified the optimal placement in the decoupled Orr–Sommerfeld and Squire systems with $Re = 10^4$, $\alpha = 1$, and $\beta = 0$ using eigenmodes.

Simulations with flow disturbances show that the dynamics propagate v waves from near the boundaries to the center of the domain, and ω waves from the center to the boundaries. This intuition supports the optimal placements shown in Fig. 5(a). Sec. 4.3 also investigated the optimal placement in the Orr–Sommerfeld/Squire system with $Re = 10^4$, $\alpha = 0$, and $\beta = 1$ using impulse responses. Simulations (see Fig. 4(a)) show that ω tends to be largest roughly near $y = \pm 0.5$; this intuitively supports the placement of actuators and sensor near these regions, as shown in Fig. 5(b).

Other Remarks. We briefly remark that in some optimal placements, certain actuators and sensors primarily exhibit feedforward behavior. This is the case, for instance, with the ω-actuator and the v-sensor in the globally optimal solution in Fig. 5(b) and Fig. 12. This is not necessarily detrimental, as [20] describes. These feedforward actuators and sensors have authority over physical regions where other actuators and sensors do not, thereby improving the control performance.

Finally, we point out that it is common for the topography of $\|\boldsymbol{G}\|_2$ with respect to certain actuator and sensor placements to be relatively flat. This is the case, for instance, with the sensor placements in Fig. 5(c). This affords the control designer additional flexibility in effective hardware placement.

5 Conclusion

This manuscript described several characteristics of effective actuator and sensor positions for the feedback control of fluid models. The $\mathscr{H}_2$ optimal control of the linearized Ginzburg–Landau and the Orr–Sommerfeld/Squire models served as test beds for this investigation.

Successful feedback control always requires actuator and sensor positions that allow the plant to stabilize and detect unstable eigenmodes. In addition, placements that cause small right-half-plane zeros, or large time lags or delays, typically lead to poor control. The placement of actuators and sensors using leading eigenmodes can be effective, but only if the dynamics are close to normal. For non-normal dynamics with nearly collocated actuators and sensors, the sensitivity region better predicts advantageous placements. Optimal growth and disturbances, on the other hand, can predict optimal placements in non-normal systems without assuming collocation. Impulse responses also reveal the individual actuators' and sensors' authority over the state. Finally, physical intuition can often predict the same results that these mathematical tools reveal.

There currently does not exist a single, unifying method that is universally able to predict optimal actuator and sensor positions. The heuristics and observations in this manuscript provide some knowledge that may be useful in approaching this goal, however. The prediction of optimal placements remains a topic of future research.

Acknowledgements. This work was supported by the National Science Foundation's Graduate Research Fellowship Program and grant CMMI-0932928.

References

1. van de Wal, M., de Jager, B.: A review of methods for input/output selection. Automatica 37(4), 487–510 (2001)
2. Morris, K.: Linear-quadratic optimal actuator location. IEEE Trans. Autom. Control 56(1), 113–124 (2011)
3. Darivandi, N., Morris, K., Khajepour, A.: An algorithm for LQ optimal actuator location. Smart Mater. Struct. 22(3) (2013)
4. Hiramoto, K., Doki, H., Obinata, G.: Optimal sensor/actuator placement for active vibration control using explicit solution of algebraic Riccati equation. J. Sound Vib. 229(5), 1057–1075 (2000)
5. Güney, M., Eşkinat, E.: Optimal actuator and sensor placement in flexible structures using closed-loop criteria. J. Sound Vib. 312(1-2), 210–233 (2008)
6. Roussopoulos, K.: Feedback control of vortex shedding at low Reynolds numbers. J. Fluid Mech. 248, 267–296 (1993)
7. Gillies, E.A.: Multiple sensor control of vortex shedding. AIAA J. 39(4), 748–750 (2001)
8. Lauga, E., Bewley, T.R.: The decay of stabilizability with Reynolds number in a linear model of spatially developing flows. Proc. R. Soc. A 459(2036), 2077–2095 (2003)
9. Giannetti, F., Luchini, P.: Structural sensitivity of the first instability of the cylinder wake. J. Fluid Mech. 581, 167–197 (2007)
10. Chen, K.K., Rowley, C.W.: H_2 optimal actuator and sensor placement in the linearised complex Ginzburg–Landau system. J. Fluid Mech. 681, 241–260 (2011)
11. Strykowski, P.J., Sreenivasan, K.R.: On the formation and suppression of vortex 'shedding' at low Reynolds numbers. J. Fluid Mech. 218, 71–107 (1990)
12. Åkervik, E., Hœpffner, J., Ehrenstein, U., Henningson, D.S.: Optimal growth, model reduction and control in a separated boundary-layer flow using global eigenmodes. J. Fluid Mech. 579, 305–314 (2007)
13. Bagheri, S., Henningson, D.S., Hœpffner, J., Schmid, P.J.: Input-output analysis and control design applied to a linear model of spatially developing flows. Appl. Mech. Rev. 62, 020803 (2009)
14. Chen, K.K., Rowley, C.W.: Fluid flow control applications of $\mathscr{H}_2$ optimal actuator and sensor placement. In: Proceedings of the American Control Conference (ACC), Portland, OR, USA. American Automatic Control Council (2014)
15. Skogestad, S., Postlethwaite, I.: Multivariable Feedback Control: Analysis and Design, 2nd edn. John Wiley & Sons Ltd., West Sussex (2005)
16. Weideman, J.A.C., Reddy, S.C.: A MATLAB differentiation matrix suite. ACM Trans. Math. Softw. 26(4), 465–519 (2000)
17. Cossu, C., Chomaz, J.M.: Global measures of local convective instabilities. Phys. Rev. Lett. 78(23), 4387–4390 (1997)
18. Chomaz, J.M.: Global instabilities in spatially developing flows: non-normality and nonlinearity. Annu. Rev. Fluid Mech. 37, 357–392 (2005)
19. Schmid, P.J., Henningson, D.S.: Stability and Transition in Shear Flows. Applied Mathematical Sciences, vol. 142. Springer, New York (2001)
20. Belson, B.A., Semeraro, O., Rowley, C.W., Henningson, D.S.: Feedback control of instabilities in the two-dimensional Blasius boundary layer: The role of sensors and actuators. Phys. Fluids 25, 054106 (2013)

An Interconnected Systems Approach to Flow Transition Control

Saulat S. Chughtai and Herbert Werner

Institute of Control Systems, Hamburg University of Technology, Germany
{saulat.chughtai,h.werner}@tuhh.de

Abstract. This paper presents the application of an interconnected systems approach to the flow transition control problem. The control system consists of a network of local controllers with dedicated actuators and sensors. The measured feedback signals used by the controllers are the local changes in wall shear force, and the generated control action is a local change in fluid wall normal velocities. The approach presented here does not require periodicity of the channel as was required by most earlier approaches. Thus, in contrast to previously proposed control schemes which operate in Fourier domain, this approach works in physical domain. Secondly it does not restrict the number of interacting units, thus allowing the application of MEMS arrays. In order to synthesize such controllers, first the dynamics of the fluid is converted into interconnected system form. The model is validated using a non-linear simulation environment developed in FLUENT, and by analysing the growth in the transient energy. The model is then used to synthesize an interconnected controller. The simulated closed-loop response shows that the controller can delay the transition by reducing the transient energy.

1 Introduction

Control of flow patterns to improve efficiency or performance is of immense technological importance. The potential benefits of flow control include improved fuel efficiency and environmental compliance. In such schemes the control action is applied to delay/advance the transition, to suppress/enhance turbulence or to prevent/provoke separation. These phenomena result in reduction of drag, enhancement of lift and reduction of flow-induced noise [1]. In this paper we consider a simple case of control of transition from laminar flow to turbulence in Poiseuille flow. The control scheme considered is as shown in Fig. 1.

In such a scheme the surface in contact of the fluid is populated with control units which interact with each other. Each control unit has its own actuator and sensor. The control algorithm not only uses the local information, obtained from the sensor, but also uses infromation from its nieghboring units. Due to this structure of the complete control scheme it is known as interconnected systems. Such control schemes can be implemented using micro-electro-mechanical-systems (MEMS) arrays. In this paper we have shown how the approach of interconnected system can be used to spatially shift the transition from laminar to turbulance.

R. King (ed.), *Active Flow and Combustion Control 2014*, Notes on Numerical Fluid Mechanics and Multidisciplinary Design 127, DOI: 10.1007/978-3-319-11967-0_9

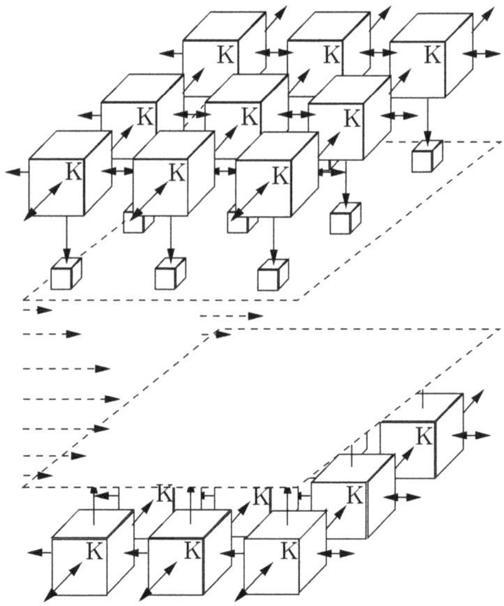

Fig. 1. Control architecture

In [2] we have shown the application of the same approach to plane Poiseuille flow. A similar approach has been used in [3]. However, there are some key differences between the two approaches. These are:

1. The model presented in [2] is based on a velocity-vorticity formulation, which has numerical advantages as discussed in [4], [5].
2. The introduction of spatial shift operators in [2] results in a model which represents an infinitely long channel. In [3], on the other hand, the authors have found a multi-input-multi-output (MIMO) model the order of which increases with the number of finite elements used.

In this paper, we have extend the approach presented in [2] to the 3D case.

Experimental studies have shown that the phenomenon of transition may occur at $Re < 5000$ in 3D case which is less than what is expected in the 2D case. In [6] the authors propose that this discrepancy between theoretical and experimental results is due to large transient energy which results in invalidation of the linear approximation, as it causes amplification of span-wise streaks. Later in [7] the authors showed that this large transient energy growth is due to the non-normality of the resulting state matrix. Hence, the cause of transition in the 3D case is non-normality of the eigenvector matrix rather than instability of eigenvalues as is the case in 2D. In [8] it is shown that the transient energy can be expressed as an LQG cost function. Based on this observation [8], [9] applied optimal control techniques for controller synthesis, where LQG controllers are synthesized. A detailed analysis of the NSE in terms of transient energy growth is presented in [10]. However, all these approaches assume periodicity of the channel. Thus, control algorithms are designed for a single spatial wave number or in Fourier domain. The control algorithm designed in this paper works directly in the physical domain thus avoids any conversions.

Dynamics of the flow is governed by the Navier-Stokes equations (NSE), which are nonlinear coupled partial differential equations. Due to the nonlinear nature of the phenomenon much research effort has been spent on nonlinear control approaches; a survey about the application of neural networks and chaos theory can be found in [1], while the application of model predictive control for flow control is presented in [11], and an approach based on adaptive control using back-stepping is presented in [12]. Some other advances in the analysis of different flows and its implication on control are surveyed in [13] and [14]. On the other hand this paper make use of linearized NSE.

The use of linearized NSE results in a singular system. In order to avoid this singularity, three possible formulations can be used [15]: a velocity-vorticity formulation, a stream wise function formulation and a velocity-pressure formulation. These linearized equations can be reduced to finite dimension either by converting these into spectral domain or by using a finite difference approach. In [16] one such model is presented. Such linearized models are used extensively and are reviewed in [17], [18] and [19]. A similar approach for modeling is used in this paper.

The paper is organized as follows: In section 2 mathematical models are presented. The models are validated using the known facts and non-linear simulation, and the concept of transient energy is used in section 3. Section 4 deals with the synthesis of a simple distributed state feedback scheme. Finally conclusions are drawn and future directions are presented in section 5.

2 Interconnected Model for the Flow Transition Control Problem

Consider a three-dimensional steady state plane channel flow with centerline velocity U_0 and channel half-width δ as shown in Fig. 2.

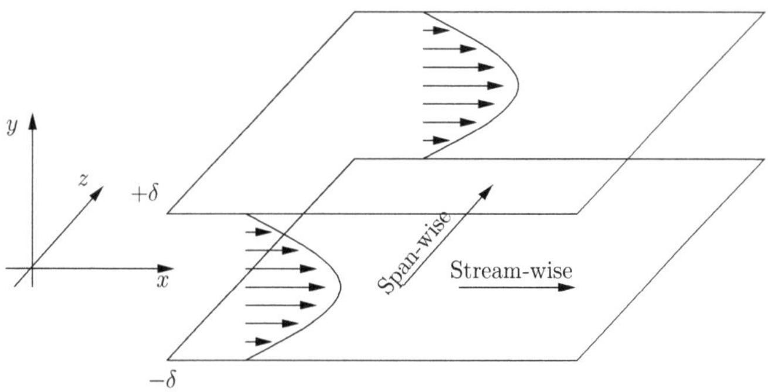

Fig. 2. Poiseuille flow, 3D case

The equations governing small, incompressible, three-dimensional perturbations $\{u, v, w, p\}$ are then approximated by the linearized NSE and the continuity equations. By eliminating p and defining a new variable ω (vorticity) we arrive at the following equivalent formulation of the NSE, which is also known as velocity-vorticity formulation

$$\Delta\dot{v} = \left\{-U\frac{\partial}{\partial x}\Delta + \frac{dU}{dy}\frac{\partial}{\partial x} + \frac{\Delta^2}{R_e}\right\} v \tag{1}$$

$$\dot{\omega} = \left\{-\frac{dU}{dy}\frac{\partial}{\partial z}\right\} v + \left\{-U\frac{\partial}{\partial x} + \frac{\Delta}{R_e}\right\} \omega \tag{2}$$

with homogeneous boundary conditions. Where Δ is the Laplacian, a dot $(\dot{})$ represents $\partial/\partial t$, $U = 1 - y^2$. The wall normal vorticity is defined as $\omega = \partial u/\partial z - \partial w/\partial x$

$$v(x, y = \pm 1, z, t) = 0 \quad \omega(x, y = \pm 1, z, t) = 0 \tag{3}$$

$$\frac{\partial v(x, y = \pm 1, z, t)}{\partial y} = 0. \tag{4}$$

Note that the above two equations have one-directional coupling, i.e the first equation does not depend on ω. The control inputs are applied at the boundaries ($y = \pm 1$), which then changes the homogenous boundary conditions into non-homogeneous ones. However, the problem can be converted into a non-homogeneous PDE with homogeneous boundary conditions by a change of variables. Let

$$v(x, y, z, t) = \phi(x, y, z, t) + q(t)g(x, z)f(y) \tag{5}$$

where $f(y)$ is any function satisfying the condition

$$f(-1) = 1, \quad f(1) = 0, \quad \frac{df(\pm 1)}{dy} = 0, \tag{6}$$

and $g(x, z)$ defines the distribution of control force in the xz-plane. Using this new variable ϕ we can recover the homogeneity in the boundary conditions. Thus the modified boundary-value PDEs are

$$\begin{aligned}\Delta\dot{\phi} + \dot{q}\Delta(gf) = -qU\frac{\partial}{\partial x}\Delta(gf) + qf\frac{dU}{dy}\frac{\partial g}{\partial x} + q\frac{\Delta^2}{R_e}(gf)\\ -U\frac{\partial}{\partial x}\Delta\phi + \frac{dU}{dy}\frac{\partial \phi}{\partial x} + \frac{\Delta^2}{R_e}\phi\end{aligned} \tag{7}$$

$$\dot{\omega} = \left\{-qf\frac{dU}{dy}\frac{\partial}{\partial z}(g) - \frac{dU}{dy}\frac{\partial \phi}{\partial z}\right\} + \left\{-U\frac{\partial}{\partial x} + \frac{\Delta}{R_e}\right\}\omega \tag{8}$$

$$\phi(x, y = \pm 1, z, t) = 0 \tag{9}$$

$$\frac{\partial \phi(x, y = \pm 1, z, t)}{\partial y} = 0. \tag{10}$$

Discretization of Span and Stream Wise Direction. The x-direction is discretized into regularly spaced segments, using the following symmetric finite difference approximation for the partial derivatives $\frac{\partial^k}{\partial x^k}$

$$\begin{aligned}
\left\{\frac{\partial}{\partial x}\right\} &= \frac{S_1 - S_1^{-1}}{2}\\
\left\{\frac{\partial^2}{\partial x^2}\right\} &= S_1 + S_1^{-1} - 2\\
\left\{\frac{\partial^3}{\partial x^3}\right\} &= \frac{S_1^2 - 2S_1 + 2S_1^{-1} - S_1^{-2}}{2}\\
\left\{\frac{\partial^4}{\partial x^4}\right\} &= S_1^2 - 4S_1 - 4S_1^{-1} + S_1^{-2} + 6.
\end{aligned} \tag{11}$$

Here S_1 is the forward spatial shift operator while S_1^{-1} is the backward shift operator along the stream-wise direction (x) . Similarly for the span wise direction (z), S_2 and S_2^{-1} will be used.

Wall Normal Discretization. Discretization in wall normal direction is done using Chebyshev polynomials with finite basis functions of degree N. Thus, at any grid point ij in the xz-plane, $\phi^{i,j}$ is given as

$$\phi^{i,j}(y,t) = \sum_{m=0}^{m=N} a_m(t)\Gamma_m(y) \tag{12}$$

$$\omega^{i,j}(y,t) = \sum_{m=0}^{m=N} b_m(t)\Gamma_m(y) \tag{13}$$

where $\Gamma_m(y)$ is the mth Chebyshev basis function. For the present work we use the basis functions proposed by McKernan [20], which are given as

$$\begin{aligned}
\Gamma_1^M &= \Gamma_1\\
\Gamma_2^M &= \Gamma_2\\
\Gamma_3^M &= \Gamma_3 - \Gamma_1\\
\Gamma_4^M &= \Gamma_4 - \Gamma_2\\
\Gamma_{m<4,odd}^M &= (\Gamma_m - \Gamma_1) - \frac{(m-1)^2(\Gamma_{m-2} - \Gamma_1)}{(m-3)^2}\\
\Gamma_{m<4,even}^M &= (\Gamma_m - \Gamma_2) - \frac{((m-1)^2 - 1)(\Gamma_{m-2} - \Gamma_2)}{((m-3)^2 - 1)}
\end{aligned} \tag{14}$$

These are shown in Fig. 3.

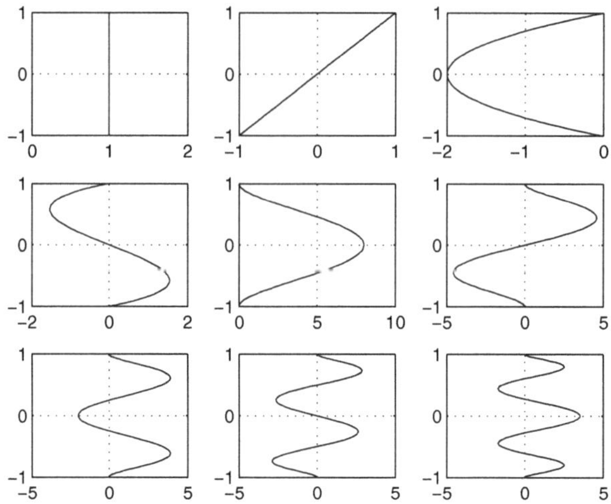

Fig. 3. Chebyshev basis functions for $m = 1, \ldots, 9$

Note that the first four basis functions do not fulfil the Dirichlet and Neumann boundary conditions. This suggests that we should eliminate columns corresponding to these in (12), while for vorticity we only require Dirichlet boundary conditions which result in elimination of the first two basis functions.

Further, let us define

$$a = \begin{bmatrix} a_4 \\ \vdots \\ a_N \end{bmatrix} \tag{15}$$

$$\Gamma = \begin{bmatrix} \Gamma_4(y_2) & \ldots & \Gamma_N(y_2) \\ \vdots & \vdots & \vdots \\ \Gamma_4(y_{N-2}) & \ldots & \Gamma_N(y_{N-2}) \end{bmatrix}, \Gamma' = \begin{bmatrix} \frac{\partial \Gamma_4}{\partial y}|_{y_2} & \cdots & \frac{\partial \Gamma_N}{\partial y}|_{y_2} \\ \vdots & \vdots & \vdots \\ \frac{\partial \Gamma_4}{\partial y}|_{y_{N-2}} & \cdots & \frac{\partial \Gamma_N}{\partial y}|_{y_{N-2}} \end{bmatrix} \tag{16}$$

$$b_1 = \begin{bmatrix} \frac{\partial^2 f}{\partial y^2}|_{y_2} \\ \vdots \\ \frac{\partial^2 f}{\partial y^2}|_{y_{N-2}} \end{bmatrix} \qquad b_2 = \begin{bmatrix} \frac{1}{Re}\frac{\partial^4 f}{\partial y^4}|_{y_2} \\ \vdots \\ \frac{1}{Re}\frac{\partial^4 f}{\partial y^4}|_{y_N} \end{bmatrix} \tag{17}$$

These basis functions can be discretized along the y-direction by making a grid of $N+1$ Chebyshev-Gauss-Lobatto points y_k, where $y_k = \cos(\pi k/N), \forall\ k = 1, ..., N$. In order to make the resulting matrices square we can eliminate the two rows close to each of the upper and lower walls, as these will be small for velocity, and one row corresponding to the vorticity next to the wall.

$$(S_1+S_1^{-1}-2)+\frac{\partial^2}{\partial y^2}+(S_2+S_2^{-1}-2))\Gamma\dot{a}+\dot{q}\Delta(gf)=$$
$$q\left(f\frac{dU}{dy}\frac{\partial g}{\partial x}-U\frac{\partial}{\partial x}\Delta(gf)+\frac{\Delta^2}{R_e}(gf)\right)-U(\frac{S_1^2-2S_1+2S_1^{-1}-S_1^{-2}}{2}+\frac{S_1-S_1^{-1}}{2}$$
$$\frac{\partial^2}{\partial y^2}+\frac{S_1S_2+S_1S_2^{-1}-2S_1-S_1^{-1}S_2-S_1^{-1}S_2^{-1}+2S_1^{-1}}{2})\Gamma a+\frac{dU}{dy}\frac{S_1-S_1^{-1}}{2}\Gamma a+$$
$$\frac{1}{R_e}(S_1^2-4S_1-4S_1^{-1}+S_1^{-2}+6)+(S_2^2-4S_2-4S_2^{-1}+S_2^{-2}+6)+\frac{\partial^4}{\partial y^4})\Gamma a$$

$$\Gamma\dot{b}=-q\frac{dU}{dy}\frac{\partial}{\partial z}(gf)-\frac{dU}{dy}\frac{S_2-S_2^{-1}}{2}\Gamma a-U\frac{S_1-S_1^{-1}}{2}\Gamma b+$$
$$\frac{1}{R_e}((S_1+S_1^{-1}-2)+\frac{\partial^2}{\partial y^2}+(S_2+S_2^{-1}-2))\Gamma b \tag{18}$$

Substituting (11), (12), (13) into (7) we obtain (18), where we have introduced a new function $g(x,z)$ which can be used to specify the shape of control action within a segment. In this study we used $g(x,z)=1$

Measurements. We assume that measurements taken at the wall provide information about the stream-wise and span-wise skin friction, from which we can determine the four quantities

$$y_1=-\frac{1}{Re}\Delta\frac{\partial u}{\partial y}\bigg|_{y=1} \qquad y_2=\frac{1}{Re}\Delta\frac{\partial u}{\partial y}\bigg|_{y=-1}$$
$$y_3=-\frac{1}{Re}\Delta\frac{\partial w}{\partial y}\bigg|_{y=1} \qquad y_4=\frac{1}{Re}\Delta\frac{\partial w}{\partial y}\bigg|_{y=-1} \tag{19}$$

Using the continuity equation and the definition of ω, it can be shown that these outputs can be represented as

$$y_1=-\frac{1}{Re}(-\frac{\partial^3 v}{\partial x\partial^2 y}+\frac{\partial^2\omega}{\partial z\partial y})\bigg|_{y=1}$$
$$y_2=\frac{1}{Re}(-\frac{\partial^3 v}{\partial x\partial^2 y}+\frac{\partial^2\omega}{\partial z\partial y})\bigg|_{y=-1}$$
$$y_3=-\frac{1}{Re}(-\frac{\partial^3 v}{\partial z\partial^2 y}-\frac{\partial^2\omega}{\partial x\partial y})\bigg|_{y=1}$$
$$y_4=\frac{1}{Re}(-\frac{\partial^3 v}{\partial z\partial^2 y}-\frac{\partial^2\omega}{\partial x\partial y})\bigg|_{y=-1}. \tag{20}$$

Using (5), (11), (12) and (13), the measurement equations can be written as

$$\begin{aligned}
y_1 &= \frac{1}{Re}((S_1 - S_1^{-1})\Gamma''a - (S_2 - S_2^{-1})\Gamma'b)\Big|_{y=1} \\
y_2 &= -\frac{1}{Re}((S_1 - S_1^{-1})\Gamma''a - (S_2 - S_2^{-1})\Gamma'b)\Big|_{y=-1} \\
y_3 &= \frac{1}{Re}((S_2 - S_2^{-1})\Gamma''a + (S_1 - S_1^{-1})\Gamma'b)\Big|_{y=1} \\
y_4 &= -\frac{1}{Re}((S_2 - S_2^{-1})\Gamma''a + (S_1 - S_1^{-1})\Gamma'b)\Big|_{y=-1}
\end{aligned} \tag{21}$$

where $y = 1$ corresponds to the first row of Γ' and Γ'', while $y = -1$ corresponds to the last row of Γ' and Γ''.

Remark 1: Estimating the Laplacian of skin friction can be replaced by applying spatial integration, using a suitable low pass spatial filter on (21).

3 Comparison of Models

In this section we compare the model obtained in the previous section against some known features of Poiseuille flow and against a nonlinear simulation. The model is obtained for $N = 20$ that results in total of 17 temporal and 102 spatial states when the model is reduced to 2D.

It has been observed that the input to output gain grows in a band of $\omega \in [0.17, 0.27]$, with a large resonant peak at 0.2 and 0.23 rad/sec for $R_e = 2000$ and 5000, respectively. This is exactly what is predicted by Orr-Sommerfeld.

3.1 Nonlinear Simulation of 2D Model

The dynamics of the underlying system are highly non-linear. In order to estimate the mismatch between the linearized model (18) and the nonlinear system, a simulation environment is developed using FLUENT.

The simulation is carried out at $Re = 2000$; only NS-equations are simulated without using any turbulence model in 2D. Other flow parameters used in the simulation are presented in Table 1. Here the under relaxation factors are used to make the simulation stable. Hence, if Ξ is a variable then its update is calculated by

$$\Xi = \Xi_{old} + \beta\epsilon$$

where Ξ_{old} is the previous value of Ξ and ϵ is the change in the variable, while β is the under relaxation coefficient. For time, the marching first order implicit formulation is used. For details on these see [21].

Table 1. Parameters used in simulation

Parameter	Values
Simulation type	Unsteady
Under relaxation parameters	Default values
Maximum iterations per time step	80
Velocity-pressure coupling	Simple
Time step	1×10^{-3} *sec.*
U_c	$1\ m/s$
Density	$1\ kg/m^3$
Viscosity	$1 \times 10^{-3}\ kg/m - s$

Input Signal. The input signal $q(t)$ is applied only at the center of the lower plate, which results in a spatially impulsive signal. However, it is temporally sinusoidal.

$$q(t) = \begin{cases} 0.1\sin(\omega t), & t \in [90, 90 + 6\pi/\omega] \\ 0, & \text{elsewhere}, \end{cases}$$

$$g(x) = \begin{cases} 1, & t \in [-0.166, 0.166] \\ 0, & \text{elsewhere}, \end{cases} \tag{22}$$

Here $\omega = 0.1, 0.2, 1\ rad/sec$ are considered. This range of frequencies is chosen since Tollmien-Schlichting waves, which are considered as the first stage of transition in 2D, have frequencies in this band. The time of simulation is set to 3 times the period of the sinusoid with 90sec of pre-calculations.

Output Calculation. At each instant of time, the software calculates streamwise (U) and wall normal velocities (V). Using (19) and a first order approximation for the differential operator, the change in wall shear force is estimated at the lower wall at different locations.

Calculation of the System Gain. Once the change in wall shear force is calculated, it is converted into Fourier domain, and the amplitude of each frequency component is calculated using the fast Fourier transform. An analysis of this reveals that for low frequencies the flow behaves nearly linearly. However, at large frequencies, like $\omega = 1 rad/sec$, the behavior is more nonlinear, as can be seen in Figure 4. It shows that the outputs at all channel locations have sinusoidal components of frequencies other then the input frequency, which is due to the nonlinearity of the system.

Deviations at other frequencies can be attributed to number of different factors: like the non-linearity of the dynamics, errors due to discretization etc. One approach to reduce these effects is to simulate a perturbation model rather then the full NS-equations. This however is not possible with FLUENT.

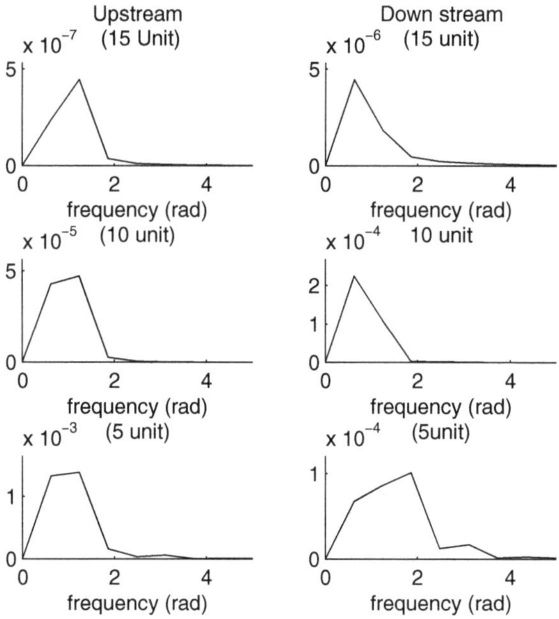

Fig. 4. Fourier transform of output for $w = 1$ rad/sec

3.2 Transient Energy of 3D Finite Difference Model

In this section we will estimate the maximum transient energy to validate the model. The transient energy is defined as [6]

$$E(t) = \frac{1}{V} \int_{-1}^{1} \int_{-\frac{1}{2}}^{\frac{1}{2}} \int_{-\frac{1}{2}}^{\frac{1}{2}} \frac{u^2 + v^2 + w^2}{2} dzdxdy \qquad (23)$$

where mass density of the fluid is assumed to be one. Since $V = 2$ and both v and $\frac{dv}{dy}$ are assumed to remain constant in x and z-direction in a single element, we obtain, using the continuity equation and the definition of ω

$$E(t) = \frac{1}{4} \int_{-1}^{1} v^2 + (\frac{\partial v}{\partial y})^2 + \omega^2 dy. \qquad (24)$$

We approximate the integral by a weighted sum as [22]

$$\int_{-1}^{1} v^2 dy = \sum_{n=0}^{N} w_n v(y_n),$$
$$\int_{-1}^{1} (\frac{\partial v}{\partial y})^2 dy = \sum_{n=0}^{N} w_n (\frac{\partial v}{\partial y}\Big|_{y_n})^2.$$
$$\int_{-1}^{1} \omega^2 dy = \sum_{n=0}^{N} w_n \omega(y_n) \tag{25}$$

where $\forall n = 1, \ldots, N-1$

$$w_n = \frac{2}{N}\sqrt{1-y_n^2}\sum_{m=1}^{N-1}\frac{1}{m}\sin(m\pi n/N)(1-\cos(m\pi)) \tag{26}$$

and $w_0 = w_N = 0$. Substituting (25) in (24) and arranging the terms in matrices we obtain

$$E(t) = \frac{\Lambda^T \Psi \Lambda + \Lambda_y^T \Psi \Lambda_y + \Lambda_\omega^T \Psi \Lambda_\omega}{4} \tag{27}$$

where

$$\Psi = \begin{bmatrix} w_0 & & 0 \\ 0 & \ddots & 0 \\ 0 & & w_N \end{bmatrix}$$

$$\Lambda = \begin{bmatrix} v_0 \\ \vdots \\ v_N \end{bmatrix}, \quad \Lambda_y = \begin{bmatrix} \frac{\partial v}{\partial y}\Big|_{y_0} \\ \vdots \\ \frac{\partial v}{\partial y}\Big|_{y_N} \end{bmatrix}, \quad \Lambda_\omega = \begin{bmatrix} \omega_0 \\ \vdots \\ \omega_N \end{bmatrix}. \tag{28}$$

In terms of the temporal state vector $(x^t)^T = [a^T, b^T]$ of the linearized model (18), (27) can be written as

$$E(t) = (x^t)^T \left(\frac{T^T \Psi T + T_y^T \Psi T_y}{4h^2}\right) x^t \tag{29}$$

where $T = diag(\Gamma, \Gamma)$ and $T_y = diag(\Gamma', 0)$ together with the deleted rows corresponding to $n = [0, 1, N-1, N]$, respectively. If the system is diagonalizable then

$$x^t = X_0 e^{\Phi t}$$

where, Φ is a diagonal matrix consisting of all the eigenvalues at its principal diagonal. Substituting this in (29) will result in a matrix whose maximum singular value at any time t will be the maximum transient energy ($E_m(t)$) starting from any initial state.

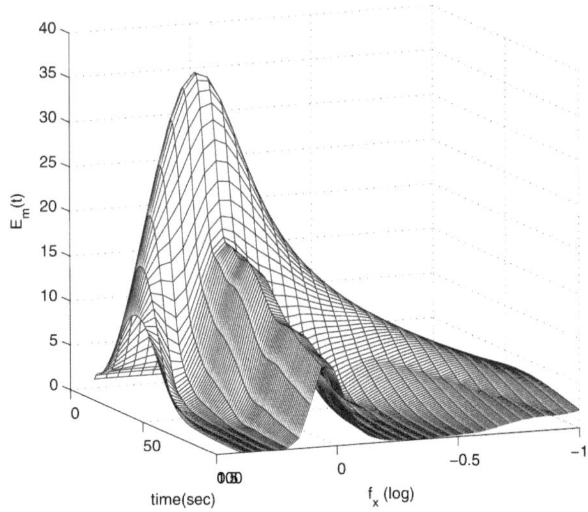

Fig. 5. $E_m(t)$ for $Re = 5000$, $f_x = [0.1, \ldots, 2.5]$ and $f_z = 0$

First $E_m(t)$ for $Re = 5000$, $f_x = [0.1, \ldots, 2.5]$ and $f_z = 0$ are calculated, where f_x is the spatial frequency in x-direction and f_z is the spatial frequency in z-direction, as shown in Fig. 5. It turns out that the system is stable, since the transient energy decreased after the initial peak. However, for a spatial frequency of nearly 1 the decay in the energy is less than for other frequencies. This is because of a pole pair which is close to the imaginary axis and for $Re > 6000$ becomes unstable. The same behavior is reported elsewhere, e.g. [16].

Next, $E_m(t)$ for $Re = 5000$, $f_x = 0$ and $f_z = [0.1, \ldots, 3]$ are calculated. These are plotted in Fig. 6. It shows a large peak of $E_m(t) \approx 4500$ at $f_z \approx 2$ at $t \approx 380$. This shows that the system is highly non-normal at this frequency. The values obtained here are in agreement with the ones found by other researchers, see e.g. [6], [23]. who found the peak of $E_m(t) = 4897$ at $t = 379$ at the same spatial frequency. The results presented above show that the model does capture the dominating features of the problem.

4 Controller Synthesis

Once the interconnected model is constructed and validated the controller can be synthesized using the approach presented in [2]. The approach searches for the controller that results in the minimum closed-loop induced $\mathcal{L}_2$-norm. Where $\mathcal{L}_2$-norm is the maximum gain over all time and space. In [24] it has been shown that the non-normality results in rise in the $\mathcal{L}_2$-norm. Hence, the approach is used to synthesize a state feedback controller with minimum $\mathcal{L}_2$-norm. The closed-loop

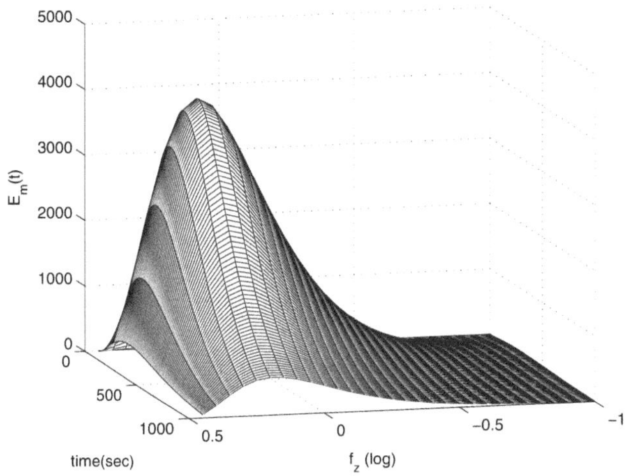

Fig. 6. Maximum transient energy $E_m(t)$ for $Re = 5000$, $f_x = 0$ and $f_z = [0.1, \ldots, 3]$

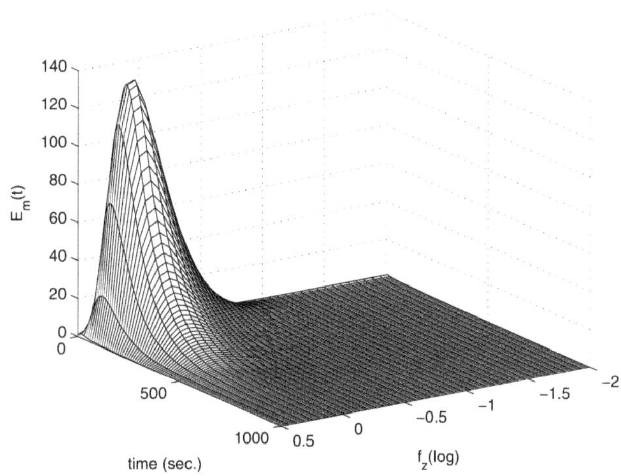

Fig. 7. Closed-loop maximum transient energy $E_m(t)$ for $Re = 5000$, $f_x = 0$ and $f_z = [0.1, \ldots, 3]$

maximum transient energy $E_m(t)$ is plotted in Fig. 7. It shows that the controller has reduced the peak transient energy from 4500 to 130. This reduction in transient energy indicates that the normality of the closed-loop system has improved.

5 Conclusion

In this paper we have shown the application of an interconnected systems approach to the flow transition control problem. The control structure consists of a network of controllers. The controllers take measurements of local changes in wall shear force and the infromation from their neighbours to generate control signals. The control signal results in changing the wall normal velocities.

To apply the approach first an interconnected model is developed which is validated using a nonlinear simulation. The Fourier transform of the output signal reveals that as we move away from the point of injection of the input signal the system behaves “more linearly”, suggesting that linear approximation can be used for controller synthesis.

Next, the interconnected model is analyzed w.r.t the growth in transient energy. It has been demonstrated that the model presented here does capture the dominating feature of 3D NSE, which are highly non-normal. The interconnected controller is then synthesized by minimizing the $\mathcal{L}_2$-norm.

The work presented here can be extended in two main directions. One is to apply the approach to problem of shifting the transition in boundary layers. Secondly, the control scheme presented here requires considerable communication between neighbours and it is difficult to consider complex boundary conditions. This can be addressed by either developing or estimating appropriate models with reduced interactions and spatially varying dynamics, using e.g the approach of [25]. Furthermore, the problem of sensor and actuator placement poses further challenges. Development and application of techniques that provide solutions to these problems will be explored in the future.

References

1. el Hak, M.G.: Modern developments in flow control. Applied Mechanics Review 49, 365–379 (1996)
2. Chughtai, S., Werner, H.: Transition control of plane Poiseuille flow - a spatially interconnected model. In: Proc. of 47th IEEE Conference on Decision and Control, Cancun, Mexico (2008)
3. Baramov, L., Tutty, O., Rogers, E.: H_∞ control of nonperiodic two-dimensional channel flow. IEEE Transactions on Control Systems Technology 12(1), 111–122 (2004)
4. Fasel, H.: Investigation of the stability of boundary layers by a finite-difference model of the Navier-Stokes equations. Journal of Fluid Mechanics 78, 355–383 (1976)
5. Daube, O.: Resolution of the 2D Navier-Stokes equations in velocity-vorticity forms by means of an influence matrix technique. Journal of Computational Physics 103, 402–414 (1992)
6. Butler, K., Farrell, B.: Three dimensional optimal disturbances in viscous shear flow. Physics of Fluids 4(8), 1637–1650 (1992)
7. Trefethen, L., Trefethen, A., Reddy, S., Driscoll, T.: Hydrodynamic stability without eigenvalues. Science 261(30), 578–584 (1993)

8. Bewley, T., Liu, S.: Optimal and robust control and estimation of linear paths to transition. Journal of Fluid Mechanics 365(12), 305–349 (1998)
9. McKernan, J., Whidborne, J., Papadakis, G.: Linear quadratic control of plane Poiseuille flow-transient behaviour. International Journal of Control 80(12), 1912–1930 (2007)
10. Jovanovic, M., Bamieh, B.: Componentwise energy amplification in channel flows. Journal of Fluid Mechanics 534, 145–183 (2005)
11. King, R., Aleksic, K., Gelbert, G., Losse, N., Muminovic, R., Brunn, A., Nitsche, W., Bothien, M., Moeck, J., Paschereit, C., Noack, B., Rist, U., Zeng, M.: Model predictive flow control. In: Proc. of 4th Flow Control Conference, Seattle, USA (2008)
12. Krstic, M., Smyshlyaev, A.: Boundary control of PDEs: A course on backstepping design. SIAM (2008)
13. el Hak, M.G., Tsai, H.: Transition and turbulence control. Lecture notes. World Scientific Publishing Co. Ltd. (2006)
14. King, R.: Active Flow Control. NNFM, vol. 95. Springer, Heidelberg (2007)
15. Peyret, R.: Spectral methods for incompressible viscous flow. Applied Mathematical Science. Springer, Berlin (2002)
16. Orszag, S.: Accurate solution of the Orr-Sommerfeld stability equation. Journal of Fluid Mechanics 50(4), 689–703 (1971)
17. Kim, J., Bewley, T.R.: A linear systems approach to flow control. Ann. Rev. Fluid Mech. (39), 39–383 (2007)
18. Sipp, D., Marquet, O., Meliga, P., Barbagallo, A.: Dynamics and control of global instabilities in open-flows: a linearized approach. Appl. Mech. Rev. 3(63), 30801 (2010)
19. Bagheri, S., Henningson, D.S.: Transition delay using control theory. Philos. Trans. R. Soc. (369), 1365–1381 (2011)
20. McKernan, J.: Control of plane Poiseuille flow: A theoretical and computational investigation. PhD thesis, Cranfield University, Dept. of Aerospace Siences, School of Engineering (2006)
21. Fluent, I.: FLUENT 6.3 User's Guide. FLUENT Inc. (2006)
22. Boyd, J.: Chebyshev and Fourier Spectral Methods. Dover Publications Inc., New York (2001)
23. Schmid, P., Henningson, D.: Stability and transition in shear flows. Springer, New York (2001)
24. Trefethen, L.: Pseudospectra of linear operators. SIAM Review 39(3), 383–406 (1997)
25. Ali, M., Ali, A., Chughtai, S., Werner, H.: Consisitent identification of spatially interconnected systems. In: Proc. American Control Conference, San Francisco, CA, USA, pp. 3583–3588 (2011)

Direct Numerical Simulation of Heat Transfer of a Round Subsonic Impinging Jet

Robert Wilke and Jörn Sesterhenn

Technische Universität Berlin, Fachgebiet Numerische Fluiddynamik,
Müller-Breslau-Str. 11, 10623 Berlin, Germany
robert.wilke@tnt.tu-berlin.de, joern.sesterhenn@tu-berlin.de
http://www.cfd.tu-berlin.de/

Abstract. This paper concentrates on the investigation of heat transfer of a confined round impinging jet. A direct numerical simulation was performed at a Reynolds number of $Re = 3300$ using a grid size of $512 \times 512 \times 512$ points. It is shown that the dissipative scales are well resolved. This enables the examination of the impact of the jet's turbulent flow field on the heat transfer of the impinged plate. In this study the distribution of the local Nusselt number is presented and related to the instantaneous flow field of the jet.

Keywords: Direct Numerical Simulation, Round Jet, Heat Transfer, Forced Convection, Impingement, Unsteady, Turbulent.

1 Introduction

Impinging jets provide an effective cooling method for various applications such as the cooling of turbine blades of aircraft. An increase of efficiency not only reduces the required cooling air mass flow and consequently the fuel consumption, but also enables new combustion concepts with even higher cooling demands to be applied in the future.

Heat transfer due to forced convection of a jet impinging on a flat plate has been studied for decades. General information including schematic illustrations of the flow fields as well as distributions of local Nusselt numbers for plenty of different geometrical configurations and Reynolds numbers Re can be found in several reviews, such as [9], [10] ,[14], [15] based on experimental and numerical results. Since experiments cannot provide all quantities of the entire flow domain spacially and temporally well resolved, the understanding of the turbulent flow field requires simulations. Existing publications of numerical nature use either turbulence modelling for the closure of the Reynolds-averaged Navier-Stokes (RANS) equations, e.g. [16], or large eddy simulation (LES), e.g. [4]. Almost all available direct numerical simulations (DNS) are either two-dimensional, e.g. [3], or do not exhibit an appropriate spatial resolution in the three-dimensional case, e.g. [8]. Recent investigations come from Dairay et al [5]. He conducted a DNS of a round impinging jet. Since those computations require tremendous

R. King (ed.), *Active Flow and Combustion Control 2014*, Notes on Numerical Fluid Mechanics and Multidisciplinary Design 127, DOI: 10.1007/978-3-319-11967-0_10

amounts of computing time, their presented results are preleminary and consider the Nusselt number and the friction coefficient only.

This study deals with a direct numerical simulation of a turbulent round impinging jet. According to Hrycak [9], for free jets as well as for impinging jets four different states dependent on the Reynolds number exist. Those states characterise the jet with regard to turbulence, ranging from dissipated laminar ($Re < 300$) to fully turbulent ($Re > 3000$). To ensure simulating a fully turbulent impinging jet, a Reynolds number of $Re = 3300$ is chosen.

2 Numerical Method

2.1 Governing Equations and Computational Domain

The governing Navier-Stokes equations are formulated in a characteristic pressure-velocity-entropy-formulation, as described by Sesterhenn [13] and solved directly numerically:

$$\frac{\partial p}{\partial t} = -\frac{\rho c}{2}\left(X^+ + X^- + Y^+ + Y^- + Z^+ + Z^-\right) + \frac{p}{c_v}\left(\frac{\partial s}{\partial t} + X^s + Y^s + Z^s\right) \tag{1}$$

$$\frac{\partial u}{\partial t} = -\left[\frac{1}{2}\left(X^+ - X^-\right) + Y^u + Z^u\right] + \frac{1}{\rho}\frac{\partial \tau_{1j}}{\partial x_j} \tag{2}$$

$$\frac{\partial v}{\partial t} = -\left[X^v + \frac{1}{2}\left(Y^+ - Y^-\right) + Z^v\right] + \frac{1}{\rho}\frac{\partial \tau_{2j}}{\partial x_j} \tag{3}$$

$$\frac{\partial w}{\partial t} = -\left[X^w + Y^w + \frac{1}{2}\left(Z^+ - Z^-\right)\right] + \frac{1}{\rho}\frac{\partial \tau_{3j}}{\partial x_j} \tag{4}$$

$$\frac{\partial s}{\partial t} = -\left(X^s + Y^s + Z^s\right) + \frac{1}{\rho T}\left(-\frac{\partial q_i}{\partial x_i} + \Phi\right) \tag{5}$$

The three Cartesian coordinates are x, y, z and the corresponding velocity components read u, v, w. Pressure, density, entropy and temperature are expressed as p, ρ, s and T. Equations for the acoustic wave amplitudes $(X/Y/Z)^{\pm}$, the entropy waves $(X/Y/Z)^s$, the transport terms $(X/Y/Z)^{(u/v/w)}$, the stress tensor τ_{ij} and Φ are given in the appendix. This formulation has advantages in the fields of boundary conditions, parallelization and space discretisation. No turbulence modelling is required since the smallest scales of turbulent motion are resolved, see section 3.1. The spatial discretisation uses 6th order compact central schemes of Lele [11] for the diffusive terms and compact 5th order upwind finite differences of Adams et al. [1] for the convective terms. To advance in time a 4th order Runge-Kutta scheme is applied.

The present simulation is conducted on a numerical grid of size $512 \times 512 \times 512$ points for the computational domain sized $L_x \times L_y \times L_z = 12D \times 5D \times 12D$, where D is the inlet diameter, see Figure 1. A confined impinging jet is characterised by the presence of two walls, the impinging plate and the orifice plate. The grid is refined in those wall-adjacent regions in order to ascertain a maximum value in time and space of the dimensionless wall distance y^+ of the closest grid point to the wall smaller than one for both plates. For the x- and z-direction a slight symmetric grid stretching is applied which refines the shear layer of the jet. The refinements use hyperbolic tangent respectively hyperbolic sin functions and lead to minimal and maximal spacings of $0.017D \leq \Delta x = \Delta z \leq 0.039D$ and $0.0017D \leq \Delta y \leq 0.016D$. The maximum change of the mesh spacing is 0.5% respectively 1%. For the circumferential averaging an equidistant grid of size $n_\theta \times n_r = 1024 \times 512$ is applied for each slice at constant height. Table 1 shows the physical parameters of the simulation. The wall temperature of 100°C and the total temperature of 20°C are chosen in order to be realisable for experiments, where the impinging plate can be kept isothermal using steam. Since the Nusselt number is commonly known as a function of the Reynolds and Prandtl number, no strong influence is expected from the Mach number. A pressure ratio of $p_t/p_\infty = 1.5$ is chosen since it leads to a high subsonic Mach number of about 0.78. This reduces the amount of required computing time related to lower Mach numbers.

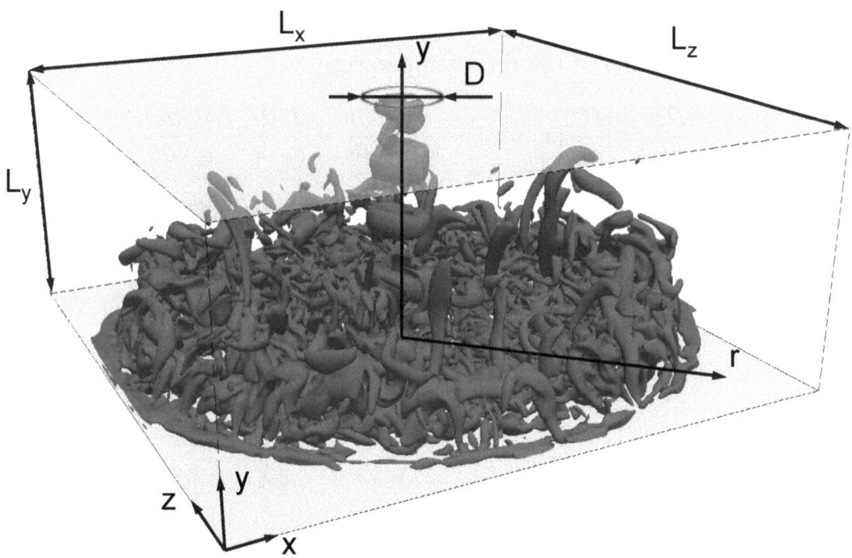

Fig. 1. Computational domain with instantaneous iso-surface of Q-criterion $Q = \frac{1}{2}\left(P^2 + \omega_{ij}\omega_{ij} - S_{ij}S_{ij}\right)$, $P = -\frac{\partial u_i}{\partial x_i}$, $\omega_{ij} = \frac{1}{2}\left(\frac{\partial u_i}{\partial x_j} - \frac{\partial u_j}{\partial x_i}\right)$, $S_{ij} = \frac{1}{2}\left(\frac{\partial u_i}{\partial x_j} + \frac{\partial u_j}{\partial x_i}\right)$ at $Q = 1000 m^2 s^{-4}$

Table 1. Physical parameters of the simulation. $p_o, p_\infty, T_o, T_W, Re, Pr, \kappa, R$ denote the total pressure, ambient pressure, total temperature, wall temperature, Reynolds number, Prandtl number, ratio of specific heats and the specific gas constant.

p_o/p	p_∞	T_o	T_W	Re	Pr	κ	R
1.5	$10^5\,Pa$	293.15 K	373.15 K	3300	0.71	1.4	287 J/(kg K)

2.2 Boundary Conditions

The computational domain is delimited by four non-reflecting boundary conditions, one isothermal wall which is the impinging plate and one boundary consisting of an isothermal wall and the inlet. The walls are fully acoustically reflective. To impose boundary conditions, the waves entering the computational domain have to be set. The isothermal wall is described in [13]. In order to derive a consistent description of the orifice plate including the inlet a new boundary condition is presented which describes the entire face with one set of equations. Therefore the time derivation of the velocity

$$\frac{\partial v_i}{\partial t} = \frac{\partial v_i}{\partial t}(x_i, t) \tag{6}$$

and the time derivation of the total temperature

$$\frac{\partial T_o}{\partial t} = \frac{\partial T_o}{\partial t}(x_i, t) = \frac{\kappa - 1}{\kappa}\frac{T}{p}\frac{\partial p}{\partial t} + \frac{T}{c_p}\frac{\partial s}{\partial t} + \frac{v_i}{c_p}\frac{\partial v_i}{\partial t} \tag{7}$$

are imposed as a function of time and space. In the specific case of the upper boundary, the entering acoustic wave Y^- can be obtained directly from (3) and the transport terms Y^u and Y^w subsequently from equation (2) and (4). The entropy wave Y^s can be derived from the equations (1), (5) and (7), viz.:

$$\begin{aligned} Y^s = -\left(X^s + Z^s\right) + \kappa\frac{\sigma}{\rho T} - \frac{c_p}{T}\frac{\partial T_0}{\partial t} + \\ + \frac{v}{T}\frac{\partial v}{\partial t} - \frac{1}{2}\frac{\kappa - 1}{\kappa}\frac{c_p c \rho}{p}\left(X^+ + X^- + Y^+ + Y^- + Z^+ + Z^-\right) \end{aligned} \tag{8}$$

The initial conditions of the present simulation are $v_i = 0$, $p = p_\infty$ and $T = T_W$, where p_∞ is the ambient pressure and T_W is the wall temperature. For the areas of the wall both derivatives are set to zero: $\frac{\partial v_i}{\partial t}\big|_W = 0$, $\frac{\partial T_o}{\partial t}\big|_W = 0$. At the inlet the derivative of the total temperature is set so that the total temperature decreases within a given time of $\Delta t = 10^{-3}$s from the initial value to the stationary value T_o. Respectively the wall normal velocity reaches the time dependent target velocity in order to maintain the total pressure of p_o. The described method has the advantage that it is very easy to define one or more inlet areas (nozzles) by multiplying the target velocity with a space function

$f(r, \theta)$ and consequently switching between inlet and wall. The location of the nozzle is defined using a hyperbolic tangent profile with a thin laminar annular shear layer described by a thickness parameter b. The parameter g moves the shear layer adjacent to the wall of the artificial nozzle:

$$f(r) = \frac{1}{2} - \frac{1}{2}\tanh\left[\left(\frac{r}{D} - \frac{D}{2} + \tanh^{-1}(1-g)\right) b\right] \quad . \tag{9}$$

Similar profiles have been uses by Freund [7] for subsonic free jets. He applied random perturbations of b in order to accelerate the transition of the jet. In agreement with his investigation, the impinging jet presented in this paper remains laminar and symmetric unless a perturbation is added as well. Therefore the thickness parameter varies in time:

$$b = e^{-\tau} b_{n-1} + \left(1 - e^{-\tau}\right)\left(b_0 + \frac{1}{N}\sum_{n=1}^{N} A_n cos(2\pi \frac{u_\infty}{D} Sr_n t + \varPhi_n) cos(\theta + \varPsi_n)\right) \quad . \tag{10}$$

$A, Sr, \varPhi$ and $\varPsi$ are random variables. This approach is a combination of those by Davidson [6] and Freund [7]. The used parameters are: $b_o = 26.47, g = 10^{-2}, N = 100, \tau = 10^{-2}$. The random variables vary as follows: $A = 0..20, Sr = 0..20, \varPhi = 0..2\pi, \varPsi = 0..2\pi$ and lead to a random distribution of the thickness parameter b with $\varDelta b_{max}/b_o = 8.87\ 10^{-4}, \sigma_{\varDelta b/b_o} = 1.30\ 10^{-4}$ per time step and $26.31 \leq b \leq 26.63$.

The non-reflecting boundary condition sets the unknown incoming wave to zero. Additionally a sponge region is applied for the outlet area $r > 11D$, that smoothly forces the values of pressure, velocity and entropy to reference values. This destroys vortices before leaving the computational domain. The reference values at the outlet were obtained by a preliminary large eddy simulation of a greater domain.

3 Results and Discussion

3.1 Energy Spectrum and Kolmogorov Scales

The scales of turbulent motion span a huge range from the size of the domain to the smallest energy dissipating ones. Since the turbulent kinetic energy is transferred downwards to smaller and smaller scales, the smallest ones have to be resolved by the numerical grid in order to obtain a reliable solution of the turbulent flow. They are given with the kinematic viscosity ν and dissipation rate ϵ by

$$l_\eta \approx \left(\frac{\nu^3}{\epsilon}\right)^{\frac{1}{4}} \tag{11}$$

and are valid for isotropic turbulence, that occurs at sufficiently high Reynolds numbers. The ratio of the mesh width to the Kolmogorov length scale $(h_x h_y h_z)^{1/3}/l_\eta$ is shown in Figure 2b and reaches a maximum value of 4.2 at the lower wall. For supersonic turbulent boundary layers, Pirozzoli et al. [12] showed that the typical size of small-scale eddies is about $5..6 l_\eta$. A strongly different behaviour for the present boundary layer of high subsonic Mach number is not expected. The maximum ratio in the area of the free jet is $(h_x h_y h_z)^{1/3}/l_\eta = 1.5$ and of the wall jet is $(h_x h_y h_z)^{1/3}/l_\eta = 2.5$.

Figure 2a shows the energy spectrum of the impinging jet. The inertial subrange is represented by a slope of $-5/3$. Since this slope is tangent to the spectrum in x- and z-direction the inertial subrange is unincisive. This means the Reynolds number of 3300 can be considered as low or *not sufficiently high* to obtain isotopic turbulence. Therefore the Kolmogorov microscales are not crucial for the assessment of the grid resolution. They provide a conservative clue. A lack of resolution means that the dissipating scales are not well resolved, which would lead to a pile-up at high wave numbers. The plot shows that the energy decreases with the scales monotonously in the dissipation area and no increase of energy at high wave numbers is present. In conclusion the applied computational grid is appropriate and the simulation can be referred to as a direct numerical simulation.

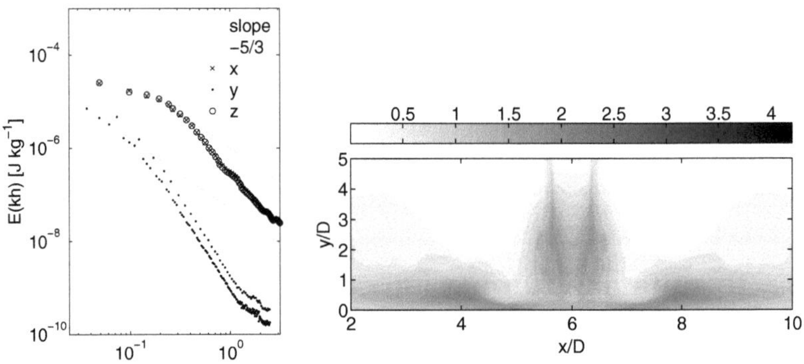

(a) Energy spectrum $E(kh)$ as function of wave number k and grid spacing h

(b) Grid spacing related to Kolmogorov length scale $(h_x h_y h_z)^{1/3}/l_\eta$ on a plane through the centre of the jet. Values below one indicate a finer grid resolution than the Kolmogorov length scale.

Fig. 2. Energy spectrum and resolution of Kolmogorov length scale

3.2 Boundary Layer

In addition to the criterion due to turbulent motion of the jet, the boundary layer including the viscous sub-layer also has to be resolved appropriately in order to achieve reliable results of the heat transfer at the impinging plate. The maximum dimensionless wall distance

$$y^+ = \frac{u_\tau y}{\nu} \tag{12}$$

of the present simulation occurs at $r/D = 0.914$ and reaches a value of $y^+ = 0.64$. The minimum number of points in the viscous sub-layer $y^+ \leq 5$ is seven for the entire domain. Figure 3 shows the velocity- and temperature boundary layer profile for different distances from the stagnation point. u^+, u_τ and τ_W are the dimensionless radial velocity, the friction velocity and the wall shear stress:

$$u^+ = \frac{u}{u_\tau} \quad , \quad u_\tau = \sqrt{\frac{\tau_W}{\rho}} \quad , \quad \tau_W = \eta \left(\frac{\partial u}{\partial y} \right)_W \quad . \tag{13}$$

The dimensionless temperature is given by

$$T^+ = \frac{T_W - T}{T_\tau} \quad , T_\tau = \frac{q_W}{\rho c_p u_\tau} \quad , \tag{14}$$

where T_τ is the friction temperature and q_W the wall heat flux.

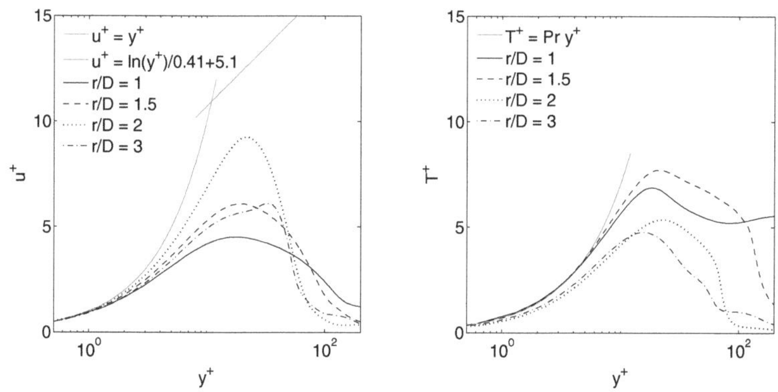

(a) Dimensionless Velocity $u^+(y^+)$ **(b)** Dimensionless Temperature $T^+(y^+)$

Fig. 3. Mean boundary layer profiles for different radial positions r/D

3.3 Heat Transfer

The main flow characteristics can be compared with experimental results from literature. Figure 4 shows the mean distributions of the radial velocity and temperature. The maximal mean radial velocity increases from the stagnation point,

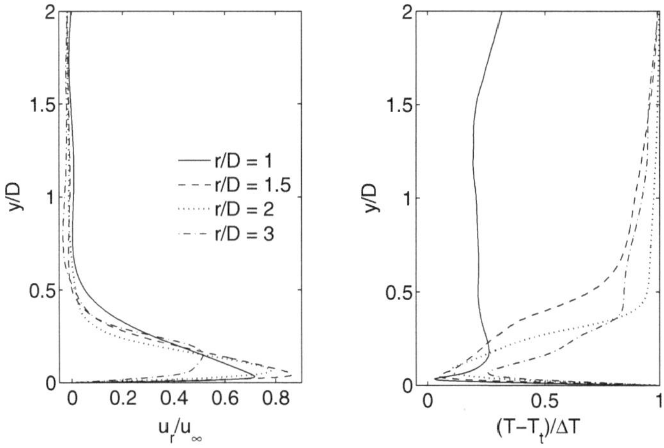

(a) Radial velocity u_r

(b) Temperature difference $T - T_t$ relative to $\Delta T = T_W - T_o$

Fig. 4. Mean distributions for different radial positions r/D

reaches its maximum value at $r/D = 1.52$ and then decreases. The maximal temperature difference $T - T_t$ increases monotonously with greater radii as a consequence of turbulent mixing of the cold impinging jet and the hot environment. For the same reason, the profile at $r/D = 1$ which is located in the shear layer of the jet is at lower temperatures than the other ones.

The heat transfer at the impinging plate is strongly related to the vortical structures of the turbulent flow field. In the shear layer of the jet (primary) ring vortices develop and grow until they collide with the wall and then stretch and move in radial direction. As soon as the primary toroidal vortex passes the deceleration area of the wall jet $r/D > 1.52$, the flow separates and forms a new secondary counter-rotating ring vortex that enhances the local heat transfer, directly followed by a likewise annular area of poor heat transfer due to separation. Travelling downstream the vortex pair becomes unstable and breaks down into smaller structures that rise. As a consequence the two rings at the wall of very high and low heat transfer vanish and the cycle restarts.

Figures 5 and 6 show the birth of those vortex rings on a x-y plane through the center of the jet. The background pictures the Mach number. Thereon contour lines of same values of the Q-criterion are drawn. The two vertical lines show the position of the maximal radial velocity. In addition, the Nusselt number is represented on the x-z plane at the wall. The first picture of the series is taken at a time step where the primary ring vortex reaches the deceleration area of the wall jet. Beside the stagnation area no strong peak in heat transfer is present. In the next three pictures, the secondary counter-rotating ring vortex is born and travels downstream. A strong annular peak in the Nusselt number can be

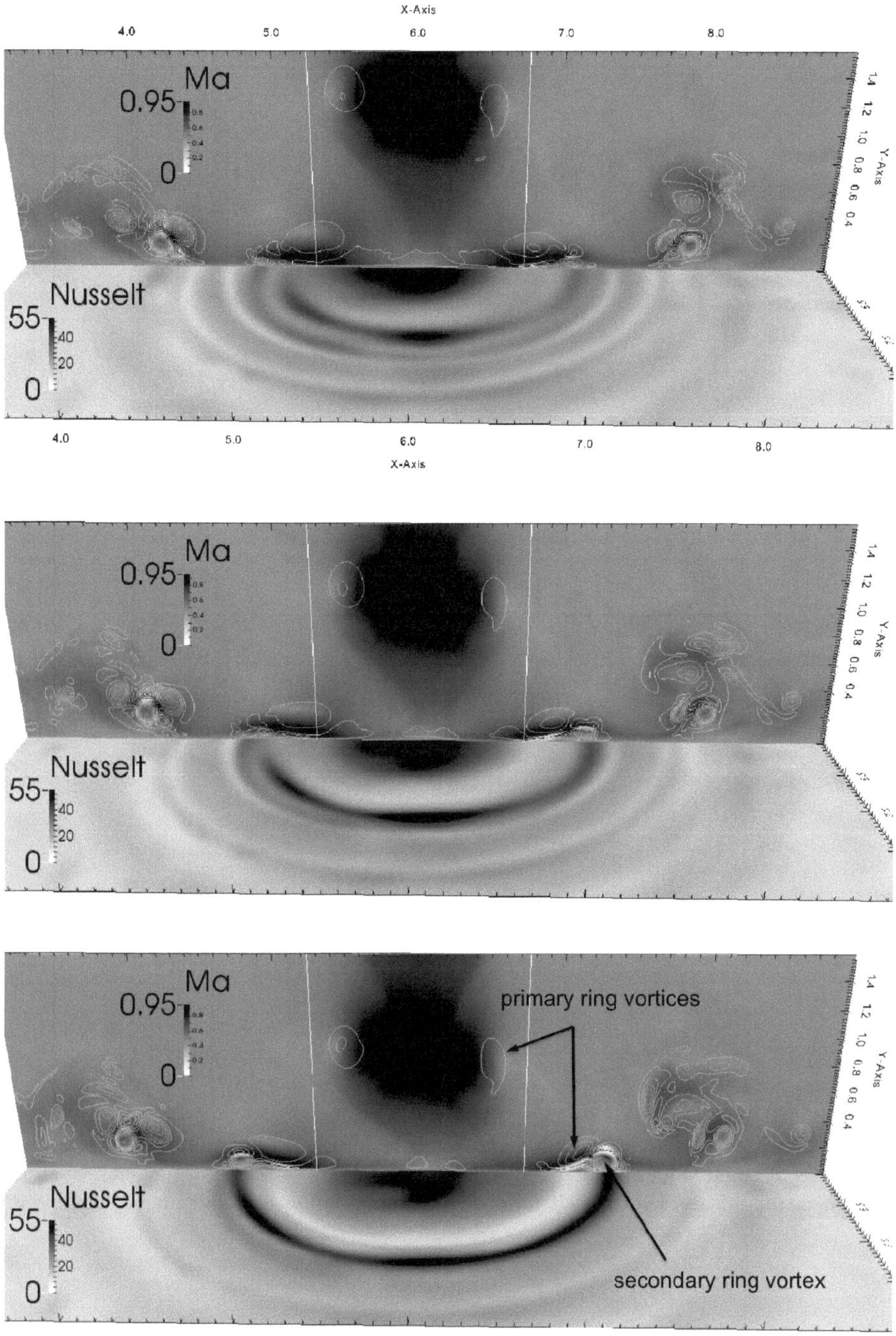

Fig. 5. Impact of the secondary vortex on the local Nusselt number illustrated by contours of equal values of Q above the Mach number Ma on a x-y-plane through the center of the jet and the Nusselt number at the impinging plate advancing in time (1)

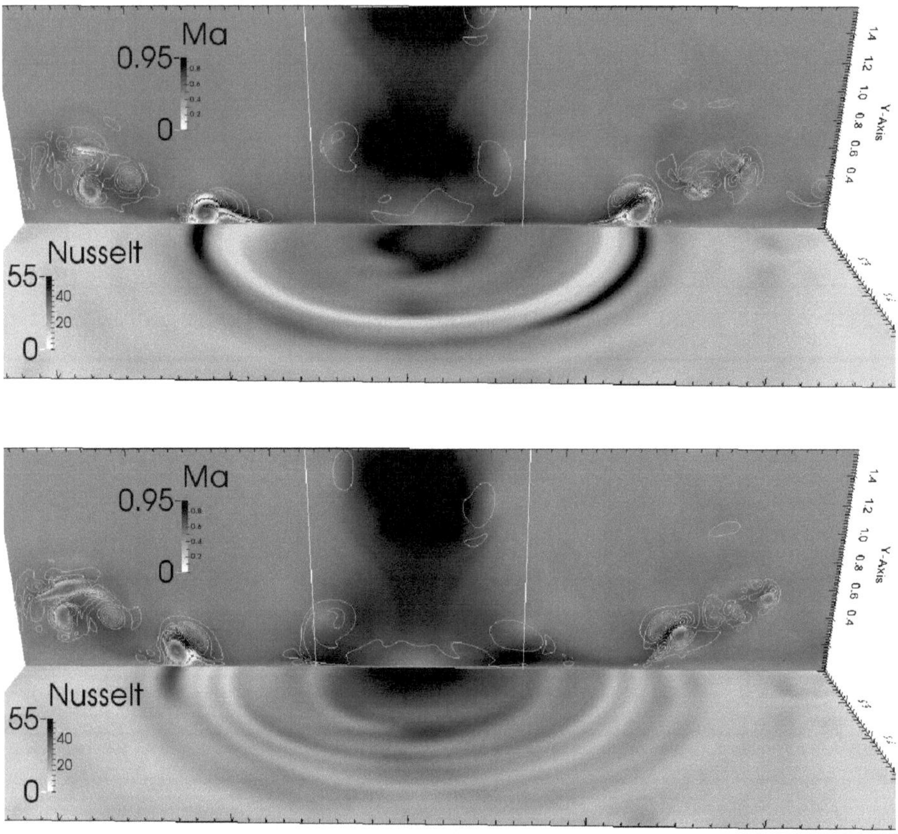

Fig. 6. Impact of the secondary vortex on the local Nusselt number illustrated by contours of equal values of Q above the Mach number Ma on a x-y-plane through the center of the jet and the Nusselt number at the impinging plate advancing in time (2)

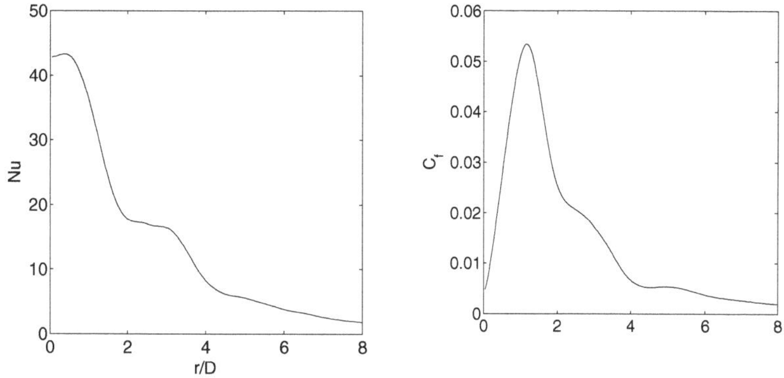

(a) Nusselt number (b) Skin friction coefficient

Fig. 7. Mean radial distributions for the local Nusselt number Nu and the skin friction coefficient $C_f = \frac{2\tau_W}{\rho u_\infty^2}$

observed which travels together with the vortex pair until they collapse and the strong heat transfer disappears.

The investigated life cycle leads to a secondary area located at $r = 2..3$ of high values of the radial Nusselt number

$$Nu(r) = -\frac{D}{\Delta T}\frac{\partial T}{\partial y}(r)\bigg|_W \tag{15}$$

that is shown in Figure 7a. The primary maximum exists at $r/D = 0.38$.

Buchlin[2] reported the inner or primary maximum at $r/D = 0.7$ and the outer or secondary at $r/D = 2.4$ for experimental data at $Re = 60000, h/D = 1$ with the remark, that the strength of both maxima decrease and the inner one moves to the stagnation point with increased nozzle to plate distances h/D and decreased Reynolds number. Since direct numerical simulations presently cannot reach high Reynolds numbers like $Re = 60000$ of the experiments, no direct comparison between the numeric values can be performed. Nevertheless Buchelins conclusions concerning the phenomenon of secondary vortex rings, the movement of the secondary maximum towards the stagnation point and the loss of strength of both maxima match the results of the present simulation well.

4 Conclusion

This paper presents a direct numerical simulation of a subsonic confined impinging jet. The chosen grid size of 512^3 points matches the requirements for reliable results on turbulent heat transfer. The Nusselt number distribution is presented and features a week primary and secondary maximum which is caused by the occurrence of secondary vortex rings that locally increase the heat transfer in an annular shape. Those first results agree with experimental data of [2]. In future work a DNS at $Re = 8000$ will be performed and directly compared to experimental results of our partner project within the Collaborative Research Centre 1029 in order to increase the practical importance of the investigations for gas turbines.

Acknowledgments. The simulations were performed on the national supercomputer Cray XE6 at the High Performance Computing Center Stuttgart (HLRS) under the grant number GCS-NOIJ/12993.

The authors gratefully acknowledge support by the Deutsche Forschungsgemeinschaft (DFG) as part of collaborative research center SFB 1029 "Substantial efficiency increase in gas turbines through direct use of coupled unsteady combustion and flow dynamics".

Appendix: Abbreviations Used in the Governing Equations

$$X^{\pm} := (u \pm c)\left(\frac{1}{\rho c}\frac{\partial p}{\partial x} \pm \frac{\partial u}{\partial x}\right) \tag{16}$$

$$Y^{\pm} := (v \pm c)\left(\frac{1}{\rho c}\frac{\partial p}{\partial y} \pm \frac{\partial v}{\partial y}\right) \tag{17}$$

$$Z^{\pm} := (w \pm c)\left(\frac{1}{\rho c}\frac{\partial p}{\partial z} \pm \frac{\partial w}{\partial z}\right) \tag{18}$$

$$X^{v} := u\frac{\partial v}{\partial x} \quad , \quad X^{w} := u\frac{\partial w}{\partial x} \quad , \quad X^{s} := u\frac{\partial s}{\partial x} \tag{19}$$

$$Y^{u} := v\frac{\partial u}{\partial y} \quad , \quad Y^{w} := v\frac{\partial w}{\partial y} \quad , \quad Y^{s} := v\frac{\partial s}{\partial y} \tag{20}$$

$$Z^{u} := w\frac{\partial u}{\partial z} \quad , \quad Z^{v} := w\frac{\partial v}{\partial z} \quad , \quad Z^{s} := w\frac{\partial s}{\partial z} \tag{21}$$

$$\tau_{ij} := 2\nu\rho\left(s_{ij} - \frac{1}{3}s_{kk}\delta_{ij}\right) \tag{22}$$

$$s_{ij} := \frac{1}{2}\left(\frac{\partial u_i}{\partial k_j} + \frac{\partial u_j}{\partial x_i}\right) \tag{23}$$

$$\Phi := \tau_{ij}s_{ij} \tag{24}$$

References

1. Adams, N.: A High-Resolution Hybrid Compact-ENO Scheme for Shock-Turbulence Interaction Problems. Journal of Computational Physics 127, 27–51 (1996)
2. Buchlin, J.: Convective heat transfer in impinging-gas-jet arrangements. Journal of Applied Fluid Mechanics 4(3) (2011)
3. Chung, Y.M., Luo, K.H.: Unsteady heat transfer analysis of an impinging jet. Journal of Heat Transfer 124(6), 1039–1048 (2002)
4. Cziesla, T., Biswas, G., Chattopadhyay, H., Mitra, N.: Large-eddy simulation of flow and heat transfer in an impinging slot jet. International Journal of Heat and Fluid Flow 22(5), 500–508 (2001)
5. Dairay, T., Fortune, V., Lamballais, E., Brizzi, L.E.: Direct numerical simulation of the heat transfer of an impinging jet. In: Department of Fluid Flow, Heat Transfer and Combustion, Institute PRIME. CNRS - Universite de Poitiers ENSMA, 14th European Turbulence Conference (September 2013)
6. Davidson, L.: Inlet boundary conditions for embedded les. In: First CEAS European Air and Space Conference (2007)
7. Freund, J.B.: Noise sources in a low-reynolds-number turbulent jet at mach 0.9. Journal of Fluid Mechanics (438), 277–305 (2001)
8. Hattori, H., Nagano, Y.: Direct numerical simulation of turbulent heat transfer in plane impinging jet. International Journal of Heat and Fluid Flow 25(5), 749–758 (2004); Selected papers from the 4th International Symposium on Turbulence Heat and Mass Transfer
9. Hrycak, P.: Heat transfer from impinging jets. a literature review. Technical report, New Jersey Institute of Technology (1981)
10. Jambunathan, K., Lai, E., Moss, M., Button, B.: A review of heat transfer data for single circular jet impingement. International Journal of Heat and Fluid Flow 13(2), 106–115 (1992)
11. Lele, S.K.: Compact finite difference schemes with spectral-like resolution. Journal of Computational Physics 103(1), 16–42 (1992)
12. Pirozzoli, S., Bernardini, M., Grasso, F.: Characterization of coherent vortical structures in a supersonic turbulent boundary layer. Journal of Fluid Mechanics 613, 205–231 (2008)
13. Sesterhenn, J.: A characteristic-type formulation of the navier-stokes equations for high order upwind schemes. Computers & Fluids 30(1), 37–67 (2001)
14. Viskanta, R.: Heat transfer to impinging isothermal gas and flame jets. Experimental Thermal and Fluid Science 6(2), 111–134 (1993)
15. Weigand, B., Spring, S.: Multiple jet impingement - a review. Heat Transfer Research 42(2), 101–142 (2011)
16. Zuckerman, N., Lior, N.: Impingement heat transfer: Correlations and numerical modeling. Journal of Heat Transfer 127(5), 544–552 (2005)

Part II
Combustion

Low Temperature Gasoline Combustion – Potential, Challenges, Process Modeling and Control

Bastian Lehrheuer, Bastian Morcinkowski, Stefan Pischinger, and Martin Nijs

Institute for Combustion Engines, RWTH Aachen University, Germany

Abstract . Worldwide efforts in combustion development focus on reducing carbon dioxide emissions. Besides improving conventional combustion systems, also alternative systems have to be considered as a measure to achieve that goal. The present work aims to increase the usability of gasoline controlled auto-ignition (GCAI) with its potential benefit regarding fuel consumption of up to 30 %.

The operating range of low temperature gasoline combustion is limited by instability and high sensibility regarding the thermodynamic state. In a first step, experimental investigations on a 1-cylinder research engine with Electromechanical Valve Train (EMVT) were carried out to identify measures that extend the GCAI operating range. In particular the injection strategy appears to be an important control value. So, a distinct amount of fuel injected during the intermediate compression by using combustion chamber exhaust gas recirculation helps to decrease the feasible load by up to 62.5 % (at 2000 min^{-1}). Increased charge dilution realized by charging or external exhaust gas recirculation influences the maximum cylinder pressure gradient and allows raising the examined load point from IMEP = 4.6 bar to IMEP = 6.7 bar.

The experiments reveal that the auto-ignition process is affected by comparable high combustion variability with negative influence on stability and controllability. Hence, numerical investigations were carried out to gain a deeper understanding of the combustion fluctuations. Using an 1D gas exchange model in combination with computational fluid dynamics (CFD) the origin of those fluctuations can be determined. On the one hand auto-correlations between consecutive combustion cycles are exposed. This demonstrates the high sensibility of the GCAI process for the thermodynamic state. Furthermore, a superimposed stochastic fluctuation caused by a distinct intake pressure noise explains the combustion variations.

The precise process model is used to develop a model-based predictive controller. To that end the complexity of the model is reduced based on 1D gas exchange calculations and the Watson approach. The reduced process model is capable to reproduce the above mentioned fluctuations.

Keywords: gasoline controlled auto-ignition, GCAI, simulation, in-cylinder flow, combustion, combustion variability, correlation.

Introduction

With potential fuel consumption benefits of up to 30 % and a significant reduction of nitrogen oxide emissions (NO_x) the low temperature combustion represents an

R. King (ed.), *Active Flow and Combustion Control 2014*, Notes on Numerical Fluid Mechanics and Multidisciplinary Design 127, DOI: 10.1007/978-3-319-11967-0_11

attractive alternative process for internal combustion engines. Contrary to the premixed turbulent combustion in common gasoline engines, which is controlled by the turbulence structure, Gasoline Controlled Auto-Ignition (GCAI) is driven by chemic kinetics. Hence, combustion characteristics are highly dependent on thermodynamic state, ambient conditions and fuel properties. Temperature and pressure level of the cylinder charge and stratification of internal residual gas fraction come into consideration as main control variables for the auto-ignition process. Investigations on an 1-cylinder research engine with Electromechanical Valve Train (EMVT) address those influencing variables and limitations of low temperature gasoline combustion.

Fig. 1. 1-cylinder research engine

Table 1. Specifications of 1-cylinder research engine

Engine specifications		
displacement	499	cm³
stroke s/ bore D	90 / 84	mm
s / D	1.07	1
compression ratio	12 (& 9.5)	1
cylinder	1	-
valves per cylinder	4	-
injection	piezoelectric actuated hollow cone injector	-
valvetrain	electromechanical valve train	-
engine setups	thermodynamic & optical investigations	-

The experiments reveal that low temperature combustion possesses cyclic fluctuations with negative influence on combustion stability. To gain a deeper understanding of the origin detailed numerical investigations are carried out including a verified 1D gas exchange model as well as computational fluid dynamics (CFD) coupled to chemical kinetics. A further step combines the numerical and experimental findings in strategies to extend the GCAI operation map, in which reduced process models beforehand used for the numerical investigations serve in a model based control software.

Gasoline Controlled Auto-ignition

The worldwide reduction of carbon dioxide emissions (CO_2) caused by the transport sector is a major focus of engine development. On that matter, reduced displacement in combination with turbocharging, valve train variability and direct injection is already a well-established measure in the field of gasoline combustion development. High potential for a further reduction in fuel consumption is assigned to the 2nd generation of lean stratified combustion ([1],[2]). Examples can already be found in serial production. Due to the lean combustion process the use of the conventional three-way catalyst for the reduction of NO_x emissions is not possible and a high cost exhaust after treatment system has to be added. An alternative of comparable CO_2 savings at up to 99 % reduced NO_x raw emissions is offered by GCAI. A decentralized ignition leads to a low global temperature in the combustion chamber with low intensity of dissociation and thus a reduction of NO_x and soot emissions [3].

Previous work shows potential fuel consumption benefits of up to 11.7 % for the New European Driving Cycle (NEDC) when compared to conventional spark-ignited (SI) operation [4]. [5] estimates the potential CO_2 savings even with up to 30 %.

Combustion Process

Contrary to the premixed, flame-propagating combustion in conventional processes, GCAI is determined by chemical kinetics. Hence, there is no propagating flame front, but multiple ignitable spots at the end of compression disengaging a chain reaction by creating favourable combustion conditions in means of temperature and pressure all over the combustion chamber. Almost simultaneous combustion leads to comparable short burn duration. Thereby the GCAI combustion is closer to the corresponding ideal isochoric cycle than the common gasoline combustion (compare Figure 2). Contrary, towards higher loads this leads to unacceptable high pressure gradients limiting the GCAI operating range.

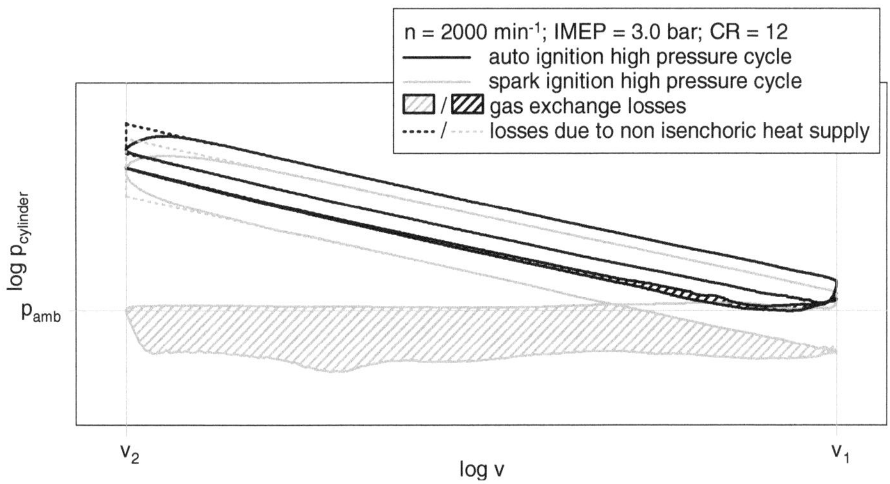

Fig. 2. Indicator diagram for GCAI and SI cycle

The required temperature to initiate auto-ignition is provided by a large amount of internal residual gas. In the case shown in Figure 2 the necessary residual gas fraction is realised by negative valve overlap as it is evident by the intermediate compression during the gas exchange cycle. Comparing the gas exchange cycles, another advantage of GCAI becomes obvious: lean operation with a high amount of residual gas allows dethrottling and thereby reduced gas exchange losses. On the other hand high residuals have negative influence on combustion stability, when the enthalpy of the exhaust gas increases at lower loads. Measures to overcome the mentioned limitations are discussed in the following.

Extended Operation Map

Using the GCAI benefits in a larger range is accompanied by an increase in CO_2 savings. Therefore, the possible influencing parameters have to be identified and classified with regard to their significance to the maximum cylinder pressure gradient $(dp/d\alpha)_{max}$, combustion stability which is characterized by the coefficient of variance COV_{IMEP} of the indicated mean effective pressure and air/fuel ratio λ.

The influencing parameters are divided into measures to influence the global temperature and pressure levels, stratification effects of recirculated exhaust gas, fuel and fresh air as well as the fuel properties such as oxidation reactions or the enthalpy of vaporization. To assess the latter, in addition to conventional gasoline fuels (RON95 and RON98) also various alcohols are tested. The influence of local stratification effects is investigated in terms of spark assistance as well as various injection strategies. With respect to the influence on the global temperature and pressure level variations of the following parameters are examined:

- internal residual gas fraction via variable valve timing
  - Combustion Chamber Recirculation (CCR)
  - Exhaust Port Recirculation (EPR)
- compression ratio via piston height
- intake manifold pressure via charging
- intake manifold temperatur
- cooled, external residual gas fraction

Due to the complexity of the investigations only the most important measures shall be discussed following. Figure 3 shows the comparison of the two above mentioned strategies to trap internal residual gas at an engine speed of $n = 2000\ min^{-1}$ and moderate load of $IMEP = 3\ bar$. In case of exhaust port recirculation the exhaust valve is kept open beyond Top Dead Center (TDC). Thereby the exhausted gas is sucked back in after TDC. The low cylinder pressure gradient respectively the slower auto-ignition process, as it is evident by the cylinder pressure trace and flame light measurements (Figure 3 below), are preferable to extend the GCAI operating range towards higher loads. Contrary, this strategy is accompanied by a high COV of IMEP when operated at lower loads like the example shown in Figure 3 ($COV_{IMEP,EPR} = 7.43\ \%$).

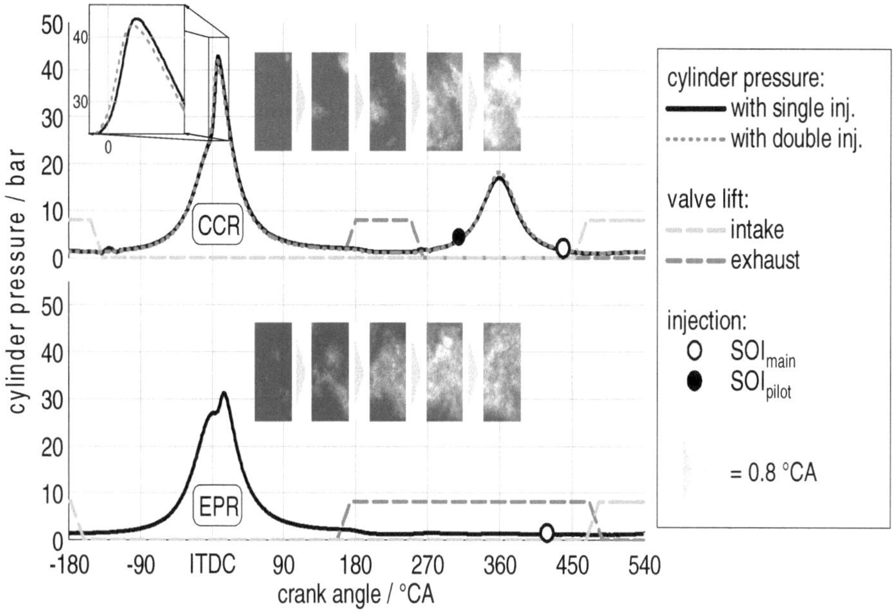

Fig. 3. Cylinder pressure, valve timing and flame light measurements at auto-ignition for combustion chamber recirculation with and without pilot-injection in comparison to exhaust port recirculation; $n = 2000\ min^{-1}$, IMEP = 3 bar, CR = 12

Combustion chamber recirculation, on the other hand, is characterized by a high negative valve overlap, thereby recompressing the exhaust gas during TDC. This leads to a comparably high and homogeneously distributed temperature of the residual gas and in consequence faster fuel conversion (Figure 3 top). Hence, a lower amount of residual gas is necessary to achieve the same load, what enables extended operation towards lower loads with increased stability ($COV_{IMEP,CCR} = 4.93\ \%$). Additionally, early exhaust valve closing in combination with direct injection allows (cylinder individual) pilot-injections during the intermediate compression. Already small injection quantities ($m_{fuel,pilot} = 2.17\ mg\ /\ m_{fuel,main} = 4.12\ mg$) can have a major influence on the auto-ignition process during main compression. The example shows a decreased COV of IMEP of $COV_{IMEP,CCR+pilot} = 1.73\ \%$. Whereas higher amounts of pilot-injection are accompanied by increased COV_{IMEP} . An almost equal intermediate compression indicates no significant fuel conversion at this point. For higher fuel amounts during intermediate compression fuel conversion can be detected what overrides the positive, stabilizing effect. Responsible for this effect might be that the ignition process depends on chemical reaction kinetics; in particular the prompt decomposition of hydrogen peroxide (H_2O_2) into hydroxyl radicals (OH·) at temperatures above $1000 - 1200$ K [14]. Small, not entirely converted fuel amounts ($T < 1000$ K) increase the number of H_2O_2 molecules for the main compression auto-ignition. Hence, pilot-injection and in consequence the amount of H_2O_2 molecules offer a promising control value to reduce cyclic fluctuations and a further measure to extend the operation range especially towards low loads. With regard to a multi-cylinder GCAI application, cylinder-specific corrections can be realized as well.

Starting from a GCAI operation with single injection during suction stroke, various strategies have been combined in order to extend the operational range towards higher and lower loads. Figure 4 shows exemplary the results at an engine speed of $n = 2000\ \text{min}^{-1}$ for both CCR and EPR. For the investigations the maximum cylinder pressure gradient was set to $(\text{dp}/\text{d}\alpha)_{\max} < 5\ bar/°CA$. The combustion stability is reflected in the standard deviation of the indicated mean effective pressure that was limited on a common value for the partial load range to $\sigma_{\text{IMEP}} < 0.15$. A further limiting value for the results as shown below is represented by the air-fuel ratio, which should be in an over-stoichiometric region for GCAI operation ($\lambda > 1$).

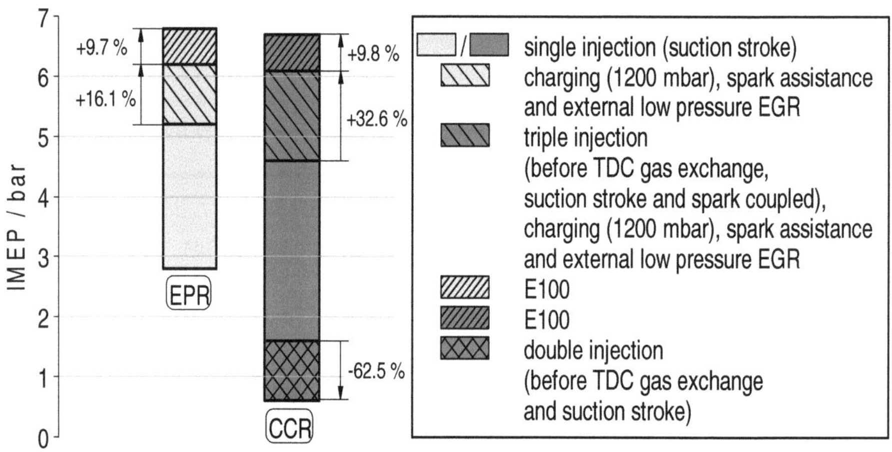

Fig. 4. Strategies for load extension; $n = 2000\ \text{min}^{-1}, \text{CR} = 12$; Boundary conditions for GCAI: $\sigma_{\text{IMEP}} < 0.15$, $(\text{dp}/\text{d}\alpha)_{\max} < 5\ bar/°CA$, $\lambda > 1$

As described before CCR features advantages towards lower loads. Even for a single injection strategy CCR enables load points down to $\text{IMEP} = 1.6\ \text{bar}$, whereas with EPR only an indicated mean effective pressure of $\text{IMEP} = 2.8\ \text{bar}$ is feasible. In combination with pilot injection it is possible to reduce the minimum load by a further 62.5 %.

The increasing maximal pressure gradient restricts the GCAI operation at higher loads. Hence, it is necessary to adopt strategies that slow down the reaction progress like increased charge dilution via charging or external exhaust gas recirculation. Also the previously described injection before top dead center gas exchange can be used to further lean out the mixture, as less residuals are necessary to achieve a certain combustion phasing. In combination with a third injection that is coupled to a supporting spark the maximum achievable load is increased by 32.6 % in case of CCR. Supporting spark leads to a stratified combustion in the vicinity of the spark plug thereby initiating the auto-ignition process. The high enthalpy of vaporization and octane number of alcohol-based fuels slow combustion and reduce the maximum cylinder pressure rise. Thus, they offer further potential for expansion of the GCAI operational range towards higher loads. With ethanol a further increase in the load of 9.8 % can be achieved. Those fuels also have the advantage to significantly reduce NO_x emissions compared to conventional RON95 fuels in the high loads due to their comparably high enthalpy of vaporization and especially their lower adiabatic flame temperature. Still NO_x emissions rise with increased engine loads. In consequence it might be necessary to switch from lean to stoichiometric conditions.

Considering the same boundary conditions, the compression ratio influences directly the initiation of the auto-ignition process by increased respectively decreased pressure at end of the compression stroke. Investigations of two compression ratios ($\mathrm{CR} = 12\ \&\ CR = 9.5$) at an engine speed of $\mathrm{n} = 1500\ \mathrm{min}^{-1}$ and an indicated mean effective pressure of IMEP = 3 bar reveal that a higher compression ratio allows to increase air-fuel ratio by 14.8 %, since there is a lower amount of residual gas necessary to achieve the same center of combustion (CA50 = 8° CA aTDC). In consequence a higher compression ratio is advantageous to cover a broad GCAI operation area. Considering the usage of both GCAI and conventional combustion in one engine a variable compression ratio is advisable.

Cyclic Fluctuations and Combustion Stability

The above mentioned investigations reveal that already small variations in ambient or combustion chamber conditions can have a major influence on combustion stability. On the one hand this provides a set of important control values for the combustion process like pilot injection, external residual gas or charging. On the other hand the auto-ignition process is very sensitive to incidental fluctuations of the global temperature and pressure level. The analysis of 200 consecutive cycles regarding the fluctuation of IMEP shows a relation ($k = -0.56$) between two sequent cycles (see Figure 5).

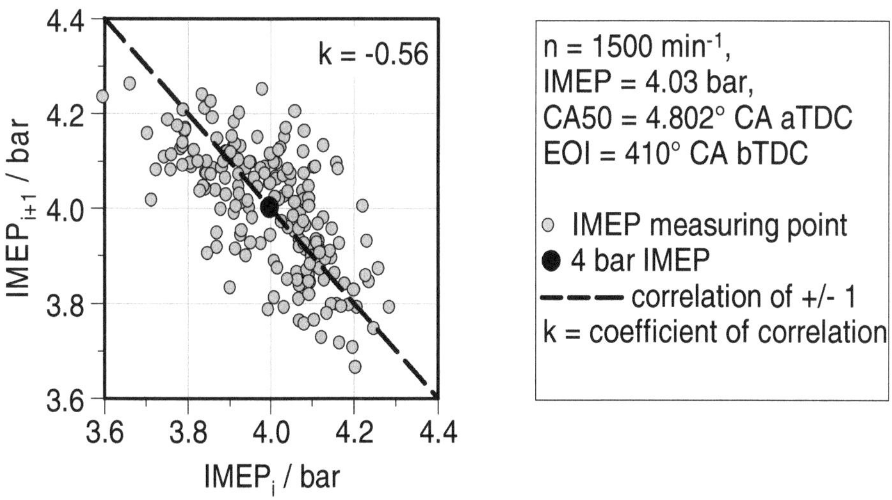

Fig. 5. Return maps for IMEP of 200 consecutive cycles at $\mathrm{n} = 1500\ \mathrm{min}^{-1}$ and $\mathrm{EOI} = 410°\ \mathrm{CA\ bTDC}$

According to that, a good combustion is followed by a worse combustion and vice versa. This relation indicates the significance of the exhaust gas enthalpy of the previous cycle for the combustion of the current cycle. Among others, this observation is reason for the numerical investigations aiming for a suitable combustion control as described in the following.

Gcai Process Modelling

At GCAI operation, the engine control becomes challenging. Therefore, an implementation of an intelligent combustion control is beneficial, where detailed knowledge of the combustion characteristics is utilized. This also implies the knowledge of cyclic fluctuations, which will be discussed in the simulation results.

Based on the understanding of effects leading to cyclic fluctuations, an efficient modelling approach of this complex combustion system is derived.

Evaluation Approach to Identify Sources of Cyclic Fluctuations

In the following an approach is introduced which divides cyclic fluctuations into two parts: An auto-correlation part analysed by 1D gas exchange modelling and a superimposed stochastic part analysed by 3D-CFD.

In a first step, a reduced 1D gas exchange model is used in which the detailed chemistry of the trapped gas mass is neglected. Therefore, the heat release rates of consecutive cycles are calculated by three-pressure analyses (TPA) [6] using measured consecutive in-cylinder, intake-port and exhaust-port pressures. These heat release rates and their order serve as input for a 1D gas exchange model. This model is able to calculate the correct mass flow over intake and exhaust valves, so the correct temperature and composition of fresh air, residual gas and fuel can be predicted.

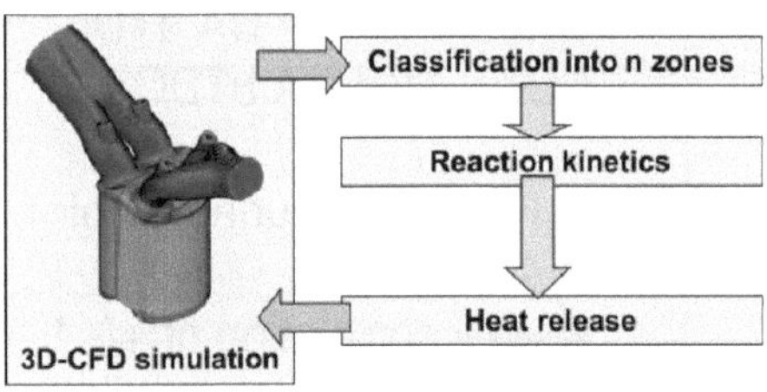

Fig. 6. Representative chemistry

Based on the solution of the transient 1D model, this work shows an approach to correlate the mean thermodynamic in-cylinder state at the end of combustion to a heat release rate of the next cycle. Subsequently to the 1D analyses, computational fluid dynamics (CFD) coupled to chemistry calculations is used to identify the sources of cyclic fluctuations in detail. For in-cylinder flow and combustion simulation, the Large Eddy Simulation (LES) approach is applied. The LES approach is a numerical technique to calculate the equations of turbulent flows by using a spatial filtering [7]. In the LES approach larger turbulent structures which are called eddies are directly calculated in a space- and time-accurate manner, while eddies that are smaller than a mesh-based filter length are modelled using a sub-grid scale model [8]. By applying the LES approach, cyclic combustion variability can be resolved [9]. In order to describe the auto-ignition process with sufficient accuracy, a combustion model based on [10] is utilized. This model solves the reaction kinetics in a discretization of the thermodynamic state instead of a spatial discretization. The discretization is performed via two dimensional classifications in the state variable mass fraction of temperature and fuel mixture fraction. Such a representative chemistry approach segregates the solving of the gas from the integration of the chemical reactions, see Figure 6.

Sources of Cyclic Fluctuations

In the following an investigation of an exemplary operation point is shown to describe the way to model the complex GCAI combustion. The engine operating point is at an engine speed of $n = 2000\ \text{min}^{-1}$ and an engine load of $\text{IMEP} = 3\ \text{bar}$. The injection strategy is one single injection with an end of injection at EOI = 80° CA after gas exchange top dead center.

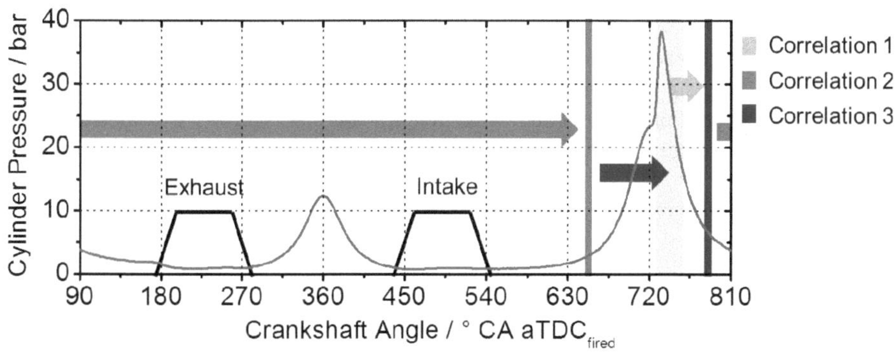

Fig. 7. Three Correlations during one cycle beginning with the end of combustion of the preceding cycle

The 1D process calculations help to find correlations between the end of combustion of the preceding cycle and the heat release of the current cycle. Figure 7 shows three possible correlations during the engine cycle that are presented in detail in the following.

The combustion can be characterized by the 50 % fuel mass burned point (MFB50). Slow burning cycles have a retarded MFB50 and fast burning are characterized by earlier MFB50. Figure 8a shows the result of a heat release analysis. The self-correlation of consecutive cycles is clearly visible. However, the Pearson's correlation coefficient [15] of k = -0.57 is not strong.

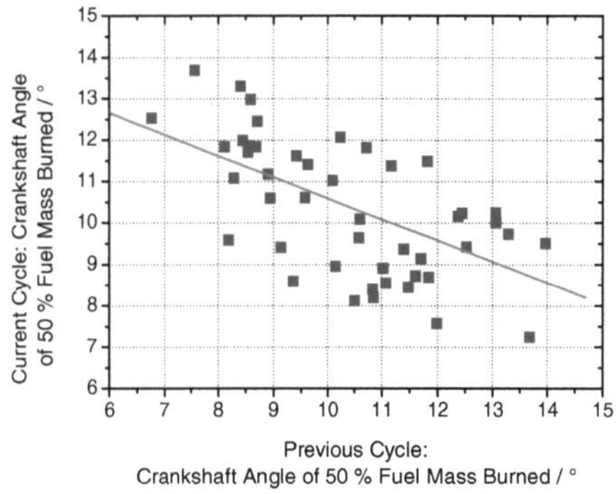

a) Auto-correlation of 50 % MFB; k = -0.57

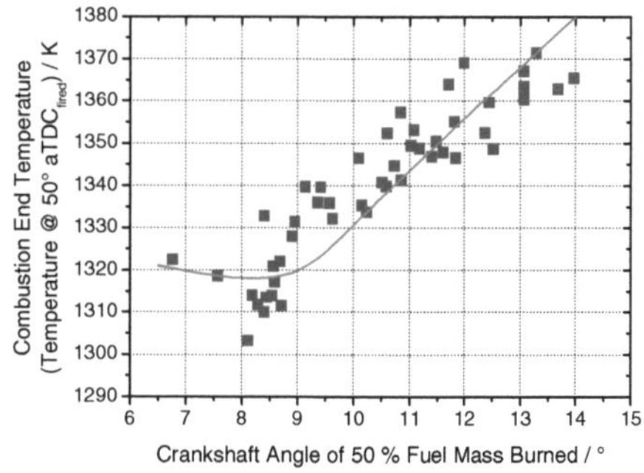

b) Correlation 1: 50 % MFB and combustion end temperature; k = -0.91

Fig. 8. a)-b): Correlations during the GCAI engine cycle

It is quite evident that the combustion quality influences the temperature during expansion. The strong correlation between MFB50 and combustion end temperature is shown in Figure 8b. Cycles with retarded combustion show decreased combustion efficiencies. That leads to higher temperatures during the expansion. The temperature minimum is located where the combustion efficiency has its optimum. For gasoline engines at part load operating conditions that optimum is typically assumed around 8° crankshaft angle after top dead center.

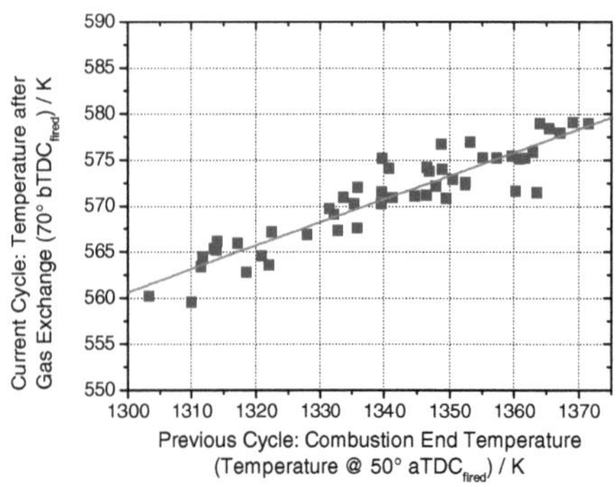

c) Correlation 2: Temperature after combustion of the preceding cycle and temperature after gas exchange; k = -0.92

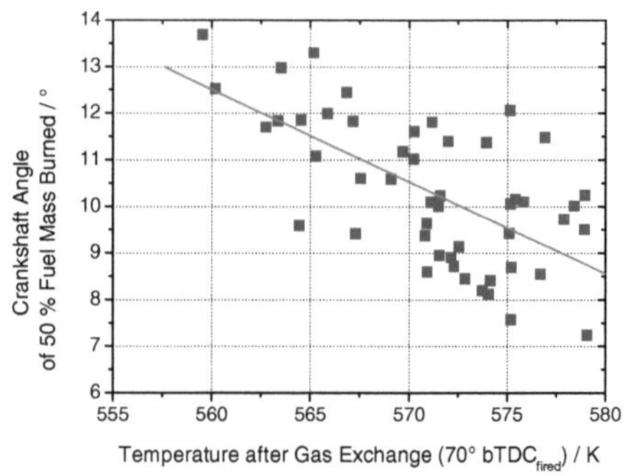

d) Correlation 3: 50 % MFB and temperature after gas exchange (resp. temperature before main compression); k = -0.61

Fig. 8. c)-d): Correlations during the GCAI engine cycle

The correlation in Figure 8c proves that the temperature state can be conserved during the gas exchange and the intermediate compression. As expected, the temperature before compression stroke correlates with the temperature after combustion and accordingly the combustion efficiency of the preceding cycle.

Figure 8d shows the correlation between mean in-cylinder temperatures and MFB50. Significant deviations from an ideal relationship are obvious. These deviations are of the same order than the auto-correlation of MFB50 (Figure 8a). This is in agreement to the common understanding that reaction chemistry is mainly influenced by the temperature state. As stated before, the fluctuations of air/fuel ratio and residual gas are too small in comparison to temperature. Thus, the fluctuating temperature state is predicted to be the source of auto-correlated cyclic combustion variability.

The discrepancy to an ideal auto-correlation is assumed to have a stochastic character due to fluctuations of stratification. In the following the effect of fluctuating in-cylinder stratifications on combustion variability is discussed by using CFD. Figure 9 shows the resolved in-cylinder flow field kinetic energies for ten cycles. LES resolves cyclic fluctuations of the kinetic energies, which are a measure for the in-cylinder flow fluctuations. These fluctuations are produced by very slight intake pressure noise beginning at 460° CA aTDC when the intake valves open.

Flow field fluctuations affect the mixture formation and will lead to fluctuations in quantities which influence reaction chemistry: the stratification of residual gas mass fraction, air-fuel ratio and temperature. Since initial conditions are the same no differences in the mean quantities are noticed.

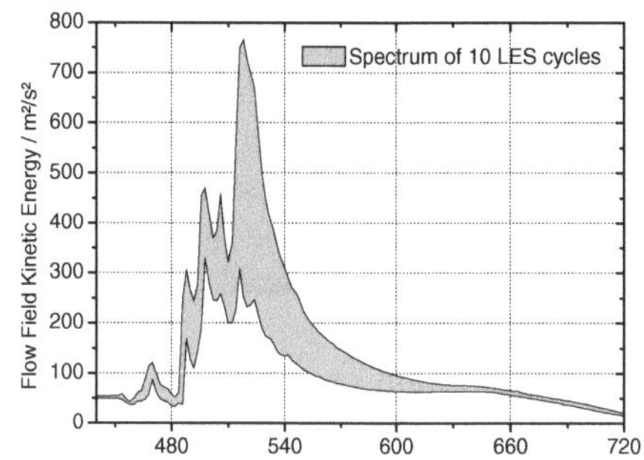

Fig. 9. Spectrum of resolved flow field kinetic energies for 10 LES cycle

Figure 10 shows the stratification of air-fuel ratio in case of early and late combustion. During homogenization (630° CA aTDC and 710° CA aTDC) significant differences of stratification can be noticed.

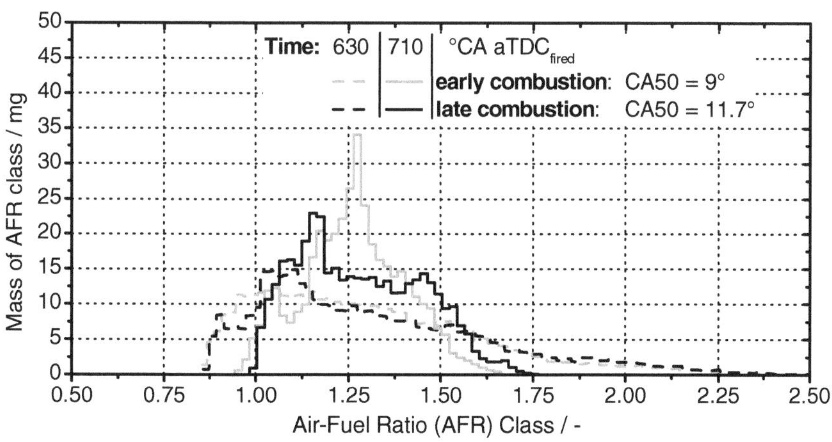

Fig. 10. Histograms of air-fuel ratio for early and late cycle at 630° CA aTDC and 710° CA aTDC

In the following a quantity is introduced, which accounts for the stratification of all variables relevant for reaction kinetics. Based on the results of the in-cylinder flow and mixture formation simulation, calculations using the RANS and LES coupled to reaction chemistry are performed.

Figure 11 shows the fired in-cylinder pressure traces calculated by CFD in comparison to the pressure traces measured at the test bench. The pressure curves show that LES resolves cyclic combustion variability. Furthermore, the measured spectrum of pressure curves is reproduced by LES. When averaging the LES cycles, the agreement of the pressure gradient and peak pressure to the measured average curve is very good.

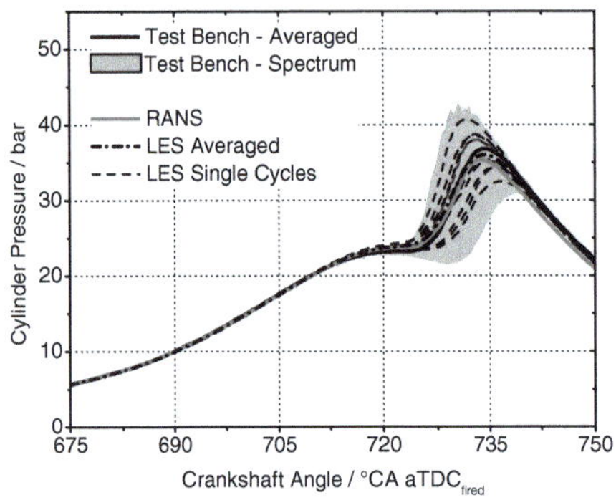

Fig. 11. Simulated LES and RANS in-cylinder pressures in comparison to measurement

The ignition behaviour for different engine cycles are shown in Figure 12 by visualisation of the locally resolved 50 % accumulated heat release iso-surfaces. It is seen that rather fine structures exist, and a specially distributed ignition occurs simultaneously at different locations in the combustion chamber. This indicates strongly that fluctuations of stratification lead to fluctuating combustions.

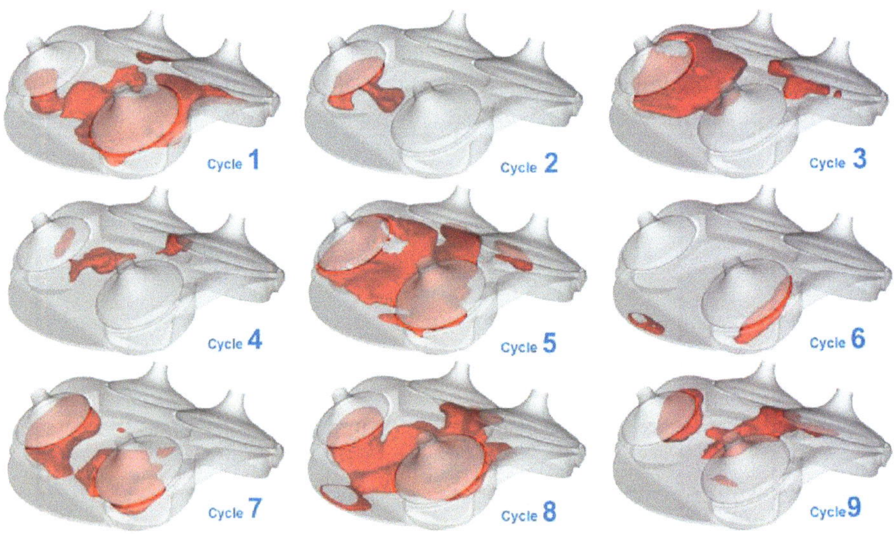

Fig. 12. Cyclic fluctuations of heat release: Distribution of 50 % heat release at 728°CA $aTDC_{fired}$

As stated before, a quantity is extracted which represents the influence of stratification on heat release. H_2O_2 is a suitable quantity. This species is produced from pre-reactions and initiates the auto-ignition at GCAI operating conditions [11].

The status of pre-reactions correlates well with the 10%, 50% and 75% fuel burned points which leads to high correlation coefficients of k > -0.92, Figure 13. As the averaged thermodynamic states of all LES cycles are equal, the only source of fluctuating H_2O_2 contents are stratification fluctuations of residual gas, air-fuel ratio and temperature. These stochastic fluctuations are primary induced by a small noise on the intake pressure.

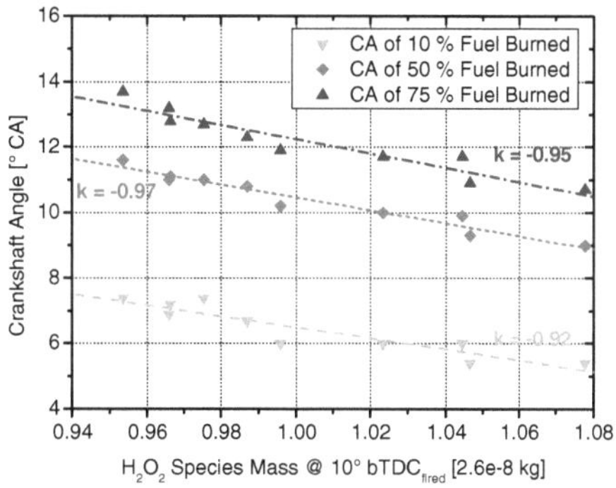

Fig. 13. Correlation of H_2O_2 in-cylinder content and combustion

Derivation of a GCAI Process Model with Reduced Calculation Time

In order to minimize the computational effort for the simulation of auto-ignition and combustion, a reduced combustion model is introduced. The reduced formulation is based on a 1D gas exchange calculation to calculate the thermodynamic state and mixture composition in the combustion chamber. Figure 14 shows the sequence of calculation with a reduced formulation to calculate combustion. The heat release x_b is used to describe the global progress of combustion.

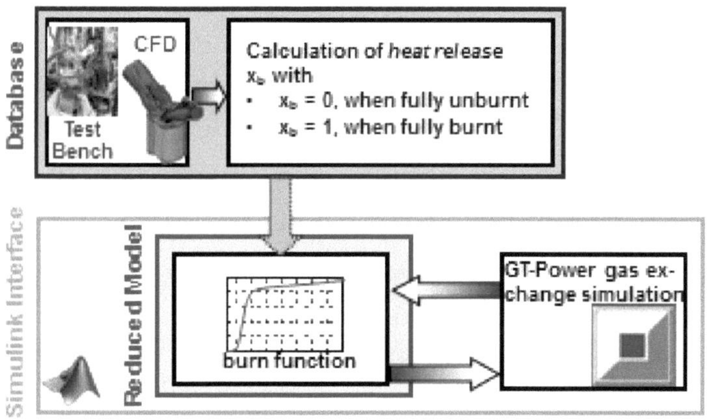

Fig. 14. Sequence of calculation with reduced approach to describe GCAI combustion

The heat release x_b is calculated by Watson approach [12] which accounts for different characteristics at early and late combustion:

$$x_b(\alpha) = \begin{cases} 0.5(1-(1-\tau^{c_{p11}})^{c_{p12}}) & \text{if } x_b(\alpha) \leq 0.5 \\ 0.5+(x_{b,max}-0.5)(1-(1-\tau^{c_{p21}})^{c_{p22}}) & \text{if } x_b(\alpha) > 0.5 \end{cases} \quad (1)$$

$$\tau = \frac{\alpha - \alpha_{xb,begin}}{\Delta\alpha} \quad (2)$$

All the parameters c_{p11}, c_{p12}, c_{p21}, c_{p22}, $x_{b,max}$, $\alpha_{xb,begin}$ and $\Delta\alpha$ depend on engine operating, like injection strategy, injection timings and valve-timing. Hence, for all conditions appropriate parameters are chosen on basis of the TPA introduced before and saved into a database.

Cyclic fluctuations are mainly represented by the variable τ. As stated before auto-correlations are primary sourced by temperature fluctuations during the cycle. Therefore, for most engine operating conditions expectation values for temperatures are also stored in the database. The calculated temperatures from the 1D gas exchange are online compared to these values. Adapted from the difference, $\alpha_{xb,begin}$ and $\Delta\alpha$ are adjusted.

In addition, the tabulated knowledge about stochastic deviations from 150 CFD simulations depending on operating conditions are incorporated to the parameters $\alpha_{xb,begin}$ and $\Delta\alpha$.

Figure 15 shows the validation with test bench measurements by applying the corresponding valve and injection timings to the process model. The abscissa describes the current cycle within the engine run. The transient behaviour of a GCAI engine with three load changes is shown. The introduced model is able to describe the transient behaviour correctly. Though, the model over predicts sudden load variations by predicting to reach a new load level within one cycle while the test bench measurements show that the combustion requires more cycles.

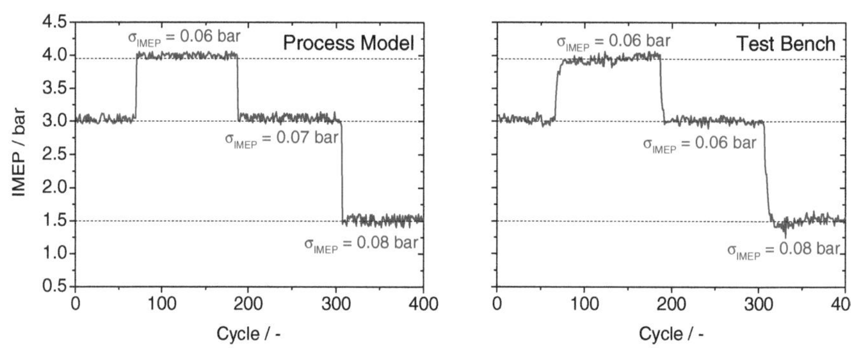

Fig. 15. Comparison of process model results and measurements of cyclic fluctuations

Cyclic fluctuations are resolved properly, so standard deviations of IMEP are in a good agreement.

An outlook to that approach is to replace 1D gas exchange modelling by a zero-dimensional air-path model. Suitable parts of this work could be used in model-based predictive controllers. Additionally a coupling of this model to a controller design environment for synchronous simulation and control is beneficial for the controller development.

Model Based Predictive Control

To control the GCAI process, a structure presented in [13] could be employed. The central element of this structure is the model-based predictive controller (MPC) which computes the actuated variables. Along with the reference value and the engine speed also an Extended Kalman-Filter, a feed-forward command governor and a block for the model parameter assignment determine inputs for this MPC.

The knowledge of this work enables to provide a controller with very good control results and minimized time for application.

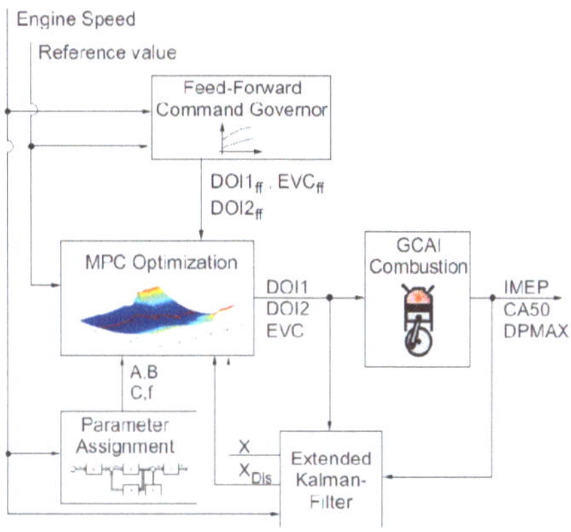

Fig. 16. Control Structure for a GCAI [13]

Conclusions

The present work reveals an approach to harness the advantages of low temperature combustion for gasoline engines. First, measures to overcome the limitations caused by an increasing maximum cylinder pressure gradient at higher loads and low stability of the process at lower loads are displayed. Thereby, especially the injection strategy including a pilot injection in combination with combustion chamber recirculation turned out to be a crucial variable to control the auto-ignition process. Higher loads are feasible through improved charge dilution. In the given example at an engine speed of $n = 2000\ min^{-1}$, the former operating range from $IMEP = 1.6 - 4.6$ bar could be extended to $IMEP = 0.6 - 6.7$ bar. Despite the comparable low general temperature level it increases with load and in consequence NO_x emissions rise. Therefore, stoichiometric GCAI-operation is advisable for higher loads to ensure a conversion in a conventional three-way catalyst.

The experimental investigations are affected by fluctuations that complicate the combustion control. Hence, a detailed process model has been implemented to provide a deeper understanding of the present correlations. 1 dimensional gas exchange simulation proves the thermodynamic state in the cylinder to be the source of auto-correlation. The auto-correlation is overlain by stochastic fluctuations. Using

LES the origin of those fluctuations can be traced back to unsteady in-cylinder state of flow and in consequence to a distinct intake pressure noise. With this knowledge a precise model including auto-correlation and stochastic fluctuation could be derived.

In order to use this development in a model-based predictive controller the computational effort had to be reduced. In a first approach, a predictive model based on 1D gas exchange simulation was shown that already reaches a high accuracy. Further improvement in terms of computational effort is expected by reducing the 1D gas exchange model by a zero dimensional one. This, finally, enables the usage of the model in a model-based predictive controller to overcome combustion fluctuations and thereby further increase the usability of low temperature combustion in gasoline engines.

Acknowledgements. The authors would like to thank the German Society of Research (DFG) for the partial financial funding of this work in the Collaborative Research Centre SFB 686 - Model-Based Control of Homogenized Low-Temperature Combustion.

References

1. Langen, P., Melcher, T., Missy, S., Schwarz, C., Schünemann, E.: Neue BMW Sechs- und Vierzylinder-Ottomotoren mit High Precision Injection und Schichtbrennverfahren. In: 28. Internationales Wiener Motorensymposium (2007)
2. Waltner, A., Lückert, P., Schaup, U., Rau, E., Kemmler, R., Weller, R.: Die Zukunftstechnologie des Ottomotors: strahlgeführte Direkteinspritzung mit Piezo-Injektor, In: 27. Internationales Wiener Motorensymposium (2006)
3. Zhao, H.: Overview of CAI/HCCI Gasoline engines, HCCI and CAI Engines for the Automotive Industry (2007)
4. Brassat, A.: Betriebsstrategien der kontrollierten Selbstzündung am aufgeladenen direkteinspritzenden Ottomotor, Dissertation, RWTH Aachen University (2013)
5. Kulzer, A., Sauer, C., Fischer, W., Karrelmeyer, R., Wintrich, T., Benninger, K.: Kontrollierte Selbstzündung beim Ottomotor CO2-Einsparpotenziale. MTZ-Motortechnische Zeitschrift 70(1), 50–57 (2009)
6. Goerg, K.-A., Morel, T.: Use of TPA (Three-Pressure Analysis) to Obtain Burn Rates and Trapped Residuals. In: GT-SUITE Conference (2005)
7. Launder, B.E., Spalding, D.B.: The numerical computation of turbulent flows. In: Computer Methods in Applied Mechanics and Engineering (1974)
8. Sone, K., Patel, N., Menon, S.: Large-eddy simulation of fuel-air mixing in an internal combustion engine, AIAA Paper (2001)
9. Smagorinsky, J.: General Circulation Experiments with the Primitive Equations. Monthly Weather Review (1963)
10. Babajimopoulos, A., Assanis, D.N., Flowers, D.L., Aceves, S.M., Hensel, R.P.: A Fully Integrated CFD and Multi-zone Model with Detailed Chemical Kinetics for the Simulation of PCCI Engines. In: 15th International Multidimensional Engine Modeling User's Group Meeting (2005)
11. Westbrook, C.K.: Chemical Kinetics of Hydrocarbon Ignition in Practical Combustion Systems. Proceedings of the Combustion Institute (2000)

12. Watson, N., Pilley, A., Marzouk, M.: A Combustion Correlation for Diesel Engine Simulation, SAE Technical Paper 800029 (1980)
13. Albin, T., Zweigel, R., Heßeler, F., Morcinkowski, B., Brassat, A., Pischinger, S., Abel, D.: Controlling GCAI (Gasoline Controlled Auto Ignition) in an Extended Operating Map. IFAC-ECOSM (2012)
14. Westbrook, C.K.: Chemical kinetics of hydrocarbon ignition in practical combustion systems. In: Sick, V., de Goey, L.P.H. (eds.) Proceedings of the Combustion Institute, vol. (28), pp. 1563–1577
15. Pearson, K.: Notes on regression and inheritance in the case of two parents. Proceedings of the Royal Society of London 58, 240–242 (1895)

Potential and Challenges of MILD Combustion Control for Gas Turbine Applications

Thivaharan Albin[1], Aline Aguiar da Franca[1,2], Emilien Varea[3], Stephan Kruse[3], Heinz Pitsch[3], and Dirk Abel[1,*]

[1] Institute of Automatic Control, RWTH Aachen University, Germany
T.Albin@irt.rwth-aachen.de
[2] CAPES Foundation, Ministry of Education of Brazil, Brasilia - DF 70040-020, Brazil
[3] Institute of Technical Combustion, RWTH Aachen University, Germany

Abstract. Moderate and Intense Low Oxygen Dilution (MILD) combustion is characterized by substantial reduction of high temperature regions and thereby reduction of thermal NO_x emissions. Beside the application in furnaces, the MILD combustion seems to be also an auspicious concept in stationary gas turbines for simultaneous reduction of NO_x and CO emissions. Nonetheless, the maintenance of MILD combustion for gas turbine relevant conditions, as the high temperature, at different operating points poses a challenge. The reason is the necessary fast mixing of recirculated burnt gases with fresh air and fuel in the combustion chamber. In this work the application of control for dealing with this task is investigated. As research approach the pulsation of the fresh gas is examined. On the one side the potential of using control for reducing the emissions level is evaluated. On the other side it is analyzed which challenges have to be solved in future concerning the control algorithm, if control shall be applied.

Keywords: Gas Turbine Combustion, Combustion Control, Moderate and Intense Low Oxygen Dilution (MILD), Flameless Oxidation (FLOX).

1 Introduction

Combustion technologies still need to be improved since regulations on pollutant emissions are becoming more and more stringent. Gas turbines require this development for achieving low emission levels combined with high combustion efficiencies. The gas turbine efficiency correlates with the turbine inlet temperature, with higher temperatures yielding higher efficiencies [1]. However, there exists

* The authors gratefully acknowledge the contribution of the Collaborative Research Center 686 Model-based control of homogenized low-temperature combustion supported by the German Research Foundation (DFG) at RWTH Aachen University, Germany, and Bielefeld University, Germany, see www.sfb686.rwth-aachen.de for details. Aline Aguiar da Franca gratefully acknowledges financial support from CAPES scholarship No. BEX 8999/13-9.

R. King (ed.), *Active Flow and Combustion Control 2014*, Notes on Numerical Fluid Mechanics and Multidisciplinary Design 127, DOI: 10.1007/978-3-319-11967-0_12

a trade-off between efficiency and emissions. For high efficiencies high temperatures are necessary, but NO_x emissions increase exponentially with temperature and linearly with residence time. Concerning CO emissions, increasing the residence time is favorable for CO oxidation to CO_2, and reducing the combustion temperature can increase carbon monoxide due to local extinction.

Combustion with high exhaust gas recirculation rates and preheated reactive gases can substantially reduce combustion peak temperatures and thereby NO_x emissions, while simultaneously showing good stability behavior, and very low noise (small temperature rise) as well as low CO emission levels. This principle is known as Moderate and Intense Low oxygen Dilution combustion (MILD) [2]. This technology has been successfully applied to furnace systems under atmospheric pressure [3], and is also an auspicious concept for reducing emissions in stationary gas turbines. The combustion conditions in gas turbines are characterized by high thermal intensity, high pressure, and high burnt gas temperatures. The higher temperatures lead to increased reactivity and, to maintain low levels of emissions, fast mixing between the reactants and the burnt gases has to be achieved. In this case, the ratio of characteristic time scales between mixing (turbulence) and chemistry (Damkoehler number) is of great importance and should be much lower than unity (well stirred reactor).

This fast mixing has to be achieved over a wide range of operating conditions, as stationary gas turbines are applied for balancing the fluctuation of energy supply provided by renewable energy sources. In the past, gas turbines were designed and optimized for one operation point. Nowadays, a high flexibility in operating conditions with low emission levels is required. In order to ensure low emissions levels for the operating conditions, control of the mixing process, the most important parameter of emissions formation, is of great interest.

Considering cold flow jets, studies have reported that jet pulsation (various amplitudes, frequency ranges, waveforms) is an effective way to enhance mixing or entrainment at the nozzle exit. Different pulsation patterns have been proposed, for example weak air excitation [4] or high frequency and amplitude excitation [5]. Parikh et al. [6] reported mixing to be enhanced with pulsation. Applying large amplitudes of pulsation at relatively low frequencies, 10 to 25 Hz, Bremhorst et al. [7] observed at $x/d = 17$, where x is the axial direction and d the nozzle exit diameter, that the level of entrainment increased by a factor of three. Vermeulen et al. [8] were able, through resonant acoustic excitation at low frequencies, to extend the effect of pulsation up to $x/d = 70$. The entrainment rate was doubled. Bremhorst et al. [9] showed the Reynolds stress to be in agreement with the mixing and entrainment enhancement, making a link with the level of turbulence. Some researchers have pointed out that the waveform of the pulsation affects the entrainment. Vermeulen et al. [8] observed sinusoidal waveforms to be efficient. Musculus [10] reported an increase in entrainment with decelerating jets, which was three times higher than that of non-pulsated jets.

For reactive flows, using a pulsating unit is not straightforward since the flame may suffer from the instabilities created by the flow pulsations. However, studies reported a better flame stabilization or combustion efficiency when properly excited [11]. Concerning the effect of pulsation on NO_x and CO emissions, Poppe et al. [12] observed a reduction by 40% of NO_x emissions through pulsation. Chao et al. [11] reported a reduction in NO_x emissions using acoustic excitations at resonant frequencies. The reduction achieved in NO_x emissions does not affect those of CO, and hence the flame stability. They argue that reduction of emissions is due to mixing improvement leading to better homogeneity in the temperature field.

For the MILD combustion, the research conducted so far concerned the passive possibilities of influencing the mixing process. As example, the influence of nozzle exit velocities and pressure on the emissions formation in MILD combustion systems was investigated by Lueckerath et al. [13]. At RWTH Aachen University experimental and simulation work has been carried out for the MILD combustion process. These studies report either the development of numerical tools for accurate simulation of MILD combustion or the characterization of the MILD combustion process (emissions, flame stabilization zones), at atmospheric or elevated pressure (see e.g. [14]). The present paper summarizes the research activities conducted at RWTH Aachen university concerning the active control of MILD combustion, which is an innovative possibility to enhance the mixing process, compared to the passive ones, i.e. changing the nozzle diameter. The aim of the present study remains twofold: 1) to quantify the potential of the control of MILD combustion concerning reduction of NO_x and CO emissions at high combustion temperatures, with the intention to solve the trade-off between emissions and efficiency, 2) to analyze the arising challenges when control shall be applied for an application in a gas turbine concerning the control algorithm. In the following sections, first the experimental apparatus is described, which includes the chamber design and the measurement techniques for emissions (CO and NO_x), OH* chemiluminescence and Particle Image Velocimetry (PIV). In the third section, the concept of pulsating unit used, is presented and analyzed with respect to limitations of the actuator. In the forth section, the effect of pulsations on MILD combustion is evaluated. Both the influence on the emissions level as well as the influence on the flame topology is illustrated. In the fifth section the challenges concerning the control algorithm is pointed out, by studying the system dynamical properties.

2 Experimental Setup

For the studies, a MILD combustion chamber based on a reverse flow configuration is used. This design ensures high recirculation rates of burnt gases. The experimental setup consists of three parts: ignition chamber, main combustion chamber, and recuperator, as shown in Fig. 1. The process of preheating the combustion chamber occurs in the ignition chamber located at the top of the setup.

When the temperature in the main combustion chamber reaches about 1100 K, air and fuel supplies are switched from the ignition to the main chamber. Fuel and air enter the main combustion chamber through two concentric nozzles located at the bottom of the chamber, resulting in a partially premixed stream,

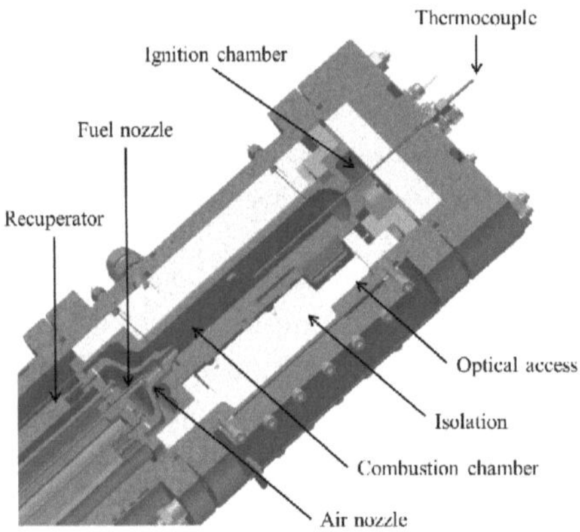

Fig. 1. Schematic of the experimental setup consisting of recuperator, combustion chamber, and ignition chamber

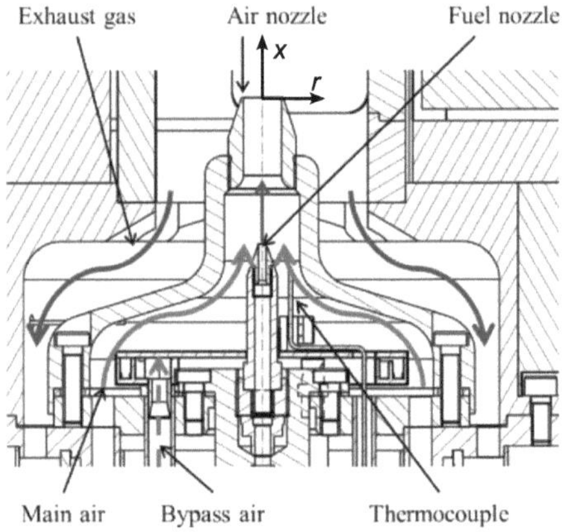

Fig. 2. Nozzle design highlighting the reverse flow configuration

as shown in Fig. 2. The diameter of the fuel nozzle is $d_{fuel} = 2$ mm. The air nozzle diameter is $d_{air} = 10$ mm. The air nozzle exit is 35 mm downstream of the fuel nozzle exit. The main combustion chamber has a square cross-section with a side length of 58 mm and is 200 mm long. Spatially adjustable quartz glass windows (40×50 mm^2) guarantee an optical access along the centerline of the chamber.

The main combustion chamber operates in a reverse flow configuration, ensuring good mixing between burnt gases, fresh air, and fuel. For deflecting the gas streams, the top end of the chamber has a spherical shape. The exhaust gas exits the combustion chamber in the nozzle plane and flows through the outer part of the recuperator to the exhaust tube. The incoming air enters the recuperator in the opposite direction and flows through the inner part of the recuperator, as shown in Fig. 2. To control the inlet temperature, bypass pipes are located in the center of the recuperator. The amount of the non-preheated bypass air which is mixed with the hot air of the recuperator upstream of the air nozzle allows to set the air inlet temperatures with good accuracy.

The setup is equipped with thermocouples of type S and K, located at different positions in the chamber. This allows to check the air inlet temperature, the wall temperatures, and the exhaust gas temperature as well as the reacting mixture temperatures at the center of the chamber.

The composition of the exhaust gases (NO_x, CO and O_2) is measured with a sensor (potentiometric cells) in the exhaust tube. The O_2 concentration is essential to correlate the emissions to 15% O_2 in the exhaust gas according to, [15] with $X_{15,i} = X_i \frac{20.9-15}{20.9-X_{O_2}}$. Hereby X_i is the measured mole fraction of species i, $X_{15,i}$ is the normalized species mole fraction at 15%, and X_{O_2} is the measured O_2 mole fraction in percent in the exhaust gas.

Since the OH* radical is a marker of the reaction region under atmospheric pressure [16], the flame structure is investigated with OH* chemiluminescence. An ICCD camera (pco. dicam) equipped with a Nikkor UV-lens (f/4.5, f = 105 mm) and a band pass filter (CWL = 310 nm, FWHM = 10 nm) is adjusted to detect OH* radicals in an area ranging from $6.5 < x/d < 9.0$ downstream of the nozzle exit. The position is measured in terms of the axes as shown in Fig. 2, $x = 0$ mm and $r = 0$ mm correspond to the center of the air nozzle exit. The OH* chemiluminescence images are averaged over 500 single shot images and normalized to the highest intensity.

To investigate the flow topology, Particle Image Velocimetry (PIV) is performed under cold flow conditions. The experimental setup consists of a double pulse Nd:YAG laser (Quantel EverGreen 200) and a double imaging camera (pco. sensicam). The laser sheet is formed by associating one cylindrical lens with a focal length fl = -25 mm and one spherical lens with a focal length fl = 250 mm. A 105-mm macro-Nikkor lens (f/2.8) is mounted to the camera, allowing a field of view of 43.5×32.6 mm^2. The magnification ratio is of 14.71 pixel/mm. The flow is seeded with olive oil droplets. The inlet flow conditions are set up such that the Reynolds numbers in cold flows correspond to those for reactive flows. The velocity calculations are performed with the LaVision software (DaVis 7.2) by standard FFT-correlation with a multi-pass scheme consisting of one pass with interrogation window size of 32×32 pixel^2 followed by one pass with 16×16 pixel^2 and overlaps of 50%. The mean velocity fields are averaged over 500 and 5000 single shot vector fields for non-pulsated and pulsated flows, respectively.

3 Actuation Strategy

3.1 The Pulsating Unit

In the following section, the setup of the pulsating unit is introduced, which is used as actuator for controlling the MILD combustion. Fig. 3 shows a schematic diagram containing the different components. Mass Flow Controllers (MFCs) for air and fuel allow a precise adjustment of the equivalence ratio. In order to avoid any back coupling on the MFCs during valve operation, a buffer tank is integrated between the MFCs and the control valve.

The control valve is a servo valve characterized by very low response times and integrated closed-loop control for the opening position. Due to these properties, it is possible to set the opening position in a precise and fast manner. This allows setting a time characteristic for the valve opening with high degree of freedom using an electronic control unit (ECU). With this overall setup, it is possible to pulse the mass flow entering the combustion chamber. Since the air mass flow rate, and thus also the effect of air pulsation is much higher than that of the fuel, in this study, only the air flow is pulsated. As ECU, a dSpace MicroAutoBox 2 is set up. For obtaining minimal sampling times, the FPGA-Board from the ECU is used to actuate the valve and sample the measurements. The sampling time is set to 1 ms.

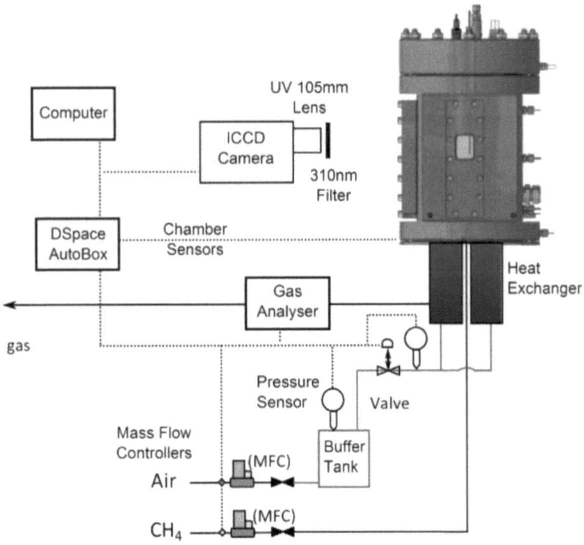

Fig. 3. Pulsating unit integrated in the MILD combustion setup

The pulsation can be modified by changing the trajectory of valve opening over time, which is even possible at runtime. For reducing the degrees of freedom, only three parameters of the pulsation specifics have been varied: waveform, frequency f, and amplitude. The waveform characterizes the specific profile of valve opening. Therefore, triangular, pulse and sinusoidal shapes are investigated. In addition to the waveform, the frequency of the pulsation can be changed. The frequency of the pressure signal can directly be changed by modifying the frequency of the actuated signal. For variation of the pressure pulsation amplitudes (without changing the frequency), the pulsation ratio PR is used. Hereby PR is defined as the ratio between the amplitude of the valve opening which is pulsated and the maximum possible amplitude w.r.t. to the valve opening.

To quantitatively check the functionality of the pulsating system, preliminary tests are done on a pulsated cold air flow. Pressure fluctuations at the nozzle exit are recorded by a piezoelectric pressure transducer (Kistler 4045A2). The pressure pulsations caused by the pulsating unit are shown in Fig. 4. The resulting pressure signal is depicted for a sinusoidal shape pulsation with $PR = 0.75$ and $PR = 0.25$, along with the actuated signal for the valve. It can be noted that pressure pulsations at the nozzle exit have the same sinusoidal shape as the valve opening signal. Additionally, the amplitude can be modified by adjusting the PR value.

Fig. 5 shows the influence of the input frequency on the pressure at the nozzle exit. The frequency is varied from $f = 1$ Hz to $f = 4$ Hz. The frequency of the pressure pulsation can effectively be adjusted by changing the frequency of the actuated signal. Additionally, it is observed that increasing frequency lowers the amplitude due to damping effects. This result shows that the pressure

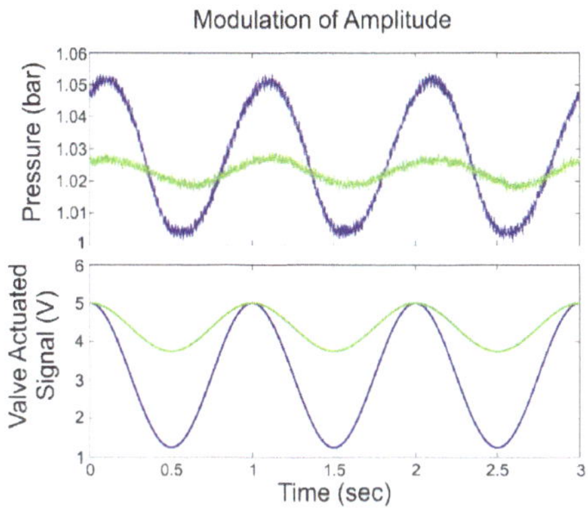

Fig. 4. Amplitude modulation of pulsated cold flow. $PR = 0.75$ (blue) vs. $PR = 0.25$ (green). Frequency $f = 1$ Hz. TOP: flow response, bottom: actuated signal

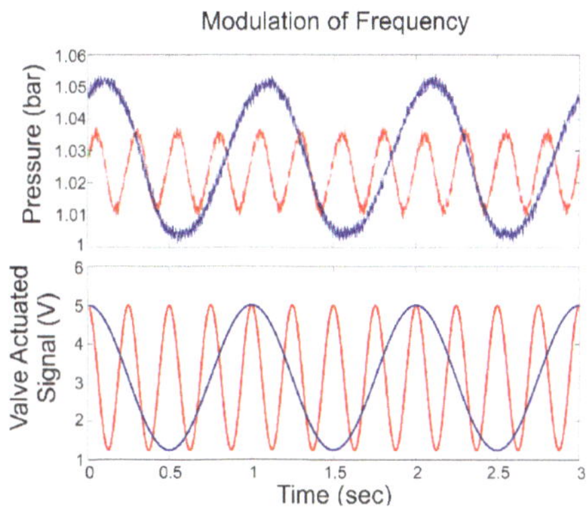

Fig. 5. Frequency modulation of pulsated cold flow. $f = 1$ Hz (blue) vs. $f = 4$ Hz (red). $PR = 0.80$. TOP: flow response, bottom: actuated signal.

pulsation amplitude achievable with this setup decreases with increasing frequency. At low frequencies, maximum pressure fluctuations of +/- 0.03 bar can be achieved. The smallest pressure fluctuation that could be recorded was limited to 1 mbar, which was measured for a frequency of 150 Hz. Thus, the actuation was applied until the frequency of 150 Hz.

Fig. 6 shows the distribution of the mean radial velocity with and without pulsation obtained by PIV measurements. The non-pulsated flow is essentially symmetric around $r/d = 0$. The pulsated flow is not fully symmetric at the nozzle exit. This results from the position of the actuator that is mounted perpendicular to the recuperator, implying a non-symmetric pulsated flow inside the recuperator, and hence at the nozzle exit. Without pulsation, the jet starts its expansion at $x/d \approx 0{,}60$, indicated to the transition to non-zero velocity at this point, zone 1 in Fig. 6. Coherent structures from Kelvin-Helmholtz instabilities are highlighted in black dashed circles. When pulsation is turned on, the jet expansion starts closer to the nozzle exit (zone 1'). In the core jet, downstream of the nozzle exit, a large recirculation zone can be observed (zone 2'), as well as a strong expansion zone (zone 3'). The existence of these different structures is the result of the pulsation. When the valve is fully opened, the jet expands and has a similar behavior as a free jet. However, when the valve starts closing, the deceleration induces a pressure drop along the centerline and a large amount of surrounding gases is recirculated in the core region. This phenomenon enhances the level of turbulence, hence the mixing, without changing the average flow nor the nozzle diameter. To sum up the PIV measurements show that for cold flow, the flow is strongly influenced by the pulsation. The radial velocity maps are completely different showing an important radial mass and momentum transfer with pulsation. Beside the radial velocity, also the axial velocity was investigated, but the mean axial velocity is similar for the pulsed and non-pulsed cases.

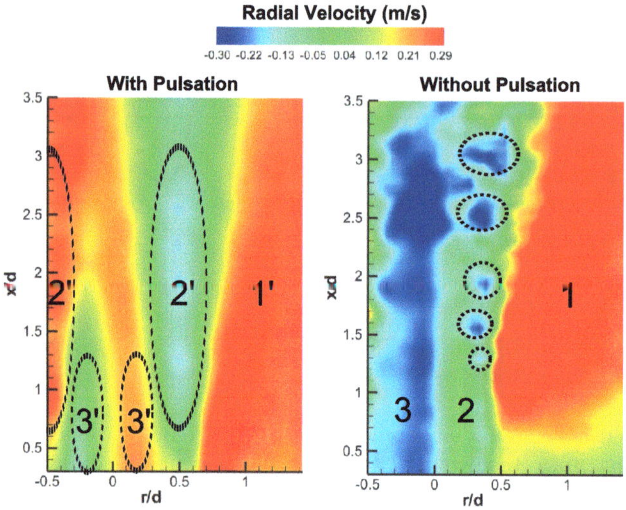

Fig. 6. Mean radial velocity for pulsated and non-pulsated flow. Cold flow measurements for the corresponding condition of $\phi = 0.67$

4 Influence of Actuation on MILD Combustion

In the following section the effect of pulsation on MILD combustion is evaluated concerning NO_x and CO emissions at stationary conditions. Finally, OH* chemiluminescence results describe the flame behavior.

4.1 Effect of Pulsation on Emissions

As reference values, the emissions are recorded for the non-pulsated case as a function of the equivalence ratio ϕ, see Fig. 7. The experiments are carried out such that the fuel flow is kept constant (8 nl/min) and the air flow rate is adjusted to match the required equivalence ratio. The procedure carried out to estimate the uncertainties is based on the average and the corresponding RMS (σ) of five experiments, performed at different days. The depicted emissions correspond to the steady state values - the emissions are averaged over a time interval of 10 minutes. For non-pulsated conditions, the results indicate that for equivalence ratios higher than $\phi = 0.6$, the NO_x emissions (black square dot line) are above 40 ppm. With decreasing equivalence ratio, *i.e.* low combustion temperature, the NO_x emissions decrease as well, until ultra-low emissions are achieved for lean conditions with $\phi < 0.50$. Concerning CO emissions (blue triangle dot line), the opposite trends are observed. For equivalence ratios $\phi > 0.40$, no traces of CO are found as they are completely oxidized to CO_2. However, approaching the extinction limit at $\phi < 0.30$, a rapid increase of CO emissions is measured. The two emissions curves indicate a very low NO_x and CO emissions region ranging from $\phi = 0.30 - 0.50$.

Due to the high temperatures required for better gas turbine efficiencies, high equivalence ratios ($\phi > 0.5$) are of great interest. However, the NO_x emissions are drastically increased in this region. As a consequence, this work studies the capability of pulsation to reduce NO_x emissions for equivalence ratios ranging from $0.57 < \phi < 0.80$.

By using design of experiments (DOE) different types of waveform, namely sinusoidal, triangle, and pulse, have been tested for the most appropriate actuation. In the preliminary work the triangular waveform (with same slope for rise and fall) has approved to be well suited. The analysis shows that both pulsations with pulse and sinus lead to a decrease of NO_x emissions, while maintaining low CO emissions levels, but the reduction is highest for the triangle waveform. This is in agreement with Vermeulen et al. ([8]), where the waveform of the pulsation was identified as an important factor for flow control. According to Musculus et al. ([10]), a decelerating jet would increase the recirculation ratio, and hence, the mixing process, which could be found for a triangle waveform. Considering this triangular waveform, experimental investigations are carried out for equivalence ratios of $\phi = 0.80$, 0.67 and 0.57 with different frequencies and pulsation ratios. The pressure is set to 1 bar and the inlet temperature is 873 K. Results for MILD combustion with pulsation are shown in Fig. 7. Under specific conditions of pulsation, described hereafter, emissions can be lowered up to 55% for the operating point of $\phi = 0.57$. For all investigated operation points the CO emissions stay below the detection limit (2 ppm). For higher equivalence ratios, $\phi = 0.80$ and $\phi = 0.67$, NO_x emissions are decreased by 13% and 28% respectively. When the pulsating unit is switched on, uncertainty increases up to 7 ppm, which is higher than that observed for non-pulsated flows.

An important finding from control side of view, is that the optimal actuation concerning pulsating ratio and frequency is very dependent on the recent operating point. For the three equivalence ratios (depicted in Fig. 7), the optimal PR and frequencies are listed in Tab. 1. The corresponding Reynolds numbers (Re) are provided as well. One can observe the strong nonlinear behavior, especially in regards to the frequency. An explanation for this operating point dependency can be found in the Strouhal number, which changes in dependence of the equivalence ratio and thus also the optimal actuation frequency. For gas turbine applications, combustion processes occur at high pressure, up to 15-30 bars. For these conditions, the time scale of chemistry is decreased so that the Damkoehler number increases and get larger than unity. Hence, the regime gets different from the well-stirred reactor regime. However, as shown in this study, pulsation can increase the mixing process by decreasing the turbulent time scale. Hence, we can expect that even at high pressure, MILD combustion can be enhanced due to pulsation.

4.2 Effect of Pulsation on Flame Topology

To clarify the effect of the pulsation on flame topology, the results of OH* measurements for one equivalence ratio are presented in Fig. 8. The images are averaged over 500 single shots and normalized by the maximum intensity. OH*

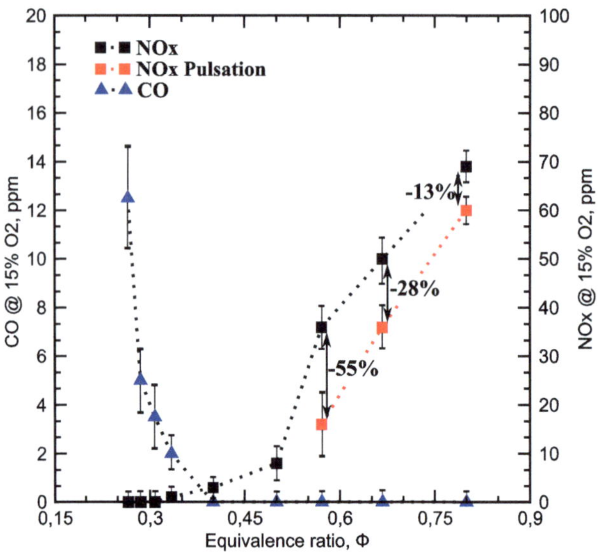

Fig. 7. Dependence of NO_x and CO emissions on ϕ at T_{in} = 873 K, and p = 1 bar,with and without pulsation

Table 1. Pulsation ratio (PR), Frequency and Reynolds number showing the lowest emissions with pulsation at different equivalent ratios

Equivalence Ratio	PR	Frequency (Hz)	Re
0.80	0.675	1.90	6001
0.67	0.750	1.07	7200
0.57	0.700	116	8400

images are recorded at the flame position ranging from $6.5 < x/d < 9.0$ downstream of the nozzle exit. OH* measurements are compared at the same location for three cases: a) Without pulsation b) with pulsation frequency of f=116 Hz, where NO_x is reduced c) with pulsation frequency of f=4 Hz, where NO_x is increased. According to the OH* measurements, Fig. 8, the flame location is influenced by the pulsation, which affects the NO_x emissions level. Depending on the pulsation conditions, the effect can increase or decrease the emissions level. When pulsation leads to a decrease in NO_x emissions, the reaction zone is more homogeneous, which is indicated by a decrease in the reaction intensity. The flame location moves downstream, which might be caused by an enhanced entrainment of burnt gases. In this case, the combustion is closer to a homogeneous combustor. If the pulsation conditions are chosen such that emissions increase, the reaction zone is less homogenous and the reaction intensity increases. The flame location moves upstream, which might be caused by an impaired entrainment of burnt gases.

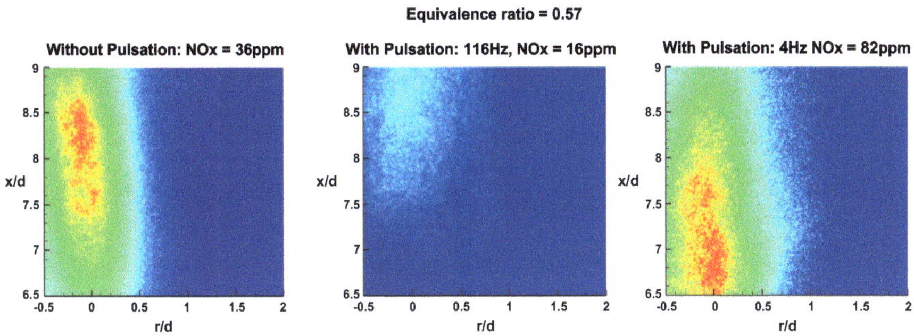

Fig. 8. OH* measurements for given conditions

5 Challenges for MILD Combustion Control

The last section has shown that the usage of pulsation can decrease the NO_x emissions in steady state by up to 55% while CO emissions are kept low. In this section the arising challenges concerning the control algorithm are discussed. For this reason, different system dynamical properties are investigated.

5.1 Transient Behavior

For investigating the transient behavior a step is applied to the actuated signal. Starting with no actuation (valve fully open) a step-wise change is applied towards a fixed frequency $f = 1.07$ Hz and triangle waveform with $PR = 0.75$. The step response can be seen in Fig. 9. After activation of the pulsation, an excessive overshoot in NO_x emissions can observed. Afterwards the NO_x emissions decrease in steady state to a level with around 45% less NO_x emissions compared to the steady state value without pulsation. Approximately 100 seconds are necessary for the NO_x emissions to reach steady state. The CO emissions remain constant at a low level during the whole experiment.

The system behavior is comparable to that of a non-minimum phase system. The reason for this dynamics is that directly after actuation, the equivalence ratio increases first as the buffer tank pressure has a slower dynamics than the valve. Afterwards the equivalence ratio and in consequence also the NO_x emissions decrease again.

A comparable transient behavior is present for the case of load variations. For gas turbines with high demands on the dynamic behavior it is beneficial to decrease the emissions arising in the transients. This means that the excessive overshoots have to be avoided and also the steady state should be reached within less time. A common way to avoid the non-minimum phase behavior is to use feedforward control. One way to tackle this issue, is to activate the actuation gradually. As consequence, the time scales of the pressure increase in the buffer tank and the time scale of the valve opening dynamics are equalized. The resulting behavior can be seen in Fig. 10 for $\phi = 0.67$. As in Fig. 9, the same

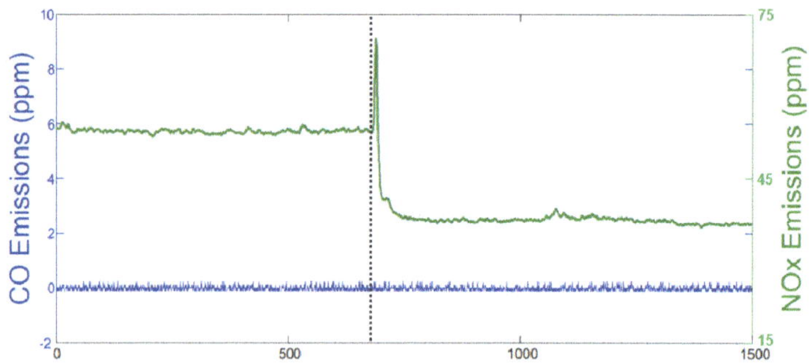

Fig. 9. Transient emissions behavior of MILD combustion when activating pulsation for $\phi = 0.67$, $PR = 0.75$, and $f = 1.07$ Hz. Stepwise activation of pulsation, start of pulsation is indicated by the vertical dashed line

frequency $f = 1.07$ Hz and a triangle waveform is used. But now the pulsating ratio is slowly increased until $PR = 0.75$ is reached as final value. One can observe that the excessive NO_x emissions are avoided and the time until steady state is drastically reduced.

5.2 Mode Switching between MILD and Conventional Combustion

For compensation of the non-minimum phase behavior in the transients feedforward control can be used. Nonetheless, other process characteristics make feedback control unavoidable, which is shown in the following. In Fig. 11, an operating point is chosen ($\phi = 0.57$, $PR = 0.75$) and the dependence of NO_x emissions on pulsation frequency variations for this specific operating point is studied. Starting with a frequency $f = 115$ Hz with NO_x emissions of 16 ppm, the frequency is increased stepwise and the steady state emissions levels are recorded. The solid line in Fig. 11 is followed until reaching the point of $f = 125$ Hz, where emissions start to strongly increase. Afterwards, the pulsation frequency is decreased again, but instead of decreasing again, the emissions are still increasing. By decreasing the frequency until the starting point of $f = 115$ Hz, a NO_x emissions level of about 34 ppm is reached. This level of emissions is similar to that observed for the operating point without pulsation. This result indicates that for a given operating condition and pulsation pattern, at least two stable modes exist, and lead to two different level of emissions. One corresponds to the MILD combustion and one to the conventional combustion. Thus, just actuating with one frequency in a feedforward manner, i.e. depending on the operating point, is not enough for maintaining a stable MILD combustion. Instead, the way the actuation has been carried out and also disturbance variables have to be accounted for in the control algorithm. For this reason a closed-loop control is necessary in order to (in best case) directly avoid the switching into the conventional combustion mode or if necessary to steer the process to the MILD

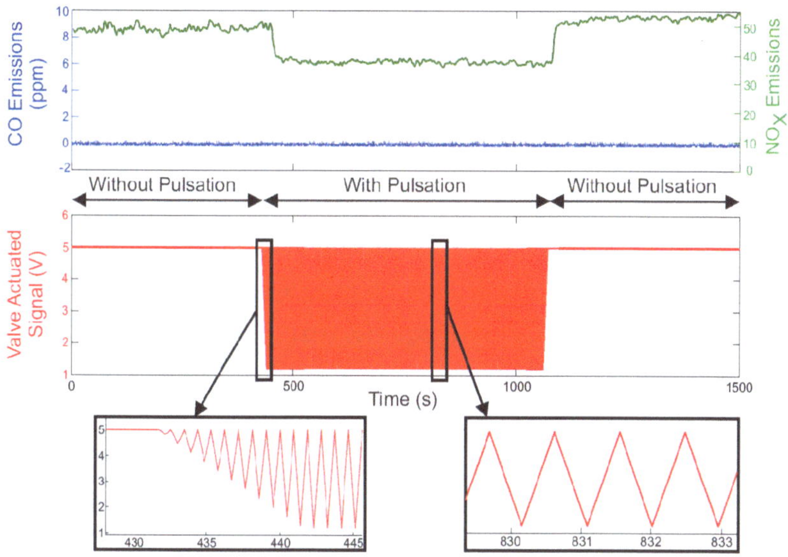

Fig. 10. Transient emissions behavior of MILD combustion with gradual activation of pulsation for $\phi = 0.67$, $PR = 0.75$ and, $f = 1.07$ Hz

combustion. Thus, all in all, feedforward control is not sufficient and instead closed-loop control is necessary. The control algorithm has to be aware of the different modes of combustion and needs to be able to perform mode switches. For this reason it seems reasonable to evaluate model-based predictive control (MPC) methods, see [17]. For the MILD combustion especially the hybrid MPC seems promising. In hybrid MPC a model can be incorporated where discrete switches between different modes can be taken account of. The process model has to reproduce the mode switch in dependence of the actuation, arising disturbance variables and load transients. The aforementioned compensation of the transient behavior with non-minimum phase behavior can be taken care of in the MPC framework as well.

5.3 Incorporation of Different Operating Points

So far, it was shown that closed-loop control is necessary for maintaining stable MILD combustion. In the following another challenge for the control algorithm is depicted. Depending on the frequency and pulsation ratio different emissions levels are present, even in the case of stable MILD combustion. As depicted in the previous section the most effective pulsation ratio and frequency depend on the recent operating point. Beside the equivalence ratio also other parameters as fuel type and wall temperature have an influence on the optimal actuation parameter. In consequence, the process behavior can be improved, if the control algorithm actuates with the most effective parameters, depending on the

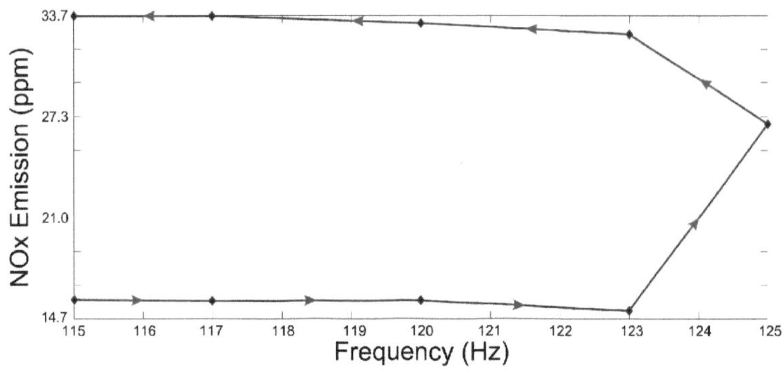

Fig. 11. NO_x emissions as a function of frequency for $\phi = 0.57$, $PR = 0.75$

recent surrounding conditions. In consequence the overall optimal control strategy would be capable of maintaining stable MILD combustion and is on top also capable of maintaining lowest emissions levels, by actuating with effective parameters. The latter task can be tackled with extremum seeking control strategies. The extremum seeking control has to be combined with the stabilizing controller. A possibility would be to further research on multiscale control algorithms, where on one time scale the stabilization is achieved and on the other time scale the extremum seeking.

6 Conclusion and Outlook

In this paper the improvement of MILD combustion for gas turbine applications concerning the emissions level by means of control is investigated. For the experiments a reverse flow configuration is used, where a new pulsating unit is introduced. This pulsating unit offers the possibility to modify the flow rate conditions at the nozzle exit. The use of the pulsating unit allows for reductions of the NO_x emissions by up to 55% while keeping the CO emissions at a low level. OH* measurements have been conducted, that show that the pulsation of the fresh air has an effective influence on the mixing process.

Based on the experiments, to evaluate the potential of emissions reduction by control of MILD combustion, further investigations have been carried out to evaluate the open challenges concerning the control algorithm. For this reason several crucial system dynamical properties are identified. The most important ones are the switching of modes (MILD combustion vs. conventional combustion) in transient operation, non-minimum phase behavior and the strong dependence of the emissions level on the actuation frequency and the surrounding conditions as the equivalence ratio.

The future work will focus on the research of closed-loop control algorithms which ensure the stable operation of the MILD combustion and simultaneous reduction of the emissions by finding the optimal actuation parameters. For achieving this task a multiscale control approach seems promising. In one time scale the stabilization could be achieved by a hybrid model predictive controller and on another time scale the emissions reduction could be achieved by an extremum seeking approach.

References

1. Lieuwen, T., Chang, M., Amato, A.: Stationary gas turbine combustion: Technology needs and policy considerations. Journal Combustion and Flame 160, 1311–1314 (2013)
2. Cavaliere, A., De Joannon, M.: Mild Combustion. Journal Progress in Energy and Combustion Science 30, 329–366 (2004)
3. Levy, Y., Sherbaum, V.: Basic thermodynamics of FLOXCOM, the low-NO_x gas turbine adiabatic combustor. Journal Applied Thermal Energy 24, 1593–1605 (2004)
4. Crow, S.C., Champagne, F.H.: Orderly structure in jet turbulence. Journal of Fluid Mechanics 48, 54–591 (1971)
5. Binder, G., Favre-Marinet, M., Kueny, J., Craya, A., Laty, R.: Jets Instationnaires. Journal Laboratoire de Mecanique des Fluides. University de Grenoble, France (1971)
6. Parikh, P., Moffat, R.: Mixing Improvement in a Resonantly Pulsed, Confined Jet. Fluid Mechanics of Combustion Systems, pp. 25–256 (1981)
7. Bremhorst, K., Harch, W.H.: Near Field Velocity Measurements in a Fully Pulsed Subsonic Air Jet. Journal Turbulent Shear Flows 1, 3–54 (1979)
8. Vermeulen, P., Ramesh, V., Miller, D., Yagle, P., Bender, E.: Ejector Pumping Effectiveness Through Pulsing Primary Flow. Journa. In: ASME Turbo Expo 2002, pp. GT-2002-30007 (2002)
9. Bremhorst, K., Hollis, P.G.: Velocity Field of an Axisymmetric Pulsed, Subsonic Air Jet. Journal AIAA 28, 2043–2049 (1990)
10. Musculus, M.P.: Entrainment waves in decelerating transient turbulent jets. Journal of Fluid Mechanics 638, 117–140 (2009)
11. Chao, Y.C., Yuan, T., Jong, Y.C.: Measurments of the stabilization zone of a lifted jet flame under acoustic excitation. Journal Experiments in Fluids 17, 381–389 (1994)
12. Poppe, C., Sivasegram, S., Whitelaw, J.: Control of NO_x Emissions in Confined by Oscillations. Journal Combustion and Flame 113, 13–26 (1998)
13. Lueckerath, R., Meier, W., Aigner, M.: FLOX®Combustion at High Pressure With Different Fuel Compositions. Journal Eng. Gas Turbines Power 130 (2008)
14. Kruse, S., Kerschgens, B., Berger, L., Varea, E., Pitsch, H.: Experimental and Numerical Study of Flameless Oxidation for Gas Turbine Applications. In: Proceedings of the European Combustion Meeting (2013)
15. Baukal, C.E.: The John Zink Combustion Handbook, p. 190. CRC Press (2001)
16. Dandy, D.S., Vosen, S.R.: Numerical and Experimental Studies of Hydroxyl Radical Chemiluminescence in Methane-Air Flames. Journal Combustion Science and Technology 82, 131–150 (1992)
17. Bemporad, A.: Model Predicitve Control Design: New Trends and Tools. In: Proceedings of the 45th IEEE Conference on Decision and Control, pp. 6678–6683 (2006)

Towards a Compressible Reactive Multiscale Approach Based on One-Dimensional Turbulence

Zoltan Jozefik[1], Alan R. Kerstein[2], and Heiko Schmidt[1]

[1] Brandenburg Technical University Cottbus-Senftenberg, Germany
[2] Consultant, Danville, CA, USA

Abstract. Due to its huge complexity, progress in understanding and prediction of turbulent combustion is extremely challenging. In principle, progress is possible without improved understanding through direct numerical solution (DNS) of the exact governing equations, but the wide range of spatial and temporal scales often renders it unaffordable, so coarse-grained 3D numerical simulations with subgrid parameterization of the unresolved scales are often used. This is especially problematic for multi-physics regimes such as reacting flows because much of the complexity is thus relegated to the unresolved small scales. One-Dimensional Turbulence (ODT) is an alternative stochastic model for turbulent flow simulation. It operates on a 1D spatial domain via time advancing individual flow realizations rather than ensemble-averaged quantities. The lack of spatial and temporal filtering on this 1D domain enables a physically sound multiscale treatment which is especially useful for combustion applications where, e.g., sharp interfaces or small chemical time scales have to be resolved. Lignell et al. recently introduced an efficient ODT implementation using an adaptive mesh. As all existing ODT versions it operates in the incompressible regime and thus cannot handle compressibility effects and their interactions with turbulence and chemistry which complicate the physical picture even further. In this paper we make a first step toward an extension of the ODT methodology towards an efficient compressible implementation. The necessary algorithmic changes are highlighted and preliminary results for a standard non-reactive shock tube problem as well as for a turbulent reactive case illustrate the potential of the extended approach.

1 Introduction

Predictive methods based on fundamental principles to model the turbulence chemistry interactions are important in turbulent reacting flow simulations to improve combustion efficiency and to reduce emissions. The existence of a wide range of length and time scales in high Reynolds number flows representative of practical applications, and the number of chemical species involved in combustion of hydrocarbon fuels, makes Direct Numerical Simulations (DNS) computationally intractable [17].

A key requirement for robust turbulent combustion modeling is that the model must be able to access a sufficient portion of the chemical-state manifold [18].

R. King (ed.), *Active Flow and Combustion Control 2014*, Notes on Numerical Fluid Mechanics and Multidisciplinary Design 127, DOI: 10.1007/978-3-319-11967-0_13

PDF models are advantageous in this regard, but are subject to significant limitations because they do not resolve flame structure. Flamelet models provide such resolution, but they rely on low-dimensional chemical manifolds. Thus, neither of these leading approaches to turbulent combustion modeling is fully satisfactory.

ODT resolves flame structure without compromising chemical-state accessibility, and achieves major cost reduction relative to DNS through reduced spatial dimensionality. ODT is a fully resolved, unsteady stochastic simulation model that emulates the Navier-Stokes turbulence. ODT has two key features. First, the properties of the flow reside on a one-dimensional domain. This 1D formulation allows full resolution of the interaction between large scales and molecular transport scales within computationally affordable simulations. Second, because vortical overturns cannot occur on a 1D domain, turbulent advection is represented using mapping events whose occurrences are governed by a random process. Unlike the Reynolds-averaged Navier-Stokes (RANS) model and large-eddy simulation (LES), which model the small scale phenomena and retain the 3D representation of the flow, ODT resolves all the scales of motion but models 3D turbulence. Hence ODT cannot capture geometrical effects and coherent flow structures, other than the so-called eddy events of ODT. In ODT, velocity components are transported and are used to determine the eddy frequency and eddy-size distribution, thereby providing a phenomenologically sound basis for driving turbulence.

As a stand-alone model, ODT has been successfully used to simulate homogeneous turbulent non-reacting [12,1,13,25,24,26,26] and reacting flows [3,21,10,9,16]. However, for stand-alone modeling of turbulent flows using ODT, one must define the dominant direction of mean property variation. For complex flows which may not have a single dominant direction, ODT has been used as a sub-grid scale model in both RANS [21,22] and LES [2] to provide closure for reacting scalars in combustion. An alternative multi-dimensional approach called ODTLES is discussed in [4,7].

To date, ODT has been formulated in most cases for zero Mach number, with constant pressure assumed, except when hydrostatic equilibrium is invoked in some applications to buoyant stratified flow. A variant of the existing ODT model in an Eulerian reference frame with compressible formulation was developed by Punati et al. [19,20] and applied to non-reacting and reacting jets. Differences between the formulation used in [19,20] and ours will be discussed in detail elsewhere. The goal of this manuscript is to extend the scope of the Lagrangian formulation of ODT, especially for reacting flows, by introducing a compressible formulation here.

This paper is organized as follows. Section 2.1 summarizes the numerical model, starting from a zero-Mach-number formulation and indicating the changes needed to introduce finite-Mach-number effects. Section 2.2 gives a short overview of ODT. For further depth on ODT, the reader is referred to [5,12,14]. Section 2.3 discusses the compressible formulation and section 2.4 introduces the model representation of the Darrieus-Landau instability caused by unsteady dilatational flow. Section 3 presents non-reacting shock-tube results with a comparison to the analytic solution and autoignition at constant volume under turbulent conditions to illustrate the potential of the developed method.

2 Computational Set Up

2.1 Numerical Model

The starting point for compressible ODT is the set of variable density zero-Mach-number equations in one spatial dimension in a Lagrangian frame work on an adaptive grid. In all equations, x is the ODT line direction. In what follows, the terms compressible and incompressible refer to finite-Mach-number and zero-Mach-number formulations respectively. Owing to constant-pressure dilatation caused by exothermic chemical reactions, incompressible does not mean solenoidal in this context.

Following the formulation in [15], we begin by writing the continuity equation in integral form for a control volume V that encloses the mixture mass. In Lagrangian formulation, the system boundary moves with the mass-average velocity so that no mass crosses the control volume boundary. Since there is no mass source term, the Reynolds transport theorem is written for the continuity equation as

$$\frac{d}{dt}\int_V \rho dV = 0, \tag{1}$$

where ρ is the density. For uniform properties inside the control volume, and in 1D, the equation reduces to

$$\frac{d}{dt}(\rho dx) = 0, \quad \Rightarrow \quad \rho dx = \text{constant}, \tag{2}$$

where dx is a Lagrangian interval. This shows, that during a time advancement of the partial differential equations, the total mass in a given grid cell is constant. The balance equations for momentum, species mass fractions, and enthalpy are

$$\frac{d}{dt}(u_i) = \frac{1}{\rho}\frac{\partial}{\partial x}\left(\mu \frac{\partial u_i}{\partial x}\right), \tag{3}$$

$$\frac{d}{dt}(Y_s) = \frac{\dot{\omega}_s}{\rho} - \frac{1}{\rho}\frac{\partial j_s}{\partial x}, \tag{4}$$

$$\frac{d}{dt}(h) = -\frac{1}{\rho}\frac{\partial q}{\partial x} + \frac{1}{\rho}\frac{\partial p}{\partial t}, \tag{5}$$

with $s = 1, ..., n_s$ and n_s is the number of different species in the gas mixture. Here, μ is the dynamic viscosity of the mixture, u_i with $i \in \{1, 2, 3\}$ are the three ODT velocity components, Y_s the mass fraction of species s, $\dot{\omega}_s$ is the chemical source term of species s, h is the enthalpy of the mixture, and p the pressure. j_s is the species diffusive flux given by

$$j_s = -\rho D_s \left(\frac{\partial Y_s}{\partial x} + \frac{Y_s}{\overline{M}} \frac{\partial \overline{M}}{\partial x} \right), \tag{6}$$

where D_s is the diffusion coefficient of species s and $\overline{M}$ is the mean molecular weight. q is the heat flux given by

$$q = -\lambda \frac{\partial T}{\partial x} + \sum_{s=1}^{n_s} h_s j_s, \tag{7}$$

where h_s is the enthalpy of species s including the heats of formation, λ is the thermal conductivity and T is the temperature. For the equation of state of a mixture of ideal gases we have

$$p = \rho T \frac{R}{\overline{M}}, \tag{8}$$

with R denoting the ideal gas constant.

In eq. (5), the pressure may vary in time for a constant volume system but also in space for the compressible formulation. We first show the formulation of pressure for the incompressible case, with spatially constant pressure, and then modify the equation for a spatially varying pressure field. In [15], the formulation for the pressure equation is described in detail. Here we only recapitulate the key parts. We begin by rewriting the equation of state in the form $p = \rho RT \sum_{s=1}^{n_s} Y_s/M_s$, where M_s is the molecular weight of species s. Taking the material derivative of this equation, and noting that for Lagrangian formulation $D/Dt = d/dt$, we obtain

$$\frac{dp}{dt} = \frac{RT}{\bar{M}} \frac{d\rho}{dt} + \frac{\rho R}{\bar{M}} \frac{dT}{dt} + \rho RT \sum_{s=1}^{n_s} \frac{1}{M_s} \frac{dY_s}{dt}. \tag{9}$$

The material derivatives on the right hand side of eq. (9) are evaluated using the transport equations for density, species mass fraction and temperature:

$$\frac{d\rho}{dt} = -\rho \frac{\partial u}{\partial x}, \tag{10}$$

$$\frac{dY_s}{dt} = \frac{\dot{\omega}_s}{\rho} - \frac{1}{\rho} \frac{\partial j_s}{\partial x}, \tag{11}$$

$$\frac{dT}{dt} = -\frac{1}{\rho c_p} \frac{\partial q}{\partial x} + \frac{1}{\rho c_p} \frac{dp}{dt} + \frac{1}{\rho c_p} \sum_{s=1}^{n_s} h_s \left(\frac{\partial j_s}{\partial x} - \dot{\omega}_s \right). \tag{12}$$

Plugging in eqs. (10, 11, 12) into eq. (9), and solving for dp/dt yields

$$\frac{dp}{dt} = -\gamma p \frac{\partial u}{\partial x} + \gamma p \Psi, \tag{13}$$

where gamma is the ratio of specific heats, $\gamma = c_p/c_v$, and Ψ is given by

$$\Psi = \frac{1}{\rho c_p T}\left(-\frac{\partial q}{\partial x} + \sum_{s=1}^{n_s} h_s \left(\frac{\partial j_s}{\partial x} - \dot{\omega}_s\right)\right) - \frac{\bar{M}}{\rho}\sum_{s=1}^{n_s}\frac{1}{M_s}\left(\frac{\partial j_s}{\partial x} - \dot{\omega}_s\right). \tag{14}$$

Equation (13) can be rewritten in terms of velocity divergence to give

$$\frac{\partial u}{\partial x} = -\frac{1}{\gamma p}\frac{dp}{dt} + \Psi. \tag{15}$$

Assuming uniform pressure, eq. (15) can be integrated over each cell to give the difference between cell face velocities for a given control volume, or it can be integrated over the entire domain to give the total dilatation. Taking the integral over the entire domain gives

$$u_{x=R} - u_{x=L} = -\frac{1}{p}\frac{dp}{dt}\int_{x=L}^{x=R}\frac{1}{\gamma}dx + \int_{x=L}^{x=R}\Psi dx, \tag{16}$$

where $x = L$ and $x = R$ denote left and right boundaries of the domain respectively. It follows, that the result for dp/dt for the incompressible formulation is

$$\frac{dp}{dt} = \left((u_{x=R} - u_{x=L}) - \int_{x=L}^{x=R}\Psi dx\right)\frac{-p}{\int_{x=L}^{x=R}\frac{1}{\gamma}dx}. \tag{17}$$

For the compressible case, with nonuniform pressure, we go back to eq. (13) and perform operation splitting. Here, we first set $\frac{\partial u}{\partial x} = 0$ and perform an integration over each cell to find the local rise in pressure due to the chemical source and heat and species flux terms in eq. (14).

$$\frac{dp}{dt} = \left(-\int_{x=l}^{x=r}\Psi dx\right)\frac{-p}{\int_{x=l}^{x=r}\frac{1}{\gamma}dx}, \tag{18}$$

where small l and r denote left and right boundaries of the cell respectively. The coupling between pressure and velocity is then discussed in section 2.3.

Time advancement of eqs. (3 - 5) is solved numerically using standard first-order finite-difference discretization and is advanced at a diffusive CFL constraint. Spatial discretization is second order on a uniform grid and formaly first order on the currently used non-uniform grid. The integration of the stiff set of chemical source terms is performed using the most recent version of the DAE solver IDA of the SUNDIALS package [11]. Thermodynamic and transport properties as well as reaction rates are calculated using the C++ interface of the CANTERA software package [8]. In this study we use the hydrogen combustion mechanism proposed in [23], that contains 22 reactions and 11 species.

2.2 Turbulence Model

In ODT, the turbulent motions that accelerate mixing are modeled through a series of stochastic rearrangement events. These events may be interpreted as the model analogue of individual turbulent eddies which are referred to as 'eddy events' or simply 'eddies'. Each eddy event interrupts the time advancement of other processes and an instantaneous transformation to the property profiles over some spatial interval $(x_0, x_0 + l)$ is applied, where x_0 represents the eddy starting location and l is the eddy length.

Eddy Events. The eddy event is central to the ODT modeling approach. It models the effects of a three-dimensional eddy using a 1D rearrangement. Eddy events are qualitatively similar to turbulence in that they have the effect of increasing gradients by redistributing the fluid elements along the 1D domain. Each eddy event consists of two mathematical operations. One is a triplet map representing the fluid displacements associated with a notional turbulent eddy and the other is a kernel transformation. The functional form chosen for the triplet map is the simplest of a class of mappings that satisfy the physical requirements of measure preservation, continuity, and scale locality over the eddy interval. The triplet map is conveniently represented by its inverse $f(x)$, such that the map moves fluid at location $f(x)$ to location x, where $f(x)$ is of the form [12]

$$f(x; x_0, l) \equiv x_0 + \left\{ \begin{matrix} 3(x - x_0) & \text{if } x_0 \leq x \leq x_0 + \frac{1}{3}l \\ 2l - 3(x - x_0) & \text{if } x_0 + \frac{1}{3}l \leq x \leq x_0 + \frac{2}{3}l \\ 3(x - x_0) - 2l & \text{if } x_0 + \frac{2}{3}l \leq x \leq x_0 + l \\ x - x_0 & \text{otherwise} \end{matrix} \right\}. \tag{19}$$

This mapping takes a line segment $[x_0, x_0 + l]$ shrinks it to a third of its original length, and then places three copies on the original domain. The middle copy is reversed, which ensures that property fields remain continuous and introduces the rotational folding effect of turbulent eddy motion. All quantities outside the $[x_0, x_0 + l]$ interval are unaffected. The triplet map is augmented by a kernel transformation to implement pressure-induced energy redistribution among velocity components while obeying energy and momentum conservation laws. This enables the model to simulate the tendency of turbulent eddies to drive the flow toward isotropy. The kernel function used is that of the vector formulation of ODT in [13,1].

Eddy Rate Distribution. The ODT velocity profiles evolve through the specification of the occurrences of eddy events. Conversely, the velocity profile supplies information that determines the size, location and frequency of these events. The eddy selection process is stochastic and follows the variable density formulation of Ahsurst and Kerstein [1]. The local rate of an eddy is taken to be $\lambda(x_0, l) = 1/l^2\tau$, and the total rate of all eddies is $\Lambda = \int\int \lambda(x_0, l) dx_0 dl$. Hence, the joint PDF of eddy parameters x_0 and l is $P(x_0, l) = \lambda(x_0, l)/\Lambda$. Eddy occurrences are sampled from a Poisson distribution with mean rate Λ, with x_0 and l parameters sampled from $P(x_0, l)$. Similar to dimensional relationships applied to fully developed turbulence, for eddy events in ODT, a relationship can be formulated between an eddy's size, it's associated energy and a time scale. The eddy time scale τ, is used to specify the eddy acceptance probability, and is computed as

$$\frac{1}{\tau} = C\sqrt{\frac{2K_0}{\rho_0 l^3}(E_{kin} - ZE_{vp} + E_{pe})}, \tag{20}$$

where E_{kin} is a measure of kinetic energy as in [1], $\rho_0 = \int \rho K^2(x)dx$, and $K_0 = \int K^2(x)dx$ where $K(x)$ is the kernel function as in [1]. E_{vp} is a viscous penalty defined using scaling arguments as $E_{vp} = \frac{1}{2}\bar{\mu}^2/\bar{\rho}l$, where $\bar{\rho}$ and $\bar{\mu}$ are the average density and viscosity in the eddy region and E_{pe} is a potential energy of the eddy defined in section 2.4. The adjustable model parameter C represents turbulence intensity and Z represents a viscous penalty factor. The evaluation of τ depends on the instantaneous flow state, so eddy occurrences are responsive to unsteadiness resulting from transient forcing or statistical fluctuations inherent in the eddy-sampling process. The eddy occurrences thus depend on the effects of prior eddies and affect future eddy occurrences. These dependencies induce spatio-temporal correlations among eddy events, leading to a physically based representation of turbulence intermittency.

2.3 Extension to Compressibility

In ODT, equations for enthalpy, species and ODT velocity components u, v and w are time advanced according to eqs. (3-5). After time advancing these equations, the extension to compressibility is made by relating pressure, density and enthalpy to a volume change by using velocity component u to advect cell faces and then introducing a feedback of the updated pressure to the u velocity. The relationships we use are based on the assumptions that the gas is an ideal gas and that the expansion process is reversible and adiabatic.

The complete time advancement follows in two steps. Starting from state n, enthalpy, species, and velocity are advanced a time step Δt through equations (3-5) to state n^*. The compressibility extension then follows by first interpolating cell center velocities u to cell faces on the non-equidistant grid,

$$u_{i-1/2}^{n^*} = u_{i-1}^{n^*} + (x_{i-1/2}^{n} - x_{i-1}^{n}) \cdot \frac{u_i^{n^*} - u_{i-1}^{n^*}}{x_i^n - x_{i-1}^n}, \tag{21}$$

where the subscript $i - 1/2$ and $i + 1/2$ denote cell face values on the left and right side of cell center, respectively. Cell face positions are then displaced

$$x_{i-1/2}^{n+1} = x_{i-1/2}^{n} + u_{i-1/2}^{n^*}\Delta t, \tag{22}$$

and the new cell sizes are calculated as

$$\Delta x_i^{n+1} = x_{i+1/2}^{n+1} - x_{i-1/2}^{n+1}. \tag{23}$$

Note, that u^{n^*} up to this point has not yet seen the pressure effect induced by chemical energy changes. Density is updated to time $n + 1$ through eq. (2) as

$$\rho_i^{n+1} = \rho_i^{n^*} \frac{\Delta x_i^n}{\Delta x_i^{n+1}}. \tag{24}$$

Based on the aforementioned assumptions, a relation between pressure and volume change can be derived and is written as $p\Delta x^\gamma = constant$. The update of the pressure is then solved for as

$$p_i^{n+1} = p_i^{n^*} \left(\frac{\Delta x_i^n}{\Delta x_i^{n+1}} \right)^\gamma . \tag{25}$$

In the momentum equation, a pressure flux is introduced as a source term and the new cell face velocities are calculated as

$$u_{i-1/2}^{n+1} = u_{i-1/2}^{n^*} - \Delta t \cdot \left(\frac{\partial p}{\partial x} \right)_{i-1/2}^{n+1} \frac{1}{\rho_{i-1/2}^{n+1}}. \tag{26}$$

Cell face velocities are then interpolated back to cell center velocities $u_i^{n+1} = (u_{i-1/2}^{n+1} + u_{i+1/2}^{n+1})/2$. Next, we want to calculate the change in enthalpy due to changes in pressure and kinetic energy. The 1st Law of Thermodynamics written in terms of enthalpy and kinetic energy is

$$dh = \frac{1}{\rho} \Delta x \cdot dp - \frac{1}{2} d(u^2). \tag{27}$$

Using eq. (27) and the equation $d(p\Delta x^\gamma) = 0$, a relation for the enthalpy update can be derived and is written as

$$h_i^{n+1} = h_i^{n^*} - \frac{p_i^{n^*}}{\rho_i^{n^*}} \frac{\gamma}{1-\gamma} \left(\left(\frac{\rho_i^{n^*}}{\rho_i^{n+1}} \right)^{1-\gamma} - 1 \right) - \frac{(u_i^{n+1})^2 - (u_i^{n^*})^2}{2}, \tag{28}$$

where the last term is the induced kinetic energy change from expansion and it is subtracted from the enthalpy.

As stated before, the time advancement is at a diffusive CFL number. However, the pressure-velocity coupling update has the additional constraint that a cell compression/expansion can not be larger than a specified factor. This is done to insure that a cell does not collapse on itself, and also for stability reasons. Therefore, the pressure-velocity update can occur in several steps, such that in each step the maximum compression/expansion is limited and then followed by mesh adaption.

2.4 Darrieus-Landau Instability Model

Planar flames are intrinsically unstable due to acceleration of the variable-density fluid caused by thermal expansion across the burning front. This is the Darrieus-Landau instability, analogous to the Rayleigh-Taylor instability that develops when heavy fluid is above light fluid in a gravitational field. This analogy allows an existing ODT representation of the Rayleigh-Taylor instability [6] to be modified in order to incorporate the Darrieus-Landau instability mechanism into ODT.

Namely, a formal analog of gravitational potential energy is introduced. In our case, the constant acceleration of the gravity is replaced by the varying time rate of change of the advecting velocity u and defined as $a(x) = \partial u(x)/\partial t$. The potential energy is then defined as

$$E_{pe} = \frac{8}{27} \int_{x_0}^{x_0+l} a(x)K(x)(\rho - \bar{\rho})dx, \tag{29}$$

where the factor 8/27 arises due to the variable density formulation and $\bar{\rho}$ is a reference density defined as the average density over the interval $[x_0, x_0+l]$. This potential energy is nonzero only where the density varies, as it is the interaction of the dilatation-induced pressure gradient and the density gradient that is the cause of this instability mechanism. E_{pe} is not a potential energy in the same sense as in a buoyant flow, because it is not based on an external energy source. For this reason, it is only used to effect the probability of acceptance of an eddy, but it is not included in the energy redistribution during an eddy event. It is however, a formal analog to the treatment of energy in the buoyant flow, and therefore a tunable coefficient is not required.

The Darrieus-Landau instability is not inherently a finite-Mach-number effect, so a representation of the instability should be incorporated into any ODT formulation involving unsteady dilatation within the ODT domain, irrespective of Mach number. It is described here because it has not previously been incorporated into ODT.

3 Results

3.1 Sod Shock Tube Problem

A useful problem to test a compressible flow code is the shock tube, in which a long straight cylinder is divided into two compartments by a central diaphragm. On the left side of the diaphragm is a gas with density $\rho = 1.15\,\mathrm{kg/m^3}$ and pressure $p = 1.0\,\mathrm{atm}$, while on the other side the gas has density $\rho = 1.7\,\mathrm{kg/m^3}$ and pressure $p = 1.5\,\mathrm{atm}$. The calculation begins by removing the diaphragm with the gases at rest. The pressure difference in the two gases drives a shock into the less dense gas, while a rarefaction moves into the denser gas. In Fig. 1 the pressure, density and velocity profiles are compared with the analytic solution at time $t = 40^{-3}\,\mathrm{ms}$. The comparison shows that the methodology is stable and that shock speed is captured accurately. Additionally, the shock profile is also well depicted.

3.2 Turbulent HCCI

To demonstrate the potential of the methodology developed, we simulate autoignition at constant volume under turbulent conditions. This is a low Mach

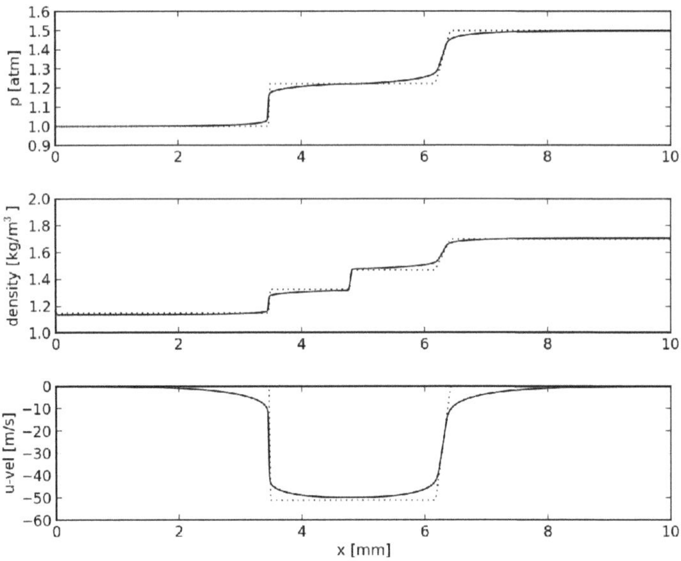

Fig. 1. Comparison of pressure, density and velocity simulation results, solid black line, with an exact Riemann solver, dotted black line, for the Sod shock tube problem at 40^{-3} ms and 300 K

number combustion case. A high Mach number combustion case will be reported in the future. Homogeneous-charge compression-ignition (HCCI) engines have potentially higher thermal efficiencies and lower NO_x and soot emissions than conventional spark-ignition and Diesel engines, respectively. HCCI engines operate under lean conditions with fuel and air well mixed before entering the cylinder. As such, we choose a lean hydrogen-air mixture with equivalence ratio 0.1, at initial pressure $p_0 = 1.0$ atm, temperature $T_0 = 950$ K and a domain length of 5 mm. Temperature and velocity are initially perturbed to create turbulent conditions with a root-mean-square of $T' = 42$ K and $u' = v' = w' = 0.8$ m/s respectively. No flux boundary conditions are applied, hence velocity at the boundaries is zero. ODT viscous penalty parameter Z and eddy frequency parameter C are set to 0.01 and 1000 respectively. Results for three different cases are compared: 1) a laminar case in which ODT eddy generation is turned off, 2) ODT eddy generation is turned on but the Darrieus-Landau instability model is excluded and 3) eddy generation including the Darrieus Landau instability model. All three cases start from the same perturbed initial condition.

Fig. 2 shows temperature, velocity, pressure, and normalized pressure profiles at different times for the case with the Darrieus-Landau model. From the velocity plots, we see that this is a low Mach number combustion. At time $t = 0.0012$ ms, large initial temperature fluctuations induce a series of small eddies that begin to homogenize the mixture. The velocity profile at the same time step shows large

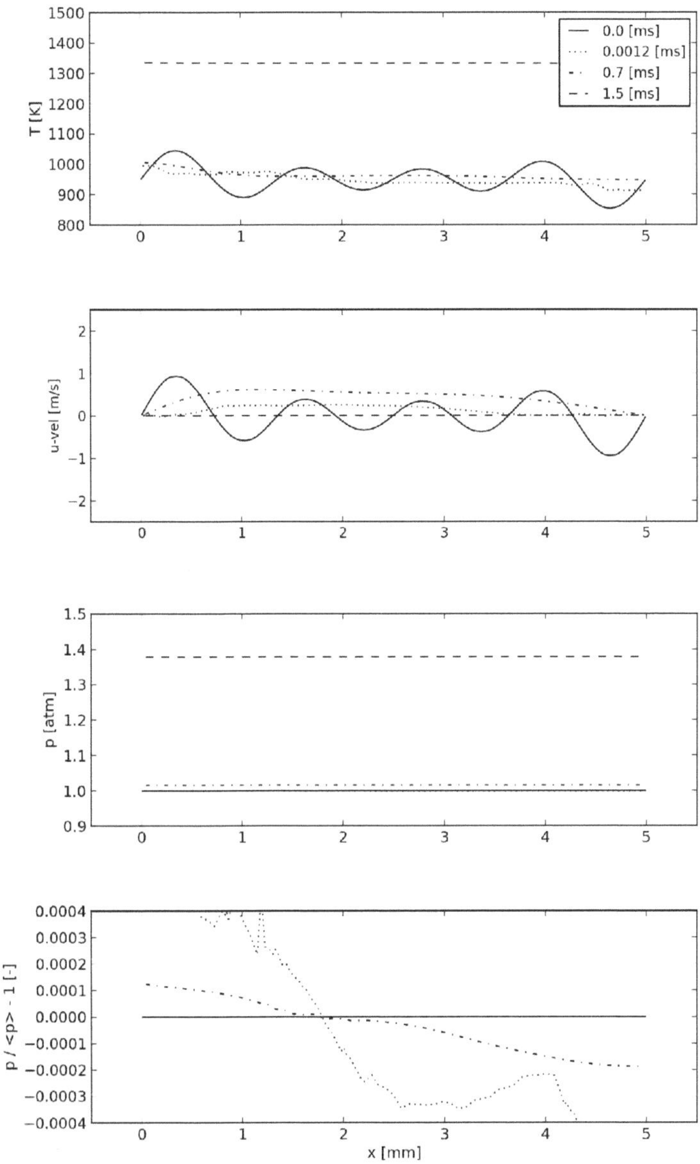

Fig. 2. For the HCCI configuration, temperature, velocity, pressure and normalized pressure profiles at different times with Darrieus-Landau model

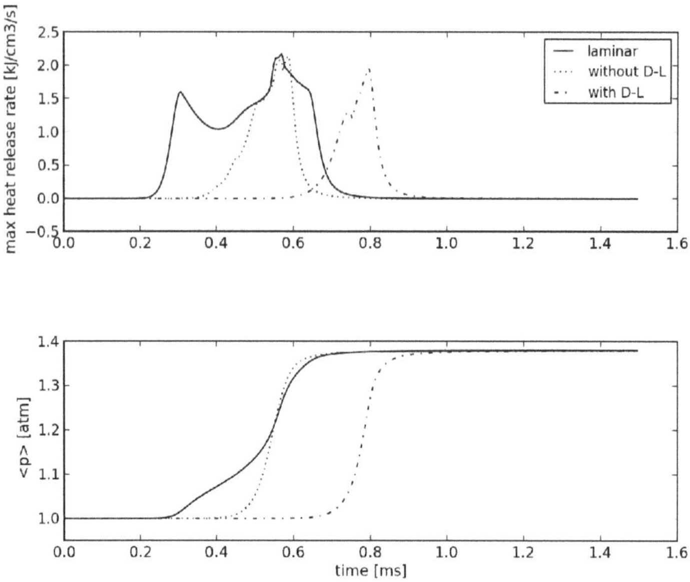

Fig. 3. For the HCCI configuration, heat release rate (top) and spatially averaged pressure (bottom) over time.

accompanying oscillation. By time $t = 0.7$ ms, the mixture begins to ignite as seen by the elevated temperature, and it is homogenized as seen by the relatively flat temperature profile. Velocity oscillations have settle down and by $t = 1.5$ ms, the reaction has completed and velocity oscillations desist. Two pressure plots are shown, the non-normalized pressure shows the quasi spatially uniform rise in pressure as the combustion proceeds, whereas the normalized pressure shows the spatial fluctuations occurring.

Fig. 3 shows the main influence of turbulent mixing with comparison to autoignition for all three cases. Starting from the same initial condition, the development of spatially averaged pressure $< p >$ and maximum heat release rate over time is shown. For our initial condition, a faster burnout of the laminar case can be seen. In fig. 2, the temperature profile at $t = 0.0012$ ms shows that turbulent mixing has lowered local peak temperatures. Turbulence has homogenized the mixture, which then reacts more slowly until it gets hot enough to react quickly. This can be further seen in fig. 3. The turbulent cases have larger ignition delay times and the pressure gradient during reaction is steeper for the turbulent cases, indicating a more homogeneous mixture than in the laminar case. In all cases, at the completion of the reaction, the same final pressure of 1.38 atm is reached, demonstrating the conversion of chemical energy into pressure rise.

Fig. 4 compares eddy time histories for the two turbulent cases, with and without the Darrieus-Landau instability mechanism, superimposed on the heat release rate. The small dashes represent individual eddies implemented at their

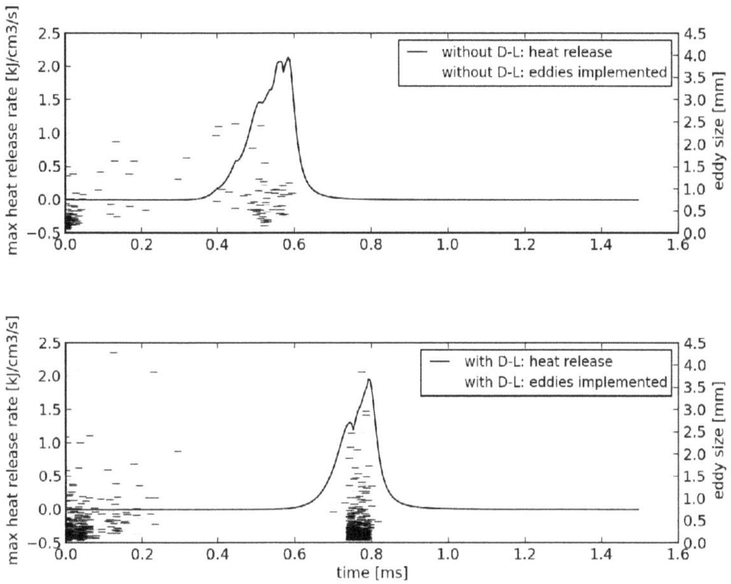

Fig. 4. For the HCCI configuration, heat release rate with superimposed time history of eddy events without (top) and with (bottom) the Darrieus-Landau instability mechanism.

depicted times, with vertical placement according to size, and the heat release rate curves are retaken from fig. 3. A suggested correlation between heat release rate and eddy acceptance for the Darrieus-Landau model is clearly seen. There are two large bursts of eddies, first, during the homogenization of the initial turbulent condition, and second, during high heat release. This indicates the ability of the model to capture the unstable nature of thermal expansion, and depicts the wrinkling of the flame surface through the occurrence of eddy events that increase the wrinkling of 1D property profiles. For the turbulent case without the Darrieus-Landau model, eddy acceptance is driven solely by the kinetic energy associated with velocity fluctuations. After the initial perturbation has been homogenized, this kinetic energy, and with it the number of accepted eddies, dwindle. There is a small second burst of eddies that are associated with heat release at approximately 0.5 ms. This however, occurs not due to the Darrieus-Landau model, but due to the induced velocity field from the heat release through the compressible formulation. Further testing is however needed to draw definitive conclusions.

4 Conclusions

In summary, an extension of the incompressible ODT methodology towards an efficient compressible implementation has been presented. Additionally, a mechanism for capturing the instabilities caused by thermal expansion across a burning front, the Darrieus-Landau instability, in ODT is developed. First, results for a standard non-reactive shock tube problem are shown and compared with analytic calculations. Second, results for autoignition under HCCI-like conditions are shown.

Results for the shock tube problem show that despite the simple compressible methodology developed, the formulation is capable of capturing pressure wave propagation. Sharp interfaces are however slightly diffused and exact profile capturing is not achieved. Further numerical testing is planned to isolate the cause.

Results for the reactive HCCI configuration show comparisons for three cases: a laminar case with ODT eddy generation turned off, and a turbulent case with and a turbulent case without the Darrieus-Landau model. In all cases, the same final elevated pressure is reached demonstrating the transfer of enthalpy change to pressure change for the closed volume. Results showing the time history of eddy events superimposed on the heat release rate for the two turbulent cases indicates that the Darrieus-Landau model is able to capture the unstable nature of thermal expansion and depict it through the occurrence of eddy events. Eddy events for the turbulent case without the Darrieus-Landau model are based solely on kinetic energy considerations and show limited correlation to thermal expansion. The HCCI application is a demonstration of the compressible numerical method with the Darrieus-Landau model for a low Mach number combustion case, results for a higher Mach number combustion case will be reported in the future.

Acknowledgements. Financial support for several stays of Alan Kerstein at BTU Cottbus-Senftenberg was provided by the Deutsche Forschungsgemeinschaft (DFG) within the SPP 1276 MetStröm project.

References

1. Ashurst, W.T., Kerstein, A.R.: One-dimensional turbulence: Variable-density formulation and application to mixing layers. Phys. Fluids 17, 25107 (2005)
2. Cao, S., Echekki, T.: A low-dimensional stochastic closure model for combustion large-eddy simulation. J. Turbul. 9, 1–35 (2008)
3. Echekki, T., Kerstein, A.R., Dreeben, T.D.: One-dimensional turbulence simulation of turbulent jet diffusion flames: model formulation and illustrative applications. Combust. Flame 125, 1083–1105 (2001)
4. Glawe, C., Schulz, F.T., Gonzalez-Juez, E.D., Schmidt, H., Kerstein, A.R.: ODTLES simulations of turbulent flows through heated channels and ducts. In: Symposium on Turbulence and Shear Flow Phenomena (2013)
5. Gonzalez-Juez, E.D., Kerstein, A.R., Lignell, D.O.: Fluxes across double-diffusive interfaces: a one-dimensional-turbulence study. J. Fluid Mech. 677, 218–254 (2011a)
6. Gonzalez-Juez, E.D., Kerstein, A.R., Lignell, D.O.: Reactive Rayleigh-Taylor turbulent mixing: a one-dimensional-turbulence study. Geophysical and Astrophysical Fluid Dyn., 1–20 (2012)
7. Gonzalez-Juez, E.D., Schmidt, R.C., Kerstein, A.R.: ODTLES simulation of wall-bounded turbulent flows. Physics of Fluids 23, 125102 (2011)

8. Goodwin, D.: CANTERA: Object-oriented software for reacting flows, `http://www.cantera.org`
9. Hewson, J.C., Kerstein, A.R.: Stochastic simulation of transport and chemical kinetics in turbulent CO/H2/N2 flames. Combust. Theory Model. 5, 669–697 (2001)
10. Hewson, J.C., Kerstein, A.R.: Local extinction and reignition in nonpremixed turbulent CO/H2/N2 jet flames. Combust. Sci. Technol. 174, 35–66 (2002)
11. Hindmarsh, A.C.: SUNDIALS: Suite of nonlinear and differential/algebraic equation solvers. Technical Report UCRL-JRNL-200037, Lawrence Livermore National Laboratory (2004)
12. Kerstein, A.R.: One-dimensional turbulence: model formulation and application to homogeneous turbulence, shear flows, and bouyant stratified flows. J. Fluid Mech. 392, 277–334 (1999a)
13. Kerstein, A.R., Ashurst, W.T., Wunsch, S., Nilsen, V.: One-dimensional tubulence: vector formulation and application to free shear flows. J. Fluid Mech. 447, 85–109 (2001)
14. Kerstein, A.R., Wunsch, S.: Simulation of a stably stratified atmospheric boundary layer using one-dimensional turbulence. Bound. -Lay. Meteorol. 118, 325–356 (2006)
15. Lignell, D.O., Kerstein, A.R., Sun, G., Monson, E.E.: Mesh adaption for efficient multiscale implementation of One-Dimensional Turbulence. Theor. Comput. Fluid Dyn. 27, 273–295 (2013)
16. Lignell, D.O., Rappleye, D.S.: One-dimensional turbulence simulation of flame extinction and reignition in planar ethylene jet flames. Combust. Flame 159, 2930–2943 (2012)
17. Poinsot, T., Veynante, D.: Theoretical and Numberical Combustion, Toulouse, 3rd edn. (2011)
18. Pope, S.B.: Small scales, many species and the manifold challenges of turbulent combustion. Proc. Combust. Inst. 32, 1–31 (2013)
19. Punati, N., Sutherland, J.C.: Application of an Eulerian One-Dimensional Turbulence Model to simulation of turbulent jets. U.S. Joint Sections of the Combustion Institute, Ann Arbor, MI (May 2009)
20. Punati, N., Sutherland, J.C., Kerstein, A.R., Hawkes, E.R., Chen, J.H.: An evaluation of the One-Dimensional Turbulence Model: Comparison with Direct Numerical Simulations of CO/H2 jets with extinction and reignition. Proc. Comb. Inst. 33, 1515 (2011)
21. Ranganath, B., Echekki, T.: An ODT-based closure model in non-premixed combustion. Prog. Comput. Fluid Dynam. 6, 409–418 (2006)
22. Ranganath, B., Echekki, T.: One-dimensional turbulence-based closure with extinction and reignition. Combust. Flame 154, 23–46 (2008)
23. Saxena, P., Williams, F.A.: Testing a small detailed chemical-kinetic mechanism for the combustion of hydrogen and carbon monoxide. Combust. Flame 145, 316–323 (2005)
24. Schmidt, R.C., Kerstein, A.R., McDermott, R.J.: Odt: A multi-scale model for 3d turbulent flow based on one-dimensional turbulence modeling. Comput. Methods Appl. Mech. Eng. 199, 865–880 (2010)
25. Schmidt, R.C., Kerstein, A.R., Wunsch, S., Nilsen, V.: New-wall les closure based on one-dimensional turbulence modeling. J. Comp. Phys. 186, 317–355 (2003)
26. Wunsch, S., Kerstein, A.R.: A stochastic model for high-Rayleigh-number convection. J. Fluid Mech. 528, 173–205 (2005)

Investigation of Equivalence Ratio Fluctuations on the Dynamics of Turbulent Lean Premixed Methane/Air Flames with a Linear-Eddy Model

Michael Oevermann[1], Christina Schrödinger[2], and Christian O. Paschereit[2]

[1] Chalmers University of Technolog, Department of Applied Mechanics, Division of Combustion, 41296 Gothenburg, Schweden
michael.oevermann@chalmers.se

[2] Technische Universität Berlin, Institut für Strömungsmechanik und Technische Akustik, Müller Breslau Str. 8, 10623 Berlin, Germany
{christina.schroedinger,oliver.paschereit}@tu-berlin.se

Abstract. Heat release fluctuations generated by equivalence ratio fluctuations may interact with the acoustics of a gas turbine combustion chamber leading to unwanted combustion instabilities, which remains a critical issue in the development of low emission, lean premixed gas turbine combustors. The present article addresses this topic by numerical investigations of one-dimensional lean premixed methane/air flames subject to prescribed sinusoidal equivalence ratio fluctuations. Compared to previous investigations, we focus on turbulent conditions and emission predictions using the one-dimensional linear-eddy model (LEM) and detailed chemistry. Within the limitations of the one-dimensional LEM the approach allows to investigate the fully non-linear regime of flame response to equivalence ratio fluctuations under turbulent conditions. Results for different forcing amplitudes and turbulence levels indicate a strongly non-linear behavior for high forcing amplitudes.

1 Introduction

Gas turbines for power generation and propulsion have long been using diffusion flame combustors for their reliable performance and stable combustion characteristics. However, diffusion flames usually exhibit high levels of soot and thermal NOx which are unacceptable under strict regulations for pollutant emissions. Among the modern combustion concepts for clean and high efficient gas turbines lean premixed combustion is a promising technology for practical systems which is already in use by the gas turbine manufactures. In lean premixed combustion fuel and oxidizer are premixed upstream of the combustion zone and burned at lean conditions leading to low temperature levels and therefore low NO_x emissions. A major challenge for operating gas turbine combustors at lean premixed conditions is their susceptibility to combustion instabilities [11,25,44]. During an instability, heat release fluctuations interact with acoustic pressure oscillations in a resonant way leading to high amplitude pressure and velocity oscillations. Heat release fluctuations can be driven by various effects and their

R. King (ed.), *Active Flow and Combustion Control 2014*, Notes on Numerical Fluid Mechanics and Multidisciplinary Design 127, DOI: 10.1007/978-3-319-11967-0_14

interaction [11], e.g., acoustic motion in the combustor [34], coherent and vortical flow structures [29,30,31], flame surface variations [34,48], flame extinction and re-ignition, and equivalence ratio oscillations [9,23,24].

Fluctuations in equivalence ratio may arise either from incomplete mixing of fuel and air or from fluctuations of fuel and/or air mass flow rates at the injection point. They have been identified as one of the major sources of heat release fluctuations. This can be explained by the strong non-linear dependence of the chemical reaction rate on the equivalence ratio, especially at lean conditions [23,24,44]. Compared to stoichiometric conditions, at lean conditions even small variations in the equivalence ratio lead to large fluctuations of the heat release which can interact in a resonant way with the acoustics of the combustion chamber.

The impact of equivalence ratio fluctuations on the dynamics of premixed flames have been investigated both experimentally [3,20,40,43] and numerically or analytically [2,5,9,23,24,35,38,44].

Kim et al. [20] perform an experimental study of the response of swirl-stabilized lean premixed flames to equivalence ratio fluctuations. They found that the response of the flame can be either linear or non-linear depending on the magnitude of the equivalence ratio oscillation. The non-linear response becomes more pronounced as modulation frequencies increase and the mean equivalence ratio decreases. For large perturbation amplitudes they also observed a non-linear response of the heat release with abrupt changes in the waveform. They conclude that the overall characteristics of the flame response to equivalence ratio fluctuations are similar to the flame response to velocity fluctuations, although the mechanisms for non-linearities are different.

Balachandran et al. investigate in [3] the response of lean premixed turbulent bluff-body stabilized flames to imposed velocity perturbations with forcing frequency 160 Hz via flame transfer function measurements. It was found that the heat release increased non-linearly for inlet velocity amplitudes exceeding 15%. For the investigated turbulent flame regimes the authors conjecture in [3] that flame area modulation is the most important phenomenon affecting the global heat release fluctuations whereas local fluctuations of strain rate and curvature have smaller effects. In a more recent study [4], however, they report for compact flames (low frequencies) that the heat release fluctuations mainly depend on equivalence ratio fluctuations and for higher frequencies it is dominated by flame-vortex interactions.

The influence of equivalence ratio fluctuations in a frequency range of 20 to 150 Hz on a turbulent premixed methane/air flame in the corrugated/thick flames combustion regime is investigated in [40]. The authors observe strongly non-linear behavior of the heat release rate which is attributed to a combined effect of equivalence ratio oscillations directly influencing heat release rate and burning velocity and turbulent velocity fluctuations. For high levels of turbulence only high amplitude and low frequency equivalence ratio forcing amplitudes reach the flame front and increase the heat release rate fluctuations.

Richardson et al. present in [35] a computational study of lean methane-air flame response to spatial and temporal equivalence rate variations in an axisymmetric laminar counterflow configuration. Investigating stratified steady flames they showed that under lean conditions back supported flames (with positive gradients of equivalence ratio) propagate faster, have reduced flame thickness and enhanced heat release rate than undisturbed flames under similar conditions, while front supported flames showed opposite trends. They also showed that the propagation speed of premixed flames also depends on unsteady effects due to strain rate and equivalence ratio fluctuations. Zhou and Hochgreb [50] showed in a counterflow configuration that the linear relation between the scalar dissipation rate and the heat release rate under different strain rates is valid also under stratified conditions. For low strain rates they concluded that flame structures under lean conditions show insignificant unsteady effects.

In [2] some of the experiments by Balachandran et al. [3] are simulated with an unsteady Reynolds-averaged Navier-Stokes (URANS) CFD approach using a modified Bray-Moss-Libby (BML) approach as a turbulent combustion model. For the frequencies considered the calculated location and distribution of heat release agree well with experiments and the CFD captures the transition from linear to non-linear behavior.

König et al. [21] considered the response of laminar stretched premixed methane-air counterflow flames at stoichiometric conditions to periodical perturbations of mixture composition at the inflow. Although the main results are of little importance for lean premixed flames, they observed that perturbations reach the flame only under certain conditions. The amplitude of perturbation reaching the flame decreases with increasing frequency due to diffusive processes and might even be completely damped before reaching the flame zone.

Equivalence ratio fluctuations have been identified in [6,44] to act via three different mechanisms on heat release fluctuations: They directly influence the heat of reaction and the flame speed and indirectly disturb the flame front via flame speed oscillations leading to a pertubation of the burning area. In [44] it is also concluded that all three processes can – and usually do in turbulent flames – interact in a nonlinear way.

The cited references with numerical investigations on the influence of equivalence ratio fluctuations on the dynamics of lean premixed flames do either investigate laminar flames or use turbulent combustion models which do not resolve the inner structure of the flames. However, gas turbine combustors usually operate in the thin reaction zones regime [33]. In this regime the smallest turbulent eddies may penetrate the preheat zone, thereby modifying the flame structure. For the simulation and investigation of unsteady flame phenomena in this regime it might be beneficial to have a model which i) resolves all scales (spatial and temporal), ii) directly simulates flame turbulence interactions, and iii) has the potential of being regime independent and can be applied in the entire parameter space with some confidence. One such model is the linear-eddy model (LEM) of Kerstein [18,19] which has been, e.g., successfully applied for premixed flames in the thin reaction zones regimes before [36].

In the present work we numerically investigate the influence of periodic equivalence ratio oscillations on lean, premixed methane/air flames that are strongly turbulent using detailed chemistry. We compare the results to laminar flames under the same conditions and evaluate pollutant formations of NO_x and CO and heat release rate fluctuations. The flame response is determined for a large range of oscillation frequencies and amplitudes. Compared to previous studies we focus on one-dimensional, turbulent flames using the LEM approach. The LEM resolves all relevant physical and chemical length and time scales and is in principle applicable to the whole range of combustion regimes. Like in direct numerical simulations (DNS), LEM resolves all scales but on a one-dimensional domain and is therefore capable of resolving directly the interaction of turbulence and chemistry at a much lower cost permitting parametric investigations.

Based on the comments made by one of the reviewers it is worth discussing some limitations of the LEM regarding effects of flame curvature and coherent flame wrinkling on the dynamics of premixed flames. It has been shown in numerous studies that LEM correctly captures flame surface generation by turbulence [27,36,49], has shown good validation against DNS [49], and, e.g., exactly obeys the expected scaling of the turbulent burning velocity in the strong-turbulence limit [49]. However, perturbations of flame surface area caused by flame propagation of a curved laminar flame or flame speed variations along the flame surface (e.g. due to equivalence ratio fluctuations and/or coherent wrinkling [10]) are not captured by the one-dimensional stand-alone LEM. That implies that flame dynamics related to such flame surface perturbations are not captured by the model in its presented form. This includes, e.g., the predictions of acoustic flame-instabilities. However, as we are focussing here on cases with strong turbulence we expect flame wrinkling due to turbulence (and its interaction with equivalence ratio perturbations) to be the leading order mechanism. This mechanism is well represented by the LEM. In a recent work Jozefik et al. [12] have developed a Darrieus-Landau instability model for the One-Dimensional Turbulence (ODT) model which can be regarded as a generalisation of the LEM and could be used in the future to address some of the mentioned limitations.

2 Numerical Method

2.1 Linear-Eddy Model

The linear-eddy mixing model was proposed by Kerstein in [13] for non-reacting flow and extended to reactive flow in [16]. It has been discussed in detail in the literature, e.g. [13,14,16,15,17] and, therefore, is only briefly summarized here. The overall LEM concept for turbulent reactive flow consists of two concurrent processes representing the respective influences of dilatation-induced advection, molecular diffusion, chemical reactions and turbulent transport. The first process is time advancement of the reactive zero-Mach-number equations on a one-dimensional domain resolving all spatial and temporal scales. The second process, turbulent transport, is implemented using a stochastic sequence of statistically independent stirring events.

LEM Diffusion and Chemical Reactions. The balance equations for species mass fractions Y_s and mixture enthalpy h are formulated in a Lagrangian (mass fixed) formulation and are solved on a one-dimensional domain resolving all spatial and temporal scales:

$$\begin{aligned} \rho\frac{\mathrm{d}Y_s}{\mathrm{d}t} &= -\frac{\mathrm{d}j_s}{\mathrm{d}x} + M_s\,\dot{\omega}_s, \\ \rho\frac{\mathrm{d}h}{\mathrm{d}t} &= \frac{\mathrm{d}p}{\mathrm{d}t} - \frac{\mathrm{d}q}{\mathrm{d}x} - \sum_s j_s\,\frac{\mathrm{d}h_s}{\mathrm{d}x} - \sum_s h_s\,M_s\,\dot{\omega}_s, \end{aligned} \tag{1}$$

where c_p is the heat capacity at constant pressure, p the pressure, q the heat flux, h_s the enthalpy of species s including the heats of formation, and $\dot{\omega}_s$ the chemical source term of species s. The equation of state for a mixture of ideal gases is $p = \rho T \sum_s Y_s R_m/M_s$ and the caloric relation is given by $h_s(T) = \Delta h_s^0 + \int_{T^0}^{T} c_{p,s}(t)\mathrm{d}T$, where Δh_s^0 is the standard heat of formation of species s. For the flames considered here we assume a constant pressure in time. The continuity equation in a Lagrangian formulation is $\rho\Delta x = \text{const.}$ and determines via the equation of state the expansion (or contraction) of each grid cell.

Linear-Eddy Mixing / Turbulent Transport. In the LEM concept, turbulent advection is implemented explicitly by stochastic eddy events. Each eddy event involves a rearrangement of all scalar quantities using so-called 'triplet maps' which mimics the influence of turbulent vortices. The effect of a triplet map is a three-fold compression of the scalar fields in a selected spatial interval whose size is denoted by l, see Figure 1 for an illustration. This map increases the scalar gradients within the selected interval, analogous to the effect of compressive strain in turbulent flow, without creating discontinuities.

Assuming the Kolmogorov inertial-range scaling the size distribution is given by [14]

$$f(l) = (5/3)\; l^{-8/3}/[\eta^{-5/3} - l_t^{-5/3}].$$

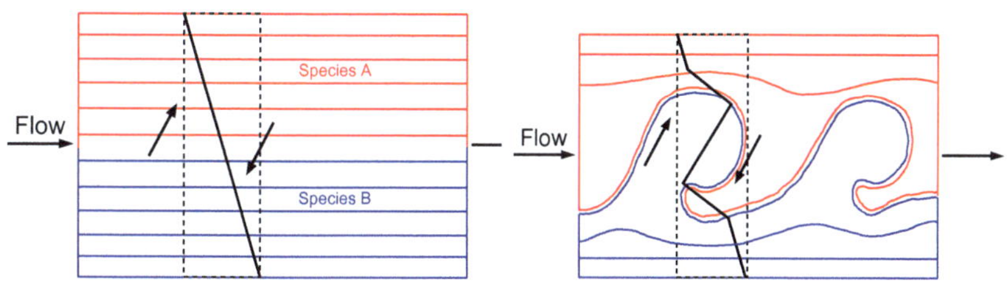

Fig. 1. Qualitative picture of a triplet map in LEM. Left: before mapping, right: after mapping.

Using the turbulent Reynolds number

$$Re_t = \frac{u' l_t}{\nu},$$

where ν is the kinematic viscosity, and u', l_t are the integral velocity and length scale, respectively, the Kolmogorov scale η is determined from the inertial scaling law $\eta = N_\eta \, l_t \, Re_t^{-3/4}$. Here, N_η is an empirical constant. The event frequency per unit length is determined by [14]

$$\lambda = \frac{54}{5} \frac{\nu Re_t}{C_\lambda l_t^3} \frac{\left[(l_t/\eta)^{5/3} - 1\right]}{\left[1 - (\eta/l_t)^{4/3}\right]},$$

where C_λ and N_η are model constants. The LEM parametrisation corresponds to a turbulent diffusivity $D_t = u' l_t / C_\lambda$.

Three quantities are needed to specify an eddy event: eddy time, eddy size l and eddy location within the domain. The eddy time is sampled under the assumption of a Poisson process with a mean eddy occurrence time of $\Delta t_{\text{eddy}} = (\lambda \, L)^{-1}$, where L is the size of the domain. The eddy location is randomly sampled from a uniform distribution, and the eddy size is randomly sampled assuming Kolmogorov inertial-range scaling, see [13] for details.

The required model constants are adopted from the literature and have been set to $C_\lambda = 15$ and $N_\eta = 10.76$ [46].

u', l_t, Re_t, and D_t (orχ) are the defining parameters of the linear-eddy model. In a coupled CFD-LEM simulations they are passed as time-dependent parameters from the CFD side to the LEM.

The linear-eddy model has been used both as a stand-alone tool and (mainly by the group of Suresh Menon at Georgia Tech) as a LES sub-grid model for the prediction of scalar mixing and turbulent combustion under various flow and combustion conditions [32,36,37,41,42,47].

It is well known that flame stretch and flame curvature have dominant effects on laminar and turbulent flames and one might ask if the LEM can somehow capture these inherently multi dimensional effects on a one-dimensional domain. One might argue that a triplet map hitting the flame or the reaction zone basically creates three copies of the flame structures which can be interpreted as a one-dimensional analog of increasing the flame area, i.e. flame stretch. Although possible, the potential of explicitly modeling flame stretch and curvature in a more elaborate way within the LEM model has not been fully explored yet [39].

The governing equations (1) are solved with an adapted code for one-dimensional turbulence (ODT) from Lignell et al. [26] which is based on a Lagrangian formulation of the advection and a central discretization of diffusion terms. The stiff chemical source term and all thermodynamic and transport data are evaluated with the software package Cantera [8]. Whereas advection is treated exactly via the Lagrangian formulation, diffusion and chemical source terms are integrated in time between two eddy events with the second order operating splitting method of Strang. The stiff chemical source term is integrated with an implicit BDF method [7].

2.2 Chemistry Model

Due to the resolution of all spatial and temporal scales in 1D, the LEM is computationally quite demanding and chemistry acceleration and/or reduced chemical mechanisms are desirable. The chemistry model used in this study is a reduced chemical mechanism based on the GRI-3.0 mechanism [45]. The reaction pathway analysis program Mixmaster of the Cantera [8] software package has been used to identify relevant reaction rates and reaction pathways for lean fuel-air mixtures with equivalence ratios ranging from 0.55 to 0.95 and to remove those species from the mechanism with a minor influence on the optimization parameters. The optimization parameters have been mass fractions of major species, NO_x and CO concentrations, and laminar flame speeds under lean conditions. The reduced mechanism consists of 23 species and 76 reactions. Fig. 2 shows some comparisons of the reduced mechanism with the original GRI-3.0 mechanism. Although the derived mechanism shows good agreement with the full GRI-3.0 mechanism (and other reduced mechanisms not shown here) for the relevant conditions and parameters in this study, we note that further comparisons

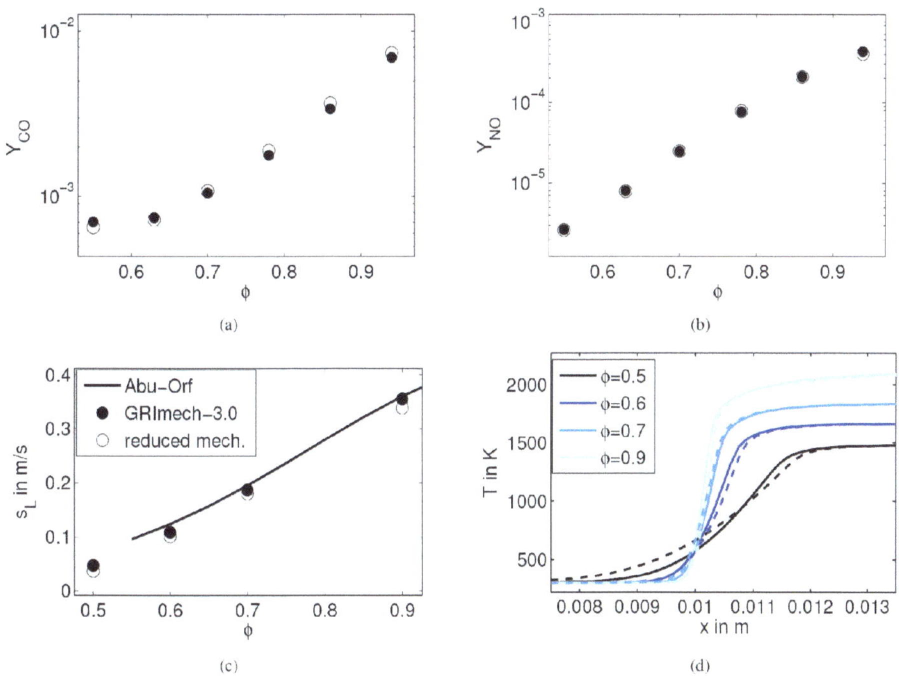

Fig. 2. Comparison between the reduced methane/air with the full GRI-3.0 mechanism [45] and experimental data of Abu-Orf & Cant [1]: (a) CO mass fractions vs. equivalence ratio, (b) NO mass fractions vs. equivalence ratio, (c) laminar flame speed vs. equivalence ratio, (d) temperature profiles for different equivalence ratios (solid lines: GRI-3.0, dashed lines: reduced mechanism).

under unsteady conditions, e.g. unsteady forcing of laminar flames, should be performed.

3 Results

In the following paragraphs we present several simulations with parameter variations of a turbulent lean premixed turbulent methane/air flame using the one-dimensional LEM. The turbulence parameters are chosen close to measurements from an atmospheric single burner test rig [22].

3.1 Comparison between Laminar and Turbulent Flame

To demonstrate some features of one-dimensional flame simulations with the LEM we compare in Fig. 3 the structure and the heat release rate of a laminar methane/air flame with a turbulent flame obtained with the LEM in the thin reaction zones regime close to the broken reaction zones regime. The freely propagating flame is kept within the computational domain by constantly adjusting the inflow velocity to the propagation speed of the flame. The equivalence ratio has been set to $\phi = 0.6$ and the unburned gas temperature is $T_u = 425$ K. The integral length scale, the Kolmogorov length scale, and the turbulent velocity fluctuation are taken from measurements in a combustion test rig [22] and set to $l_t = 2.5e - 2$ m, $\eta = 1e - 4$ m, and $u' = 10$ m/s, respectively. The laminar flame thickness is approx. 1 mm and the laminar flame speed $s_l = 0.22$ m/s. The instantaneous temperature profile of the turbulent flames in Fig. 3 shows characteristic features of a premixed flame in the thin reaction zones regime on a one-dimensional domain: multiple flame structures with temperatures between unburnt and burnt temperature as a one-dimensional analog to the increased flame front area in a turbulent flame and turbulent eddies of various sizes. The (averaged) turbulent flame or consumption speed, s_D, obtained via [28]

$$s_d = \frac{M_s}{\rho_u(Y_{s,b} - Y_{s,u}} \int \dot{\omega}_s \, \mathrm{d}x, \tag{2}$$

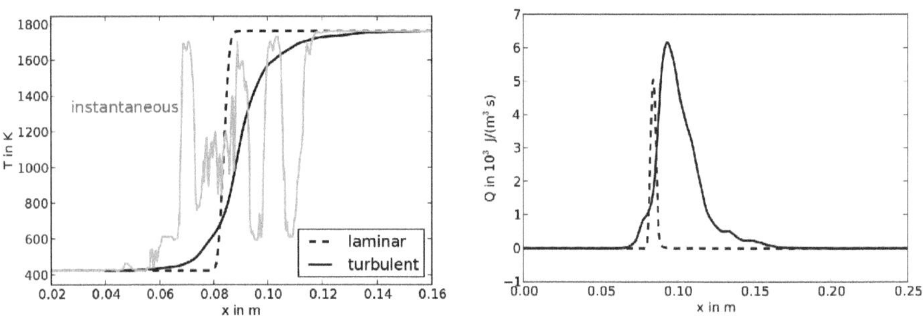

Fig. 3. Laminar and turbulent one-dimensional flame structures for a lean premixed methane/air flame at equivalence ratio $\phi = 0.6$: Temperature profiles (laminar, instantaneous and time-averaged turbulent) on the left and heat release rate on the right

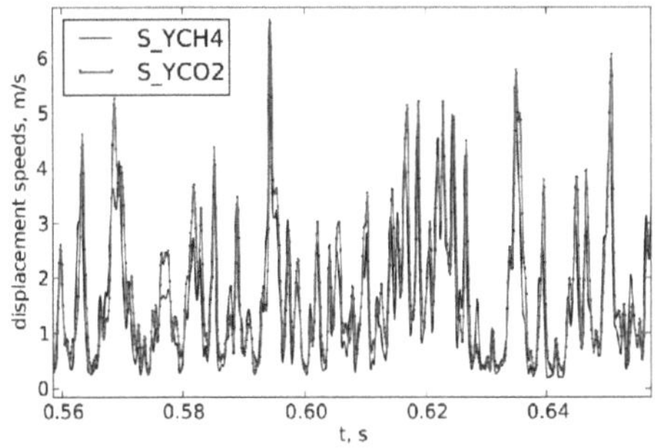

Fig. 4. Time history of the displacement speed evaluated for species CH_4 and CO_2

where indices s, u, and b denote the species index, the unburnt, and the burnt state, is 1.7 m/s, i.e. the ratio of the turbulent to the laminar flame speed is $s_d/s_l = 7.7$. The averaged profiles in Fig. 3 show the thickened turbulent flame structure and an increased heat release rate of the turbulent flame compared to the laminar flame.

Figure 4 shows a typical time history of the displacement speed, evaluated for $s = CH_4, CO_2$ in eq. (2), obtained with the LEM. For different major species such as CH_4 and CO_2 the results should be similar which is confirmed in the figure.

3.2 Influence of Equivalence Ratio Fluctuations

In this section we present results on the influence of equivalence ratio fluctuations on the dynamics of a turbulent lean premixed methane/air flame. Turbulence parameters, mean equivalence ratio, and unburnt temperature of the flame are identical with the conditions in the previous section. Equivalence ratio fluctuations are generated by sinusoidal variations of the fuel mass fraction at the inflow. The simulations are conducted for forcing frequencies ranging from 20 Hz to 150 Hz with relative amplitudes between 16% and 99%. Due to viscous and turbulent diffusion, however, the amplitudes reaching the flame front are much smaller than the imposed amplitudes at the inlet. Table 1 summarized the relative amplitudes at the flame front at given forcing frequency and amplitude at the inlet for laminar flames. Compared to the amplitudes observed in [9] at different locations of a laminar slot burner flame we investigate here much higher amplitudes entering the flame zone. As in [9] we observe strongly decreasing amplitudes at the flame front with increasing forcing frequencies. This can be explained by the steeper spatial gradients in species mass fractions leading to increased fuel–air mixing rates and hence, decreasing fluctuations at the flame front. For the investigated forcing amplitudes and frequencies in this study we did not encounter local flame quenching. For comparisons with laminar flames we adjust the flame position such that the amplitudes of the equivalence ratio

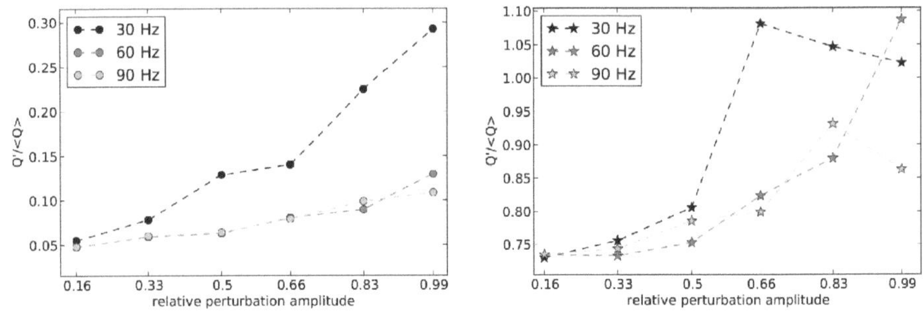

(a) laminar (b) turbulent

Fig. 5. Influence of the equivalence ratio perturbation amplitude on heat release rate fluctuations for different forcing frequencies

Table 1. Relative amplitudes at the flame for different frequencies and relative amplitudes at the inlet for laminar flames and periodic forcing

Amplitude	f in Hz		
at inlet	30	60	90
0.16	0.016	0.0021	$4.3{\cdot}10^{-4}$
0.33	0.032	0.0044	$8.9{\cdot}10^{-4}$
0.50	0.047	0.0067	0.0014
0.66	0.064	0.0088	0.0018
0.83	0.081	0.011	0.0022
0.99	0.096	0.013	0.0027

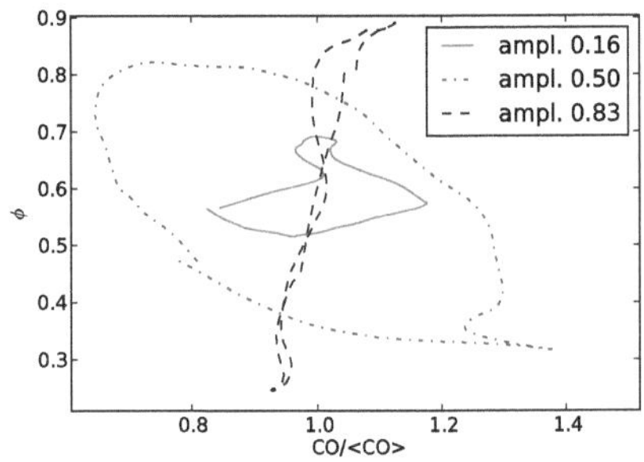

Fig. 6. Phase averaged relative CO concentrations 10 cm behind the flame front for the turbulent case at different forcing amplitudes with 30 Hz forcing frequency

fluctuations reaching the flame are comparable for the laminar and the turbulent flame at the same forcing frequency.

Fig. 5 compares the influence of the perturbation amplitude on heat release rate fluctuations for forcing frequencies of 30, 60, and 90 Hz in the laminar and the turbulent case. The laminar flames show monotonically increasing heat release rate fluctuations with increasing perturbation amplitude with a large difference between 30 and 60 Hz forcing frequency. This can be directly related to – again – the effective equivalence ratio fluctuation entering the flame front. As shown in Table 1 the relevant amplitudes of the equivalence ratio fluctuation reaching the flame front are much higher for low forcing frequencies than for high forcing frequencies. The results for 60 Hz and 90 Hz indicate that for those forcing frequencies the response to equivalence ratio fluctuations is almost linear whereas a weak convex non-linear response can be observed for the low frequency forcing of 30 Hz. The turbulent flames shows a qualitatively similar behavior for relative perturbation amplitudes up to 0.5. For perturbation frequencies of 30 Hz and 90 Hz we observe decreasing heat release rate fluctuations for amplitudes higher than 0.66 and 0.83, respectively. Forcing with 60 Hz leads to monotonically non-linear increasing heat release rate fluctuations as in the laminar case. The qualitatively different response for increasing forcing amplitudes at different forcing frequencies can be related to the phase interaction between the spatially oscillating turbulent flame brush and the prescribed sinusoidal equivalence ratio oscillations at the inlet may lead to an enhanced (or reduced) damping of the equivalence ratio fluctuations entering the flame zone. As an indicator for this claim Fig. 6 shows phase averaged relative CO concentration for the turbulent case at different forcing amplitudes with forcing frequency of 30 Hz. The figure clearly indicates a non-monotonic response of CO to forcing amplitudes. Compared to the laminar cases, the turbulent flames exhibit substantially larger relative heat release rate fluctuations. The increased heat release rate fluctuations can be related to the structure of the turbulent flame as shown in Fig. 3: In the laminar case heat release fluctuations are only due to smooth variations of equivalence ratio fluctuations entering the flame zone. In the turbulent case we see a highly disturbed unsteady flame structure leading to high turbulent fluctuations of the heat release rates.

Fig. 7 (a) shows the influence of equivalence ratio fluctuations with a relative forcing amplitude of 0.66 on heat release rate fluctuations for the laminar and the turbulent case as a function of the forcing frequency. Both laminar and turbulent flames show qualitatively similar results in the sense that the heat release rate fluctuations decrease for increasing perturbation frequencies up to 60 Hz. Above 60 Hz the heat release fluctuations remain almost constant with increasing perturbation frequencies. With increasing perturbation frequencies viscous damping increases and only small equivalence ratio amplitudes enter the flame zone. As explained above, the turbulent flame features much higher heat release rate fluctuations than the laminar flame. The almost constant values for frequencies above 60 Hz indicate that turbulence contributes most to the heat release rate fluctuations at high forcing frequencies. For all forcing frequencies the heat release rate fluctuations are in the same order of magnitude as the mean heat release rate for the turbulent flame. In comparison, the heat release fluctuations for the laminar flame start at approx. 25% of the mean heat release rate for 20 Hz and reach lower than 10% at higher frequencies.

Fig. 7 (b) displays the impact of the perturbation amplitude on the mean heat release rate for different forcing frequencies. The figure shows almost no influence for the laminar case and decreasing mean heat release rates with increasing perturbation amplitudes for the turbulent case. As the heat release rate is a linear function of the equivalence ratio for lean mixtures and small perturbation amplitudes we expect a negligible influence for the laminar case with sinusoidal equivalence ratio fluctuations entering the flame zone. As expected the turbulent flames exhibit much higher heat release rates than the laminar flames. The decreasing heat release rates for increasing pertubation amplitudes in our simulations are due to decreasing mass flow rates with increasing

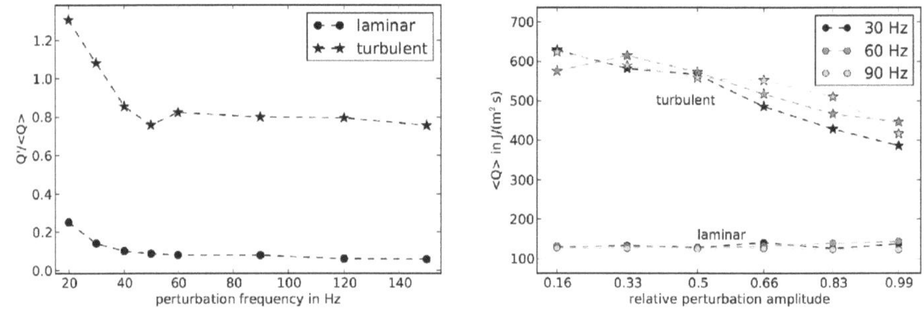

(a) heat release rate fluctuation (b) heat release rate

Fig. 7. Influence of the equivalence ratio fluctuations on the heat release rate fluctuations and the heat release rate

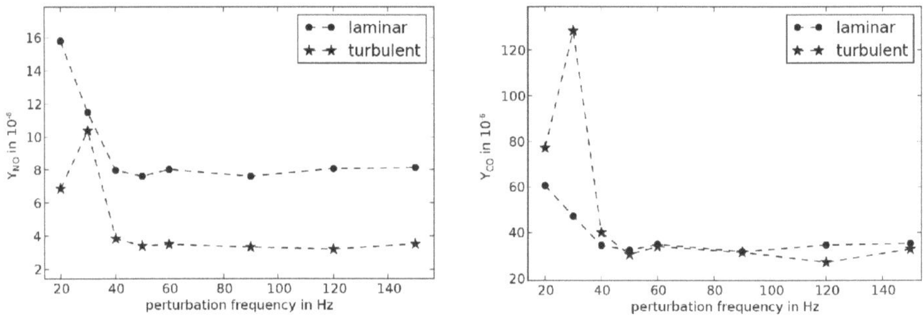

Fig. 8. Influence of the forcing frequency on NO and CO emissions at a relative forcing amplitude of 0.66

perturbation amplitudes to keep the turbulent flame brush statistically on a fixed spatial position.

Some results on the influence of equivalence ratio fluctuations on NO and CO emissions are summarized in Fig. 8. Both NO and CO equilibrium concentrations exhibit exponential growth with increasing equivalence ratio under lean conditions. NO reaches its maximum close to stoichiometric conditions and decreases with increasing equivalence ratios in the rich area due to decreasing temperatures. CO increases further in the rich area and also at very lean conditions due to quenching effects. CO and NO are both convex functions of the equivalence ratio under lean conditions. Fig. 8 shows the frequency dependence of NO and CO for a laminar and a turbulent flame with a relative forcing amplitudes of 0.66. For frequencies higher than 50 Hz NO and CO mass fractions reach a constant level. Compared to CO the NO emissions are lower in the turbulent case than in the laminar case. For the laminar flame we see increasing NO and CO values with decreasing frequencies. Due to the convex dependence of NO and CO on the equivalence ratio under lean conditions, one can expect increased values when a sinusoidal equivalence ratio fluctuation passes the flame front. For the turbulent flame we see a sharp peak for both, NO and CO, at a perturbation frequency of 30 Hz. This can be explained by the fact that at 30 Hz the equivalence ratio amplitudes observed at the flame front have the highest values. For higher frequencies we observe increased damping of the amplitudes in front of the flame. Due to nonlinear interactions of equivalence ratio oscillations and flame front propagation the amplitudes are smaller again for 20 Hz forcing frequency. A more detailed investigation of the impact of equivalence ratio fluctuations on emissions will be presented in an upcoming publication.

4 Summary and Conclusions

In this work we investigate the influence of equivalence ratio fluctuations on the dynamics of turbulent lean premixed methane/air flames under typical conditions for gas turbine combustors. The simulations are performed with a one-dimensional model where turbulent mixing is modeled via the LEM. Compared to previous analytical and numerical investigations, the LEM resolves all spatial and temporal scales and therefore allows to simulate directly the interaction of turbulence and chemistry. This is particularly important in situations where local interactions of turbulent eddies with the flame lead to disturbance or destruction of flame structures, e.g. in the thin reaction zones or distributed reaction zones regime of premixed flames, and unsteady effects such as quenching and re-ignition. Under such highly turbulent conditions flame surface variations due to curvature effects and coherent wrinkling – which are not included in the LEM – are likely to be of minor importance relative to flame surface variations due to turbulence. However, a confirmation of this statement is left for further studies.

The presented results demonstrate the capabilities of the one-dimensional LEM approach to simulate qualitatively and to some extend quantitatively the non-linear response of turbulent premixed flames to equivalence ratio fluctuations. The results show partly substantially qualitatively different results for laminar and turbulent flames for the same forcing frequencies and amplitudes reaching the flame zone. Almost all results indicate that at high forcing frequencies viscous damping significantly decreases the effective equivalence ratio fluctuations entering the flame zone and that turbulence dominates the effect of increased heat release rate fluctuations. However, for low forcing frequencies we observe a substantial influence of equivalence rate fluctuations on heat release rate fluctuations.

Further studies with the model will focus on the impact of equivalence ratio fluctuations on emissions and the coupling with an acoustic model to simulate equivalence ratio fluctuations driven by the interaction with acoustic modes leading to flame instabilities. Current work is also focused on using the LEM as a sub-grid combustion model for large eddy simulations.

Acknowledgments. The authors gratefully acknowledge the financial support of the German Science Foundation (DFG). We also thank Alan Kerstein and David Lignell for their support and many fruitful discussions.

References

1. Abu-Orf, G.M., Cant, R.S.: Joint Meeting of the Portuguese, British, Spanish and Swedish Sections of the Combustion Institute, April 1-4, Madeira, Portugal (1996)
2. Armitage, C.A., Balachandran, R., Mastorakos, E., Cant, R.S.: Combustion and Flame. Combustion and Flame 146, 419–436 (2006)
3. Balachandran, R., Ayoola, B.O., Kaminski, C.F., Dowling, A.P., Mastorakos, E.: Combustion and Flame 143, 37–55 (2005)
4. Balachandran, R., Dowling, A.P., Mastorakos, E.: Proceeding of the European Combustion Meeting (2011)
5. Birbaud, A.L., Ducruix, S., Durox, D., Candel, S.: Combustion and Flame 154, 356–367 (2008)
6. Cho, J.H., Lieuwen, T.: Combustion and Flame 140(1-2), 116 – 129 (2005)
7. Cohen, S.D., Hindmarsch, A.C.: Computers in Physics 10(2), 138–143 (1996)
8. Goodwin, D.: Cantera: An object-oriented software toolkit for chemical kinetics, thermodynamics, and transport processes (2009), `http://code.google.com/p/cantera`
9. Hemchandra, S.: Combustion and Flame 159, 3530–3543 (2012)
10. Hemchandra, S., Peters, N., Lieuwen, T.: Proceedings of the Combustion Institute 33, 1609–1617 (2011)
11. Huang, Y., Yang, V.: Progress in Energy and Combustion Science 35, 293–364 (2009)
12. Jozefik, Z., Kerstein, A.R., Schmidt, H.: To be presented at the "Active Flow and Combustion Control" Conference, Berlin, Germany (September 2014)
13. Kerstein, A.R.: Combustion Science and Techology 60, 391–421 (1988)
14. Kerstein, A.R.: Combustion and Flame 75, 397–413 (1989)
15. Kerstein, A.R.: Journal of Fluid Mechanics 231, 361–394 (1991)
16. Kerstein, A.R.: Combustion Science and Technology 81, 75–96 (1992)
17. Kerstein, A.R.: Journal of Fluid Mechanics 240, 289–313 (1992)
18. Kerstein, A.R.: Combustion and Flame 75(3-4), 397–413 (1989)
19. Kerstein, A.R.: Journal of Fluid Mechanics 231, 361–394 (1991)
20. Kim, K.T., Lee, J.G., Quay, B.D., Santavicca, D.: Journal of Engineering for Gas Turbines and Power 133(2), 021502 (2010)
21. König, K., Bykov, B., Maas, U.: Flow, Turbulence and Combustions 83, 105–129 (2009)
22. Lacarelle, A., Moeck, J.P., Tenham, A., Paschereit, C.O.: 47th AIAA Aerospace Sciences Meeting and Exhibit, Orlando, Florida. AIAA paper, January 5-8, pp. 2009–2986 (2009)
23. Lieuwen, T., Neumeier, Y., Zinn, B.T.: Combustion Science and Technology 135, 193–211 (1998)
24. Lieuwen, T., Zinn, B.T.: 27th Symposium (International) on Combustion/The Combustion Institute, pp. 1809–1816 (1998)
25. Lieuwen, T.C., Yang, V.: AIAA, Inc. (2005)
26. Lignell, D.O., Kerstein, A.R., Sun, G., Monson, E.I.: Theoretical and Computational Fluid Dynamics 27(3-4), 273–295 (2013)
27. Menon, S., Kerstein, A.R.: Twenty-Fourth Symposium (International) on Combustion, pp. 443–450 (1992)
28. Oevermann, M., Schmidt, H., Kerstein, A.: Combustion and Flame 155, 370–379 (2008)

29. Paschereit, C.O., Gutmark, E., Weisenstein, W.: Combustion Science and Technology 138, 213–232 (1998)
30. Paschereit, C.O., Gutmark, E., Weisenstein, W.: AIAA Paper 99-0711 (1999)
31. Paschereit, C.O., Gutmark, E.J., Weisenstein, W.: AIAA Journal 38(6), 1025–1034 (2000)
32. Patel, N., Menon, S.: Combustion and Flame 153, 228–257 (2008)
33. Peters, N.: Turbulent Combustion. Cambridge University Press (2000)
34. Poinsot, T., Candel, S.M.: Combustion Science and Technology 61, 121–151 (1988)
35. Richardson, E.S., Granet, V.E., Eyssartier, A., Chen, J.H.: Combustion Theory and Modelling 14(6), 775–792 (2010)
36. Sankaran, V., Menon, S.: Proceedings of the Combustion Institute 28, 203–209 (2000)
37. Sankaran, V., Menon, S.: Proceedings of the Combustion Institute 30, 575–582 (2005)
38. Sattelmayer, T.: Journal of Engineering for Gas Turbines and Power 125, 11–19 (2003)
39. Schmidt, H., Oevermann, M., Bastiaans, R.J.M., Kerstein, A.R.: Technical Report 09-09, Konrad-Zuse-Zentrum Berlin, ZIB (2009)
40. Schrödinger, C., Nolte, D., Oevermann, M., Paschereit, C.O.: Proceedings of the European Combustion Meeting (2013)
41. Sen, B.A., Menon, S.: Combustion and Flame 157, 566–578 (2010)
42. Sen, B.A., Menon, S.: Combustion and Flame 157, 62–74 (2010)
43. Seo, S., Lee, S.-Y.: Flow, Turbulence and Combustion 85, 95–112 (2010)
44. Shreekrishna, Hemchandra, S., Lieuwen, T.: Combustion Theory and Modelling 14(5), 681–714 (2010)
45. Smith, G.P., Golden, D.M., Frenklach, M., Moriarty, N.W., Eiteneer, B., Goldenberg, M., Bowman, C.T., Hanson, R.K., Song, S., Gardiner Jr., W.C., Lissianski, V.V., Qin, Z.: GRI-Mech 3.0, The Gas Research Institute (1999), `http://www.me.berkeley.edu/gri-mech/`
46. Smith, T.M., Menon, S.: Combustion Science and Technology 128, 99–130 (1997)
47. Some, K., Menon, S.: Journal of Engineering for Gas Turbines and Power 125, 435–443 (2003)
48. Venkataraman, K.K., Preston, L.H., Simons, D.W., Lee, B.J., Lee, J.G., Santavicca, D.A.: Journal of Propulsion and Power 15(6), 909–918 (1999)
49. Woosley, S.E., Kerstein, A.R., Sankaran, V., Aspden, A.J., Röpke, F.K.: The Astrophysical Journal 704, 255–273 (2009)
50. Zhou, R., Hochgreb, S.: Combustion and Flame 160, 1070–1082 (2013)

Model-Based Detection of Misfirings in an Annular Burner Mockup

Sascha Wolff and Rudibert King

Technische Universität Berlin, Fachgebiet Mess- und Regelungstechnik,
Hardenbergstraße 36a, 10623 Berlin, Germany
{sascha.wolff,rudibert.king}@tu-berlin.de

Abstract. If a detonation fails in a tube of a pulsed detonation engine, special measures are needed to refill this tube with fresh air. To initiate these measures, misfirings have to be detected reliably. In this contribution, the acoustic effects of burner tubes on a down-stream plenum are exploited to determine which of multiple tubes did not produce the expected acoustic signature. This is done with a bank of Kalman filters in combination with the Bayes' rule. To develop and test these methods, a surrogate, non reacting experimental set-up is considered.

Keywords: Failure Detection, Pulse Detonation Engine, Kalman Filter, Bayes' Rule.

1 Introduction

Due to the fact that the components of modern gas turbines are well tuned to operate at high efficiency levels, new approaches are necessary if their performance is to be enhanced further. In current gas turbines the combustion is described by an isobaric process resulting in a rise of temperature at a constant pressure level. A higher efficiency could be achieved by using isochoric combustion which is accompanied by an increase of the pressure level in the combustion chamber [1]. This can be accomplished, e.g. by changing the combustion type from a deflagration to a detonation. Thus, the question arises how to get air, which is needed for the combustion, from the turbine's compressor into the combustion chamber, i.e. from a lower to a higher pressure level. One method to overcome this challenge is to use the acoustics induced by the detonations itself. For explanation, we first consider a single detonation tube. By igniting a detonation in the tube an acoustic pressure wave is generated. It is then reflected as a negative pressure wave when reaching the open end of the tube. This end is coupled to a plenum. The pressure of the reflected wave is below the mean pressure level of the combustion chamber, and, even more important, below the pressure level of the turbine's compressor. When this wave reaches the closed, upstream, end of the tube, which is connected to the compressor, it can be refilled with air from the compressor by opening the closed end at the correct time. A new detonation wave can be initiated, and the process is sustained. Hence, the crucial factor to keep the system running is the acoustic pressure wave generated by

R. King (ed.), *Active Flow and Combustion Control 2014*, Notes on Numerical Fluid Mechanics and Multidisciplinary Design 127, DOI: 10.1007/978-3-319-11967-0_15

the detonation. If the detonation fails (e.g. wrong mixture of fuel and oxidizer for the given conditions in the tube), the tube cannot be refilled because of the missing negative pressure wave. Therefore, the combustion chamber design is composed of a circular array of detonation tubes which all fire into a common plenum. For a restart of a faulty tube, we propose to use the remaining working ones. This could be achieved by a proper firing synchronisation producing a pressure minimum in the plenum at the open end of the faulty tube. To initiate such a firing sequence, it is necessary to first detect the failed detonation. As a detonation wave generates high temperatures and pressures, the application of sensors inside or near the detonation zone is delicate. Moreover, for deliberately rotating the pressure profile inside the plenum in case of a misfiring, pressure sensor information from the plenum is needed in any case. For this reason only pressure sensors in the annular plenum are considered here for fault detection.

Since the acoustic waves generated by the detonations excite acoustic modes in the annular plenum, the tubes can be considered as actuators of a dynamic system from a control engineering point of view. Misfiring, hence, can be viewed as an actuator fault. Many different methods have been used in the past for fault detection of actuators. One particular approach utilises a disturbance observer for the estimation of the actuator inputs [2],[3]. To this end, new states corresponding to the inputs are introduced into the state observer's differential equations. Thus, the actuation inputs are estimated and actuator faults can be detected. A problem with this approach is that the observability of the system is quickly lost if multiple inputs are to be estimated. Also, quick changes in the actuator signals cannot be detected satisfactorily due to measurement noise. Another way to detect actuator faults is to use so-called unknown input estimators [4],[5]. This way, tracking of fast changing actuation signals is possible, but the number of observable inputs is strictly constrained to the number of available sensors. For this reason, this contribution exploits a multiple filter concept extended by a Bayesian estimation approach to calculate the probability of an actuator fault having happened [6],[7],[8],[9],[10].

To set up the method, a mathematical model of the acoustics of the annular plenum is first derived and the model parameters are adjusted with the help of experimental data in Sect. 2. For simplicity, the experimental set-up to validate the aforementioned algorithm does not use detonation tubes but loudspeakers, which generate acoustic waves similar to the signature of a detonation. In Sect. 3, the statistical methods to detect faults are revisited. Sect. 4 gives experimental results, before some conclusions are drawn in Sect. 5.

2 Experimental Set-Up, Physical Model and System Identification

The experimental set-up is described in Sect 2.1. For model-based fault detection, a model is needed relating actuator signals with pressure readings. Starting from the equations of the homogeneous system without inputs in Sect. 2.2, transfer functions are formulated combining acoustic particle velocities of the actuators with measured pressures, i.e, describing the inhomogeneous system. To account for effects not modelled theoretically, model parameters are adapted in section 2.3.

2.1 Experimental Set-Up

The acoustic system used for the experiments is shown in Fig. 1. The experimental set-up mainly consists of two concentrically arranged aluminium tubes with different diameters covered by an aluminium circular ring plate on both sides. The ring plate on the front side at $z = 0$ is provided with twelve holes arranged in a circle for the connection of actuators mimicking the effect of detonation tubes. Valves connected to pressurized air or loudspeakers can be used for this purpose. Here, we concentrate on the use of loudspeakers.

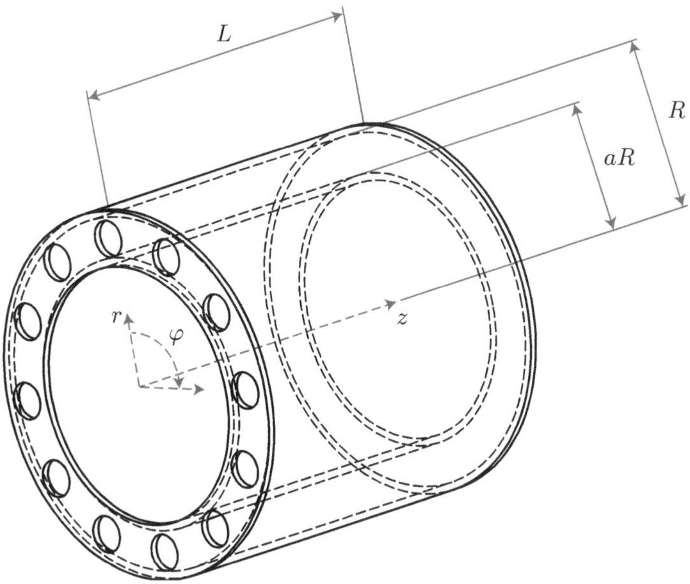

Fig. 1. Anular gap with twelve circular holes in the front plate (at $z = 0$) for the connection of surrogate firing tubes/actuators. The dimensions of the system are given by: $L = 0.6\,\text{m}$, $aR = 0.32\,\text{m}$, $R = 0.4\,\text{m}$, $a = 0.8$.

The block diagram in Fig. 2 shows the interconnection between the essential elements of the test rig. A digital signal processor (DSP), see Tab. 1, running at a sampling rate of $f_s = 10^4\,\text{Hz}$, controls the voltage of the loudspeakers which results in a movement of their respective membranes. By this, acoustic particle velocities are generated equal to the velocity of the loudspeaker's membrane, exciting the annular duct. To measure the effect, twelve microphones arranged in a circle and equally spaced are used. The microphones are calibrated with a pistonphone, see Tab. 1. The position vectors of the k-th loudspeaker and j-th microphone are given by

$$\boldsymbol{x}_k = \left[\tfrac{R(a+1)}{2},\, \tfrac{\pi}{2}(k-1),\, 0L\right]^T \ , \tag{1}$$

$$\boldsymbol{x}_j = \left[R,\, \tfrac{\pi}{2}(k-\tfrac{1}{2}),\, \tfrac{1.03}{6}L\right]^T \ . \tag{2}$$

Note that a cylindrical coordinate system $[r,\, \varphi,\, z]^T$ shown in Fig. 1 is used here. Furthermore, a laser vibrometer is used to measure the velocities of the loudspeaker membranes, and, thus, the acoustic particle velocity generated by the associated actuator.

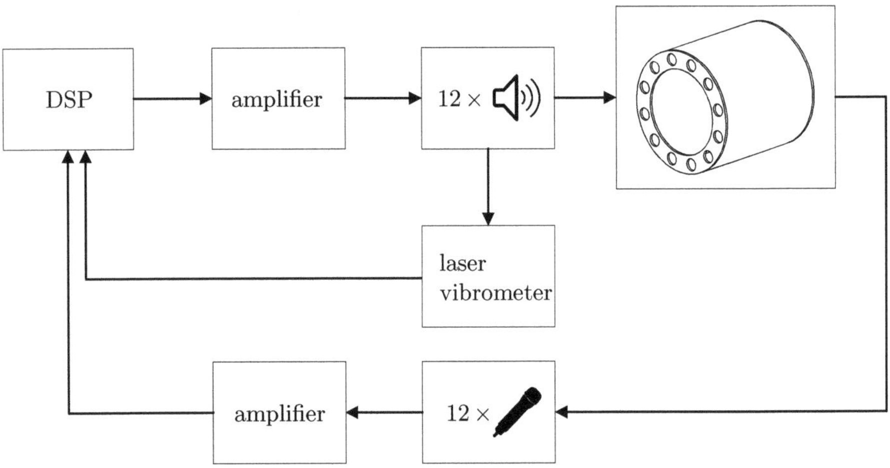

Fig. 2. Experimental set-up with DSP, amplifiers, actuators and sensors

2.2 Physical Model of the Acoustic System

Homogeneous System. As nonlinear acoustic effects cannot be produced in the experimental set-up, the calculation of the eigenfunctions and eigenfrequencies of the system is based on the solution of the homogeneous linear acoustic wave equation [11]

$$\frac{1}{c^2}\frac{\partial^2 p}{\partial t^2} - \Delta p = 0 \ , \tag{3}$$

and the linearised Euler equation

$$\rho_0 \frac{\partial \boldsymbol{u}}{\partial t} = -\nabla p \ . \tag{4}$$

Table 1. Measurement equiment

DSP	dSpace DS1006 processor board
	dSpace DS2103 DA-board
	dSpace DS2004 AD-board
loudspeakers	12 × TI-100 8 Ohm Visaton
microphones	12 × G.R.A.S. 40BP 1/4" ext. polarized pressure microphone
	12 × G.R.A.S. 26AC 1/4" standard preamplifier
laser vibrometer	Polytec CLV 1000
	Polytec CLV 700
pistonphone	G.R.A.S. 42AA pistonphon

The vector of velocities is given by $\boldsymbol{u} = [u_r, u_\varphi, u_z]^T$, where all entries as well as p are fluctuating quantities. The boundary conditions of the system for rigid walls are

$$u_r(t, r = R, \varphi, z) = 0 \ , \tag{5}$$
$$u_r(t, r = aR, \varphi, z) = 0 \ , \tag{6}$$
$$u_z(t, r, \varphi, z = 0) = 0 \ , \tag{7}$$
$$u_z(t, r, \varphi, z = L) = 0 \ . \tag{8}$$

The variables t, c, ρ_0, p, $\boldsymbol{u}$, L, R, a denote time, speed of sound, mean density of the gas, acoustic pressure, acoustic particle velocity, length, outer radius and ratio of inner to outer radius of the annular plenum. The solution of such systems is well known [11].

For the subsequent application of the mathematical model in the experimental fault detection only a limited number of eigenmodes and eigenfrequencies will be considered due to limitations in real-time computation. A small annular gap leads to very high eigenfrequencies for higher radial modes. Since high eigenfrequencies are considerably damped and thus not much involved in the response of the system, the radial dependence of the pressure is neglected. If additionally $a > 0.7$, the eigenfrequencies of the system are given by [12]

$$\omega_{lm} = c\sqrt{\left(\frac{l\pi}{L}\right)^2 + \left(\frac{2m}{R(1+a)}\right)^2} \ , \quad l, m \in \mathbb{N}_0 \ , \tag{9}$$

where l and m are the number of the axial and azimuthal mode, respectively. In combination with the eigenfunctions [12]

$$\Phi_{c,lm} = \cos\left(\tfrac{l\pi}{L}z\right)\cos\left(m\varphi\right) \ , \tag{10}$$
$$\Phi_{s,lm} = \cos\left(\tfrac{l\pi}{L}z\right)\sin\left(m\varphi\right) \ , \tag{11}$$

where the indices s and c represents the sine and cosine mode, respectively, the solution of (3) is given by

$$p(\varphi, z, t) = \sum_{l=0}^{\infty} \sum_{m=0}^{\infty} (A_{lm} \Phi_{c,lm} + B_{lm} \Phi_{s,lm}) e^{i\omega_{lm} t} \ , \tag{12}$$

with A_{lm} and B_{lm} coefficients determined by the initial conditions of the system. Note that for m unequal to 0 or a multiple of 6, the eigenfrequencies are degenerate with an algebraic and geometric multiplicity of two. The eigenfunctions (10) and (11) form an orthogonal basis of the associated two-dimensional eigenspace.

Inhomogeneous System. Each loudspeaker $k \in [1, 2, \ldots, 12]$ is modelled as an ideal point source in the acoustic particle velocity $\boldsymbol{u}_k = [u_{k,r}, u_{k,\varphi}, u_{k,z}]^T$, where $||\boldsymbol{u}_k||_2 = u_{k,z}$. Each point source is positioned in the centre of a hole of the front plate. A linear acoustic system with sources and an additional term for the dissipation is described by

$$c^2 \Delta p - \alpha \frac{\partial p}{\partial t} - \frac{\partial^2 p}{\partial t^2} - = h \ , \tag{13}$$

$$\boldsymbol{n} \cdot \nabla p = -f \ , \tag{14}$$

with h,f,$\boldsymbol{n}$ and α being a source in the volume, a source at the surface, the normal vector of the volume's surface at the position of the source and the damping coefficient, respectively [13]. Equations (13), (14) can be transformed into the frequency domain and solved using Green's function which can be composed of a linear combination of the eigenfunctions (10),(11) used as modes. If flush-mounted pressure sensors are applied to the plenum wall, a transfer function $H_{j,k}(\imath\Omega) = \frac{\hat{p}_j}{\hat{u}_{k,z}}$ can be derived relating the j-th pressure sensor information in the frequency domain $\hat{p}_j$ to the k-th actuator signal in the frequency domain $\hat{u}_{k,z}$

$$H_{j,k}(\imath\Omega) = -\imath\Omega\rho_0 c^2 A \sum_{l,m=0}^{\infty} \left(\frac{\Phi_{s,lm}(\boldsymbol{x}_j)\Phi_{s,lm}(\boldsymbol{x}_k) + \Phi_{c,lm}(\boldsymbol{x}_j)\Phi_{c,lm}(\boldsymbol{x}_k)}{\Lambda_{lm}\left(\Omega^2 - i\Omega\alpha - \omega_{lm}^2\right)} \right) . \tag{15}$$

Here, Ω, $\imath$, ρ_0 and A are the frequency, imaginary unit, mean density of the gas inside the annular gap and the area of the individual holes of the front plate. The position vectors of the k-th actuator and j-th sensor are given by (1),(2). Theoretically, the factor Λ_{lm} amounts to [13]

$$\Lambda_{lm} = \int_0^L \int_0^{2\pi} \int_{aR}^{R} \Phi_{c,lm}^2 r dr d\varphi dz = \frac{L\pi R^2(1-a^2)}{(2-\delta(l))(2-\delta(m))} \ , \tag{16}$$

$$\delta(\xi) = \begin{cases} 1, & \xi = 0 \\ 0, & \xi \neq 0 \end{cases} \ . \tag{17}$$

2.3 Reduced Model and System Identification

Reduced Model. To account for effects seen in the experiment but not described by (15), some parameters are identified experimentally. Moreover, due to the limited computing capacity available for a real-time fault detection the infinite sum in (15) has to be approximated by a finite one. Thus only summands corresponding to those eigenfrequencies which are within the range of the frequencies excited by the actuators can be regarded. This will be partly based on experimental data as well. To this end, the signal of the vibrometer, measuring the input of the system, is used to identify a transfer function relating it to the acoustic pressures as output signals. The output is measured with $n_y = 12$ microphones. Aiming for a good match between the analytical model and the experimental set-up, the system parameters ω_{lm}, Λ_{lm} will be adjusted by an identification algorithm. In addition, the damping factor α in (15) is replaced by modal damping coefficients α_{lm}, allowing for a different damping factor for every eigenfrequency. These new parameters α_{lm} will be also adjusted by the identification algorithm. This way, a better agreement between the experimental plant and the analytical model is achieved.

The signal of the acoustic particle velocity used to simulate a detonation event is shown in Fig. 3(a). The important frequency components of this signal are calculated next. For that purpose, the normalized energy contained in a specific frequency range from 0 Hz to Ω, denoted as $\Pi(\Omega)$, is defined

$$\Pi(\Omega) = \frac{\int_{0\,\mathrm{Hz}}^{\Omega} S_{uu}\, d\Omega}{\int_{0\,\mathrm{Hz}}^{f_s/2} S_{uu}\, d\Omega}, \quad \Omega \leq f_s/2 \ . \tag{18}$$

Here, $f_s/2$ is the Nyquist frequency for the given sampling rate and S_{uu} is the power spectral density calculated from the signal in Fig 3(a). For the signal from Fig 3(a), 90 % of the energy of the signal is contained in a frequency range from

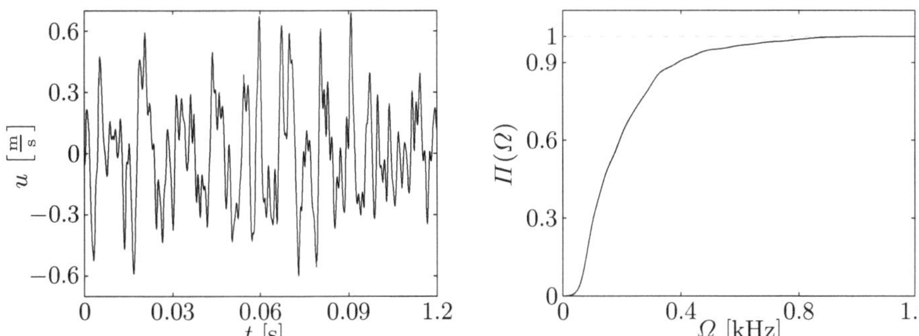

(a) Particle velocity at a membrane measured by a vibrometer during a simulated detonation event.

(b) Normalized integral of the power spectral density of the signal in Fig. 3(a).

Fig. 3. Time and frequency domain data of a proposed detonation event

0 Hz to 400 Hz, i.e. $\Pi(0.4\,\text{kHz}) \approx 0.9$, see Fig. 3(b). Using (9), the expected eigenfrequencies for the given geometry (see Fig. 1) are listed in Tab. 2. Combining these two observations, only the eigenfrequencies written in black are considered as main contributors to the behaviour of system.

The structure of the system from the k-th actuator to the j-th sensor including the seven chosen eigenfrequencies is given by a transfer function of order

$$n_x = \underbrace{5 \cdot 2 \cdot 2}_{\text{degenerated modes } (m \neq 0)} + \underbrace{2 \cdot 2 \cdot 1}_{\text{non degenerated modes} (m = 0)} = 24. \quad (19)$$

To determine the unknown parameters of these transfer functions, a state space representation

$$\dot{\tilde{\boldsymbol{x}}}(t) = \tilde{\boldsymbol{A}}\tilde{\boldsymbol{x}}(t) + \tilde{\boldsymbol{B}}\boldsymbol{u}(t) \ , \quad (20)$$

$$\boldsymbol{y}_c(t) = \boldsymbol{C}\tilde{\boldsymbol{x}}(t) \ , \quad (21)$$

of the system is used, where $\boldsymbol{u} \in \mathbb{R}^{n_u}, \tilde{\boldsymbol{x}} \in \mathbb{R}^{n_x}$, $\boldsymbol{y}_c \in \mathbb{R}^{n_y}$, $\tilde{\boldsymbol{A}} \in \mathbb{R}^{n_x \times n_x}, \tilde{\boldsymbol{B}} \in \mathbb{R}^{n_x \times n_u}$ and $\boldsymbol{C} \in \mathbb{R}^{n_y \times n_x}$, are the states of the system, the acoustic pressures measured by the microphones, the system matrix, the input matrix and the output matrix, respectively. The dimensions for the given system are $n_y = 12$, $n_u = 12$ and $n_x = 24$. Since a discrete-time state space model is needed in Sect. 3, the corresponding discrete-time system is given by

$$\boldsymbol{x}(t_{i+1}) = \boldsymbol{A}\boldsymbol{x}(t_i) + \boldsymbol{B}\boldsymbol{u}(t_i) \ , \quad (22)$$

$$\boldsymbol{y}(t_i) = \boldsymbol{C}\boldsymbol{x}(t_i) \ , \quad (23)$$

$$t_i = i \cdot T_s \ , i \in \mathbb{N} \ , \quad (24)$$

with T_s being the sampling time of the DSP.

Identification Experiment. For the identification of the adjustable parameters a linear chirp signal from 60 to 500 Hz with a peak-to-peak voltage of 10 V is used to excite the loudspeakers consecutively. As an example, the non-parametric frequency response $G1$ from u_1 to y_1, identified by using a spectral analysis [14], is shown in Fig. 4. The bode plot for the same transfer function calculated from the analytical model, denoted by $G2$, including the aforementioned

Table 2. Eigenfrequencies [Hz] of the annular gap calculated using (9)

		m-th mode						
		0	1	2	3	4	5	6
l-th mode	0	0.00	151.64	303.28	454.92	606.56	758.20	909.84
	1	285.83	323.57	416.75	537.26	670.53	810.29	953.68
	2	571.67	591.44	647.13	730.58	833.50	949.56	1074.52
	3	857.50	870.80	909.55	970.70	1050.34	1144.63	1250.24
	4	1143.33	1153.35	1182.87	1230.51	1294.27	1371.89	1461.17

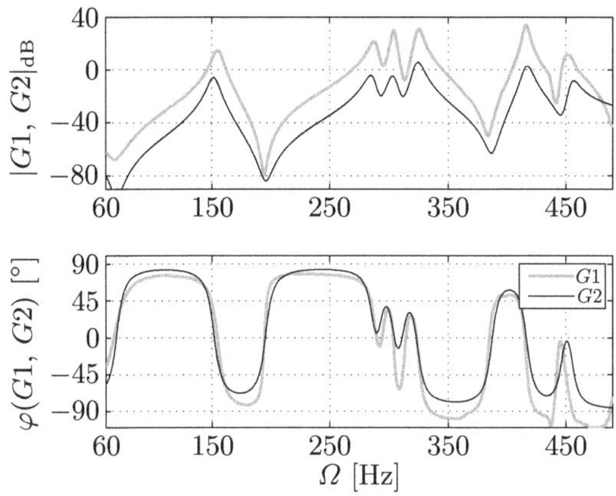

Fig. 4. Frequency responses of the experimental plant and the analytical model without parameter adjustment. $G1$ is the Frequency response from u_1 to y_1, identified by using a spectral analysis. $G2$ is the Frequency response from u_1 to y_1 caculated using (15) with $\alpha = 60$.

7 eigenfrequencies and a damping factor $\alpha_{lm} = 60, \forall\, l, m$ without any parameter adjustments is shown in the same figure. Even though non of the parameters are calibrated, there is already a rather good qualitative match between both frequency responses. This match will be improved in the next step.

Parameter Identification. For a better match of the analytical model and the real system, the aforementioned parameters ω_{lm}, Λ_{lm} and α_{lm} are calibrated by the prediction error method (PEM) [14]. To this end, all other parameters in the time continuous state space representation of the analytical model are fixed. The tunable ones are initialized with $\alpha_{lm} = 60$ and with the values calculated by (9) and (16), respectively. Figure 5 shows the same transfer functions as depicted in Fig. 4, but now with adjusted parameters resulting from the PEM algorithm. The transfer function with adjusted parameters is denoted by G_3.

Cross-Validation. The last step of the identification is the validation of the identified model with a cross-validation experiment which was not used for identification. The actuator signal used for this purpose is the one shown in Fig. 3(a). A small time frame of the real sensor data y_1 and the output of the adjusted analytical model for this experiment $\hat{y}_1$ are compared in Fig. 6. An adequate match of both signals is achieved. The fit value related to the normalised mean square error for the time frame of the whole experiment and all n_y sensors is given by

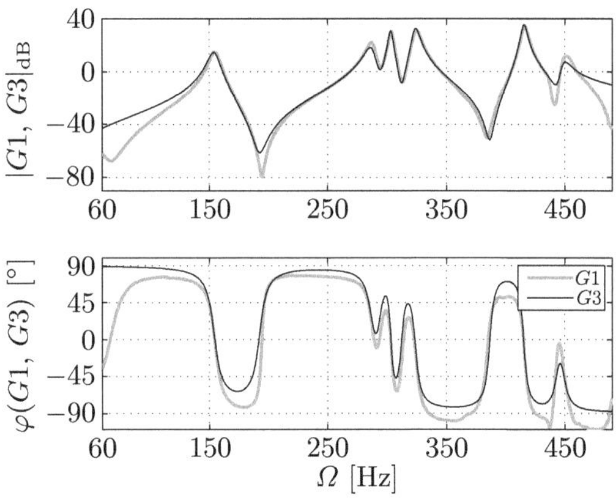

Fig. 5. Frequency responses of the experimental plant and the analytical model with adjusted parameters. G_3 is the frequency response from u_1 to y_1 caculated using (15) with adjusted parameters ω_{lm}, Λ_{lm} and α_{lm}.

$$f_{\mathrm{NMSE}} = \frac{1}{n_y} \sum_{k=1}^{n_y} \left(1 - \frac{||\boldsymbol{y}_k - \hat{\boldsymbol{y}}_k||_2^2}{||\boldsymbol{y}_k - \bar{\boldsymbol{y}}_k||_2^2} \right) = 0.74 \ , \tag{25}$$

where $\boldsymbol{y}_k = [y_k(t_i), y_k(t_{i+1}), y_k(t_{i+2}), \dots]$ and $\hat{\boldsymbol{y}}_k = [\hat{y}_k(t_i), \hat{y}_k(t_{i+1}), \hat{y}_k(t_{i+2}), \dots]$. The variable $\bar{\boldsymbol{y}}_k$ denotes the mean value of the time series data of the k-th sensor $\boldsymbol{y}_k$. In comparison, the fit value for the two signals depicted in Fig 6 amounts to 0.84. Accordingly, the identified system is expected to describe the behaviour of the experimental plant sufficiently well.

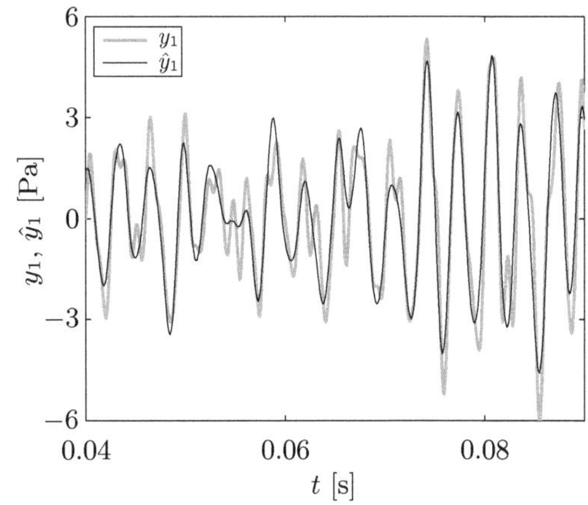

Fig. 6. Measured and modeled pressure from cross-validation experiment

3 Fault Detection

To detect whether and which actuator/detonation tube had failed, a set of models is used in parallel. A nominal model assumes that all actuators perform as expected. A first fault model describes the effect of a failure of the first actuator. An additional fault model is based on the occurrence of a second actuator fault, and so forth. Now, the predictions of all models are compared against measurements obtained in the framework of model-based estimators, i.e. Kalman filters. A Bayesian probability approach is finally used to decide which model, i.e. which fault, can describe the measurements best. In the case considered below, a single fault model has to suffice, due to computational limits. However, by respecting some restrictions on the ignition sequence, which are explained below, a reliable and fast fault detection can be implemented this way.

Bayesian Probability Approach for Linear Stochastic Models. Assume a number of $n \in \mathbb{N}$ model hypotheses, where M_k, $k \in [1, \ldots, n]$ denotes one of the n possible models. The real model is denoted by M^r. Let in addition M^r and M_k, $\forall k$ be linear stochastic models. Define the k-th hypotheses conditional probability

$$p_k(t_i) = p(M_k|\boldsymbol{Y}(t_i)) \ , \tag{26}$$

as the probability of model M_k, under the condition that $\boldsymbol{Y}(t_i)$, the measurement history until t_i, is observed. With these assumptions, $p_k(t_i)$ can be calculated with a recursive algorithm by using the Bayesian probability approach

$$p_k(t_i) = \frac{p(\boldsymbol{Y}(t_i)|M_k, \boldsymbol{Y}(t_{i-1}))p_k(t_{i-1})}{\sum_{j=1}^{n} p(\boldsymbol{Y}(t_i)|M_j, \boldsymbol{Y}(t_{i-1}))p_j(t_{i-1})} \ . \tag{27}$$

Here, $p_k(t_{i-1})$ and $p_k(t_i)$ are the a priori and a posteriori probabilities, respectively, and $p(\boldsymbol{Y}(t_i)|M_k, \boldsymbol{Y}(t_{i-1}))$ is the so-called likelihood function. The calculation of the likelihood functions is performed below by means of Kalman filters [15]. Starting the detection process for $t = t_0$, the first a priori probabilities $p_k(t_0)$ are typically initialized with a value of $\frac{1}{n}$, when n is the number of considered model alternatives. One important property of this algorithm is that, if $M^r \notin (\cup_{\forall k} M_k)$, the model with the best match will still get the highest conditional probability.

Multiple Filter Concept. The idea of the multiple filter concept is depicted in Fig. 7. A number of Kalman filters (KFs) is running in parallel [6]. All filters are updated at the sampling instants $i \cdot T_s$ by measurements $\boldsymbol{y}(t_i)$, $\boldsymbol{y} \in \mathbb{R}^{n_y}$. Each KF belongs to a different type of error that could occur in the real plant. The filters are tuned to operate well for a specific failure mode. Thus a discrimination in terms of the "correctness" of each filter's behaviour can be achieved. This is done by using the aforementioned Bayesian probability approach as will be outlined in the following.

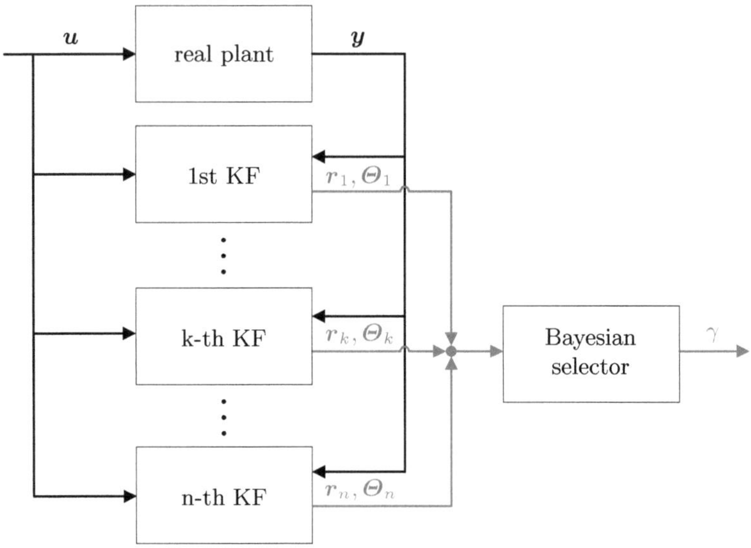

Fig. 7. Multiple filter concept for fault detection

Assume that the k-th system corresponding to the k-th error hypothesis be given by

$$\boldsymbol{x}_k(t_{i+1}) = \boldsymbol{A}_k \boldsymbol{x}_k(t_i) + \boldsymbol{B}_k \boldsymbol{u}_k(t_i) + \boldsymbol{w}_k(t_i) \ , \quad (28)$$

$$\boldsymbol{y}_k(t_i) = \boldsymbol{C}_k \boldsymbol{x}_k(t_i) + \boldsymbol{v}_k(t_i) \ , \quad (29)$$

with $\boldsymbol{x} \in \mathbb{R}^{n_{x_k}}$, $\boldsymbol{y} \in \mathbb{R}^{n_y}$ $\boldsymbol{A}_k \in \mathbb{R}^{n_{x_k} \times n_{x_k}}$, $\boldsymbol{B}_k \in \mathbb{R}^{n_{x_k} \times n_{u_k}}$ and $\boldsymbol{C}_k \in \mathbb{R}^{n_y \times n_{x_k}}$. The variables $\boldsymbol{v}_k \in \mathbb{R}^{n_x}$ and $\boldsymbol{w}_k \in \mathbb{R}^{n_y}$ are uncorrelated white noise processes with $\boldsymbol{w}_k \sim \mathcal{N}(0, \boldsymbol{R}_k)$ and $\boldsymbol{v}_k \sim \mathcal{N}(0, \boldsymbol{Q}_k)$, respectively. Thus, $\boldsymbol{Q}_k \in \mathbb{R}^{n_{x_k} \times n_{x_k}}$ and $\boldsymbol{R}_k \in \mathbb{R}^{n_y \times n_y}$ are the covariance matrices of the process noise and of the measurement noise. The error-free system is designated with $k = 1$. The k-th KF is given by the equations for the measurement update

$$\boldsymbol{\Theta}_k(t_{i+1}) = \boldsymbol{C}_k \boldsymbol{P}_k(t_{i+1}|t_i) \boldsymbol{C}_k{}^T + \boldsymbol{R}_k \ , \quad (30)$$

$$\boldsymbol{r}_k(t_{i+1}) = \boldsymbol{y}_k(t_{i+1}) - \boldsymbol{C}_k \hat{\boldsymbol{x}}_k(t_{i+1}|t_i) \ , \quad (31)$$

$$\boldsymbol{K}_k(t_{i+1}) = \boldsymbol{P}_k(t_{i+1}|t_i) \boldsymbol{C}_k{}^T \boldsymbol{\Theta}_k^{-1} \ , \quad (32)$$

$$\hat{\boldsymbol{x}}_k(t_{i+1}|t_{i+1}) = \hat{\boldsymbol{x}}_k(t_{i+1}|t_i) + \boldsymbol{K}_k(t_{i+1}) \boldsymbol{r}_k(t_{i+1}) \ , \quad (33)$$

$$\boldsymbol{P}_k(t_{i+1}|t_{i+1}) = \boldsymbol{P}_k(t_{i+1}|t_i) - \boldsymbol{K}_k(t_{i+1}) \boldsymbol{C}_k \boldsymbol{P}_k(t_{i+1}|t_i) \quad (34)$$

and the time update

$$\hat{\boldsymbol{x}}_k(t_{i+1}|t_i) = \boldsymbol{A}_k \boldsymbol{x}_k(t_i|t_i) + \boldsymbol{B}_k \boldsymbol{u}(t_i) \ , \quad (35)$$

$$\boldsymbol{P}_k(t_{i+1}|t_i) = \boldsymbol{A}_k \boldsymbol{P}_k(t_i|t_i) (\boldsymbol{A}_k)^T + \boldsymbol{Q}_k \ , \quad (36)$$

where $\hat{\boldsymbol{x}}_k$ is the estimated system state and $\boldsymbol{P}_k \in \mathbb{R}^{n_{x_k} \times n_{x_k}}$ the covariance matrices of the estimation error [15]. The variables $\boldsymbol{\Theta}_k \in \mathbb{R}^{n_y \times n_y}$ and $\boldsymbol{r}_k \in \mathbb{R}^{n_y}$

describe the innovation or measurement residual and its estimated covariance matrix, respectively. It can be shown, see for instance [15], that the likelihood functions $p(\boldsymbol{Y}(t_i)|M_k, \boldsymbol{Y}(t_{i-1}))$ needed in the Bayes' rule (27) can be calculated using these innovations $\boldsymbol{r}_k$ and covariance matrices $\boldsymbol{\Theta}_k$ by

$$p(\boldsymbol{Y}(t_i)|M_k, \boldsymbol{Y}(t_{i-1})) = \frac{1}{(2\pi)^{\frac{n_y}{2}} (\det \boldsymbol{\Theta}_k(t_i))^{\frac{1}{2}}} \exp\left(-\frac{1}{2}\boldsymbol{r}_k(t_i)^T \boldsymbol{\Theta}_k^{-1}(t_i)\boldsymbol{r}_k(t_i)\right) . \tag{37}$$

This can be interpreted in the following way: If the k-th KF is not working properly (i.e. wrong failure case, wrong system description) the innovation $\boldsymbol{r}_k(t_i)$ will not reflect its estimated covariance matrix $\boldsymbol{\Theta}_k(t_i)$. This results in a small value of the likelihood function. If, on the other hand, the k-th KF describes the plant's behaviour well, a higher value for the likelihood function can be expected. Therefore, by using (27) and (37), the highest values of the corresponding conditional probabilities $p_k(t_i)$ will be assigned to the KF most related to the actual behaviour of the experimental plant. Thus, the plant's actual error state can be determined by selecting the KF with the highest conditional probability.

4 Experimental Results and Discussion

For the application of the multiple filter concept the different types of errors that could occur have to be determined. It is assumed that the scheduled ignition of a detonation is synchronised with a sampling instant t_i. All model probabilities are initialized at this time with $p_k(t_i) = \frac{1}{n}$, with n as the number of hypotheses. For the detection of detonation faults in the k-th tube/actuator, two hypotheses are formulated:

- the detonation is successfully ignited at time t_i, corresponding to a normalised actuation signal of the k-th actuator of $u_k(t_i) = 1$,
- the detonation is not ignited at time t_i. Now, $u_k(t_i) = 0$.

For simplicity, a situation is considered in which only one tube after the other can produce a detonation. Hence, two KFs are designed to represent both hypotheses, respectively. Both exploit the very same discrete-time state space model (22),(23). The only difference is the value of the used normalised input $u_k(t_i)$ of the k-th actuator/tube. Some time later, say at t_{i+d_i}, the Bayesian selector indicates the process state with its output γ, cf. Fig. 7. The time-varying detection time is $d_i \cdot T_s$. The output of the selector is $\gamma = 1$ for a successful ignition and $\gamma = 2$ for a fault. A value of $\gamma = 0$ indicates that the Bayesian selector has not made a decision until the current point in time. The Bayesian selector is initialised with $\gamma = 0$ at $t = t_i$. Without any further measures, the KF representing the wrong hypothesis would produce increasingly inaccurate estimates over time. If some time later this hypothesis becomes the correct one, this KF would need an unacceptable long (detection) time to correct its states and to give a high likelihood for the Bayesian selector. For that reasons, all, i.e. in this simple case both, KFs and the Bayesian selector itself are initialized whenever a new tube is ignited. If normal operation ($\gamma = 1$) is detected the estimates of the fault model are initialised with the estimates of the nominal model, i.e. $\hat{\boldsymbol{x}}_2(t_i) := \hat{\boldsymbol{x}}_1(t_i)$. Likewise, $\hat{\boldsymbol{x}}_1(t_i) := \hat{\boldsymbol{x}}_2(t_i)$ in the case $\gamma = 2$.

The performance of the fault detection algorithm is tested on the experimental plant. For the test it is assumed that only one actuator/tube fires at the same time. If more than one actuator/tube should fire at the same time, more KFs are needed (each one describing a specific combination of successful and misfiring tubes). Nominally, the 12 actuators/tubes are ignited sequentially (i.e. the first tube fires first, the second tube fires second, and so forth). The time distance between two shots of two neighbouring loudspeakers is 0.01 s, resulting in a firing frequency of 100 Hz. For testing the detection algorithm, every second actuator/tube is assumed to misfire when ignited. The experimental results are presented for a time window from 0.24 s to 0.36 s relating to the point in time when the first tube is ignited the first time at t=0 s. Thus, the selected time window covers a the third full cycle of ignitions, starting with the first tube and ending with the twelfth. Figure 8 shows in the upper row the output γ of the Bayesian selector. A value of $\gamma = 1$ or $\gamma = 2$ is produced whenever the model probability is above $p_k(t_i) = 0.9$. In the left graph of the second row of Fig 8, the conditional model probabilities of the nominal model is given. As $p_2 = 1 - p_1$, the model probability of the fault model can easily be derived from this figure. This example shows that a very reliable fault detection is obtained. A close-up of the model probability $p_1(t_i)$ is given as well in the lower row of Fig. 8 highlighting the variable detection time $d_i \cdot T_s$. This delay results on the one hand from the Bayes' rule (27) when all a priori model probabilities are initialised with $\frac{1}{2}$ at the time of a scheduled ignition. Initially, both KF produce reasonable estimates, so that their respective likelihoods are similar. An additional time delay will be introduced as every detonation needs a short time to excite the acoustics inside the the plenum. For the given data the mean time needed to detect a detonation is 4.2 ms. A misfiring is detected in 2.7 ms.

A prerequisite for a successful fault detection of the actual version is the exact knowledge of the ignition timing. This can be explained by the outcome of the likelihood function of a model (37). The value obtained depends on the innovation (31), i.e. the difference between the model output and the current measurement data. A difference between the expected and the real ignition timing results in a phase shift between the related actuation signals. Thus, a phase shift between the measurement and the expected model output is generated which leads to a high innovation and hence, to a low probability of the related model. To overcome this challenge, the time delay between the measurement and the model outputs could be estimated to compensate for uncertain ignition timings. The implementation of the suggested improvement will be the next step for the robustification of the algorithm.

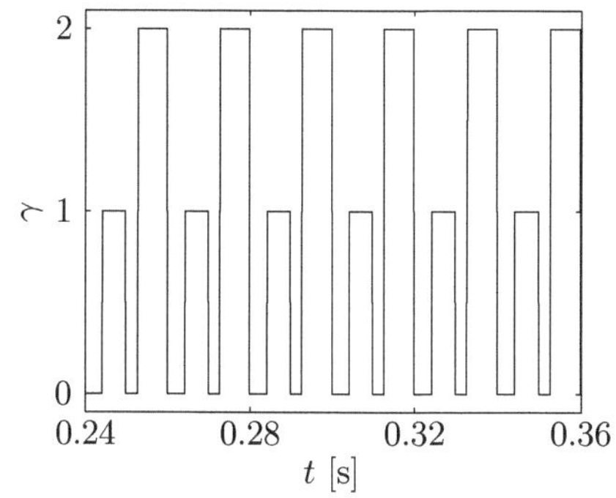

(a) Output of the Bayesian selector.

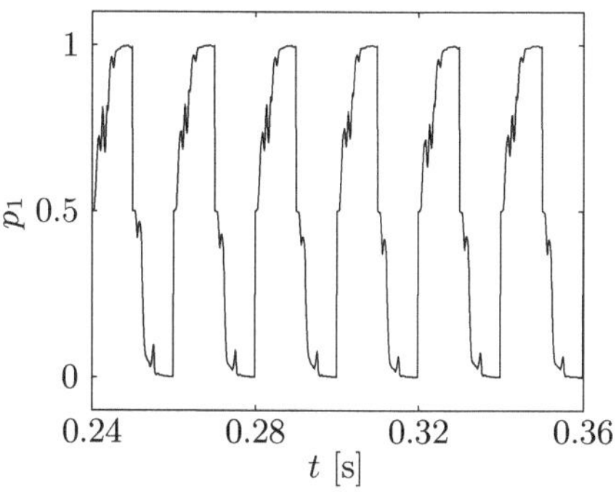

(b) Signal of the hypothesis conditional probability of the first KF, corresponding to a successfull detonation.

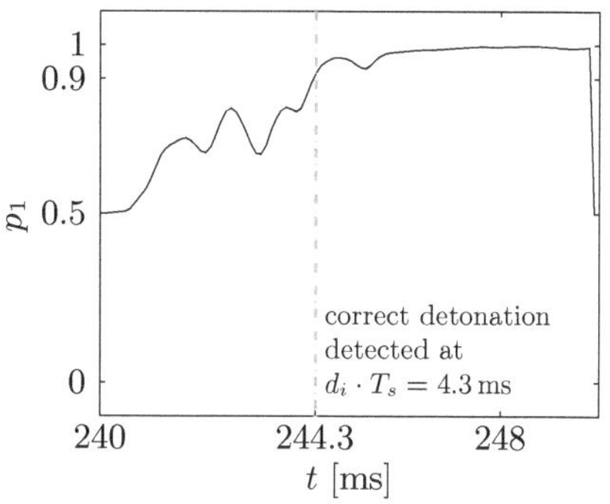

(c) Close-up of Fig. 8(b) highlighting the time when the Bayesian detector presents its decision.

Fig. 8. Time signals of the hypthesis conditional probabilitie p_1 and the γ-value calculated by the Bayesian detector

5 Conclusion

An algorithm for a model-based detection of misfiring in an annular burner mockup was derived. It was implemented on a DSP resulting in a real-time detection of misfiring. Thus, a tool providing the information needed for a restart of faulty tubes has been successfully developed.

The performance of the algorithm depends on a significant difference between the successful detonation model and the misfiring model. For the considered annular burner mockup this difference is the actuation signal, i.e. the particle velocity. Thus, it is crucial to know how the particle velocity, generated by a successful detonation and a misfiring looks like.

Acknowledgements. The authors gratefully acknowledge support by the Deutsche Forschungsgemeinschaft (DFG) as part of collaborative research center SFB 1029 "Substantial efficiency increase in gas turbines through direct use of coupled unsteady combustion and flow dynamics".

References

1. Roy, G., Frolov, S., Borisov, A., Netzer, D.: Pulse detonation propulsion: challenges, current status, and future perspective. Progress in Energy and Combustion Science 30(6), 545–672 (2004)
2. Hallouzi, R., Verdult, V., Babuska, R., Verhaegen, M.: Fault detection and identification of actuator faults using linear parameter varying models. In: Proceedings of the 16th IFAC World Congress, vol. 16(1), pp. 1821–1821 (2005)
3. Patton, R.J., Klinkhieo, S.: Actuator fault estimation and compensation based on an augmented state observer approach. In: Joint 48th IEEE Conference on Decision and Control and 28th Chinese Control Conference, pp. 8482–8487 (2009)
4. Darouach, M., Zasadzinski, M., Xu, S.J.: Full-order observers for linear systems with unknown inputs. IEEE Transactions on Automatic Control 39(3), 606–609 (1994)
5. Darouach, M., Zasadzinski, M., Onana, A.B., Nowakowski, S.: Kalman filtering with unknown inputs via optimal state estimation of singular systems. International Journal of Systems Science 26(10), 2015–2028 (1995)
6. King, R., Gilles, E.D.: Multiple filter methods for detection of hazardous state in an industrial plant. AIChE 36(11), 1697–1706 (1990)
7. Menke, T.E., Maybeck, P.S.: Sensor/actuator failure detection in the vista F-16 by multiple model adaptive estimation. IEEE Transactions on Aerospace and Electronic Systems 31(4), 1218–1229 (1995)
8. Rago, C., Prasanth, R., Mehra, R.K., Fortenbaugh, R.: Failure detection and identification and fault tolerant control using the IMM-KF with applications to the Eagle-Eye UAV. In: Proceedings of the 37th IEEE Conference on Decision & Control, vol. 4, pp. 4208–4213 (1998)
9. Zhang, Y., Li, X.R.: Detection and diagnosis of sensor and actuator failures using IMM estimator. IEEE Transactions on Aerospace and Electronic Systems 34(4), 1293–1313 (1998)
10. Maybeck, P.S.: Multiple model adaptive algorithms for detecting and compensating sensor and actuator/surface failures in aircraft flight control systems. Int. J. Robust Nonlinear Control 9, 1051–1070 (1999)
11. Rienstra, S., Hirschberg, A.: An introduction to acoustics. Technical report, Eindhoven University of Technology (2013)
12. Moeck, J.P.: Analysis, Modeling, and Control of Thermoacoustic Instabilities. PhD thesis, Technische Universität Berlin (2010)
13. Schuermans, B.: Modeling and Control of Thermoacoustic Instabilities. PhD thesis, École Polytechnique Fédérale de Lausanne (2003)
14. Ljung, L.: System Identification: Theory for the User, 2nd edn. Prentice-Hall, Englewood Cliffs (1999)
15. Gelb, A.: Applied Optimal Estimation. The M.I.T. Press (1986)

Part III
Constant Volume Combustion

Transient Mixing Enhancement of a Transverse Jet in Supersonic Cross Flow Using Pulse Detonation

Timothy Ombrello[1,*], Chung-Jen Tam[2], Willie Haw[3], and Campbell Carter[1]

[1] U.S. Air Force Research Laboratory, Wright-Patterson AFB, OH, USA
{timothy.ombrello.1,campbell.carter}@us.af.mil
[2] Taitech, Beavercreek, OH, USA
chung_jen.tam.1.ctr@us.af.mil
[3] U.S. Air Force Institute of Technology, Wright-Patterson AFB, OH, USA
willie.haw@afit.edu

Abstract. Enhanced mixing and conditioning of a transverse jet in Mach-2 cross flow was investigated through the application of a staged pulse detonation injector. NO planar laser-induced fluorescence and high-frame-rate shadowgraph imaging provided measurements for comparison to CFD modeling of the interaction. The large momentum flux of the pulse detonator led to significant redistribution of the plume from an upstream injectant for several milliseconds while simultaneously elevating its temperatures. The result was a conditioning of the transverse jet plume that could be related to increased reactivity if the upstream injectant was fuel. Typical PD exhaust times would enable potential quasi-steady conditioning of the transverse jet plume if pulsation frequencies on the order of 100 Hz are used.

1 Introduction

One of the more challenging problems in high-speed air-breathing propulsion systems, such as supersonic combustion ramjets (scramjets), is achieving sufficient mixing of fuel with the supersonic air flow within the confines of the combustor. A lack of mixing typically leads to reduced chemical heat release from the fuel and therefore significantly reduced thrust and overall efficiency. Mixing aids can be implemented by intrusions to the flow, such as by using struts or ramps, in order to distribute fuel to the core supersonic flow and create recirculation/mixing regions in their wakes.[1-4] Consequently, these intrusions can lead to overall pressure losses and thermal management issues. Therefore, the pursuit of enhanced mixing techniques is warranted, especially when considering larger-scale scramjets where mixing needs to be achieved over greater transverse distances in the supersonic flow.

The most common fuel delivery technique to supersonic cross flows relies upon the sonic transverse/oblique jet. This technique has been investigated in significant detail using a variety of injection angles, orifice geometries, and pressure ratios and can provide the near-field mixing that is necessary in a supersonic combustor.[5-14] A

[*] Corresponding author.

R. King (ed.), *Active Flow and Combustion Control 2014*, Notes on Numerical Fluid Mechanics and Multidisciplinary Design 127, DOI: 10.1007/978-3-319-11967-0_16

nominal schematic representation of a circular transverse sonic jet in a supersonic cross flow is shown in Fig. 1.[15] The underexpanded sonic jet produces a canted barrel shock terminating with a Mach disk, an upstream bow shock, and separation/recirculation regions upstream and downstream of the orifice. This type of jet is commonly characterized in terms of the jet momentum flux ratio, which is defined as the ratio of the jet momentum flux to the freestream momentum flux. Larger jet momentum flux ratios lead to more rapid near-field mixing because of the large fluidic intrusion to the supersonic cross flow, which in turn produces a large-scale counter-rotating vortex pair downstream of the jet, as well large eddies near the interface between the jet and the freestream.[8] While the near-field mixing is enhanced with larger jet momentum flux ratios, there are also more stagnation pressure losses. One way of mitigating the stagnation pressure losses is to angle the jet to the freestream.[11-14] The oblique injection technique has similar far-field mixing compared to normal injection, but the near-field mixing suffers. Therefore, one of the only methods of increasing the overall mixing of a jet in a supersonic cross flow is to enhance the near-field mixing characteristics with the introduction of significant inertial components, which is typically closely tied to having larger jet momentum flux ratios and hence more stagnation pressure losses.

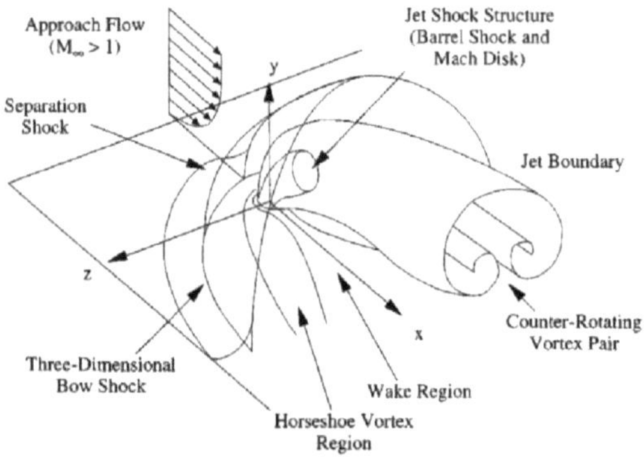

Fig. 1. Perspective view of a mean transverse injection flowfield. Reproduced with permission from Gruber et al. (1997).

In an attempt to mitigate stagnation pressure losses, staged injection or pulsed injection has been investigated. With staged injection, two (or more) jets aligned in the streamwise direction benefit from the first jet shielding the second jet to allow for increased penetration [16,17] of the second jet, as well as increased near-field mixing, [18] all while reducing stagnation pressure losses. Pulsed injection (Hz to kHz) is effective in providing a transient plume with varying jet momentum flux ratios to increase penetration and large eddy formation for enhanced mixing that would only be possible at a constant higher jet momentum flux ratio that would have large overall stagnation pressure losses.[19,20] Therefore, staging and transient plumes have the potential to increase mixing while minimizing stagnation pressure losses.

To take advantage of using large jet momentum flux ratios, but pulsing to minimize overall stagnation pressure losses, the transient pulse from a detonation can be utilized to enhance mixing of a wall jet with a supersonic cross flow. By staging a pulse detonator (PD) downstream of a jet, the high-pressure and high-velocity detonation plume has the potential to significantly increase penetration and mixing of the upstream jet with the supersonic flow. Furthermore, the high-temperature and radical rich plume can provide some conditioning of the mixed flow by elevating the temperature for increased reactivity downstream.

PD's have recently been applied to a supersonic cross flow to examine the transient exhaust plume interaction.[21] The PD plume was shown to penetrate deeply into a Mach-2 cross flow (up to 7 PD diameters within 12 PD diameters downstream) with a high momentum flux and having a total exhaust time (reaching static tunnel conditions) of 4-5 ms. While the investigation showed what the global interaction would be for a staged injection configuration, with a jet upstream of the PD, a detailed investigation of the jet plume interaction with the PD was not performed. Therefore, the focus of this paper is on understanding the degree of mixing enhancement and conditioning of an upstream jet with a supersonic cross flow using a PD. The constituents of the jet for this mixing study were chosen to be NO mixed in air instead of a fuel in order to provide a species that was more easily interrogated with planar laser-induced fluorescence (PLIF). High-frame-rate shadowgraph imaging was also applied, and collectively the results of the diagnostics were used for comparison to a CFD model of the mixing process.

2 Experimental Set-Up

Experiments were performed in a supersonic wind tunnel facility capable of operating continuously with peak stagnation conditions of 2860 kPa and 922 K at a flow rate as high as 15.4 kg/s.[22] For the current experiments a rectangular test section was used with a height (z) of 13.0 cm and a width (y) of 15.2 cm. The flow was accelerated through a nozzle to a Mach number of 2 with a total pressure of 275.8 kPa (static pressure of 35.2 kPa) and a total temperature of 300 K (static temperature of 170 K). Large windows were installed in the test section in order to provide optical access for non-intrusive diagnostic techniques.

A picture and schematic of the tunnel test section with pulse detonator (PD) installed are shown in Fig. 2. Both the jet and PD were injected along the streamwise centerline, normal to the supersonic flow from a removable floor block. The jet and PD had exit diameters of 4.76 mm and 10.3 mm, respectively, and were separated by 25.4 mm (the jet was upstream of the PD). For all experiments, the injection pressure of the jet was maintained at 345 kPa and consisted of air mixed with NO/N_2.

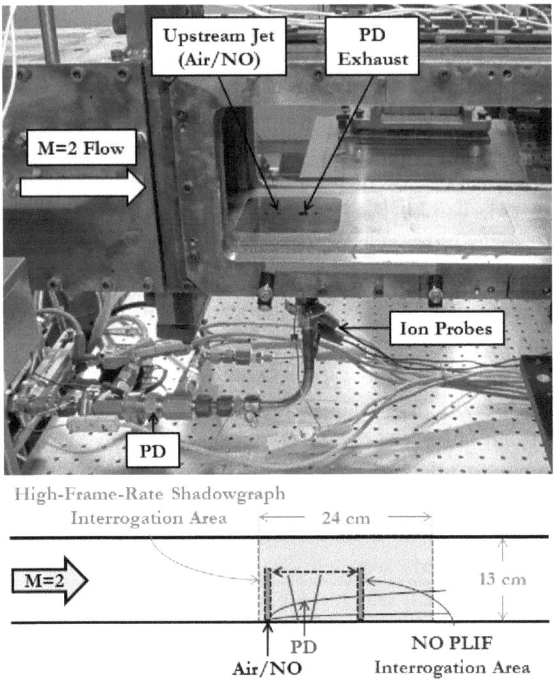

Fig. 2. Picture of the experiment in the tunnel test section and a schematic representation of the diagnostics interrogation areas.

The PD was comprised of a 61 cm long and 10.3 mm inside-diameter stainless steel tube. The reactants were injected into the closed end of the tube using automotive-style injectors sized for the pressures and flowrates desired in order to achieve an appropriate reactive mixture. The mixture was ignited with a conventional automotive spark igniter (~10 mJ/pulse) and deflagration-to-detonation transition was achieved with a Shchelkin-like spiral. Two different sets of reactants were used for the PD: C_3H_8/N_2O and ultra-high purity CH_4/O_2. These reactant pairs were chosen because of the amount of NO that they produced in their plumes; NO was the target species for the PLIF measurements, and more details on NO production and measurements will be given below. Switching between the two sets of reactants only required changing the injectors and therefore preserved the overall length and internal volume of the PD. The reactants were chosen to allow for ease of ignition and propagation of a detonation through the small inside diameter of the PD tube at the static pressure conditions of the tunnel. To ensure that a detonation was achieved for each firing, two ion probes were placed in the PD tube (just upstream of the PD exit into the test section) to measure the detonation wave speed.

High-frame-rate shadowgraphy provided an integrated view of the PD plume across the span of the test section while also capturing the development in the streamwise direction. The illumination source was an Hg-Xe lamp where the light was collimated, passed through the flow path—providing visualization of the full test section height of 13 cm and approximately 24 cm in width (see Fig. 2)—and then impinged upon a sheet of frosted glass. The shadowgraph on the frosted glass was then imaged by a Photron SA5 CMOS (complementary metal oxide semiconductor) camera at 50,000 frames per second and a resolution of 512 pixels by 272 pixels, and the high framing rate allowed for interrogation of individual PD events. Note that shadowgraphs were normalized with (i.e., divided by) a time-averaged image taken under quiescent conditions (no tunnel flow); this improved markedly the quality of the images. NO PLIF was used to image the jet and PD plume in the spanwise direction by taking "slices" at various streamwise coordinates and times in relation to the PD being fired. The laser set-up for the NO PLIF system used the second harmonic (λ=532 nm) of an injection seeded Nd:YAG (Quanta Ray GCR-170) to pump a dye laser (Lumonics HD-300B) to produce 622 nm radiation from Rhodamine 640 dye. The third harmonic of the injection seeded Nd:YAG (λ=355 nm) was mixed with the 622 nm radiation within a β-barium borate (BBO) crystal to produce approximately 4 mJ/pulse of 226 nm radiation. The dye laser was tuned to the $R_1(8.5)$ line of the NO $A^2\Sigma^+ \leftarrow X^2\Pi$ ($v'=0$, $v''=0$) band. The laser sheet that was passed through the tunnel test section in the spanwise direction was formed by a plano-concave cylindrical lens (-50 mm focal length) and a plano-convex spherical lens (1-m focal length) producing a sheet of approximately 7.5 cm in height. The resulting fluorescence was imaged off axis using a Princeton Instruments SuperBlue PI-MAX 3 intensified charge-coupled-device camera (ICCD, having a 1024 x 1024 pixel array). To mitigate read noise, the pixels were binned 2 x 2 (so that a lower digitization rate could be used). To mitigate image blur, the camera and UV lens (45-mm f/1.8 Cerco lens) were mounted to a Scheimpflug adapter (Dantec); PLIF images were then corrected for *perspective distortion* using the algorithm provided by Burger and Burge [23] within the image processing program ImageJ.[24]

While the shadowgraph only required one trigger before the PD fired in order to capture the entire exhaust process, the NO PLIF required accurate timing measurements. To synchronize the NO PLIF with the PD firing, the laser system was used as the master clock. The 200-ns intensifier gate pulse (centered in time over the laser pulse) was used to trigger the PD, which would then be imaged on the following laser pulse. Therefore, to collect 100 fluorescence images, the intensifier was gated 101 times. To then find the times when the PD began to exhaust in relation to the laser firing (and ICCD image acquisition), the ion probes were used. The time was recorded between the intensifier trigger signal and the fall in voltage of the last ion probe. Since the detonation was at Chapman-Jouguet (C-J) conditions when it reached the ion probes (confirmed by the measured wave speed), the transit time remained nearly constant between the last ion probe and exhausting into the supersonic flow. Therefore, the ion probe signal was used as a marker to vary the delay times between the PD firing and the NO PLIF image in order to acquire different time slices of the jet and PD plume interaction. This enabled averaging of multiple images from the different firings at the same time after the PD began to exhaust. For all of the images shown, there were 10-20 single-shot images averaged for each time ($\pm$ 3 μs) to provide the best comparison to the average plume from the upstream jet. More details of the NO PLIF experimental set-up can be found in reference [21].

While PLIF was adopted for imaging cross-sections of the plume, a specific NO transition was chosen which rendered relative concentration measurements nearly impossible. This was done because of the wide range of temperatures that would be present between the cold NO jet and the hot PD plume. Since the focus of this investigation was on how the cold jet injectant was being influenced by the PD plume, the $R_1(8.5)$ line of the A-X (0,0) band was chosen because of having a peak population at low temperatures with a sharp decay at high temperatures. The order of magnitude difference in the expected fluorescence signal between the jet (~200 K) and the PD (~3000 K) allowed for significant suppression of NO produced by the PD, and therefore a focus on the NO jet plume.

3 Results

3.1 Pulse Detonator Exhaust Process

The PD exhaust process into the supersonic flow is a rapid event, only lasting several milliseconds. Initially there is a high-pressure, high-temperature, and high-velocity pulse produced from the detonation emerging into the supersonic flow. The detonation is closely followed by a blow-down process with a choked and then unchoked exhaust until the tube has reached tunnel static pressure conditions. An example of the PD exhaust process is shown in Fig. 3 with snapshots from 50,000 frames per second (fps) shadowgraphy. Shadowgraph images were taken at two separate streamwise locations (the entire shadowgraph system was mounted on a translation stage that could be moved in relation to the tunnel) and were "stitched" together. Since the detonation process was repeatable, two separate firings of the PD could be matched easily to produce the images shown in Fig. 3. In order to provide some perspective of how intrusive the PD plume was, the images also show the upstream jet whose bow and barrel shock can clearly be seen at 0 μs. Since the jet constituents are only room temperature air mixed with NO, there were no significant density gradients in the plume that could be imaged with shadowgraphy. On the other hand, the elevated temperature plume from the PD could be seen easily. Within 160 μs the PD plume had penetrated far into the supersonic cross flow from the initial detonation, immediately followed by a slightly weaker high-pressure blow-down process. To better quantify the PD exhaust process, the plume height as a function of time at different downstream locations is shown in the plot of Fig. 3. The initial PD plume penetration was significant, reaching z/d_{PD}=3 within 100 μs at an x/d_{PD}=0. The plume then continued to penetrate to z/d_{PD}=5 at an x/d_{PD}=4.93 in the near-field and then started to slow in growth when approaching the far-field with z/d_{PD}=6 at an x/d_{PD}=12.32. After the detonation, the blow-down process lasted for many milliseconds with the plume at z/d_{PD}=2 at an x/d_{PD}=12.32 after 2 ms.

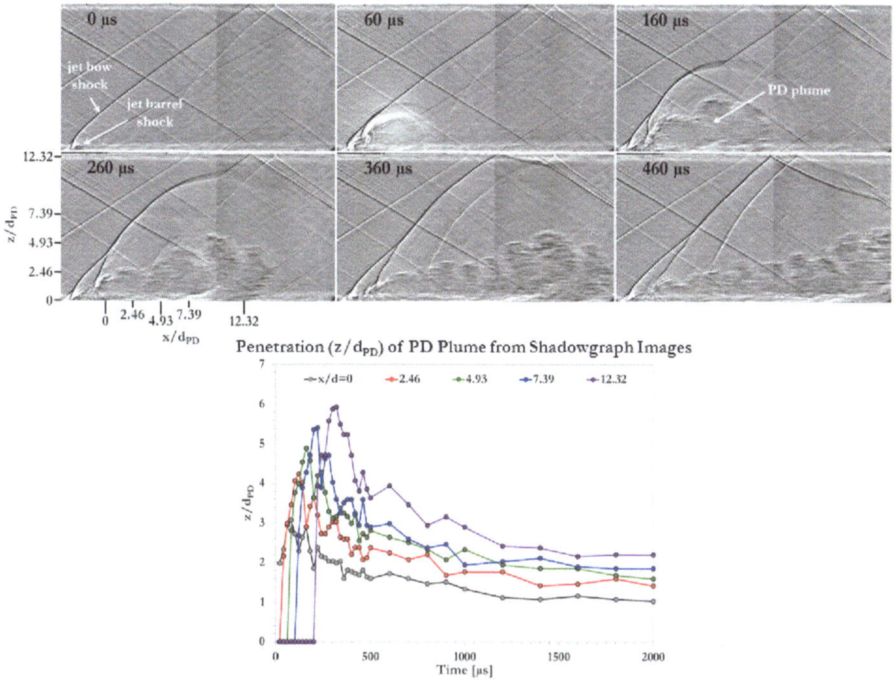

Fig. 3. Shadowgraph of PD plume and plot of its penetration (z/d_{PD}) as a function of time and downstream location

The strong pulse from the PD clearly leads to enhanced near-field mixing because of the amount of momentum that is added to the flow. The large under-expansion from the PD created a strong bow shock that was nearly normal to the flow, leading to significant separation and recirculation upstream of the PD that was followed with the weaker, but still significant high-pressure blow-down to enhance mixing. The total time for the complete exhausting of the PD to when pre-PD firing conditions were reached was approximately 4 ms. While the shadowgraph provided interrogation of the overall exhaust of the PD, it did not directly show any of the mixing characteristics that would be present between the upstream jet and PD plume.

3.2 NO Planar Laser-Induced Fluorescence (PLIF) of Jet and PD

NO PLIF allowed for imaging of the spanwise slices of both the jet and PD plume. For the jet, the air was seeded with an NO/N_2 mixture (50,000 ppm of NO in N_2), resulting in an NO mole fraction of approximately 5,000 parts per million (ppm). The high concentration provided clear PLIF images of the plume as it evolved downstream (shown in Fig. 4). Each image in Fig. 4 was the average of 100 single-shot images, and all are shown with the same look-up table in order to provide some perspective of the level of mixing. The momentum flux ratio of the jet to the freestream was fixed at approximately 1.3 and the penetration into the flow in the near-field aligned well with previous studies [8].

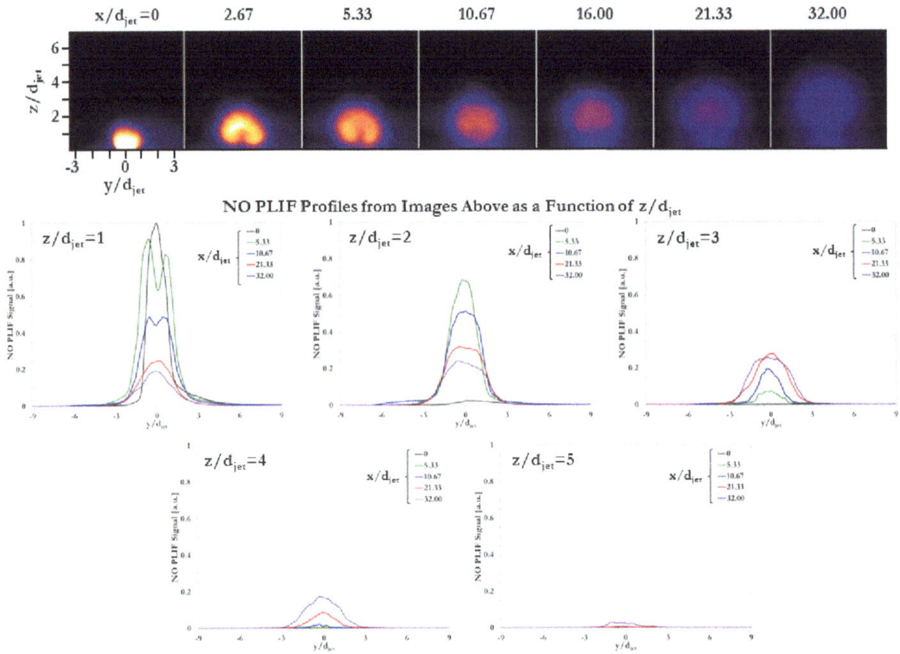

Fig. 4. NO PLIF of jet versus streamwise location (x/d_{jet}) and plots of the relative intensity profiles

The counter-rotating vortex pair is clearly shown at x/d_{jet}=2.67 and 5.33 in the near-field, with a gradual fading into a more diffuse plume by an x/d_{jet}=16 and beyond. This jet structure and behavior is common to a normal jet in a supersonic cross flow. In the plots of Fig. 4, profiles at different z/d_{jet} as a function of x/d_{jet} are shown for the NO PLIF images. The plots further emphasize how the plume evolved downstream, reaching a z/d_{jet}=5 by an x/d_{jet}=32. This provided a baseline as to the normal plume produced before interaction with the PD.

For the PD, initially the C_3H_8/N_2O reactants were adopted because of their use in a previous study.[21] Since the C_3H_8/N_2O PD was capable of producing significant concentrations of NO (on the order of 1000s of ppm),[25] the PD plume was imaged easily. This of course made it challenging, if not impossible to image NO from an upstream jet. While the $R_1(8.5)$ transition of NO helped to suppress the fluorescence signal from the PD, it was not sufficient to allow for discerning the difference between the NO from the upstream injectant and the PD. Therefore, changes had to be made to the PD in order to minimize NO production so the NO from the jet could be imaged easily in relation to the PD plume. In order to suppress the NO production, ultra-high purity CH_4/O_2 (99.999% for CH_4 and 99.999% for O_2) was substituted for the reactants, so that no nitrogen was present in the PD. While there was no nitrogen injected into the PD tube, there was still the possibility of NO production from two sources: one from any residual air in the tube upon filling with reactants from the injectors, and the other from reaction of air with the high-temperature PD plume when emerging and mixing with the tunnel flow. Nevertheless, the NO production from these two sources was found to be sufficiently suppressed to allow for better viewing of the NO from the upstream jet. To demonstrate the level of suppression of NO production, NO PLIF images of the PD plume at the same downstream location and time after firing is shown in Fig. 5. Image (a) in Fig. 5 shows the plume from the C_3H_8/N_2O PD, while (b) from the ultra-high purity CH_4/O_2 PD. Each image was the average of 10-20 single-shot images at the specific time after the PD began to exhaust (± 3 μs). Note the intensity scale of the images, where (b) is more than a factor of 30 lower than (a). Also, note that the plume height and width is preserved, even though the reactants were very different. This was expected since N_2O and O_2 behave very similarly as oxidizers in terms of detonations, and there is little variation in C-J conditions when using different hydrocarbons.

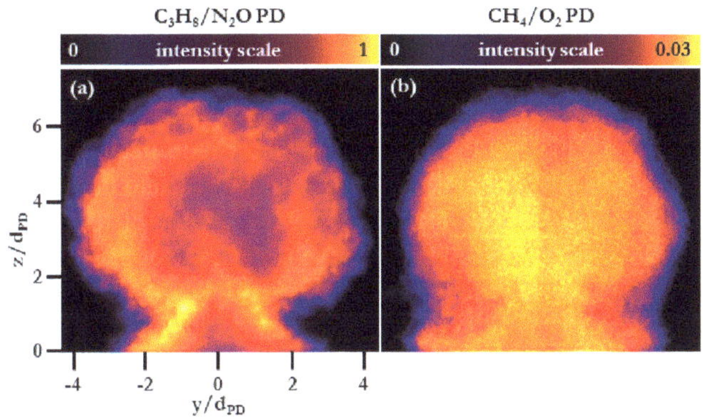

Fig. 5. Comparison of NO PLIF with different PD reactants showing suppression of NO production

Even though the NO production was significantly suppressed with the ultra-high purity CH_4/O_2 PD, there still remained some fluorescence signal that would skew the results of where the NO from the upstream jet was going in relation to the PD plume. Therefore, the PD was fired with and without the jet upstream and the images subtracted to arrive at the final image. This process is shown in Fig. 6 with NO PLIF images taken over the top of the PD exit 100 μs after it began to exhaust (± 3 μs with an average of 10-20 images). Image (a) shows the PD plume only, image (b) the PD plume with upstream NO jet, and image (c) the subtraction of (a) from (b). Again, note the relative difference in the range scale used in (a) versus (b) and (c) to show the level of suppression of NO coming from the PD. The resulting image, (c), clearly shows the NO from the upstream jet being forced around the barrel shock structure of the PD plume.

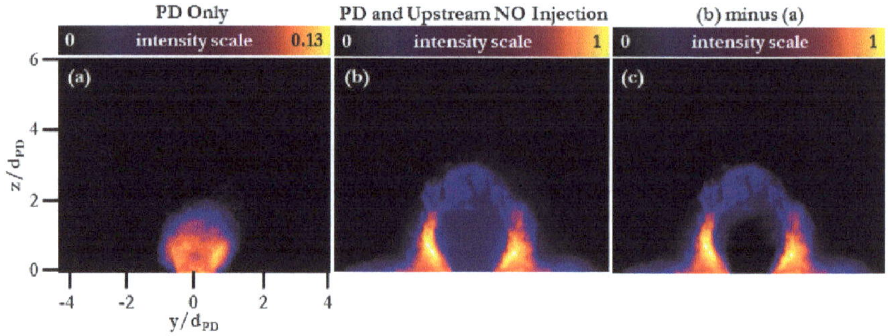

Fig. 6. Comparison of NO PLIF with and without upstream NO injection

With the established confidence of the imaging technique to examine the entrainment and distribution of NO from the upstream jet within and around the PD plume, the time resolved interaction process was examined. Since the PD exhaust was a transient process, the plume needed to be examined at various times and downstream locations. In order to capture the PD plume, the NO PLIF sheet was placed at different streamwise locations and delayed to various times after the PD began to exhaust into the test section. Figure 7 shows three different regimes identified for the interaction of the PD plume with the upstream jet of NO with the same look-up table used. Each of the three rows start with a time after the PD began to exhaust into the tunnel and then its evolution downstream by following the approximate tunnel core flow velocity (~517 m/s). In the first row at 10 μs, the PD plume is just beginning to emerge into the tunnel and pushes the NO plume from the upstream jet. As this "initial push" evolves downstream, the NO is shown to only be located around the periphery of the PD plume. This makes sense since there is no mechanism to mix the core PD plume with the upstream injected NO other than the slow diffusion process. In the second row, the PD barrel shock has grown to its greatest extent from the initial detonation by approximately 40 μs. Similar to the PD plume at 10 μs, the NO was pushed into the core supersonic flow to a larger *z/d*, but there is also the appearance of NO around the base of the barrel shock. This was the beginning of the process by which NO was being forced around the barrel shock as it flowed downstream. By 90 μs at the next downstream location, it becomes clear that the NO was starting to be entrained into the PD plume and mixed within the counter-rotating vortex structure. As the plume evolved further downstream, the NO appeared to be distributed across a significant *y/d* and *z/d* with signal levels being at the threshold of what can be seen with the look-up table range used. After the maximum barrel shock structure in the second row, the third row shows a significantly decayed barrel shock over the PD at 80 μs. The PD exhaust is now clearly in the high-pressure blow-down portion of the cycle and the NO was being forced around the barrel shock structure. The evolution downstream again shows the counter-rotating vortex structure and hence entrainment of the NO from the upstream jet but with less spanwise (*y/d*) growth. This follows the less under-expanded PD barrel shock that was narrower than in the second row at 40 μs. This "decay" process from the PD blowing-down to static tunnel conditions consumes the majority of the PD exhaust cycle. Through this part of the cycle, the NO from the upstream jet was pulled to large *z/d*'s and distributed across the span (*y/d*) of the PD plume.

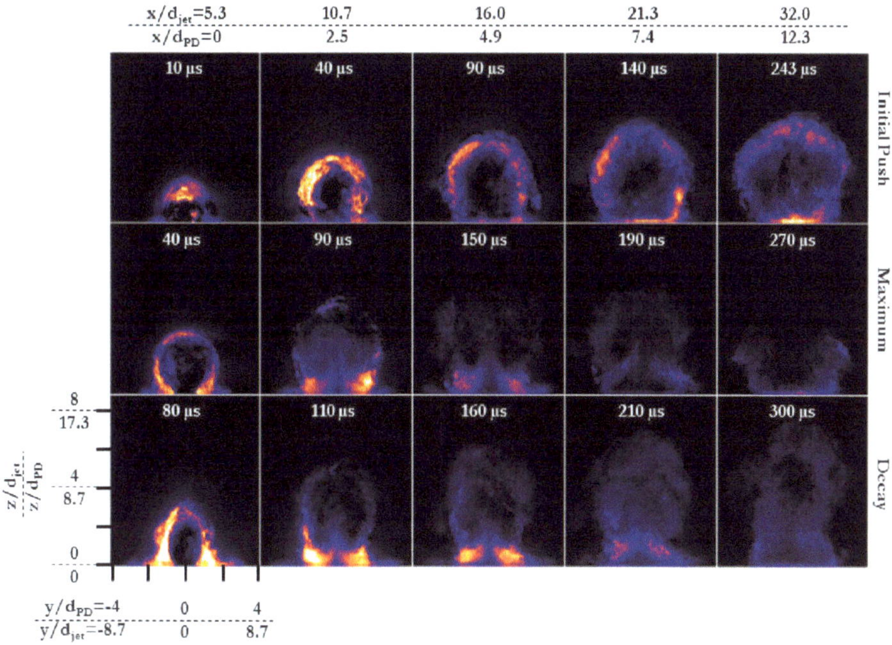

Fig. 7. NO PLIF showing the three stages of interaction between the PD and jet plume as it evolved downstream

3.3 CFD and Synthetic LIF

While the NO PLIF was able to show the influence of the PD on the upstream NO jet plume, it could not be used to quantify the concentration and/or temperature distributions. Since it was desired to know how the NO from the upstream jet was mixed and temperature elevated within the PD plume to draw some comparisons of how a fuel could be conditioned for downstream reaction, CFD modeling was employed. In a previous study, a simple model of the PD exhausting into a Mach 2 flow was established and verified through both shadowgraph and NO PLIF measurements.[21] The model attempted to replicate the transient pressure pulse from the PD with conditions grounded in ideal C-J detonation calculations for the PD mixture. The model had an initial spike from the detonation with P_{static}=834 kPa, T_{static}=3500 K, v=1500 m/s, and duration of 20 µs, and for the beginning of the blow-down process P_{static}=503 kPa, T_{static}=3000 K, and v=1105 m/s. This model was applied in three-dimensional unsteady RANS simulations using the CFD++ code.[26] CFD++ uses a finite-volume numerical framework, with multi-dimensional Total Variation Diminishing (TVD) schemes and Riemann solvers for accurate representation of supersonic flows. The Harten-Lax-van Leer-Contact (HLLC) Riemann solver with min-mod flux limiting was used in the simulations. Multi-grid acceleration was used to provide a fast and accurate solution methodology for both the steady and unsteady flows. The turbulence was modeled using the two-equation cubic κ-ε model. At solid surfaces, an advanced two-layer wall function with equilibrium and non-equilibrium blending was employed to reduce grid requirements. The turbulent Schmidt and Prandtl numbers were fixed at 0.7 and 0.9, respectively. There were no chemical reactions simulated since the focus of the study was on the flow physics between the PD and the supersonic cross-flow. This was reasonable since the majority of the chemical reactions from the PD were occurring inside the tube before exhausting into the supersonic cross-flow.

The computational domain consisted of half the width and the full height of the test section with symmetry applied along the streamwise centerline to reduce the required computational time. Thus, the PD tube also consisted of a half circular geometry. The total number of grid cells in the numerical simulations was approximately 8.5 million using a structured topology. The numerical simulations were extended from the plenum, upstream of the facility nozzle, to the entrance of the test section to provide the appropriate inflow conditions for the test section. This methodology obviates the use of *ad-hoc* profiles at the test section entrance. A no-slip, adiabatic-boundary condition was imposed on the solid walls; however, an inviscid wall assumption was imposed on the PD tube due to its small diameter. Furthermore, a zero-gradient condition was applied at the outflow plane.

The numerical approach for the PD simulation was divided into two steps. First, a steady-state solution was performed for the supersonic wind tunnel without the interaction of the PD tube. Second, the PD flow field was suddenly initiated into the wind tunnel floor, similar to a shock tube problem. At that time, the simulation was performed in a time-accurate manner to investigate the flow physics between the PD and supersonic flow. The first 3 μs of the detonation-front process was calculated with a time step of 10^{-8} s, and then it was increased to 10^{-7} s for the remainder of the process. The total time simulated was 400 μs.

The model accurately captured the PD plume as it evolved downstream and therefore could be adopted in this study to examine how the NO jet interacted with the PD plume. In order to move forward with the CFD model as a reasonable representation of the interaction, the NO PLIF images needed to be compared to the model for the current experiments. For this, synthetic laser-induced fluorescence (LIF) images were calculated from the CFD. The synthetic NO LIF signal, S_f, was calculated in the simulations using a combination of the CFD solution and laser excitation parameters via the equation for linear (weak) excitation:

$$S_f = KX_{NO}\left(\frac{P}{T}\right)g(v_L)\left(\frac{1}{A+Q_e}\right)f_B. \tag{1}$$

All of the flowfield independent parameters are grouped into K and include the detector gain/efficiency, the solid angle for detection of photons, the probe volume for each detector element, the Einstein B coefficient for absorption, the speed of light, the laser sheet fluence, the rate at which fluorescence photons are emitted, and the fluorescence efficiency factor. The other parameters are a function of the CFD solution, such as the NO mole fraction X_{NO}, the pressure, P, and the temperature, T. The term $g(v_L)$ is the laser overlap integral for laser frequency v_L, with $\int_{-\infty}^{\infty} g(v_L)\,dv_L = 1$, and is a function of the linewidths and lineshapes of the laser and the NO transition, as well as line shifts from the Doppler effect and collisions. Here, the transition linewidth is described by a Voigt function that is calculated with the routine given by Humlicek, [27] while the collisional broadening and shift coefficients are from the work of Chang, et al.[28] The laser lineshape was assumed to be Gaussian, and synthetic LIF images with a laser linewidth of $\Delta\nu_L$=0.2 cm^{-1} (FWHM) were computed. The total fluorescence decay rate, $(A+Q_e)$, is a function of the fluorescence natural decay rate, A, and the electronic quenching rate, Q_e, and was evaluated using the Settersten database.[29-31] Lastly, f_B is the Boltzmann fraction of the probed ro-vibronic state, which was calculated using simplified term-energy expressions[32] for the specific spin-split J state of NO probed by the laser (here, the J=8.5 level of $X^2\Pi_{1/2}$ (v″=0) state).

The results of comparing the PLIF images to the synthetic LIF images of the NO jet as it evolved downstream showed good agreement in the overall shape and size of the plume. To then compare to the jet plume interacting with the PD, images were chosen at times that represented the evolution of the plume downstream, starting with the large barrel shock structure over the PD at a time of 50 μs after the PD began to exhaust. A comparison of the measured and synthetic LIF images is shown in Fig. 8. The synthetic LIF agreed reasonably well with the measured values in terms of the overall size and LIF signal distribution as the plume evolved downstream. The contours in the synthetic LIF images mark the 1 ppm concentration of NO.

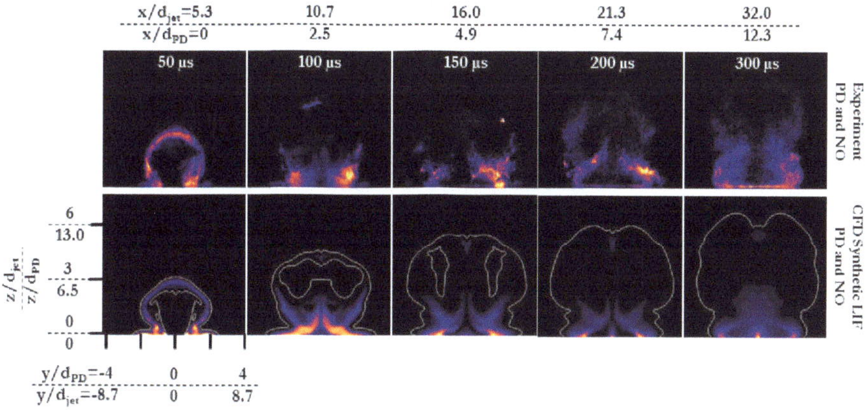

Fig. 8. Comparison of measured and computed NO PLIF as the PD plume evolved downstream

With the synthetic LIF agreeing reasonably well with the NO PLIF, the CFD solutions were used to show the relative differences between the NO jet plume without and with the PD interaction. Figure 9 shows the comparison between the NO jet plume with and without the PD interaction. In Fig. 9(a) when there was only the NO jet, the NO reached into the flow to approximately z/d_{jet}=2-3 with a localized and tightly concentrated plume. With the PD interaction (Fig. 9(b)), in this case at 160 μs as the PD blow-down process was occurring, the NO plume was bifurcated at smaller z/d_{jet} and reached into the flow to approximately a z/d_{jet}=8. Most importantly, much of the NO became elevated in temperature to between approximately 1000 K and 2000 K (Fig. 9(c)).

Looking further downstream to an x/d_{jet}=32 (x/d_{PD}=12.3) (Fig. 9(d)), the NO jet plume remained fairly localized and tightly concentrated and reached into the flow to approximately a z/d_{jet}=5. With the PD interaction (Fig. 9(e)), in this case at 300 μs as the PD was near its maximum when tracked upstream to the barrel shock, the NO plume was distributed across a significant y/d and reached an approximate z/d_{jet}=17. Again, most importantly, a significant quantity of the NO was elevated in temperature (Fig. 9(f)), with values at or above 600 K, and some locations reaching as high as 1600 K.

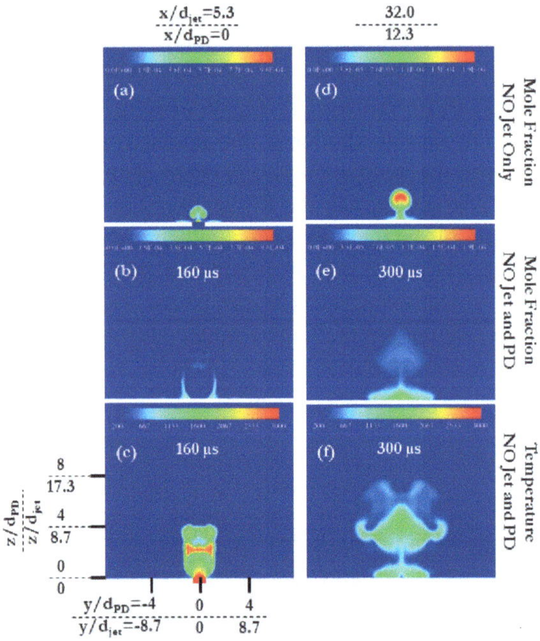

Fig. 9. Distributions of NO and temperature from CFD

3.4 Implications of Plume Conditioning

The results of the measurements and simulations showed that the jet plume cross sectional area could be increased by a factor of six with the PD and elevated to temperatures in excess of 2000 K. Therefore, it was worthwhile to put into context what the implications would be if the jet NO was replaced with fuel. Take C_2H_4 as an example as the fuel since it has approximately the same molecular weight and hence diffusivity as NO and air. If C_2H_4 was injected cold, as in the current experiments, it would require some ignition source, such as elevated temperatures to achieve chemical heat release. Therefore, the normal injection case would have a near infinitely long ignition delay time at the static pressure (35.2 kPa) and static temperature (170 K) conditions present in the current experiments. On the other hand, the PD interaction with the jet plume resulted in significantly elevated temperatures.

To provide some perspective, the ignition delay time for a stoichiometric premixture of C_2H_4/air at the static pressure condition of 35.2 kPa was calculated using SENKIN

[33] and the USC Mech II [34] chemical kinetic mechanism. While the actual jet mixing experiment is by its nature not perfectly premixed, especially near the fuel injector, the simulations of ignition delay time for a specific equivalence ratio can provide a relative understanding of the role of elevated temperature achieved from mixing with the PD plume. Therefore, in the case of C_2H_4/air for temperatures in excess of 1400 K, which is well within values provided by the PD, ignition delay times can be decreased to roughly 100 μs. This is less than the residence time within a scramjet combustor and is a fraction of the residence time within a cavity based flameholder.

Furthermore, while the conditioning of an individual fuel plume with one pulse from one PD can be very effective for mixing and conditioning for enhanced reactivity, practical implementation would require multiple PD's to interact with multiple fuel injection locations across the span of a flow. Fortunately, a PD can be branched or sent through a manifold in the same way that the fuel can before being injected. An example of this is shown in Fig. 10 where the same PD used in the supersonic mixing enhancement tests was attached to a manifold with a series of small exit holes (2.4 mm diameter) and fired into a quiescent environment with a static pressure of 101 kPa. Since the fuel/oxidizer mixture is so reactive and fills the entire manifold before detonating (note the reactants emerging from the tube at 0 μs), there are individual detonations from each small hole and the subsequent blow-down. Therefore, one PD can potentially be used to condition many fuel injection locations.

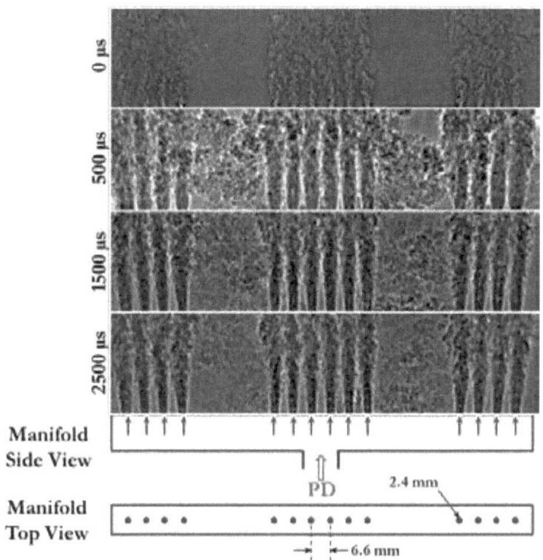

Fig. 10. Shadowgraph of PD plume when sent through a manifold (static pressure of 101 kPa)

4 Summary and Conclusions

Enhanced mixing and conditioning of a transverse wall jet in a Mach-2 cross flow was achieved through the application of a streamwise staged pulse detonator (PD).

The high momentum flux of a transient PD pulse produced a large barrel shock structure from the highly under-expanded jet exhaust. The flow of the upstream injectant around the fluidic body of the barrel shock and up into the counter-rotating vortex allowed for significant distribution into the core supersonic flow with the elevated temperature PD plume. Planar laser-induced fluorescence (PLIF) of the NO molecule and computational fluid dynamic simulations showed that the cross section of the jet plume was increased by up to a factor of six and elevated to temperatures in excess of 2000 K. The distribution of upstream injectant and the elevation of temperature can provide significant conditioning of the flow. In the case of the injectant being a fuel, the conditioning would allow for increased reactivity and, therefore, the potential for greater control of burning efficiencies within the confines of a combustor. Furthermore, the ability to send the detonation through a manifold would enable conditioning of multiple fueling locations with one PD. While only discrete pulses from the PD were considered for this mixing study, successive pulsing of the PD could provide a more quasi-continuous conditioning of a fuel plume. With typical PD exhaust times on the order of 3-10 ms, which depends upon the length/internal volume, as well as the gas density inside the tube, pulsation frequencies on the order of 100-300 Hz could provide quasi-steady conditioning. Since the fuel conditioning would typically be for a short-duration event, such as ignition or operation of a scramjet during an off-design transition, and since each pulse requires a small quantity of fuel and oxidizer, multiple pulsation events would be possible with a fairly compact system.

References

1. Bogdanoff, D.: Advanced Injection and Mixing Techniques for Scramjet Combustors. Journal of Propulsion and Power 10(2), 183–190 (1994)
2. Tomioka, S., Murakami, A., Kudo, K., Mitani, T.: Combustion Tests of a Staged Supersonic Combustor with Strut. Journal of Propulsion and Power 17(2), 293–300 (2001)
3. Doster, J.C., King, P.I., Gruber, M.R., Carter, C.D., Ryan, M.D., Hsu, K.-Y.: In-Stream Hypermixer Fueling Pylons in Supersonic Flow. Journal of Propulsion and Power 25(4), 885–901 (2009)
4. Hsu, K.-Y., Carter, C.D., Gruber, M.R., Barhorst, T., Smith, S.: Experimental Study of Cavity-Strut Combustion in Supersonic Flow. Journal of Propulsion and Power 26(6), 1237–1246 (2010)
5. Schetz, J.A., Billig, F.S.: Penetration of Gaseous Jets Injected into a Supersonic Stream. Journal of Spacecraft and Rockets 3(11), 1658–1665 (1966)
6. Papamoschou, D., Hubbard, D.G.: Visual Observations of Supersonic Transverse Jets. Experiments in Fluids 14, 468–476 (1993)
7. McDaniel, J.C., Graves, J.: Laser-Induced Fluorescence Visualization of Transverse Gaseous Injection in a Nonreacting Supersonic Combustor. Journal of Propulsion and Power 4(6), 591–597 (1988)
8. Gruber, M.R., Nejad, A.S., Chen, T.H., Dutton, J.C.: Mixing and Penetration Studies of Sonic Jets in a Mach 2 Freestream. Journal of Propulsion and Power 11(2), 315–323 (1995)
9. Kobayashi, K., Bowersox, R.D.W., Srinivasan, R., Carter, C.D., Hsu, K.-Y.: Flowfield Studies of a Diamond-Shaped Fuel Injector in a Supersonic Flow. Journal of Propulsion and Power 23(6), 1168–1176 (2007)

10. Tomioka, S., Jacobsen, L., Schetz, J.: Sonic Injection from Diamond-Shaped Orifices into a Supersonic Crossflow. Journal of Propulsion and Power 19(1), 104–114 (2003)
11. Lin, K.-C., Ryan, M., Carter, C., Gruber, M., Raffoul, C.: Raman Scattering Measurements of Gaseous Ethylene Jets in Mach 2 Supersonic Crossflow. Journal of Propulsion and Power 26(3), 503–513 (2010)
12. Mays, R.B., Thomas, R.H., Schetz, J.A.: Low Angle Injection into a Supersonic Flow. AIAA Paper, 89–2461 (1989)
13. Fuller, E.J., Mays, R.B., Thomas, R.H., Schetz, J.A.: Mixing Studies of Helium in Air at High Supersonic Speeds. AIAA Journal 30(9), 2234–2243 (1992)
14. Schetz, J.A., Thomas, R.H., Billig, F.S.: Mixing of Transverse Jets and Wall Jets in Supersonic Flow. In: IUTAM Symposium on Separated Flows and Jets, Novosibirsk, USSR, pp. 807–837 (1990)
15. Gruber, M.R., Nejadt, A.S., Chen, T.H., Dutton, J.C.: Compressibility Effects in Supersonic Transverse Injection Flowfields. Physics of Fluids 9(5), 1448–1461 (1997)
16. Hollo, S.D., McDaniel, J.C., Hartfield, R.J.: Quantitative Investigation of Compressible Mixing - Staged Transverse Injection into Mach 2 Flow. AIAA Journal 32(3), 528–534 (1994)
17. Lee, S.-H.: Characteristics of Dual Transverse Injection in Scramjet Combustor, Part 1: Mixing. Journal of Propulsion and Power 22(5), 1012–1019 (2006)
18. Takahashi, H., Masuya, G., Hirota, M.: Supersonic Turbulent Mixing Structure in Staged Injection Flowfields. In: 48th AIAA Aerospace Sciences Meeting Including the New Horizons Forum and Aerospace Exposition, AIAA-2010-753 (2010)
19. Kouchi, T., Sasaya, K., Watanabe, J., Shibayama, H., Masuya, G.: Penetration Characteristics of Pulsed Injection into Supersonic Crossflow. In: 46th AIAA/ASME/SAE/ASEE Joint Propulsion Conference and Exhibit. AIAA-2010-6645 (2010)
20. Kouchi, T., Sakuranaka, N., Izumikawa, M., Tomioka, S.: Pulsed Transverse Injection Applied to a Supersonic Flow. In: 43rd AIAA/ASME/SAE/ASEE Joint Propulsion Conference & Exhibit. AIAA-2007-5405 (2007)
21. Ombrello, T., Carter, C., McCall, J., Schauer, F., Tam, C.-J., Naples, A., Hoke, J., Hsu, K.-Y.: Enhanced Mixing in Supersonic Flow Using a Pulse Detonation Combustor. In: 50th AIAA Aerospace Sciences Meeting, AIAA-2012-0123 (2012)
22. Gruber, M.R., Nejad, A.S.: New Supersonic Combustion Research Facility. Journal of Propulsion and Power 11(5), 1080–1083 (1995)
23. Burger, W., Burge, M.J.: Digital Image Processing: An Algorithmic Introduction to Using Java. Springer (2008)
24. Rasband, W.S., Image, J.: U.S. National Institutes of Health, Bethesda, Maryland, USA (1997-2012), `http://imagej.nih.gov/ij/`
25. Hoke, J.L., Bradley, R.P., Katta, V.P., Schauer, F.R.: Emissions in a Pulsed Detonation Engine. In: 47th AIAA Aerospace Sciences Meeting, AIAA-2009-505 (2009)
26. Metacomp Technologies (2009), `http://www.metacomptech.com/index.html`
27. Humlicek, J.: An efficient method for evaluation of the complex probability function and its derivatives. Journal of Quantitative Spectroscopy and Radiative Transfer 21, 309 (1979)
28. Chang, A.Y., DiRosa, M.D., Hanson, R.K.: Temperature dependence of collisional broadening and shift in the NO A $\leftarrow$ X (0,0) band in the presence of argon and nitrogen. Journal of Quantitative Spectroscopy and Radiative Transfer 47, 375 (1992)
29. Settersten, T.B., Patterson, B.D., Gray, J.A. Temperature- and species-dependent quenching of NO A $^2\Sigma^+$ (v ′=0) probed by two-photon laser-induced fluorescence using a picosecond laser. Journal of Chemical Physics 124, 234-308 (2006)

30. Settersten, T.B., Patterson, B.D., Carter, C.D.: Collisional quenching of NO A $^2\Sigma^+$ (v ′=0) between 125 and 294 K. Journal of Chemical Physics 130, 204–302 (2009)
31. Settersten, T.B., Patterson, B.D., Humphries IV, W.H.: Radiative lifetimes of NO A $^2\Sigma^+$ (v ′=0) and the electronic transition moment of the A $^2\Sigma^+$– $X^2\Pi$ system. Journal of Chemical Physics 131, 104–309 (2009)
32. Herzberg, G.: Molecular Spectra and Molecular Structure: Spectra of Diatomic Molecules, Van Nostrand Reinhold (1950)
33. SENKIN from the CHEMKIN II package, Reaction Design, 6440 Lusk Boulevard, Suite D-205, San Diego, CA 92121, http://www.reactiondesign.com
34. Wang, H., You, X., Joshi, A.V., Davis, S.G., Laskin, A., Egolfopoulos, F., Law, C.K.: USC Mech. Version II. High-Temperature Combustion Reaction Model of $H_2/CO/C_1$-C_4 Compounds (May 2007), http://ignis.usc.edu/USC_Mech_II.htm

An Experimental Study of Different Obstacle Types for Flame Acceleration and DDT

Joshua A.T. Gray, Christian O. Paschereit, and Jonas P. Moeck

Technische Universität Berlin,
Institut für Strömungsmechanik und Technische Akustik,
Müller-Breslau-Str. 8 10623 Berlin, Germany
{joshua.gray,jonas.moeck,oliver.paschereit}@tu-berlin.de

Abstract. Harnessing detonations for energy conversion and transport applications requires methods for efficient deflagration-to-detonation transition (DDT) over short distances. The results of three different experiments, characterizing different types of obstacles for flame acceleration and DDT are reported in this work. Flame acceleration by obstacles with identical blockage ratio but different geometric details is investigated using light-sheet tomography. Small but distinct differences in propagation speeds are identified, which correspond to the various obstacle geometries. DDT experiments are carried out to investigate these configurations beyond initial flame acceleration observable with high-speed imagery. A strong effect of obstacle spacing on DDT success is observed, indicating an optimal spacing of slightly larger than two tube diameters. A so-called pseudo-orifice is considered in order to recreate the flow behind a mechanical orifice with the same blockage ratio considered in the previous experiments (0.43). The pseudo-orifice injects fluid perpendicular to the flow, creating a circumferential jet-in-crossflow configuration. Particle image velocimetry is conducted in an acrylic water test-rig in order to measure the flow field in several planes in the acrylic combustion chamber model to assess the effect of the pseudo-orifice on the flow.

Keywords: flame acceleration, deflagration-to-detonation transition, pulse detonation combustion, fluidic obstacle.

1 Introduction

Pulse detonation engines have been investigated for over fifty years mainly for unmanned propulsion applications, but only recently have pulse detonation combustors (PDCs) been considered a viable alternative for isobaric combustors presently used in gas turbines for power generation and aero-engines [1]. This idea, though it has been under research a great deal in the last decades, is not new. The use of this cycle was investigated analytically in the mid-twentieth century by Zel'dovich [2] and later independently by Jacobs [3]. The premise is

R. King (ed.), *Active Flow and Combustion Control 2014*, Notes on Numerical Fluid Mechanics and Multidisciplinary Design 127, DOI: 10.1007/978-3-319-11967-0_17

taking advantage of the thermodynamic efficiencies of a non-isobaric combustion process, in contrast to the constant-pressure process exhibited by the Brayton cycle in gas turbines (see [4] or [5]).

Due to the high amount of energy required to directly initiate a detonation [6], obstacles are typically used to introduce turbulence into the flow and accelerate a deflagratively propagating flame until the heat release rate of the turbulent flame and its propagation speed are sufficiently high to permit transition to detonation [7,8]. At this point, the SWACER (Shock Wave Amplification by Coherent Energy Release) mechanism plays a vital role [9]. The SWACER mechanism requires certain gradients in the induction time of the mixture which result in coeherent energy release. This can be achieved through turbulence to promote such gradients in temperature and mixture concentrations. The combination of the flame propagation speed, heat release rate, and induction time gradients results in small local explosions. The pressure waves from these small explosive kernels coalesce into a shock wave which propagates through the gas, causing autoignition in the unburnt mixture. Thus, a detonation can be obtained by accelerating a deflagrative flame. This process is called the deflagration to detonation transition (DDT) and requires much less ignition energy with the trade off of requiring an acceleration section of the PDC operating at suboptimal efficiencies. Therefore, shortening of the acceleration section is of utmost importance in view of fully exploiting the efficiency increase offered by detonative combustion.

The purpose of this work is twofold. First, obstacles of several geometries were investigated using high-speed laser-sheet tomography to assess their effect on initial flame acceleration. Additionally, the distance of the obstacles from the point of ignition as well as the separation distance, in the case of multiple obstacles, were investigated. Ionization probes were used on a different setup to capture front propagation at speeds at which high-speed imagery is more difficult.

The second goal of this work is to investigate the viability of using pseudo-orifices to replace mechanical orifices as "obstacles" in the flow. Instead of placing a mechanical obstacle in the flow, additional fluid is injected perpendicularly to the flow, resulting in a circumferential jet-in-crossflow configuration. Pseudo-orifices have the advantage of inducing less pressure loss in the high-velocity filling and purging phases of the detonation chamber, as the flow is not obstructed if the orifice is not activated during these phases. Another, and arguably more important, advantage is the fact that, as there is no physical obstacle in the flow, the thermal loading on vital components is drastically reduced, increasing life expectancy and reliability. Finally, this option gives the engineer another screw to turn with the goal of optimization of the DDT process. The more interaction and control can be exerted, the more such process can be improved based on injection timing, flow rates, etc. A pseudo-orifice was recently proposed and tested by Knox et al. [10] for hydrogen–air mixtures in a tube with an inner diameter of roughly 53 mm. However, DDT lengths for a single orifice in this configuration were above 1 m at atmospheric pressure. Increasing the initial pressure in the detonation tube decreased the DDT length to roughly 760 mm.

By investigating the non-reacting flow field inside of the detonation chamber, it may be determined which geometries and flow rates lead to higher turbulence intensities and simultaneously reduce DDT length. These geometries can then be tested on reacting test-rigs for verification.

2 Experimental Setup and Data Processing

In this study, experiments were conducted on three different setups. High-speed laser sheet tomography was conducted on a setup with a tube made of acrylic glass. Front propagation speeds were measured on a setup with ionization probes in a metal tube. Finally, experiments were conducted in a water test-rig in order to assess the impact of the pseudo-orifice on the flow field in the detonation chamber.

2.1 Flame Acceleration Setup Using Laser Sheet Tomography

The acrylic tube had an inner diameter of 30 mm and was filled with a stoichiometric hydrogen–air mixture whose composition was confirmed by two coriolis mass flowmeters. The mixture was given ten seconds to settle, afterwhich the contents of the tube were ignited using a sparkplug at the closed end of the tube, imparting roughly 1.2 mJ to the mixture.

The flame propagation was observed using light-sheet tomography. A fine mist of droplets (mean diameter of 1 µm) composed of silicone oil was produced by sending the combustion air through a seeding generator, introducing a small amount of oil into the air before it was directed into the flame tube. A laser sheet from a CW-laser (Quantum Laser Finesse) was aligned along the axis of the tube with a thickness of 1 mm. In light-sheet tomography, the laser illuminates the oil droplets so that the unburnt gas may be identified from the Mie-scattered light. As the flame passes through the mixture, the droplets evaporate, and the combustion products are no longer illuminated. This results in a two-dimensional view of the interface between burnt and unburnt gases [11]. Due to the light scattered from the oil particles, it was necessary to reduce the length of the tube to 350 mm in order to retain the required light intensity in the area of interest. This intensity allowed for images to be taken at 16000 frames per second by a high-speed camera installed perpendicularly to the laser sheet. Pressure was not recorded on this setup, and no visible density changes (indicating a pressure wave) were observed; therefore, interactions between pressure waves and flame were not considered here. The experimental setup may be seen in Fig. 1.

Several obstacles of various geometries having a blockage ratio of 0.43 were installed in the tube to increase turbulence, resulting in flame acceleration. These obstacles were made of hard opaque plastic, in order to allow for ease of producing complex geometries. The blockage ratio was determined to be ideal for bringing about the deflagration-to-detonation transistion [8,12]. The blockage ratio is determined as the area of the obstacle divided by the area of the cross-section of the tube. The geometries used are depicted in Fig. 2. Similar geometries were

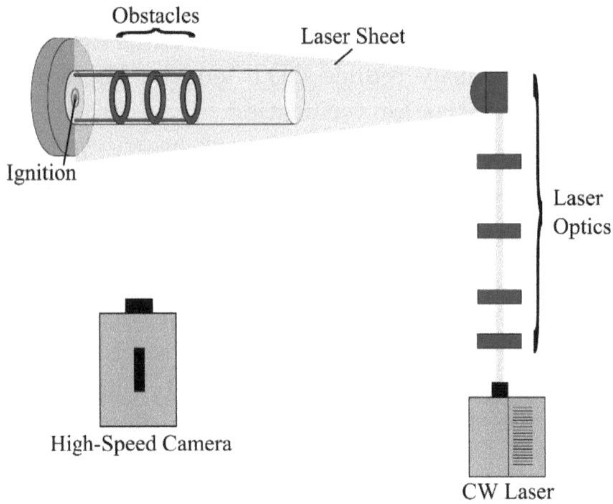

Fig. 1. Experimental setup for flame acceleration using laser sheet tomography

used by Knystautas et al. [13] The rationale behind the chosen obstacles was to vary the type of flame folding. For instance, the orifice features a flame expanding upon passing through it and vortices that interact with the wall, while the plates present an obstacle to the flame tip and produce vortices that interact with one another downstream. The obstacles were attached to two metal threaded rods outside of the plane of the laser with a diameter of 2.5 mm and held in place by nuts. This setup produces additional turbulence, but it is assumed that this is negligible compared to that produced by the obstacles themselves. Despite this disadvantage, the setup was chosen due to the flexibility of installing the orifices at arbitrary and easily adjustible positions. Several researchers have investigated orifices attached in this way (see, for example, [14,15]).

In choosing the geometries, the effect the obstacle would have on the fluctuation length scales was also considered. Plates A–D, for example, all have the same blockage ratio but have a different number of slits, decreasing the size of the trailing vortices. Additionally, Orifice Z, Plate Z, and Disc Z are similar to their geometric counterparts, however, possess serrated edges (blockage ratio is, nevertheless, held constant). It was envisaged that these edges would function as vortex generators, increasing the amount of turbulence through the introduction of streamwise vorticity.

2.2 Setup for DDT Experiments

Ignition was achieved in this setup using the same system as in the experiments in Sec. 2.1. Here, a steel tube with an inner diameter of 39 mm was used. Tube sections with smaller inner diameters could be inserted into this tube, changing the effective diameter of the combustion chamber. This resulted in inner diameters of 30 mm, 32.8 mm, and 39 mm. For the two smaller diameters, inserts cut to

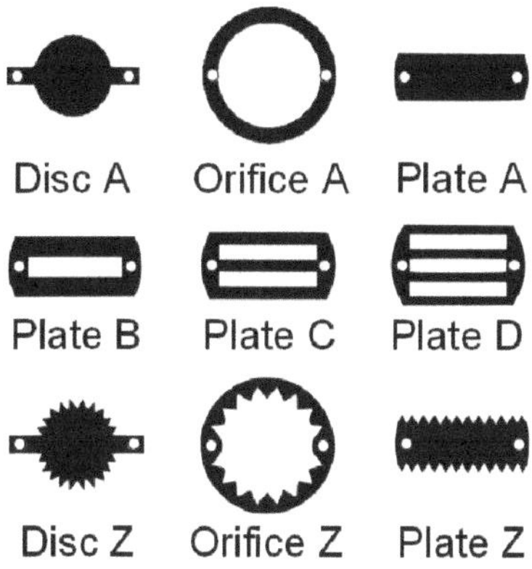

Fig. 2. Obstacle geometries used in experiments for initial flame acceleration. All obstacles have a blockage ratio of 0.43.

the correct separation lengths were used as spacers between the orifices (see Fig. 3). Up to four orifices with blockage ratios of 0.43 were used in the experiments. For the largest tube diameter (without inserts), the orifices were installed exactly as in the experiments with the acrylic tube, with two threaded rods. The steel tube consisted of a mixing section of 110 mm in length and 22.5 mm in diameter followed by the flame acceleration and measurement tube, having a length of 1500 mm. Four ionization probes were installed at the far end of the tube at 910, 1110, 1310, and 1510 mm from the point of ignition, at the back wall of the mixing section. Ionization probes operate on the principle that combustion fronts contain ionized species. If an electric potential difference is applied across two electrodes and a reaction front containing ions is passing the gap between the electrodes, then a current is induced and the potential difference is equalized. In other words, the voltage signal falls to zero at the instant the reaction front reaches the electrodes. By measuring the time of flight between the probes, the front speed can be determined. At least ten test runs were taken for each parametric combination of tube inner diameter, number of orifices, and separation distance, resulting in a total of roughly 500 test runs.

2.3 Setup for Flow Field Measurements with the Pseudo-orifice

Experiments were conducted in a water test-rig in order to assess the impact of the pseudo-orifice on the flow field in the detonation chamber. This test-rig was also constructed out of acrylic glass for optical access. The inner diameter of the tube representing the detonation chamber was 40 mm. A perforated plate was located at the inlet of the tube (see Fig. 4). This was done to be similar to the planned fired test-rig. In the fired test-rig, this perforated plate is responsible for increasing turbulent mixing of the gases as they enter the tube. It also serves to reduce the chance of flashback into the premix chamber, as the gas is injected

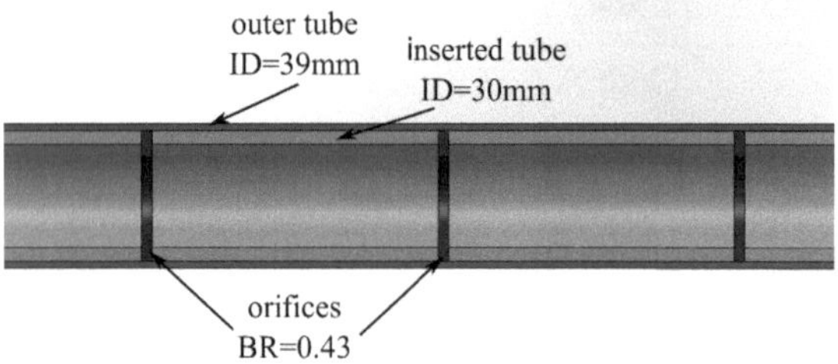

Fig. 3. DDT setup for multiple tube diameters and orifice separation distances

upstream of this plate. The circular slot is intended to purge the area directly behind the centerbody (where the spark plug is installed in the PDC). It was found in previous studies that exhaust gas accumulation in this recirculation zone frequently led to ignition failure.

A series of orifices was used to increase turbulence in the flow. The blockage ratio was 0.43, the distance between obstacles 85 mm. This was found in the previously mentioned experiments to be a separation distance at which reproducible detonations could be obtained for stoichiometric hydrogen–air mixtures (see Sec. 3.2). The thickness of the orifices was 3 mm. The first orifice was located at 55 mm downstream of the inlet. A manifold was installed at the location of the fourth orifice in which either a mechanical orifice or a pseudo-orifice could be inserted. In this study, the pseudo-orifice investigated was a ring slot measuring 0.3 mm. This created, so to say, a "sheet in crossflow" configuration.

Particle image velocimetry was conducted using a high-speed laser at 750 Hz for each image pair. The system consists of a pulsed pair of Quantronics Darwin Duo 100 W Nd:YLF lasers and a high-speed CMOS camera (Photron FastCam SA1.1). The pulse separation between each pair of images was 200 µs. The data processing was done with interrogation windows of 24×24 pixels with 50% overlap filtered with a 3×3 pixel median filter and subtraction of the background image of minimum light intensity. The laser sheet was aligned in a longitudinal plane containing the central axis of the tube. Several measurements were conducted at three axial locations (see Fig. 5). The test-rig used two separate circuits both drawing from the same water supply. The water supply was seeded with silver-coated, hollow glass spheres having a mean particle size of 15 µm. The volume flow of the main circuit was set using a gate valve to a bulk velocity of 0.7 m/s. The circuit for the pseudo-orifice was controlled with a fast-acting proportional valve, allowing for unsteady injected volume flows.

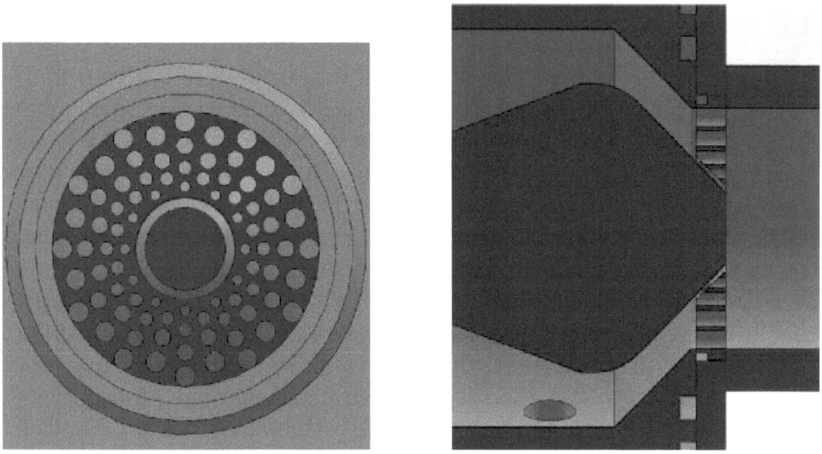

(a) Frontal view (b) Sectional side view

Fig. 4. Perforated plate between mixing chamber and detonation chamber with front view and sectional side view

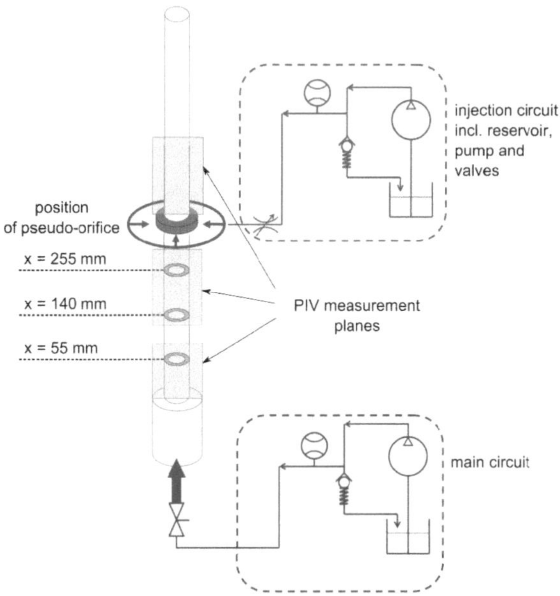

Fig. 5. Experimental setup for flow field investigations of a pseudo-orifice using PIV

3 Results and Discussion

3.1 Flame Acceleration Induced by Obstacles of Different Shapes

Individual obstacles were initially tested at a distance of 50 mm from the closed end of the tube. The locus of the interface between burnt and unburnt gases determined using laser-sheet tomography is shown in Fig. 6 for various obstacle geometries. Three experimental realizations for each case are plotted, indicating the repeatabiliy of the measurements. The locus corresponding to a tube without obstacles is added for comparison. Flame propagation without obstacles exhibits the typical "tulip flame" instability [16], accompanied by acoustic oscillations and a reversal of flame movement (at later times not shown). Mounting any of the obstacles at 50 mm results in suppression of this instability within the tube. Orifice A as well as Plates A, B, and C result in the highest flame acceleration, primarily directly after the obstacle, with a maximum propagation speed of around 120 m/s. Orifice A resulted in nearly the same flame propagation as those from the plates and with a similar final velocity, although just past the orifice, the flame shortly decelerates more than with the other obstacles. Plates B and C exhibit a somewhat weaker acceleration but shortly past the obstacles once again overtake the others, finally reaching the same speed as the other obstacles. A series of images for Orifice A and Plate A are shown in Fig. 7. Acceleration in both cases is evident from the superlinear progress of the interface with time. Although the structure of the reaction front is quite different in the two cases, the final propagation velocity is very similar.

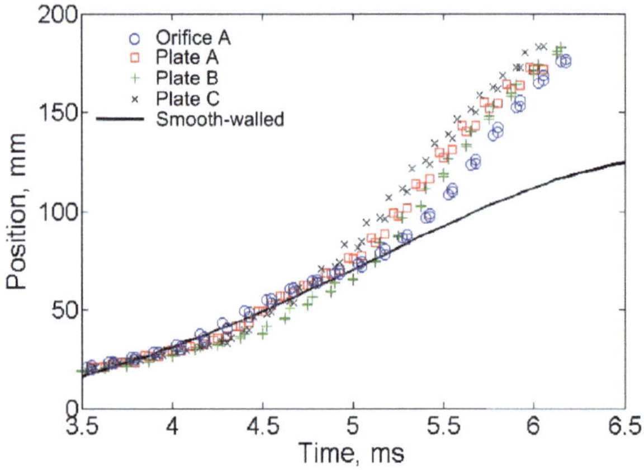

Fig. 6. Flame propagation for various obstacles and for a tube without obstacles. Axial location of the unburnt–burnt interface is plotted versus time after ignition. For cases with obstacles, results from three independent experiments are shown.

Fig. 7. High-speed images of accelerating flames using a single obstacle (Orifice A, left; Plate A, right), both having blockage ratios of 0.43. Time between two frames is 3 µs.

Plate D and Disc A resulted in flame propagation similar to that of Plates A–C. However, the propagation exhibited slightly more scatter and, thus, they are not shown in Fig. 6 for the sake of clarity. The obstacles with serrated edges proved to show no improvement in flame propagation speed. In fact, Orifice Z produced a significantly lower propagation speed than Orifice A.

Orifice A was also investigated in more detail, as such orifices are frequently used in pulse detonation combustors. Since the orifice is in contact with the wall of the combustor, it allows for heat to be more quickly transferred when cooled from the wall, decreasing thermal loading. First, the distance of the orifice from the closed end of the tube was varied from $1D$ to $3D$ in increments of $0.5D$, D being the inner diameter of the tube. Highest flame acceleration occurred when the orifice was at $2.5D$, resulting in a propagation speed of 111 m/s. The weak dependence of initial flame acceleration on the geometric details of the obstacles may be taken to indicate that the process is mainly affected by the integral fluctuation scales.

Finally, a series of two to three orifices were investigated and the distance between them was varied from $1D$ to $3D$. The first orifice was mounted at 50 mm. A series of two orifices produced maximum propagation speeds of 298 m/s with a spacing of $2.5D$. When a series of three orifices was mounted in the tube, the propagation speed increased from 266 m/s at $1D$ to 470 m/s at $2D$. With a distance between the three orifices greater than $2D$, the propagation velocity increased so drastically that it was no longer possible to detect it. Experiments with both two and three orifices suggest that the optimal distance lies somewhat above $2D$, which is consistent with results from literature, indicating that $2D$ is an optimum obstacle spacing for flame acceleration [17].

3.2 DDT Results

The experiments in the steel tube were conducted in order to confirm results seen in the previous experiments, now allowing DDT to occur in a longer tube. An exemplary result for a test run is shown in Fig. 8. The propagation speed in this case, calculated from the delay in the arrival times of the reaction front at the probes, is determined to be 1960 m/s. The first orifice was placed at 100 mm, roughly $2.5D$ for the largest tube, found to be advantageous in experiments in the acrylic glass tube. Of the three inner diameters, only that of 39 mm was able to consistently sustain a detonation. The tube with a diameter of 32.8 mm resulted in one instance of an overdriven detonation, and the tube of 30 mm produced no detonations whatsoever, although these sizes are much larger than the expected detonation cell size for stoichiometric hydrogen–air mixtures of 1–1.5 cm (see [18] and [19]).

The results of the tests with the 39 mm tube are shown in Fig. 9. Here, it can be seen that a detonation occurs only after the use of four orifices. With three orifices, the flame is able to reach the "choking" regime discussed by Guirao [8]. The test series at 70 mm and 90 mm are at the boundary of producing a detonation. For example, ten test runs with 70 mm separation distance resulted in 8 detonations at roughly 2000 m/s and two choked cases at 1000 m/s, with no values inbetween. It is interesting that the 80 mm separation distance is suddently no longer sufficient to produce a detonation, while a separation distance of 85 mm results in very stable DDT. This sensitive dependence of DDT success rate underlines the necessity of introducing a means to control and optimize the effect of the obstacle-induced flow dynamics. One possibility that offers this additional degree of freedom is the pseudo-orifice.

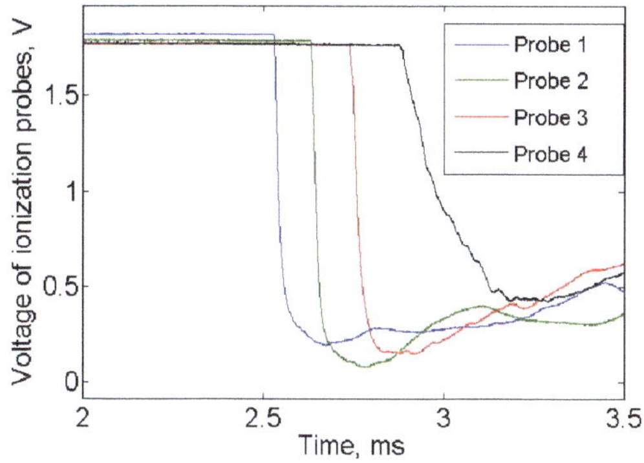

Fig. 8. Voltage signals from four ionization probes with a spacing of $\Delta x = 200\,\mathrm{mm}$

3.3 Flow Fields of the Pseudo-orifice

The results from the PIV measurements in a longitudinal plane are shown in Fig. 10. A steady-state volume flow of roughly 460 L/h (jet velocity of 3.4 m/s) was injected through the pseudo-orifice. Both the mean axial velocity as well as the turbulence intensity are plotted in the three measurement planes investigated. The entirety of the tube was not measured. In several areas, this was not possible due to the obstruction from the mechanical orifices or the manifold containing

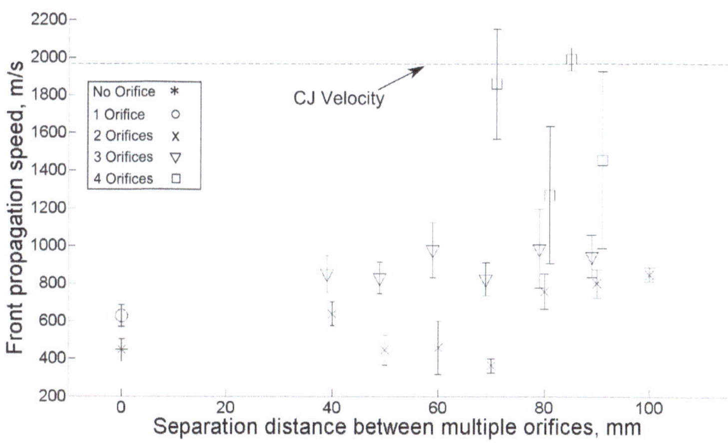

Fig. 9. Front propagation speeds with various numbers of orifices and different spacings in a 39 mm tube, determined from time-of-flight measurements using ionization probes. Error bars indicate the standard deviation of each test series. Blockage ratio for all orifices is 0.43.

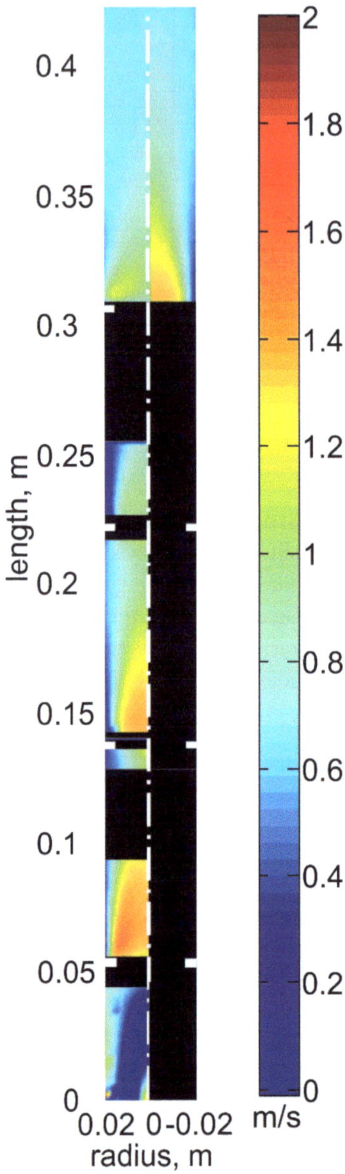

(a)

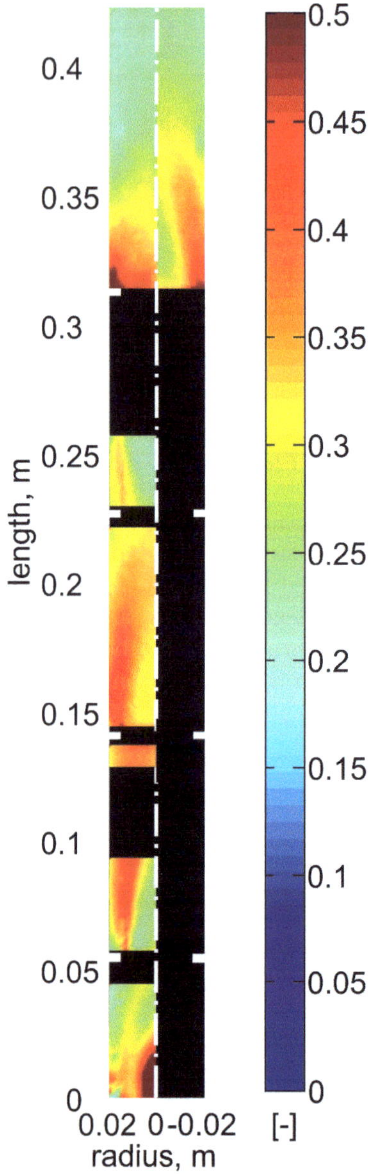

(b)

Fig. 10. Mean axial velocity (a) and turbulence intensity (b) in the detonation tube. The right side of each figure corresponds to the case with the mechanical orifice; the left side represents the case where the fourth mechanical orifice is replaced by the pseudo-orifice. Areas not recorded are shown in black.

the last orifice. The turbulence intensity was calculated from the in-plane fluctuation components (axial and radial) only. Since an orifice generates primarily azimuthal vorticity, it is assumed that the azimuthal fluctuation component is comparatively small.

All measurements conducted upstream of the fourth orifice were made only with the pseudo-orifice, while the measurements downstream of the orifice were taken with both the mechanical orifice as well as the pseudo-orifice in the fourth position. The mean axial velocity along the centerline is increased by the presence of the orifices due to the flow restriction. These velocities are on the order of two times the bulk velocity. The turbulence intensities are around 0.4, mostly in the shear layers downstream of each orifice. There are two exceptions. The first is the strong shear layer formed by the slot in the inlet. The second is behind the fourth orifice. The mechanical orifice creates a turbulent shear layer similar to the other orifices with slightly higher turbulence intensities, reaching 0.5 and above. The pseudo-orifice creates a similar region of higher turbulence fluctuations. The noticeable difference lies in the structure of the fluctuation field. Additionally, the turbulent flow from the pseudo-orifice reaches completely to the centerline of the tube. This increases the average turbulence intensity along the entire transverse plane. It achieves this at the expense of a noticeably shorter zone of higher turbulence stretching in the axial direction.

4 Summary and Conclusions

Laser-sheet tomography was conducted on a propagating flame in a tube with obstacles. All of the obstacles, with the exception of those with serrated edges, perform similarly in terms of flame acceleration. In the case of single orifices, the highest flame acceleration was observed when the orifice was mounted at $2.5D$. In the case of two orifices, the highest flame acceleration occurred at a spacing of $2.5D$. For three or more orifices, drastic flame acceleration was observed for distances between orifices above $2D$, but not at $2D$ itself. This suggests that the ideal separation distance, at least for an inner diameter of $D = 30\,\text{mm}$, lies somewhere between $2D$ and $2.5D$.

Reliable detonations were only achieved with the 39 mm tube with at least four orifices. The remaining tubes seem to be too small for the detonation to propagate. In the largest tube, four orifices at a separation distance of 85 mm produced repeatable detonations, while 70 mm and 90 mm separation distances resulted in the flames propagating in one of two regimes, that of the "choked" detonation and that of the CJ detonation. In this particular configuration, the best repeatibility was found to be around $2.2D$, though the somewhat less than optimal reproducibility of the 70 mm, 80 mm, and 90 mm separation distances makes it difficult to conclusively infer that this is the best separation distance.

The PIV measurements on the pseudo-orifice illustrate how this device helps to bring about DDT. Higher levels of turbulence are brought into the center of the detonation tube, leading to strong concentration and temperature gradients, supporting the SWACER mechanism. Additionally, the average turbulence intensity is increased in comparison to the mechanical orifice. This indicates that pseudo-orifices may be used to produce DDT, especially when considering the available optimization parameters such as timing, length scales, mass flows, velocities, injection type, and the ability to inject a different mixture, allowing for stratification of fuel concentration at the point of injection.

Acknowledgments. The authors gratefully acknowledge support by the Deutsche Forschungsgemeinschaft (DFG) as part of collaborative research center SFB 1029 "Substantial efficiency increase in gas turbines through direct use of coupled unsteady combustion and flow dynamics". Furthermore, the authors would like to thank Bernhard Bobusch, Robert Bahnweg, Steffen Terhaar, Thoralf Reichel, Richard Berg, Niklas Blum, and Niclas Hanraths for their ideas, support, and assistance.

References

1. Rasheed, A., Furman, A.H., Dean, A.J.: Experimental investigations of the performance of a multitube pulse detonation turbine system. Journal of Propulsion and Power 27(3), 586–596 (2011)
2. Zel'dovich, Y.B.: To the question of energy use of detonation combustion. Zhurnal tekhnicheskoi fiziki 10(17), 1453–1461 (1940)
3. Jacobs, S.J.: The energy of detonation. NAVORD Report 4366, U.SṄaval Ordnance Laboratory, White Oak, MD. NTIS AD113271 – Old Series (1956)
4. Wintenberger, E., Shepherd, J.E.: Thermodynamic analysis of combustion processes for propulsion systems. In: AIAA Aerospace Sciences Meeting and Exhibit, AIAA 2004-1033 (2004)
5. Wintenberger, E., Shepherd, J.E.: Thermodynamic cycle analysis for propagating detonations. Journal of Propulsion and Power 22(3), 694–697 (2006)
6. Schultz, E., Wintenberger, E., Shepherd, J.: Investigation of deflagration to detonation transition for application to pulse detonation engine ignition systems. In: 16th JANNAF Propulsion Meeting (1999)
7. Lee, J.H., Knystautas, R., Chan, C.K.: Turbulent flame propagation in obstacle-filled tubes. Symposium (International) on Combustion 20(1), 1663–1672 (1985)
8. Guirao, C.M., Knystautas, R., Lee, J.H.: A summary of hydrogen–air detonation experiments. NUREG/CR-4961, SAND87-7128. Sandia National Laboratories (1989)
9. Lee, J.H.S., Knystautas, R., Yoshikawa, N.: Photochemical initiation of gaseous detonations. Acta Astronautica 5, 971–982 (1978)
10. Knox, B.W., Forliti, D.J., Stevens, C.A., Hoke, J.L., Schauer, F.R.: A comparison of fluidic and physical obstacles for deflagration-to-detonation transition. In: AIAA Aerospace Sciences Meeting and Exhibit, AIAA 2011–587 (2011)
11. Boyer, L.: Laser tomographic method for flame front movement studies. Combustion and Flame 39(3), 321–323 (1980)
12. Dorofeev, S.B., Sidorov, V.P., Kuznetsov, M.S., Matsukov, I.D., Alekseev, V.I.: Effect of scale on the onset of detonations. Shock Waves 10, 137–149 (2000)
13. Knystautas, R., Lee, J.H., Moen, I., Wagner, H.G.: Direct initiation of spherical detonation by a hot turbulent gas jet. Symposium (International) on Combustion 17(1), 1235–1245 (1979)

14. Lee, S.-Y., Watts, J., Saretto, S., Pal, S., Conrad, C., Woodward, R., Santoro, R.: Deflagration to detonation transition processes by turbulence-generating obstacles in pulse detonation engines. Journal of Propulsion and Power 20(6), 1026–1036 (2004)
15. Porowski, R., Teodorczyk, A.: Experimental study on DDT for hydrogen–methane–air mixtures in tube with obstacles. Journal of Loss Prevention in the Process Industries 26, 374–379 (2013)
16. Clanet, C., Searby, G.: On the "tulip flame" phenomenon. Combustion and Flame 105, 225–238 (1996)
17. Theodorczyk, A., Drobniak, P., Dabkowski, A.: Fast turbulent deflagration and DDT of hydrogen–air mixtures in small obstructed channel. International Journal of Hydrogen Energy 34(14), 5887–5893 (2009)
18. Ciccarelli, G., Ginsberg, T., Boccio, J., Economos, C., Sato, K., Kinoshita, M.: Detonation Cell Size Measurements and Predictions in Hydrogen–Air–Steam Mixtures at Elevated Temperatures. Combustion and Flame 99, 212–220 (1994)
19. Tieszen, S.R., Sherman, M.P., Benedick, W.B., Shepherd, J.E., Knystautas, R., Lee, J.H.S.: Detonation cell size measurements in hydrogen–air–steam mixtures. Progress in Astronautic and Aeronautics 106, 205–219 (1986)

Investigation of Fluidic Devices for Mixing Enhancement for the Shockless Explosion Combustion Process

Bernhard C. Bobusch[1], Phillip Berndt[2], Christian O. Paschereit[1], and Rupert Klein[2]

[1] Technische Universität Berlin, Institut für Strömungsmechanik und Ttechnische Akustik, – Hermann-Föttinger-Institut – Müller-Breslau-Str. 8, D-10623 Berlin, Germany
{bernhard.bobusch,oliver.paschereit}@tu-berlin.de
[2] Freie Universität Berlin, Department of Mathematics, Geophysical Fluid Dynamics, Arnimallee 6, D-14195 Berlin, Germany
{pberndt,rupert.klein}@math.fu-berlin.de

Abstract. Fuel-air mixing is a crucial process in low emission combustion systems. A higher mixing quality leads to lower emissions and higher combustion efficiencies. Especially for the innovative constant volume combustion processes "Shockless Explosion Combustion" (SEC) the mixing of fuel and air is an important parameter, since the whole combustion process is triggered and controlled via the equivalence ratio. To enhance the passive scalar mixing, fluidic oscillators are investigated and compared to the standard jet in crossflow fuel injection configurations. The mixing quality of the different geometries is assessed in a water test-rig by making use of planar laser induced fluorescence. After a short introduction to the SEC-process, the test-rig and the different injection configurations are introduced. To verify whether the mixing quality is sufficient for the SEC-process, a numerical investigation using the experimentally determined unmixedness is conducted. It is not only shown that the fluidic oscillators are able to enhance the mixing quality and create an independence of the mixing quality from the jet in crossflow momentum, but it is also verified in a first numerical calculation that the achieved mixing quality might be good enough for the Shockless Explosion Combustion process.

Keywords: mixing, constant volume combustion, Shockless Explosion Combustion, numerical combustion.

1 Introduction

High quality mixing is one of the most crucial parameters in modern low emission combustion systems. This becomes even more crucial regarding advanced homogeneous pulsed (constant volume) combustion approaches such as the homogeneous charge compression ignition (HCCI) for internal combustion engines

R. King (ed.), *Active Flow and Combustion Control 2014*, Notes on Numerical Fluid Mechanics and Multidisciplinary Design 127, DOI: 10.1007/978-3-319-11967-0_18

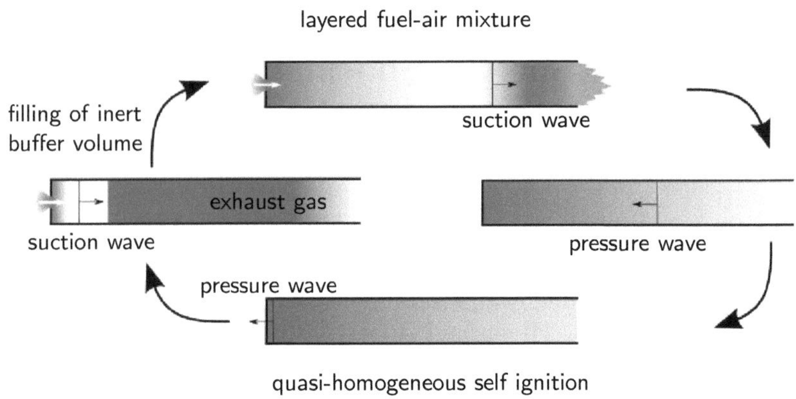

Fig. 1. Process cycle of the shockless explosion combustor

or the shockless explosion combustion (SEC) process for gas turbine combustion. Both of these systems rely on the auto-ignition of a homogeneous fuel-air-charge inside a combustor, thus, the preparation of a perfectly mixed combustion charge is the most important factor for these systems.

For the SEC-process not only is a perfectly homogeneous fuel-air mixture needed, but in addition, a stratification of the equivalence ratio in the combustor is necessary. An overview of this process is shown in Fig. 1. Like other constant volume combustion processes, the shockless explosion combustion process is based on a periodic combustion process. A standing pressure wave is established inside the combustion tube. The moment this pressure wave reduces the pressure at the tube inlet below the plenum pressure, the tube is filled with compressor air (left). After filling a volume with pure air, fuel is added to the combustion air at the inlet position until around 40% of the tube is filled with a combustible mixture (top). The air volume is needed to separate the hot flue gases of the previous cycle from the fresh fuel-air mixture. The suction wave is reflected from the open end of the tube and travels upstream to the inlet (right). Due to the hot air from the compressor, the mixture undergoes auto-ignition (bottom). The equivalence ratio inside the combustion chamber is adjusted in a way that the ignition delay matches the residence time of the mixture in the tube. Thus, the entire fuel-air volume undergoes homogeneous auto-ignition and the mixture is burnt instantaneously without any shock waves. In addition, the ignition delay is adjusted to match the oscillation period. This means that the combustion of the mixture occurs simultaneously with the pressure wave raising the pressure at the tube inlet. The pressure wave is amplified and travels to the end of the combustion tube, where it is reflected as a suction wave and restarts the process.

This process has by implication very high demands on the mixing quality:

1. The mixing needs to be very fast to avoid regions where the ignition delay is too short.
2. The mixing must be homogeneous in the radial plane.

3. The axial stratification of equivalence ratio demands for very low mixing in the axial direction.
4. The mixing must be independent from the fuel volume flow (i.e., jet in crossflow momentum) to have a broad range of possible equivalence ratios for base load, part load, and idle operation.

To create this type of mixing several different jet in crossflow configurations are investigated in this work. Namely the tested configurations are round jet, rectangular jet, slit, and spatially oscillating jet in crossflow. The jet in crossflow configuration is the easiest, most robust, and very common way to mix two fluids. Over the past decades, the round turbulent jet in crossflow was widely investigated as a steady jet (e.g., [1–4]) or as a modulated/pulsed jet (e.g., [5–7]). The rectangular jet in crossflow is not as thoroughly described as the round jet, but recently this configuration was analyzed more extensive as well (see [8, 9]). The slit injection lies somewhere between these two configurations. It can be seen as a two-dimensional jet in crossflow with infinite depth. For the spatially oscillating jet, only a few publications can be found. Nathan et al. [10] extensively investigated jet parameters and their impact on the mixing for an inline configuration. Several oscillating devices, mechanical and fluidic, were tested. It was shown that the oscillation enhances the mixing especially close to the fuel inlet. Arnaud and Paschereit [11] tested the enhancement of scalar mixing due to a spatial oscillation of a jet in crossflow in a water test-rig. They confirmed the findings of Nathan et al. [10] and in addition found out that the spatial oscillation reduces the dependency of the mixing quality on the jet in crossflow momentum. These advantages make the spatially oscillating jet in crossflow injection very promising for the demands of the shockless explosion combustion process. To assess and compare the mixing quality of the different configurations they were investigated in a water test-rig making use of a fluorescent dye.

Known from literature (e.g., [12, 13]), the mixing of water with dye is a good indicator for the fuel-air mixing in combustion systems. In addition, the results of the experiments are used in a numerical calculation to see if the mixing quality is high enough to assure a reliable SEC-process.

The remainder of the paper is organized as follows: First the experimental setup, measurement technique, and evaluation methods are presented. This is followed by the experimental results and the numerical investigations.

2 Experimental Setup

To investigate the mixing performance of several geometries, a test-rig was build employing water as the fluid. The entire geometry is based on the planned SEC combustion tube. A schematic overview of the rig can be found in Fig. 2. A more detailed cut through the dye inlet section can be found in Fig. 3. The test-rig consists of a main combustion tube made of acrylic glass with an inner diameter of 40 mm and a length of 800 mm. Upstream of this tube a 25 mm valve is installed to open and close the main mass flow. It is followed by a diffuser which increases the inner diameter up to the mentioned 40 mm of the main tube (see

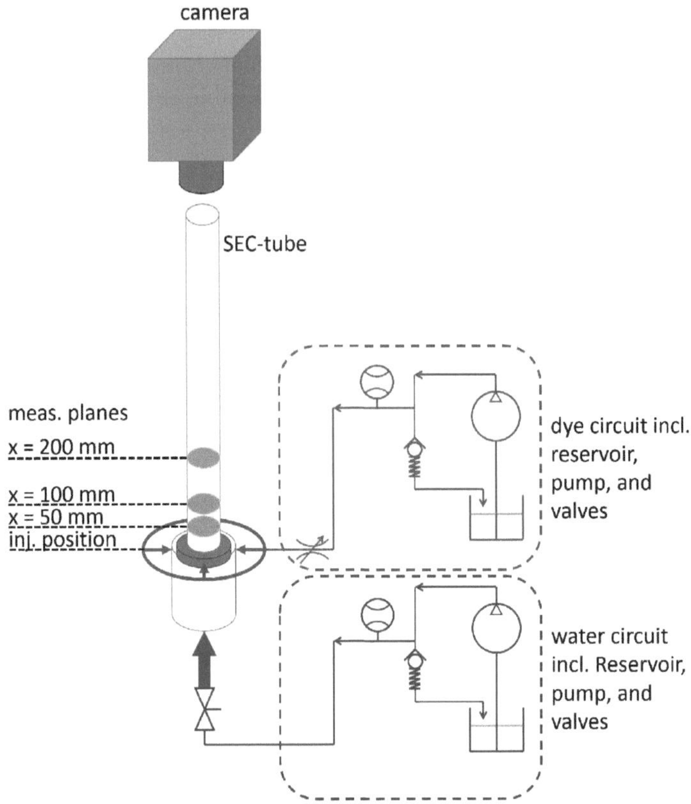

Fig. 2. Sketch of the test-rig

Fig. 3). It was assured that the flow is attached to the surface of the diffuser at all times.

Downstream of the diffuser interchangeable disks represent the fuel, in this case dye, inlet. The disk is colored yellow for a better visibility in Fig. 3. Pictures and details of the investigated geometries can be found later in this section. The flow rate of the dye can be adjusted using an electrically driven proportional valve, which was installed upstream of the inlet disc plenum.

Two separated water circuits were realized for the injection of the main and the dye flow. Both circuits are equipped with pressure driven one-way valves in the circuit to generate the needed pressure upstream of the two main valves.

To investigate the mixing quality and determine the unmixedness parameters of the different injection geometries, Planar Laser Induced Fluorenscence (PLIF) was employed in three radial planes at different axial positions downstream of the dye injection. The images were taken with a high-speed camera from the downstream side of the tube. A shutter frequency of 500Hz minimizes the blurring of the image due to long exposure times.

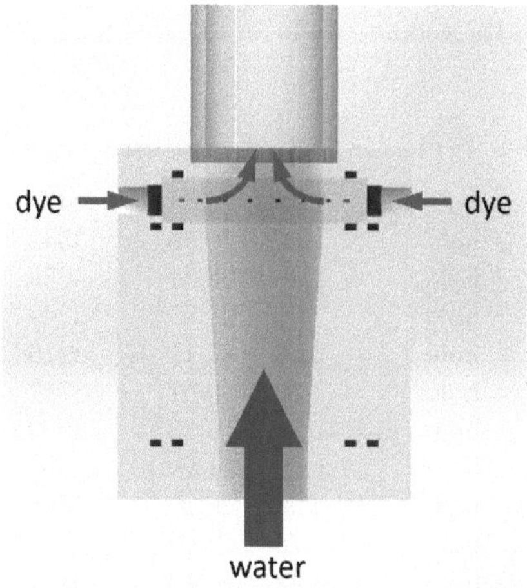

Fig. 3. Cut through the fuel inlet section of the water test-rig

21 different injection geometries including round, rectangular, and spatially oscillating (fluidic) jets, as well as slits were investigated. Different cross-section areas were employed to create a broad range of jet in crossflow momentums for all injector disks as well as different frequencies for the fluidic oscillators. The jet in crossflow momentum J is defined as the squared ratio between the jet flow velocity w_j and the main flow velocity w_0.

$$J = \left(\frac{w_j}{w_0}\right)^2 \tag{1}$$

Note that for all four different injector geometries (i.e., round, rectangle, oscillating, slit) different sizes and amounts of injection ports were tested. This was done to investigate the behavior of these injectors over a wide range of jet in crossflow momentums. Since the volume flow range for all configurations was approximately the same, the different sizes result in different ranges of J. This broad range is needed in the SEC process due to the fact that the fuel mass flow is changed within each injection cycle, as mentioned in the introduction. Additionally for distinct injection ports two different numbers of ports were investigated. A summary of the different geometries, including the main flow parameters, can be found in Tab. 1. The Reynolds number of the main flow was $Re = 31280$ and thus, a fully turbulent flow was assured. To give an idea of the different injection geometries a three-dimensional model of four of the used injector disks is shown in Fig. 4. For better visibility of the fluidic oscillator, a cut was made through the material to show the geometry.

The investigated spatially oscillating jets were created using fluidic oscillators. From literature [11], it is known that these devices can enhance the spatial

Table 1. Investigated injector geometries including the main parameters. All units in mm or mm^2 respectively.

Disk	Outlet geom.	Outlets	Type	d_h	A_{Outlet}	$Re_{jet;min}$	$Re_{jet;max}$	J
1	round	13	hole	1.40	20.01	4479	11143	16.7-103.6
2	round	13	hole	1.90	36.86	3542	8371	5.7-34.2
3	round	13	hole	2.40	58.81	3059	7519	2.7-16.0
4	round	7	hole	2.00	21.99	6249	14485	13.2-82.4
5	round	7	hole	2.60	37.17	4916	11361	5.8-31.2
6	round	7	hole	3.30	59.87	3787	9984	2.2-15.0
7	rectangle	13	hole	1.26	16.33	3909	9493	15.8-93.4
8	rectangle	13	hole	1.69	21.99	3359	7138	6.4-32.6
9	rectangle	13	hole	2.00	26.06	2548	5732	2.6-13.4
10	rectangle	7	hole	1.73	12.09	5574	12668	14.1-87.9
11	rectangle	7	hole	2.29	16.06	4183	9983	4.5-31.0
12	rectangle	7	hole	2.71	18.97	3308	8127	2.4-14.7
13	rectangle	13	fluidic osc.	1.31	16.97	3430	8092	11.2-62.5
14	rectangle	13	fluidic osc.	1.78	23.18	2943	7223	4.4-26.7
15	rectangle	13	fluidic osc.	2.17	28.24	1889	5618	1.2-10.9
16	rectangle	7	fluidic osc.	1.79	12.54	4098	11088	8.6-62.6
17	rectangle	7	fluidic osc.	2.40	16.83	3303	9726	3.1-26.7
18	rectangle	7	fluidic osc.	2.87	20.09	2516	7828	1.3-12.2
19	slit	1	—	0.40	50.19	893	2728	8.2-76.3
20	slit	1	—	0.60	75.22	867	2701	3.4-33.3
21	slit	1	—	0.80	100.21	891	2674	2.0-18.4

mixing, especially close to the injection position. The oscillatores used were described in detail in [14]. The oscillatory parameters of these devices were determined by numerical simulations, which were carried out with a validated numerical model [15]. The oscillation frequency was calculated from a time history plot of the velocity inside the mixing chamber.

For each of these 21 configurations, 8 fuel flow rates were investigated at the mentioned three axial positions. Recording 1632 snapshots for each of these tests over 822,000 pictures were taken in total. Each picture was corrected for background reflections and normalized using a homogeneous picture with the maximum dye concentration. It was assured that the dye concentration was low enough to be well within the linear regime of the fluorescence intensity. From the normalized images containing pixel values between 0 and 1, the two unmixedness paramters U_x and U_t were calculated, which represent the spatial and the temporal mixing quality repectively. Based on the work of Danckwerts [16], the parameters are defined as follows:

$$U_x = \frac{\sigma_x^2}{\sigma_0^2} = \frac{\sigma_x^2}{C_\infty^* \left(1 - C_\infty^*\right)}, \tag{2}$$

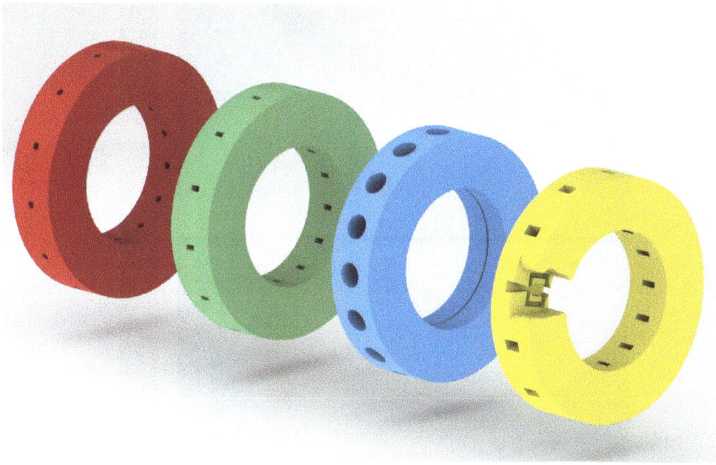

Fig. 4. Colored CAD-Models of the four different injector types (from left to right): round holes (red), rectangular holes (green), slit (blue), and fluidic oscillator (yellow)

$$\text{where} \qquad \sigma_x^2 = \frac{1}{N_i - 1} \sum_{i=1}^{N_i} \left(\overline{C^*}(i) - C_\infty^*\right)^2, \tag{3}$$

and σ_x^2 denotes the mixture variance of the temporally averaged concentration field $\overline{C^*}(i)$, which is recorded by the N_i camera pixels and is defined in Eq. 3. The variance immediately before the start of the mixing process (σ_0^2) is calculated from the dye concentration C_∞^*. The spatial unmixedness parameter U_x represents the average spatial mixing quality and gives a value between 0 (perfectly mixed) and 1 (not mixed at all). To investigate the temporal unmixedness, the parameter U_t is used, which employs the variance of all the concentration records. It is defined as:

$$U_t = \frac{\sigma_t^2}{\sigma_0^2} = \frac{\sigma_t^2}{C_\infty^* \left(1 - C_\infty^*\right)}, \tag{4}$$

$$\text{where} \qquad \sigma_t^2 = \frac{1}{N_i N_t - 1} \sum_{i=1}^{N_i} \sum_{i=1}^{N_t} \left(C^*(i,t) - C_\infty^*\right)^2. \tag{5}$$

N_t is the number of snapshots recorded during one measurement sequence (1632 in this work). According to Eq. 5, U_t captures both spatial and temporal fluctuations in the concentration. These unmixedness parameters can be used as a first measure to evaluate the mixing quality of a given injector geometry.

For the transfer of the experimental results to the numerical investigations, an additional analysis method was employed, which is presented in a next section.

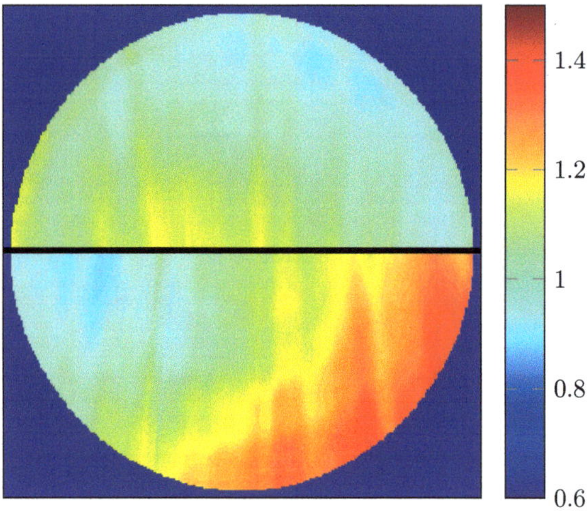

Fig. 5. Averaged, normalized spatial concentration field for disk 13 (upper half) and disk 1 (lower half)

3 Experimental Results

As a first impression of the results of the experiments, two pictures of the averaged and normalized concentration field C in the first measurement plane are shown in Fig. 5. For a better visibility, only one half of the pictures is shown for the two different configurations, whereby one of the two pictures was flipped upside-down allowing them to be plotted in one figure. The top half of the picture corresponds to disk 13 (see Tab. 1), an example for a high mixing quality. The lower half of the picture corresponds to disk 1 which shows a rather poor mixing quality due to the different injection geometry. This is clearly visible in the inhomogeneities of the concentration field. In addition to the pictures the concentration histograms for the two measurements are shown in Fig. 6. It is visible that the distribution of the pixel counts has a much sharper peak for disk 13 (see Fig. 6a) than for disk 1 (Fig. 6b). The vast amount of data (see Tab. 1) demands the selection of the injection geometries for further investigations by making use of the defined umixedness parameters. For the SEC-process, a fast, reliable and high-quality mixing is necessary in the radial plane, while almost no mixing is desired in the axial direction. From the steady injection point of view, the latter is unquantifiable since no changes in the dye volume flow are present in the current work. Accordingly, the spatial unmixedness parameter U_x is used to investigate the overall mixing parameters of the different geometries.

The range of the spatial unmixedness parameter for the different injection geometries is plotted in Fig. 7 for the first measurement plane 50 mm downstream of the injection point (see Fig. 2). For this plot, all the geometries of the same type are grouped together to show the overall performance of such a configuration, since the different injection parameters make a plot of the individual

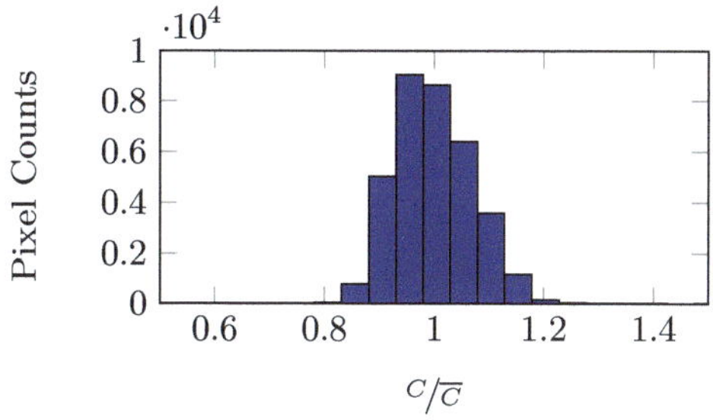

(a) Disk 13 — fluidic osc. $d_h = 1.31\,\text{mm}$

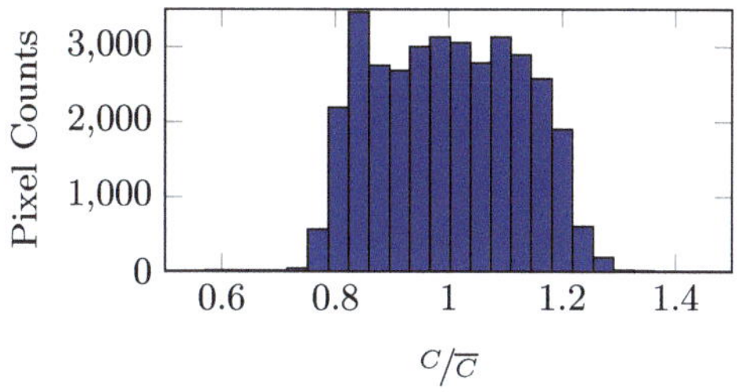

(b) Disk 1 — holes; $d_h = 1.4\,\text{mm}$

Fig. 6. Histogram of normalized concentration for (a) disk 13 and (b) disk 1 at $x = 50\,\text{mm}$

measurements rather unclear. Accordingly, disks $1 - 6$ are summed up in the black range, $7 - 12$ in the red, $13 - 18$ in the green, and finally the slits $(19 - 21)$ in the blue range. As one can see in this plot the slits and the round holes give a lower mixing quality than the rectangular holes and the fluidic oscillators. The rectangular holes show a better mixing quality (lower unmixedness parameter) on the lower side of the range than the fluidic oscillators but a worse mixing quality than the fluidic oscillators for the upper boundary of the range. Even though the mixing quality is slightly better for the rectangular holes in some cases, they fail for other injection situations. However, to assure the reliability of the mixing system over a wide range of equivalence ratios, fuel mixtures, and power levels, the fluidic oscillators show a more desirable behavior. Due to these advantages, the fluidic oscillators are chosen to be analyzed in more detail in this work. To gain a deeper insight into the mixing mechanisms of the oscillating jet, two different sizes of fluidic oscillators (disks $13 - 15$ and disks $16 - 18$) have been designed. Since the range of flow rates is defined by the dye pump, the two different oscillators were manufactured with 3 different aspect ratios each. From literature, it is known that the frequency of these oscillators scales linearly with the volumetric flow rate [14, 17]. Accordingly, the dependency of the frequency on the jet in crossflow momentum ratio J is of quadratic nature. To show the frequency ranges, J is plotted against the frequency in Fig. 8. It is clearly visible that even though the jet in crossflow momentum J is in the same range for both

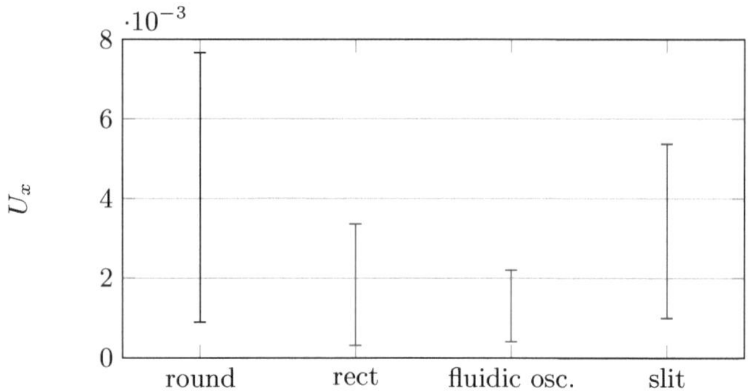

Fig. 7. Spatial unmixedness parameter U_x at measurement plane $x = 50$ mm for the four diffferent geometries: round jet, rectangular jet, fluidic oscillator, and slit)

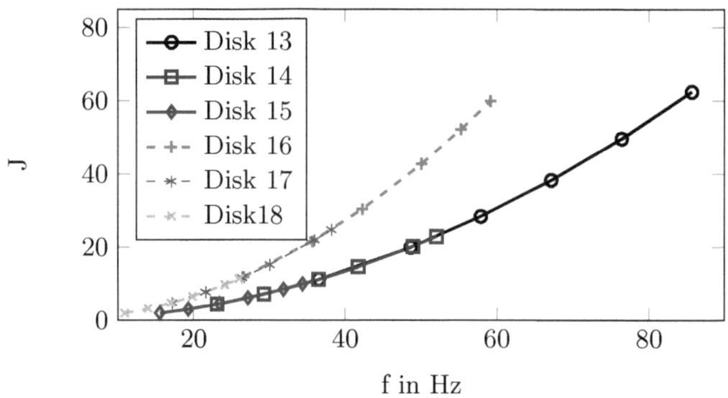

Fig. 8. Frequency of the fluidic oscillators for the different geometries and different values of J

sizes, the oscillation frequency is different. Disks 16 – 18, which consist of seven larger fluidic devices give a lower frequency range than disks 13 – 15.

Keeping this in mind, the mixing performances of disks 13 and 16 were compared. For the same J and, thus, the same volume flow, Reynolds number, and jet penetration, they create different frequencies. To see the impact of this change in injection geometry, the spatial unmixedness is plotted against J in Fig. 9. Two main conclusions can be made from these results. First, the mixing quality of the smaller oscillators with a higher frequency is better close to the point of injection. This might also be affected by the fact that for the smaller fluidic oscillators, 13 oscillating jets are present, while only 7 were employed for disk 16. However, it can be stated that a higher oscillation frequency may be favorable for a faster mixing process. The second observation from Fig. 9 is that the mixing quality of both configurations converge as the mixture is convected downstream. This confirms the findings of Lacarelle et al. [11] that the mixing enhancement of the fluidic oscillators is most pronounced close to the injection plane.

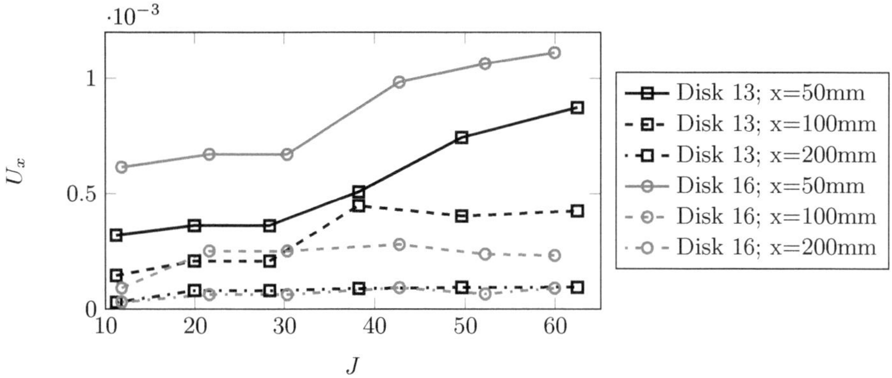

Fig. 9. Spatial unmixedness versus J for disks 13 and 16 for the three measurement planes

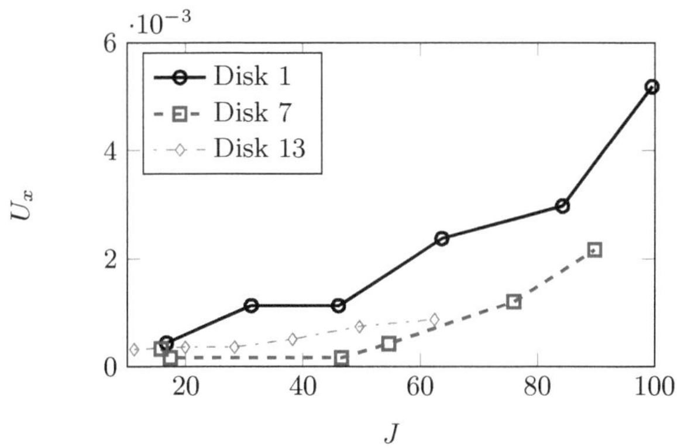

Fig. 10. Spatial unmixedness versus J for disks 1, 7, and 13 at $x = 50\,\text{mm}$

Now, knowing the influence of the oscillating frequency, the influence of the jet in crossflow momentum can be analyzed. To do so the spatial unmixedness parameters of disks 1, 7, and 13 are plotted with respect to J in Fig. 10. These disks have been selected because they have a similar hydraulic diameter and represent three different injection geometries. As already mentioned, the injection geometry for the SEC-process needs to be as independent from the jet in crossflow momentum as possible. From Fig. 10, it can be deduced that this desired characteristic is best represented by the fluidic oscillators (disk 13). Even though the rectangular holes (disk 7) have a higher mixing quality for some values of J, they show a very pronounced dependency on the jet in crossflow momentum. An even stronger dependency can be seen for the circular holes of disk 1. Keeping in mind that the frequency increases with J and that a higher frequency leads to better mixing, the independence of the fluidic injection from J is likely to stem from the increase in the oscillating frequency. These findings again confirm the results from Lacarelle et al. [11], who stated that fluidic oscillators may not only

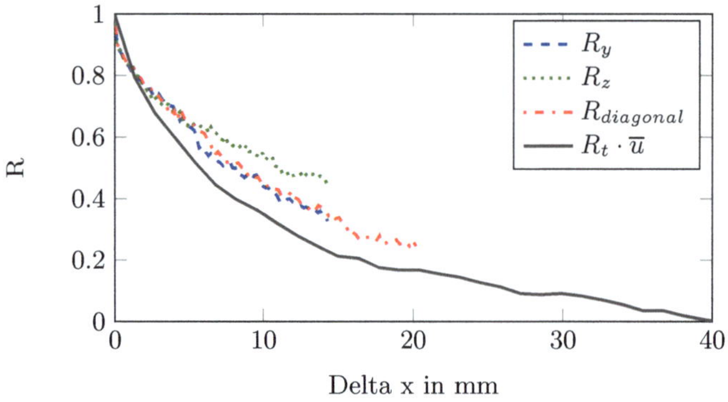

Fig. 11. Correlation coefficients for three spatial directions plus the axial direction calculated using the Taylor hypothesis

enhance the passive scalar mixing between two fluids but also make that mixing quality less dependent from the jet in crossflow momentum.

Following these observations, disk 13 is the best injection configuration of the investigated geometries. In addition to the mixing quality of the different geometries, the mixing parameters of the best configuration were included in a numerical model of the SEC process to investigate if the mixing quality of the given setup is good enough to assure a reliable Shockless Explosion Combustion.

The data of most promising configuration (injection disk 13) was selected to be investigated by the numerical simulations. Since the most upstream auto-ignition location (i.e., location of the combustible mixture) in the combustion tube is at least located a distinct distance downstream of the injection point, the data in the third measurement plane ($x = 200\,\text{mm}$) was analyzed. From the temporally averaged picture, the spatial variance was calculated according toEq. 3. Since no data was available in the axial direction, except the individually taken radial planes, the Taylor hypothesis was employed to gain an insight into the size of the axial mixing structures of the flow.

To use the Taylor hypothesis, defined as

$$\frac{\partial C}{\partial t} = -u \frac{\partial C}{\partial x}\bigg|_{x=x_0}, \tag{6}$$

which states that spatial fluctuations in an homogeneously turbulent field, where the streamwise turbulent fluctuations are low ($u' \ll \overline{u}$), can be calculated from temporal fluctuations at a given coordinate, using the mean streamwise velocity $\overline{u}$. To assure the homogeneous character of the fluctuations in the scalar mixing field, the temporal and spatial correlations where compared. Since all of the correlation curves are reasonably close to each other (see Fig. 11), the use of the Taylor hypothesis is justified. Accordingly, from the individual snapshots of concentration at different times in the measurement plane, an integral length scale of the mixing fluctuations was calculated as

$$\Lambda = \int_0^{\Delta x^*} R(\Delta x) d(\Delta x), \tag{7}$$

where Δx^* is defined as the first intersection of $R(\Delta x)$ and the x-axis. This length scale of 9.7 mm was then used together with the spatial variance to create an equivalence ratio stratification where the worst possible mixing deviations were imposed. This stratification was then used in the numerical simulation to assess the reliability of the SEC-process.

4 Numerical Investigation

The effect of Gaussian noise with the observed standard deviation from the experiment, $\sigma = 2.424 \cdot 10^{-3}$, on the homogeneous ignition of a stratified mixture has been numerically investigated by the simulation of the dynamics of a mixture with perturbed equivalence ratios.

The exact numerical setup is based on the internal state of a pipe in which the SEC process is about to enter the combustion stage; prior to this state, gas flowed from the pipe's inlet at $x = 0$ to a position x in a time τ_{flow} that is assumed to be given via a constant inflow velocity u_0. This, in turn, is given by the geometry of the pipe and fuel properties or can be arbitrarily chosen for single shot ignitions, respectively. Here, it is chosen such that the gas can flow 40 cm within the range of the ignition delay times. Since the ignition shall occur homogeneously, the ignition delay time of the gas that ends up at position x must be chosen to be $\tau_0 + \tau_{\text{flow}}$, such that the remaining ignition delay time τ at the beginning of the simulation is τ_0 everywhere. This choice is done by altering the equivalence ratio Φ between stoichiometric (1.0) and lean (0.5) values. In the numerical setup, perfect advection without diffusion is assumed, such that the initial values can be calculated by selecting Φ such that $\tau = \tau_0 + \tau_{\text{flow}}$ and then solving the mixture in an isochoric 0D reactor for $\tau_{\text{flow}} + \eta\tau_0$, where η is some fraction smaller than unity and chosen such that the assumption that the fluid is only advected is not violated due to the pressure rise of the reaction. For this simulation, $\eta = 0.7$ was chosen.

For the chemical kinetics, the latest fuel suggestion and associated reaction mechanism for a 3 bar test rig, which will also be presented at AFCC 2014 [18], namely a composition of

$$CH_3OCH_3 + 1.1\,H_2 + 0.8\,CH_4 \tag{8}$$

in air, has been used. The initial temperature was chosen to be 787.5 K.

Within the setup described above, the choices of Φ were perturbed with different noise levels on equidistant points with 9.7 mm spacing and linear interpolation in between. The resulting stratification was then placed into a 1D domain as initial conditions for a fluid dynamic simulation that solves the Euler equations with a thermally perfect equation of state and the above reaction mechanism. The in-house code is based on a second-order MUSCL extension of the HLL

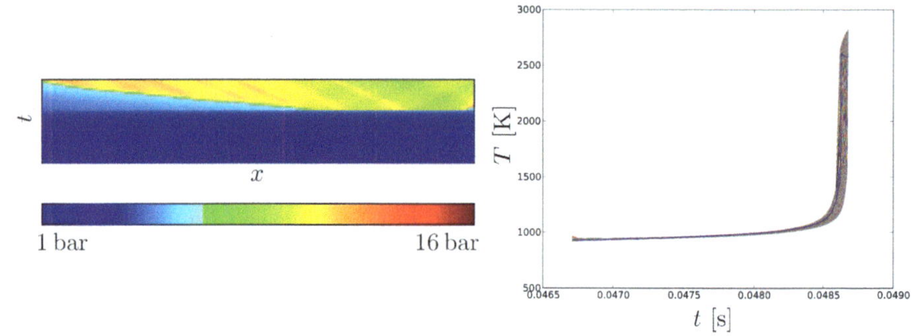

(a) Testcase 1: Φ limited to 0.75 – 1.0.

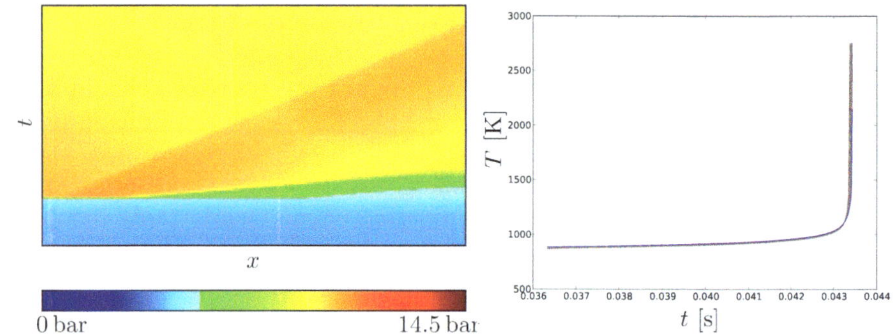

(b) Testcase 2: Noise reduced to $\pm 2\sigma$ perturbations.

Fig. 12. Ignitions for the different simulations. The heatmaps show pressure evolution, the other plot shows the temperature as a function of time overlain for each of the computational cells.

approximate Riemann solver [19] and uses Strang splitting to couple with an isochoric 0D kinetics code. Reflecting wall boundary conditions have been employed for the left boundary, and a fixed pressure plenum at 3 bar for the right one.

If the full range of equivalence ratios is used and an extremal (deterministic) noise of alternating 3σ and -3σ is applied, the different sensitivities at the extremal Φ values are too high and a detonation develops near $\Phi = 1$, traveling towards the $\Phi = 0.5$ boundary. The current quality of mixing is, therefore, insufficient and requires improvement. To assess the possibilities with the current mixing and estimate the required variation to allow usage of the full Φ range, further simulations were conducted. They show two possible ways to achieve homogeneous combustion:

By limiting the choice of Φ to either $0.5 - 0.75$ or $0.75 - 1.0$, the sensitivity is sufficiently reduced to ensure homogeneous combustion. Test case 1 in Fig. 12 shows the simulation results for the upper range. From the ignition delay plot, one can observe that the ignition takes place throughout the domain in a 0.1 ms window. In the heat map, the transition from blue to yellow/orange is the combustion. The steeper lines are pressure waves in the burnt gas. The results for the lower range of Φ values are slightly better, with the ignition time varying by only half the upper range's value, but otherwise similar.

Reducing the variance by a factor of $\frac{2}{3}$ is mathematically equivalent to a reduction of the confidence interval to a 2σ environment. A simulation with reduced variance can, therefore, be equivalently interpreted as being an improved mixing quality or accepting 5% non-SEC ignitions. In both cases, the combustion is almost completely homogeneous, as can be seen in test case 2 of Fig. 12.

5 Conclusion

The innovative constant volume combuston process SEC has very high and specific demands on the fuel-air mixing. In order to create the desired levels of mixing quality, several geometries were investigated in a water test-rig. They were analyzed employing planar laser induced fluorescence and evaluated by the unmixedness parameter U_x.

It was shown that spatially oscillating jets in crossflow created by fluidic oscillators are able to get close to the needed mixing characteristics. In contrast to a slit injection, the round, or the rectangular jet in crossflow, fluidic oscillators allow the creation of a very high mixing quality already close to the fuel injection plane. In addition, the mixing quality is independent from the jet in crossflow momentum for these types of injectors. The results indicate that this is likely to stem from the increasing oscillation frequency of the fluidic oscillators. A higher jet in crossflow momentum leads to a higher volumetric flow in the oscillator and, thus, a higher oscillating frequency. By comparing two different sizes of fluidic oscillators, it was possible to show that higher frequencies are favorable for the mixing process. However, since higher frequencies lead to smaller geometries for the fluidic oscillators the pressure loss of the fuel injection system will increase and might contradict the advantages of a higher mixing quality. Hence, the optimal oscillating frequency must to be identified for every new injection configuration.

In order to validate the achieved mixing quality in the highly demanding SEC-process, a numerical simulation was conducted. Based on these calculations it was possible to show that for an ideal filling charge of the SEC-combustor, the identified perturbations in the equivalence ratio are likely low enough to assure a reliable homogeneous auto-ignition. The experimental verification of these findings and the further investigation of the injection and ignition process will be the scope of future work. However, the presented work gives confidence that the proposed SEC-process can be achieved reliably and efficiently.

Acknowledgement. The authors gratefully acknowledge support by the Deutsche Forschungsgemeinschaft (DFG) as part of collaborative research center SFB 1029 "Substantial efficiency increase in gas turbines through direct use of coupled unsteady combustion and flow dynamics". In addition, the authors would like to thank the CONFET for assistance in the lab and helpful discussions.

References

1. Keffer, J., Baines, W.: The round turbulent jet in a cross-wind. Journal of Fluid Mechanics 15(4), 481–496 (1963)
2. Kamotani, Y., Greber, I.: Experiments on a turbulent jet in a cross flow. AIAA Journal 10(11), 1425–1429 (1972)
3. Andreopoulos, J.: On the structure of jets in a crossflow. Journal of Fluid Mechanics 157, 163–197 (1985)
4. Smith, S.H., Mungal, M.G.: Mixing, structure and scaling of the jet in crossflow. Journal of Fluid Mechanics 357, 83–122 (1998)
5. Eroglu, A., Breidenthal, R.: Structure, penetration, and mixing of pulsed jets in crossflow. AIAA Journal 39(3) (2001)
6. Johari, H., Pacheco-Tougas, M., Hermanson, J.: Penetration and Mixing of Fully Modulated Turbulent Jets in Crossflow. AIAA Journal 37(7), 842–850 (1999)
7. Johari, H.: Scaling of Fully Pulsed Jets in Crossflow. AIAA Journal 44(11), 2719–2725 (2006)
8. Weston, R., Thames, F.: Properties of aspect-ratio-4.0 rectangular jets in a subsonic crossflow. Journal of Aircraft 16(10), 701–707 (1979)
9. Plesniak, M.W., Cusano, D.M.: Scalar mixing in a confined rectangular jet in crossflow. Journal of Fluid Mechanics 524, 1–45 (2005)
10. Nathan, G., Mi, J., Alwahabi, Z., Newbold, G., Nobes, D.: Impacts of a jet's exit flow pattern on mixing and combustion performance. Progress in Energy and Combustion Science 32(5-6), 496–538 (2006)
11. Lacarelle, A., Paschereit, C.O.: Increasing the Passive Scalar Mixing Quality of Jets in Crossflow With Fluidics Actuators. Journal of Engineering for Gas Turbines and Power 134(2), 021503 (2012)
12. Clarke, A., Gerrard, A., Holliday, L.: Some experiences in gas turbine combustion chamber practice using water flow visualization techniques. Symposium (International) on Combustion 9(1), 878–891 (1963)
13. Lacarelle, A., Sebastian, G., Paschereit, C.O.: A Quantitative Link Between Cold-Flowscalar Unmixness and NOX Emissions in a Conical Premixed Burner. In: ASME Turbo Expo, Number X, ASME GT 2010–23132 (2010)
14. Bobusch, B.C., Woszidlo, R., Bergada, J.M., Nayeri, C.N., Paschereit, C.O.: Experimental study of the internal flow structures inside a fluidic oscillator. Experiments in Fluids 54(6), 1559 (2013)
15. Krüger, O., Bobusch, B.C., Woszidlo, R., Paschereit, C.O.: Numerical Modeling and Validation of the Flow in a Fluidic Oscillator. In: 21st AIAA Computational Fluid Dynamics Conference, Reston, Virginia, American Institute of Aeronautics and Astronautics (June 2013)
16. Danckwerts, P.: The definition and measurement of some characteristics of mixtures. Applied Scientific Research, Section A 3, 279–296 (1952)

17. Bobusch, B.C., Woszidlo, R., Krüger, O., Paschereit, C.O.: Numerical Investigations on Geometric Parameters Affecting the Oscillation Properties of a Fluidic Oscillator. In: Proceedings of the 21st AIAA Computational Fluid Dynamics Conference (2013)
18. Cai, L., Pitsch, H.: Tailoring fuels for shockless explosion engine (preliminary title). In: Proceedings of AFCC (2014)
19. Harten, A., Lax, P.D., Van Leer, B.: On upstream differencing and godunov-type schemes for hyperbolic conservation laws. SIAM Review, 35–61 (1983)

Tailoring Fuels for a Shockless Explosion Combustor

Liming Cai and Heinz Pitsch

Institute for Combustion Technology, RWTH Aachen University,
Templergraben 64, 52062 Aachen, Germany
{lcai,h.pitsch}@itv.rwth-aachen.de

Abstract. Shockless isochoric explosion has been proposed as an alternative to conventional constant-pressure combustion in gas turbines to improve the thermal efficiency. Fully homogeneous auto-ignition and fast burnout facilitate quasi-isochoric combustion, but the ignition delay times depend on local temperatures and mixture composition. This implies that homogeneous auto-ignition is difficult to achieve in the presence of local inhomogeneities. Therefore, in this study, the ignition delay time, its temperature dependence and the excitation time of the potential fuels are investigated. An automatic optimization technique is used to tailor fuels such that the temperature dependence of ignition delay times is minimized to realize homogeneous isochoric auto-ignition in realistic systems even in the presence of inhomogeneities and also to avoid the formation of detonation waves by considering the kinetic properties. Reaction kinetic mechanisms are developed for the chemical modeling of the relevant fuel species and tailor-made compositions of the fuel mixtures are determined by means of simulation.

Keywords: shockless explosion combustor, tailor-made fuel mixture, excitation time, chemical kinetics.

1 Introduction

In the Brayton-cycle used in gas turbines, constant-pressure combustion takes place after an adiabatic compression, and it is followed by an adiabatic expansion. Nowadays, the efficiency of gas turbines comes to around 40%, limited by achievable turbine inlet temperatures and the combustion process at constant pressure. To increase the efficiency of gas turbines by more than 10%, constant-volume combustion can be seen as an alternative. Principally, detonations can be used for nearly constant volume combustion in open systems, as for instance applied in pulse detonation engines (PDE). However, this comes at the expense of extreme thermal and mechanical loads on engine components due to strong shock waves. A shockless explosion combustor (SEC) could be another potential alternative [1]. Through controlled homogeneous auto-ignition of the appropriately stratified fuel/air mixture, a more efficient isochoric thermodynamic process can be achieved without the occurrence of detonation. In an ideal situation, the fully

R. King (ed.), *Active Flow and Combustion Control 2014*, Notes on Numerical Fluid Mechanics and Multidisciplinary Design 127, DOI: 10.1007/978-3-319-11967-0_19

homogeneous auto-ignition and the fast burnout facilitate isochoric combustion locally. In addition, the deflagration/detonation transition found in PDEs vanishes in shockless explosion engines, leading to improved efficiency.

The backbone of the SEC concept is a quasi homogeneous isochoric auto-ignition process. Since a reliable homogeneous auto-ignition process in an SEC combustor is not easy to achieve and has a very high and specific demand on fuel-air mixing quality [1], details of chemical kinetics of fuels can play a central role in achieving this combustion mode.

In addition, these kinetic details are also of particular importance to avoid the formation of detonation. Detonation develops due to an acoustic resonance between the reaction wave velocity and the speed of sound [2]. As analyzed by Bradley et al. [3] and Peters et al. [4], the wave velocity is defined as the inverse of the temperature derivative of the ignition delay time coupled locally with a negative spatial temperature gradient in direction of the front propagation. Therefore, the establishment of the resonance depends strongly on the fuel kinetics. This again calls for the investigation of the reaction kinetics. The relevant kinetic parameters for SEC engines are introduced here.

Ignition delay times. The most important kinetic parameter for isochoric homogeneous combustion is the ignition delay time. It indicates the required time for fuel/air mixtures to reach auto-ignition during the combustion process. After this induction period, the reactants deplete rapidly and the chemical heat is released within a very short time span that is several orders of magnitude smaller than the ignition delay time. Therefore, the ignition delay time denotes approximately the time period of the chemical conversion of the fuel/air mixture. As demonstrated by Bradley et al. [3], for the prevention of a detonation, the induction of fuel/air mixtures is desired to take place in a time interval which is shorter than the characteristic time of sound wave propagation in the combustion chamber. For a chamber length of 1 meter, ignition delay times are required to be smaller than 1 ms. However, regarding the mixing process, a long ignition delay time is expected to ensure a more homogeneous mixing of fuel and air. Therefore, a chemical induction time of 1 ms represents the optimal ignition delay time.

Temperature dependence of ignition delay times. The fully homogeneous auto-ignition enables the infinitely fast reaction wave velocity. As the wave velocity is much larger than the sound speed, the possibility of developing a detonation is minimized. However, the local inhomogeneities and the fluctuations of the mixture temperatures are unavoidable due to the charge cycle, the residual exhaust gas and the heat transfer through the combustor wall. All these issues make it difficult to yield the homogeneous auto-ignition, if ignition delay times are sensitive to local temperatures, which will then increase the propensity for auto-ignition. Therefore, the ignition delay times should ideally be independent on the initial temperature of the fuel/air mixtures in the SEC chamber.

Excitation time. The excitation time is defined by Meyer et al. [5] as the period from the beginning of energy release to the point of maximum power. It indicates the rate of chemical energy release and follows the induction of fuel ignition [6]. Once the fuel auto-ignites, the exothermic elementary reactions are accelerated, leading to significant chemical energy release, which ends with the depletion of the primary reactants. When the fuel/air mixture is inhomogeneously distributed in the combustion chamber, auto-ignition can first take place at several "hot spots" and the resulting pressure increase initiates sound waves. These sound waves may couple with the reaction wave along a negative temperature gradient. Once the propagation length of the sound wave during the excitation time is smaller than the length over which the temperature gradient is constant, a resonance can be established and thus the detonation wave can form [2,3,4]. Therefore, in the framework of CRC 1029, a long excitation time is desired in shockless explosion combustors to avoid the detonation waves.

In this study, the goal is to enable a quasi-homogeneous isochoric ignition process in the novel shockless explosion combustor by taking advantage of the fuel chemical kinetics. The above described kinetic properties are investigated. As numerous studies [7,8,9,10] have reported the strong temperature dependence of ignition delay times of a wide variety of fuels, for a pure fuel, it is difficult to meet the requirements on the kinetics of the potential SEC fuel. The current study therefore focuses on the development of fuel mixtures, whose temperature dependence of ignition delay times is eliminated or adjusted to promote the homogeneous isochoric combustion by taking advantage of the negative temperature coefficient (NTC) behavior of several fuels. Through the blend of fuels with NTC and without NTC behavior, temperature sensitivity is eliminated, such that the resulting mixtures when used in the SEC, are expected to ignite simultaneously everywhere in the volume, regardless of the temperature fluctuations in the combustion chamber. Moreover, the fuel mixtures are supposed to provide the desired ignition delay times at the relevant conditions. An optimization framework is developed in this work to automatically tailor the blending ratio of mixtures. Reaction mechanisms are developed for the chemical modeling of the relevant fuels.

The presentation of this paper is organized as follows. The first part provides details of the mechanism development. In the second part, the investigation of the important kinetic parameters and the fuel tailoring are described in detail.

2 Mechanism Development

Both liquid and gaseous fuel mixtures are suggested to reduce the temperature dependence of ignition delay times to enable homogeneous isochoric combustion. While a large variety of fuel species, e. g. n-heptane, n-dodecane, iso-octane and toluene, are available to compose the liquid mixture, the choice of the component of the gaseous mixtures is limited to hydrogen, methane, ethane and dimethyl ether (DME). In this section, a kinetic mechanism is developed to include the oxidation chemistry of a large number of species, which are the potential mixture components for SEC. The chemical model should on the one hand predict the oxidation of individual components accurately and is on the other hand supposed to reflect the combustion properties of surrogate mixtures correctly.

A well-validated kinetic mechanism for hydrocarbon fuels has been developed in a series of recent studies. Blanquart et al. [11] developed a chemical mechanism to describe combustion of fuels spanning from C_1 to C_8 hydrocarbon species. The detailed chemistry of the substituted aromatic species, e. g. toluene, xylene, styrene and 1-methylnaphthalene was then added by Narayanaswamy et al. [12] into that mechanism [11]. Recently, the model was further extended to include the reaction pathways of the short chain alkane n-pentane. [13], the long chain n-dodecane [14] and the gasoline surrogate components, n-heptane, iso-octane and ethanol [15]. The model has been extensively and successfully validated against experimental data measured in shock tubes, rapid compression machines, combustion vessels, perfectly stirred reactors and flow reactors. This mechanism is taken as the base mechanism in this study, as the interesting species for SEC are included, except for dimethyl ether.

The oxidation chemistry of DME is adopted from the published mechanism by Fischer et al. [16] and built as additional module that is combined with the above described base model. To reduce the computational effort, the detailed mechanism of DME [16] was reduced to a skeletal level using a multi-stage reduction strategy developed by Pepiot-Desjardins and Pitsch [17]. The reduction procedure involved the elimination of species and reactions. The database used to perform the reduction includes the conditions where experimental data have been reported [9]. This reduction process eliminated the species and reactions with negligible effect on reaction kinetics at the relevant temperature and pressure range. Numerical results using the mechanism of Fischer et al. [16] and the reduced mechanism are compared and shown in Fig. 1. Over the entire domain of interest, the reduced mechanism is found to produce negligible changes in results.

The inclusion of the DME chemistry into the base mechanism was accomplished using the interactive tool from Ref. [18]. This interactive tool automatically identifies common species and reactions from the different mechanisms. Rate conflicts detected during the mechanism combination are always resolved in favor of the rigorously validated base mechanism, therefore leaving that part of the mechanism virtually unchanged. The numerical ignition delay times using the proposed combined mechanism are compared with those computed using the detailed mechanism [16] and the experimental data from Ref. [9] in Fig. 2. It is clearly seen that the present model is in close agreement with the detailed model. The present model demonstrates the capability to predict autoignition of DME over a variety of conditions. For the sake of brevity, the validation of individual relevant species is not shown here, so the reader is directed to our previous studies [11,12,13,14,15].

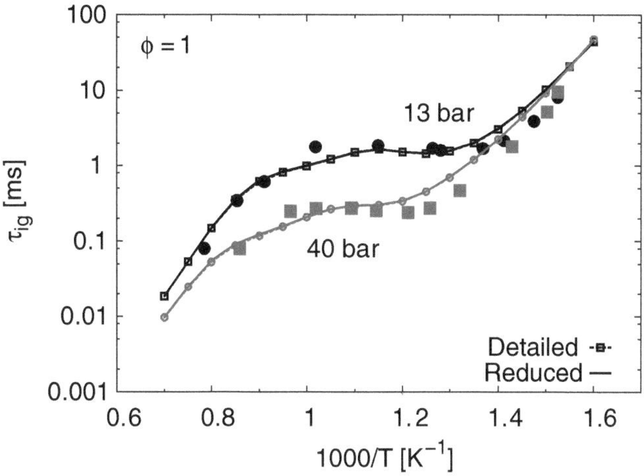

Fig. 1. Ignition delay times of stoichiometric DME/air mixture at 13 and 40 bar. Solid symbols denote the experimental data by Pfahl et al. [9]. Solid and dashed lines show the numerical results with the reduced and the detailed mechanism [16], respectively.

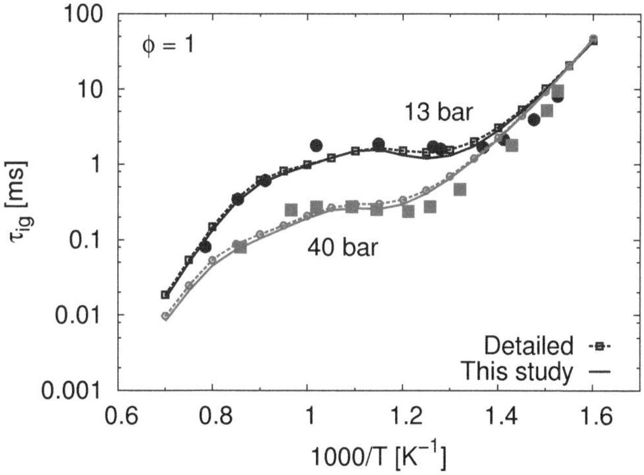

Fig. 2. Ignition delay times of stoichiometric DME/air mixture at 13 and 40 bar. Solid symbols denote the experimental data by Pfahl et al. [9]. Solid and dashed lines show the numerical results with the combined model developed here and the detailed mechanism [16], respectively.

3 Fuel Tailoring

In this section, the critical kinetic parameters for SEC are numerically investigated. Fuel mixtures are tailored to fill the requirements on the ignition delay times, their temperature dependence and the excitation times.

3.1 Computational Details and Relevant Conditions

The computational details and the relevant boundary conditions for SEC operation are explained first. The ignition delay times of the SEC fuels are computed in a homogeneous constant-volume configuration. The typical operation conditions of gas turbines are generally also relevant for the shockless explosion combustor. Typical conditions include the initial temperatures of fuel/air mixtures from 800 to 1000 K, equivalence ratios from 0.5 to 1.0, and a high pressure of 30 bar. In order to realize and demonstrate the shockless explosion at ambient pressure, the auto-ignition process at atmospheric pressure is also considered.

Numerical calculations are performed using the zero-dimensional configuration of an isochoric homogeneous reactor in the open source FlameMaster code [19]. Adiabatic premixed combustion is assumed to take place and the species and energy equations are solved. The ignition delay time is defined as the time to reach the maximum temperature change.

3.2 Ignition Delay Times and Their Temperature Dependence

Negative Temperature Coefficient (NTC) Behavior. One of the major purposes in this study is to eliminate the temperature dependence of ignition delay times by blending different fuels. This will be achieved by taking advantage of the NTC behavior of several fuels. As revealed in Fig. 1, the fuel ignition is retarded with a decrease in the initial temperatures at high and low temperatures. However, in the temperature range of approximately 750-1000 K, it can be observed that the ignition processes are accelerated when the initial temperatures decrease. This regime is hence called the NTC regime. The reason for this behavior lies on the competition between the low temperature chain branching and the high temperature chain breaking channels. For the species with NTC behavior, the well-known low temperature chain branching pathway is dominant at low temperatures. The addition of molecular oxygen to the fuel radical yields a peroxy-radical. Through the following isomerization and the second oxygen addition, a peroxy-hydroperoxide radical is produced. The peroxy-hydroperoxide radical isomerizes rapidly and then decomposes to give an OH radical and a keto-hydroperoixde. The important intermediate species, keto-hydroperoxide, can further decompose to form the second OH radical, which completes the low temperature chain branching process. At increased initial temperatures, the decomposition channel of the fuel radical is activated and competes with its oxidation pathway. As the chain branching channel is thus inhibited, the ignition delay time increases. The NTC behavior has been observed for a large number of species, e. g. *n*-heptane [22], iso-octane [23], di-*n*-butyl ether [10] and 2-methyl heptane [24]. In general, the low temperature chain branching channel is active and thus the NTC behavior can be obtained, when the molecule structure of species enables the isomerization reactions of peroxy and peroxy-hydroperoxide radicals involving 5-, 6-, 7- and 8-member transition state rings. In this sense, the short chain species, e. g. hydrogen, methane, ethane and ethanol do not show NTC behavior. As aromatic rings are difficult to open and the aromaticity causes a strong C-H bond dissociation energy of the carbon atoms in the ring, aromatic species with a short side chain also do not show the NTC behavior.

Automatic Composition Optimization. The fuels with NTC and without NTC behavior are blended to eliminate the temperature dependence of ignition delay times of SEC fuel. With the tailored blending ratio, the temperature dependence of ignition delay times is supposed to vanish in the temperature range 800-900 K. An automatic framework is developed here to optimize the blending ratio aiming at the temperature-independent auto-ignition with the desired ignition delay times.

The optimization framework is based on the solution mapping method originally developed by Frenklach [20,21]. For this study, the compositions of each component in the fuel mixture are chosen as the optimization objects that will be calibrated. They are normalized as

$$x_k = \ln\left(\frac{X_k}{X_{k,0}}\right) / \ln f_k, \tag{1}$$

where X_k is the composition of the kth fuel species in the fuel mixture and $X_{k,0}$ is set to 1/number of species here, which means the fuel components are initially mixed with a blending ratio of one. The normalized parameters are treated as uniformly distributed and bounded to the interval $\{-1, 1\}$. f_k indicates the adjustable range of the compositions

$$X_k/f_k \leq X_k \leq X_k f_k. \tag{2}$$

Here, f_k is set to four.

In the optimization approach, the normalized compositions are fitted by minimization of the least-squares difference between the model predictions $\tau_{ig,r}(\mathbf{x})$ and the desired ignition delays $\tau_{ig,r}^{req}$. This is expressed as the minimization of the following objective function

$$\Phi(\mathbf{x}^*) = \min_{x} \sum_{r=1}^{n} \left(\frac{\tau_{ig,r}(\mathbf{x}) - \tau_{ig,r}^{req}}{\sigma_r^{req}}\right)^2, \tag{3}$$

where $\mathbf{x}^*$ is the set of the optimized normalized compositions and n is the number of conditions. For the rth condition, the value of the target $\tau_{ig,r}^{req}$ and its prescribed uncertainty σ_r^{req} have to be provided. In the optimization function, the ignition delay time $\tau_{ig,r}(\mathbf{x})$ is calculated through the so-called response surface technique, which relates here the species compositions to the optimization targets in form

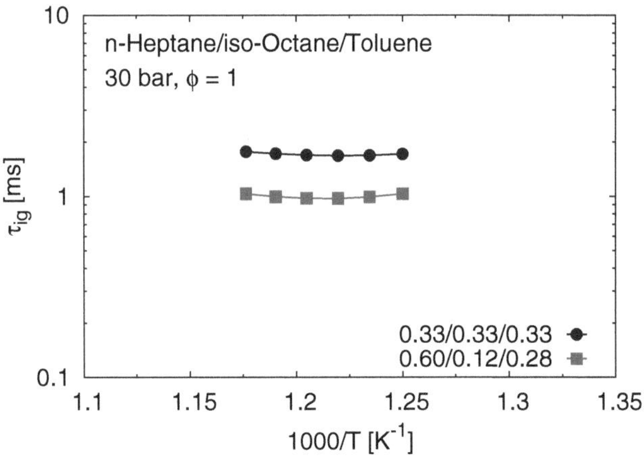

Fig. 3. Ignition delay times of stoichiometric n-heptane/iso-octane/toluene/air mixture at 30 bar. Solid lines show the numerical results.

of a second order polynomial. The composition optimization is performed here by minimizing $\Phi(\mathbf{x}^*)$ in Eq. (3). At each considered condition, the target $\tau_{ig,r}^{req}$ is set to 1 ms and the uncertainty σ_r^{req} is 0.05 ms. The feasibility of the automatic optimization is demonstrated here using a liquid mixture of n-heptane, iso-octane and toluene.

The liquid mixture is subjected to the optimization process. The initial molar blending ratio of n-heptane, iso-octane and toluene is 0.33/0.33/0.33. The numerical ignition delay times of the initial blend are shown in Fig. 3 for the temperature range from 800 to 850 K and found to be weakly sensitive to the mixture temperatures. However, the ignition delay times are twice as high as those desired. The automatic optimization framework is thus used to calibrate the blending ratio between the three components for an ignition delay time of $\tau_{ig} = 1$ ms. Six conditions between 800 and 850 K with an increment of 10 K are considered in the optimization process and the ignition delay times at these conditions are targeted with an absolute value of 1 ms. The calibrated blending ratio is determined as 0.60/0.12/0.28 and the ignition delay times of the tailored mixture are denoted in Fig. 3 by the red solid line. For all six conditions, the tailored mixture gives a satisfactory ignition delay time of 1 ms and the temperature-independent self-ignition remains unchanged.

The ignition delay times of the initial mixture in an extended temperature regime (800-900 K) are depicted in Fig. 4. With an increase in initial temperatures, the auto-ignition process decelerates slightly. In order to eliminate this temperature dependence, the automatic framework is again employed to tailor the mixture composition. Five conditions between 800 and 900 K with an increment of 25 K are considered in the calibration with an absolute ignition delay time of 1 ms as target. The results of the tailored mixture are shown in Fig. 4. While the tailored mixture ignites with an induction time of around 1 ms, the ignition delays of the tailored mixture are strongly affected by mixture temperatures. This indicates the difficulty to give satisfactory ignition behavior in all respects over a very wide temperature range by using *n*-heptane, iso-octane and toluene as the fuel mixture. Instead of adding a new fuel component, an alternative is to optimize the equivalence ratio of the fuel blend/air mixture. By taking the equivalence ratio as an additional calibration object into the automatic framework, a rich oxidation with an equivalence ratio of 2.26 is proposed from the optimization process. The blending ratio is determined as 0.26/0.63/0.11 and the computed ignition delay times of the mixture are shown in Fig. 4. The ignition delay times are equal to 1 ms at all conditions and as expected, the initial mixture temperatures are of minor importance for ignition delays. The rich oxidation leads to an acceleration of the auto-ignition and the fitted blending ratio eliminates the temperature dependence of ignition delay times. Overall, the examples shown here demonstrate the capability of the automatic framework to optimize the composition of fuel mixtures by taking account of the relevant kinetic properties.

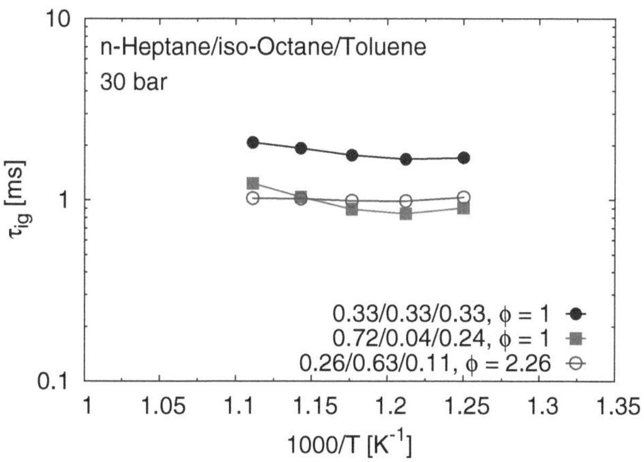

Fig. 4. Ignition delay times of stoichiometric *n*-heptane/iso-octane/toluene/air mixture at 30 bar. Solid lines show the numerical results.

Simulations for a gaseous mixture of H_2, CH_4 and DME with the blending ratio of 0.33/0.33/0.33 show ignition with an induction time of 1 ms over the temperature range of 800-900 K at 30 bar and $\phi = 1.0$ (Fig. 5). Therefore, its blending ratio is not further optimized.

Prediction Accuracy. Since the compositions of the fuel mixtures are determined and optimized by means of simulation, the model prediction accuracy of ignition delay times is of particular importance. The proposed model has

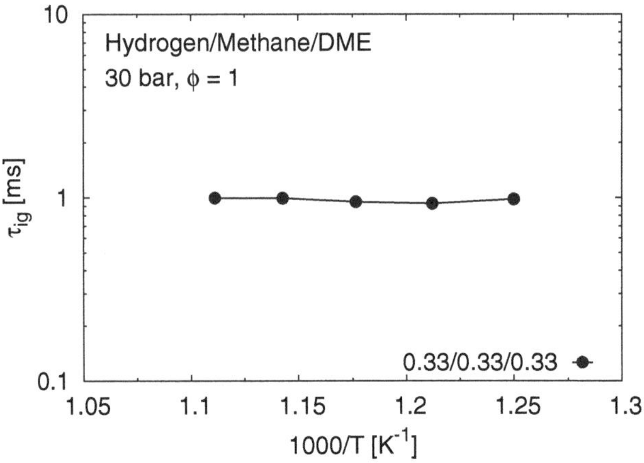

Fig. 5. Ignition delay times of stoichiometric hydrogen/methane/DME/air mixture at 30 bar. Solid lines show the numerical results.

been validated for the individual fuel species in section 2 and in earlier publications [11,12,13,14,15]. Here, the validation is performed for auto-ignition of mixtures. A liquid mixture composed of n-heptane, iso-octane and ethanol has been experimentally investigated by Fikri et al. [26]. The reported experimental data are compared with numerical results in Fig. 6. As found in the literature, the ignition delay times of these three fuels are strongly influenced by the initial temperatures of the fuel/air mixtures [22,23,25]. While n-heptane and iso-octane show a distinct NTC behavior, the ignition delay times of ethanol grow with the decrease in temperature in the entire temperature range. Here, n-heptane, iso-octane and ethanol are blended in a volume ratio of 0.18/0.62/0.20. As shown in Fig. 6, the mixture gives almost identical ignition delay times in the temperature range of 750-850 K at 10 and 30 bar, which is of particular interest for shockless explosion combustors. These data again indicate the possibility of using a fuel mixture to eliminate the temperature dependence of ignition delay times. More importantly, the numerical ignition delay times of the mixture are in close agreement with the experimental data by Fikri et al. [26], which confirms the good prediction accuracy of the proposed model.

Ignition Acceleration. The shockless explosion concept is designed for gas turbine relevant high pressures as well as for atmospheric pressure. Here, the ignition delay times of the n-heptane/iso-octane/ethanol/air mixture are calculated for a variety of pressures. The results are presented in Fig. 7. At a pressure of 3 bar, a weakly temperature-dependent auto-ignition is observed. With a further decrease in pressure, the ternary mixtures fail to ignite homogeneously regardless of the initial mixture temperatures, even though they give a satisfactory temperature dependence of ignition delay times at higher pressures. Mixtures with modified composition or additional components are required for ambient pressure.

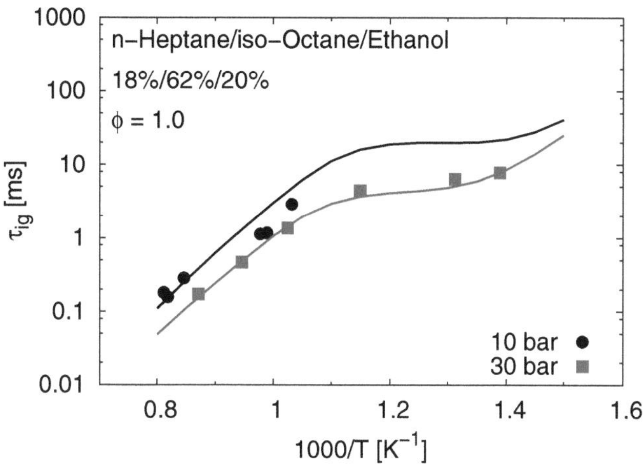

Fig. 6. Ignition delay times of stoichiometric *n*-heptane/iso-octane/ethanol/air mixture at 10 and 30 bar. Symbols denote the experimental data by Fikri et al. [26]. Solid lines show the numerical results using the proposed model.

Another critical issue shown in Fig. 7 is the absolute value of the ignition delay time. As described above, an ignition delay time of 1 ms is required to avoid the formation of detonation waves at different pressures. Several species, e. g. alkyl nitrates have been successfully added into petroleum fuels to improve the ignition propensity. These species can also be added into fuel mixtures for SEC and their effects on ignition delay times can be experimentally investigated. However, numerical calculations cannot be performed presently, because neither detailed chemical mechanisms nor validation data are available for these species.

A practical method to accelerate ignition is the manipulation of the oxygen/nitrogen ratio of the oxidizer. The numerical ignition delay times of the liquid *n*-heptane/iso-octane/ethanol mixtures with various O_2/N_2 ratios are illustrated in Fig. 8. It can be immediately seen that the ignition delays decrease with increasing oxygen ratio. The temperature independence of the induction times remains favorable. Overall, in order to obtain the required ignition delay times, the oxygen concentration in air might be optimized. Such an optimization can also be performed in the automatic framework proposed in this work.

3.3 Excitation Time

The excitation time indicates the rate of chemical energy release during the strong temperature rise period of auto-ignition and plays an important role in the formation of detonation waves. As reported in Ref. [6], the excitation times of hydrogen, methane, acetylene, ethylene, and ethane are generally two and three orders of magnitude shorter than the ignition delay times and not sensitive to temperature.

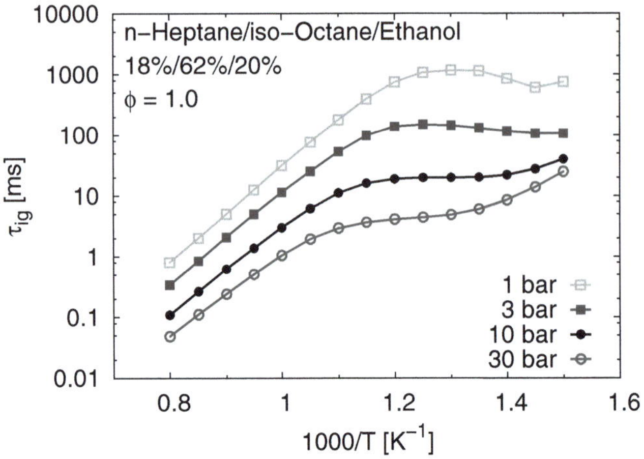

Fig. 7. Ignition delay times of stoichiometric *n*-heptane/iso-octane/ethanol/air mixture at 1 bar, 3 bar, 10 and 30 bar. Solid lines show the numerical results with the proposed model.

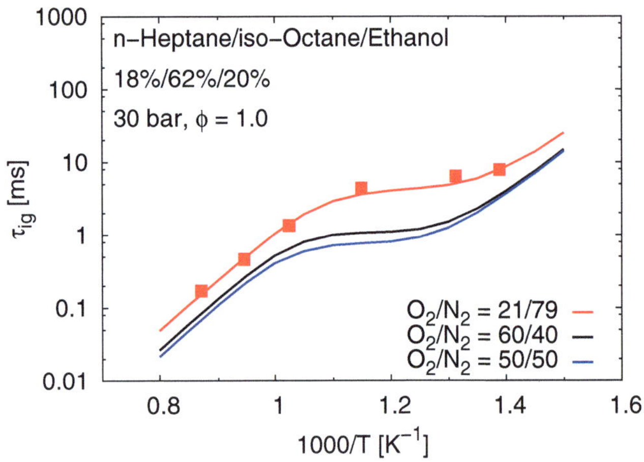

Fig. 8. Ignition delay times of stoichiometric *n*-heptane/iso-octane/ethanol/air mixture at 30 bar with varied oxygen/nitrogen ratios. Symbols denote the experimental data by Fikri et al. [26]. Solid lines show the numerical results using the proposed model.

In this study, the excitation time is further evaluated by means of the numerical calculation. The excitation time is defined as the time period from 1% up to the maximum heat release. The study by Lutz et al. [6] reported that the excitation time of stoichiometric fuel/air mixtures is a stronger function of pressure than temperature at pressures up to 10 bar. The excitation times and the ignition delay times of DME at higher pressures are shown in Fig. 9. Consistent with the observation from Lutz et al. [6], the excitation times are strongly influenced by the pressure at conditions with higher pressures as well. While the excitation times are always around two orders of magnitude shorter than the ignition delay times at various pressures, the absolute excitation times decrease with the acceleration of homogeneous auto-ignition.

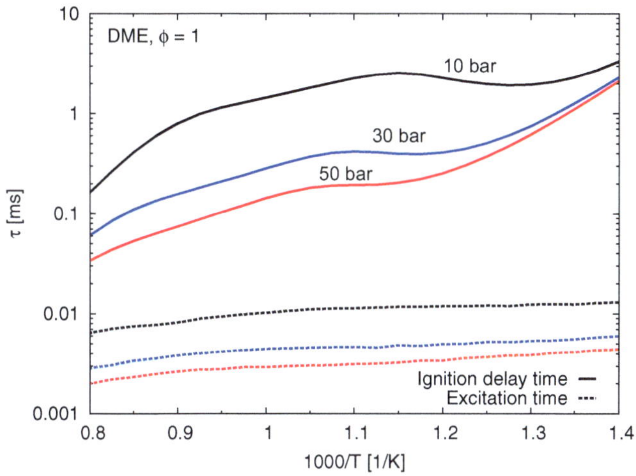

Fig. 9. Ignition delay times and excitation times of stoichiometric DME/air mixture at 10, 30 and 50 bar

The effects of equivalence ratio on the excitation time are highlighted in Fig. 10, again using the oxidation of DME as an example. In Fig. 10, the lean DME/air mixtures show longer ignition delay times and an improved reactivity is observed for rich DME/air mixtures. The ignition delays of DME/air mixtures are continuously increased by decreasing the equivalence ratio. However, this is not the case for the excitation time. While the excitation times of the lean DME/air increase significantly, the chemical energy release of rich mixtures is also found to take place in an extended time interval. It is also interesting to note that the relative changes in the excitation times are much larger than those of the induction times for lean mixtures. As depicted in Fig. 10, rich mixtures present the desired ratio between the characteristic time scales of the ignition and the burnout of fuel mixtures. The ignition accelerates, but the heat release is prolonged. This behavior of rich mixtures could be utilized to demonstrate the feasibility of the shockless explosion combustion. However, it is meaningless

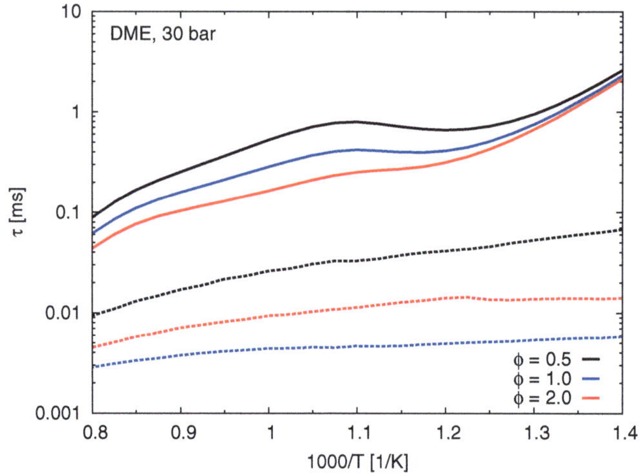

Fig. 10. Ignition delay times (solid lines) and excitation times (dashed lines) of DME/air mixture at 30 bar and $\phi = 0.5$, 1.0 and 2.0.

for the practical application in gas turbines, as modern combustion devices are typically operated at lean or stoichiometric conditions for good fuel efficiency and emission reduction. Another potential operation condition demonstrated in Fig. 10 is the lean combustion at 800 K. While the ignition delay times of lean, stoichiometric and rich mixtures are similar, the excitation time of the lean mixture is distinctly larger than those of other cases.

The induction times and the excitation times are numerically calculated for various potential fuel components and compared in Fig. 11. At the conditions numerically investigated, the ignition delay times of DME, *n*-heptane, ethanol and toluene are distinctly different, especially at lower temperatures. DME and *n*-heptane show NTC behavior, while the ignition delay times of toluene and ethanol increase continuously with temperature. Particularly interesting here is the fact that the excitation times of the four fuels are very similar despite the pronounced differences in ignition delay times. The computed excitation times depend on the rate of exothermic chemical reactions including mainly the oxidation of formaldehydes and carbon monoxides. As the specific kinetics of DME, *n*-heptane, ethanol and toluene are built upon the same base chemistry in this study, the same reaction rate for formaldehyde and carbon monoxide oxidation are employed. This leads to similar time scales of energy release even for different fuel components.

Overall, the excitation time is not sensitive to temperature, while it is a function of pressure and equivalence ratio. It is difficult to extend the heat release interval and to improve the ignition delay at the same time.

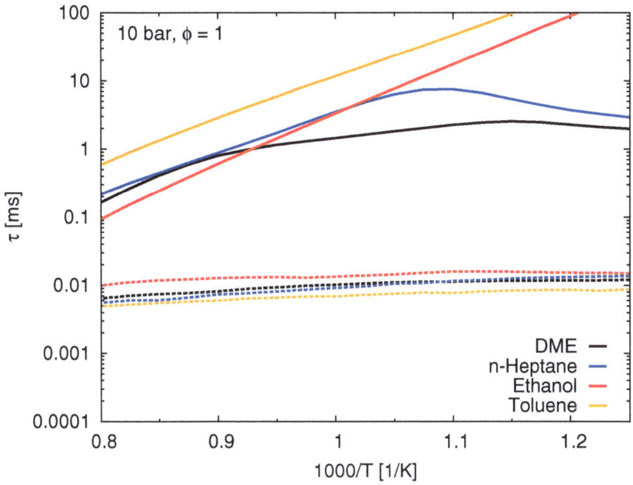

Fig. 11. Ignition delay times (solid lines) and excitation times (dashed lines) of DME/air, *n*-heptane/air, ethanol/air and toluene/air mixtures at 10 bar and $\phi = 1.0$.

4 Conclusion

In the current study, the chemical kinetics of the potential SEC fuels are investigated. In the novel shockless explosion combustor, a quasi-homogeneous isochoric auto-ignition of stratified fuel/air mixture is supposed to take place, which should lead to pressure-rise combustion and improve combustion efficiency. In order to realize the concept and also to avoid the formation of detonation waves, the ignition delay time, its temperature dependence and the excitation time are considered as optimization targets in the development of appropriate fuel blends. By appropriately blending the fuels, the temperature dependence of ignition delay times can be eliminated at the operating conditions of interest. The desired ignition delay times can be obtained by increasing the oxygen concentration in the oxidizer. However, the excitation times, which indicate the rate of the energy release, are found to be difficult to control by fuel blending, because they depend only weakly on the fuel structure.

Acknowledgments. The authors gratefully acknowledge support by the Deutsche Forschungsgemeinschaft (DFG) as part of collaborative research center SFB 1029 "Substantial efficiency increase in gas turbines through direct use of coupled unsteady combustion and flow dynamics".

References

1. Bobusch, B.C., Berndt, P., Paschereit, C.O., Klein, R.: Investigation of fluidic devices for mixing enhancement for the shockless explosion combustion process. In: Proceedings of AFCC (2014)

2. Zeldovich, Y.B.: Regime classification of an exothermic reaction with nonuniform initial conditions. Comb. Flame 39, 211–214 (1980)
3. Bradley, D., Morley, C., Gu, X.J., Emerson, D.R.: Amplified pressure waves during autoignition: Relevance to CAI engines SAE Technical Paper 2002-01-2868 (2002)
4. Peters, N., Kerschgens, B., Paczko, G.: Super-knock prediction using a refined theory of turbulence SAE. Int. J. Engines 6(2), 953–967 (2013)
5. Meyer, J.W., Cohen, L.M., Oppenheim, A.K.: Study of exothermic processes in shock ignited gases by the use of laser shear interferometry. Comb. Sci. Tech. 8, 185 (1973)
6. Lutz, A.E., Kee, R.J., Miller, J.A., Dwyer, H.A., Oppenheim, A.K.: Dynamic effects of autoignition centers for hydrogen and C1,2-hydrocarbon fuels. In: Twenty-Second Symposium (International) on Combustion/The Combustion Institute, pp. 1683–1693 (1988)
7. Peters, N., Paczko, G., Seiser, R., Seshadri, K.: Temperature cross-over and non-thermal runaway at two-stage ignition of n-heptane. Comb. Flame 128(1-2), 38–59 (2002)
8. Westbrook, C.K., Pitz, W.J., Herbinet, O., Curran, H.J., Silke, E.J.: A comprehensive detailed chemical kinetic reaction mechanism for combustion of n-alkane hydrocarbons from n-octane to n-hexadecane. Combust. Flame 156, 181–199 (2009)
9. Pfahl, U., Fieweger, K., Adomeit, G.: Self-ignition of diesel-relevant hydrocarbon-air mixtures under engine conditions. Proc. Combust. Inst. 26, 781–789 (1996)
10. Cai, L., Sudholt, A., Lee, D.J., Egolfopoulos, F.N., Pitsch, H., Westbrook, C.K., Sarathy, S.M.: Chemical kinetic study of a novel lignocellulosic biofuel: Di-n-butyl ether oxidation in a laminar flow reactor and flames. Combust. Flame 161, 798–809 (2014)
11. Blanquart, G., Pepiot-Desjardins, P., Pitsch, H.: Chemical mechanism for high temperature combustion of engine relevant fuels with emphasis on soot precursors. Combust. Flame 156, 588–607 (2009)
12. Narayanaswamy, K., Blanquart, G., Pitsch, H.: A consistent chemical mechanism for oxidation of substituted aromatic species. Combust. Flame 157, 1879–1898 (2010)
13. Cai, L., Pitsch, H.: Mechanism optimization based on reaction rate rules. Combust. Flame 161, 405–415 (2014)
14. Narayanaswamy, K., Pepiot-Desjardins, P., Pitsch, H.: A chemical mechanism for low to high temperature oxidation of n-dodecane as a component of transportation fuel surrogates. Combust. Flame 161, 866–884 (2014)
15. Cai, L., Pitsch, H.: Chemical mechanism for combustion of gasoline surrogate fuels (in preparation)
16. Fischer, S.L., Dryer, F.L., Curran, H.J.: The reaction kinetics of dimethyl ether. I: High-temperature pyrolysis and oxidation in flow reactors. International Journal of Chemical Kinetics 32(12), 713–740 (2000)
17. Pepiot-Desjardins, P., Pitsch, H.: An efficient error-propagation-based reduction method for large chemical kinetic mechanisms. Combust. Flame 154, 67–81 (2008)
18. Pepiot-Desjardins, P.: PhD. thesis, Department of Mechanical Engineering, Stanford University (2008)
19. Pitsch, H.: A C++ computer program for 0D combustion and 1D laminar flame calculations
20. Frenklach, M.: Systematic optimization of a detailed kinetic model using a methane ignition example. Combust. Flame 58, 69–72 (1984)

21. Frenklach, M., Packard, A., Seiler, P., Feeley, R.: Collaborative data processing in developing predictive models of complex reaction systems. Int. J. Chem. Kin. 36, 57–66 (2004)
22. Curran, H., Gaffuri, P., Pitz, W., Westbrook, C.K.: A Comprehensive Modeling Study of n-Heptane Oxidation. Combust. Flame 114, 149–177 (1998)
23. Curran, H., Gaffuri, P., Pitz, W., Westbrook, C.K.: A comprehensive modeling study of iso-octane oxidation. Combust. Flame 129, 253–280 (2002)
24. Sarathy, S.M., Westbrook, C.K., Mehl, M., Pitz, W.J., Togbé, C., Dagaut, P., Wang, H., Oehlschlaeger, M.A., Niemann, U., Seshadri, K., Veloo, P.S., Ji, C., Egolfopoulos, F.N., Lu, T.: Comprehensive chemical kinetic modeling of the oxidation of 2-methylalkanes from C7 to C20. Combust. Flame 158, 2338–2357 (2011)
25. Fieweger, K., Blumenthal, R., Adomeit, G.: Self-ignition of S.I. engine model fuels: A shock tube investigation at high pressure. Combust. Flame 109, 599-619 (1997)
26. Fikri, M., Herzler, J., Starke, R., Schulz, C., Roth, P., Kalghatgi, G.T.: Autoignition of gasoline surrogates mixtures at intermediate temperatures and high pressures. Combust. Flame 152, 276–281 (2008)

Unsteady Combustion Processes Controlled by Detailed Chemical Kinetics

Michael A. Liberman

Nordita, KTH Royal Institute of Technology and Stockholm University,
Roslagstullsbacken 23, 10691 Stockholm, Sweden
Moscow Institute of Physics and Technology, Russia
michael.liberman@nordita.org

Abstract. A correct description of unsteady, transient combustion processes controlled by chemical kinetics requires knowledge of the detailed chemical reaction mechanisms for reproducing combustion parameters in a wide range of pressures and temperatures. While models with fairly simplified gas-dynamics and a one-step Arrhenius kinetics in many cases makes possible to solve the problem in question in explicit analytical form, many important features of combustion can not be explained without account of the reactions chain nature, describing qualitatively a few major properties of the phenomena in question with some poor accuracy if any, often rendering misinterpretation of a verity of combustion phenomena. However, for modeling real three-dimensional and turbulent flows we have to use reduced chemical kinetic schemes, since the use of detailed reaction mechanisms consisting up to several hundreds species and thousands reactions is difficult or practically impossible to implement. In this lecture we consider the option of a reliable reduced chemical kinetic model for the proper understanding and interpretation of the unsteady combustion processes using hydrogen-oxygen combustion as a quintessential example of chain mechanisms in chemical kinetics. Specific topics covered several of the most fundamental unsteady combustion phenomena including: the regimes of combustion wave initiated by initial temperature non-uniformity; ignition of combustion regimes by the localized transient energy deposition; the spontaneous flame acceleration in tubes with no-slip walls; and the transition from slow combustion to detonation.

1 Introduction

Significant insights have been obtained during the past decades by a long series of theoretical and numerical studies in our understanding and interpretation of combustion fundamentals, yet many questions still remain unresolved or poorly understood. A one-step Arrhenius kinetics model, used for the theoretical combustion studies, in many cases makes possible to solve the problem in question in explicit analytical form. However, many important features of combustion can not be explained without account of the reactions chain nature. The range of validity of simplified schemes is necessary very limited and their applicability to the understanding and modeling of combustion phenomena must be examine with great

R. King (ed.), *Active Flow and Combustion Control 2014*, Notes on Numerical Fluid Mechanics and Multidisciplinary Design 127, DOI: 10.1007/978-3-319-11967-0_20

care. Furthermore, it became clear that the use of a one-step Arrhenius model has not always been interpreted properly, rendering misinterpretation of a verity of combustion phenomena. A one-step model does not reproduce even two distinct stages of the combustion reaction: the induction stage and exothermal one. Another fundamental difference between a one-step Arrhenius model and the chain-branching reaction is an avalanche multiplication of intermediate species, atoms and radicals, and their repetitive reactions constituting reaction chains.

Knowledge of the detailed reaction mechanisms is of special importance for understanding unsteady combustion processes, which are controlled by chemical kinetics, such as ignition and self-ignition processes (e.g. engine knock), flame acceleration, and the deflagration-to-detonation transition (DDT), and it is absolutely necessary for a correct description of kinetically controlled pollutant formation. Development and exploitation of reliable detailed chemical kinetic models, identification of the important kinetic pathways and accurate kinetic–transport models remain among the major challenges in combustion science and technology being essential for the design of efficient and reliable engines and for controlling emissions. The availability of such models is essential for gaining scientific insight into the most fundamental combustion phenomena. If, however, real three-dimensional and often turbulent flows with a large temperature and density gradients are considered, we have to use reduced chemical schemes, since the use of the detailed reaction mechanisms involves massive computing times which can be impossible to implement. Even in the combustion of simple hydrocarbons, the chemical kinetics is reproduced by relatively large reaction mechanisms, and for complex hydrocarbons the number of chemical species and elementary reactions can be up to several hundreds and up to several thousands. In principle, the detailed reaction mechanisms were developed for different combustible mixtures and today are available and can be used in the simulation of spatially homogeneous reaction systems. If, however, real three-dimensional and often turbulent flows are considered, we have to use reduced chemical schemes, since the use of detailed reaction mechanisms involves massive computing times which can be difficult and even impossible to implement. Even in the combustion of simple hydrocarbons, the chemical kinetics is reproduced by relatively large reaction mechanisms, and for complex hydrocarbons the number of chemical species can be up to several hundreds and the number of elementary reactions up to several thousands. It is known [1, 2] that in many cases both quantitative and qualitative features of the studying processes essentially depend on the choice of chemical kinetics model. Therefore, there is great interest in reduced reliable chemical reaction schemes consisting of not too large number of elementary reactions.

A quintessential example of chain mechanisms in chemical kinetics and combustion science is the H_2/O_2 mechanism, which has been a major topic of research for many decades. The number of available mechanisms are significant, yet there are uncertainties related to the elementary reactions involved [3] and available for validation experimental data is scarce, especially for low-temperature kinetics ($T<1000K$). Usually the verification of the reduced kinetic schemes covers the data on induction periods, equilibrium composition, temperature of the products, and the normal laminar flame velocity. The later are pure thermodynamic characteristics and are more or less in a good agreement with experimental data at normal pressure. To study unsteady combustion processes, such as e.g. flame acceleration, ignition, etc., which are accompanied by compression and shock waves, it is necessary to take into account parameters such as the induction period and the period of exothermal reaction, which determine the chemical time scales competing with the transport time scales in establishing the zone of energy release, which in turn determines the evolution of the flame. Even more important for a correct description of transient processes is the pressure dependence of flame parameters. It is also important to get the solution of basic problems using a model, which validation is similar to that used in the simulation of multidimensional unsteady problems.

2 Reduced Kinetic Models for Hydrogen-Oxygen Gaseous Mixture

We consider several widely used chemical kinetic schemes for gaseous H_2/O_2 and H_2-airmixtures [1, 3-9] using the full gas-dynamic transport model [10] for calculation of a laminar flame characteristics and taking into account the correlations between evolutionary and gas-dynamics parameters at different initial pressures. A widely accepted standard numerical procedures for verification of the chemical schemes are based on the solution of 0D problem and 1D eigenvalue problem, which is distinguished mathematically from the computational gas-dynamics setup used for the simulation of the unsteady combustion processes. The method of eigenvalue problem often used for verification of the chemical schemes is far distinct also from experimental setups of the ignition and combustion and in some cases it may cause considerable differences between experimental and numerically obtained data.

The governing equations used for the verification of chemical kinetic schemes are the one-dimensional version of the multidimensional time-dependent, reactive Navier-Stokes equations including the effects of compressibility, molecular diffusion, thermal conduction, viscosity and chemical kinetics for the reactive species with subsequent chain branching, production of radicals and energy release, which will be used for modeling of unsteady combustion problems.

$$\frac{\partial \rho}{\partial t}+\frac{\partial(\rho u)}{\partial x}=0 \tag{1}$$

$$\frac{\partial Y_i}{\partial t}+u\frac{\partial Y_i}{\partial x}=\frac{1}{\rho}\frac{\partial}{\partial x}\left(\rho D_i\frac{\partial Y_i}{\partial x}\right)+\left(\frac{\partial Y_i}{\partial t}\right)_{ch} \tag{2}$$

$$\rho\left(\frac{\partial u}{\partial t}+u\frac{\partial u}{\partial x}\right)=-\frac{\partial P}{\partial x}+\frac{\partial \sigma_{xx}}{\partial x}, \tag{3}$$

$$\rho\left(\frac{\partial E}{\partial t}+u\frac{\partial E}{\partial x}\right)=-\frac{\partial(Pu)}{\partial x}+\frac{\partial}{\partial x}(\sigma_{xx}u)+\frac{\partial}{\partial x}\left(\kappa(T)\frac{\partial T}{\partial x}\right)+$$

$$+\sum_k\frac{h_k}{m_k}\left(\frac{\partial}{\partial x}\left(\rho D_k(T)\frac{\partial Y_k}{\partial x}\right)\right), \tag{4}$$

$$P=R_B T n=\left(\sum_i\frac{R_B}{m_i}Y_i\right)\rho T=\rho T\sum_i R_i Y_i\,, \tag{5}$$

$$\varepsilon=c_v T+\sum_k\frac{h_k\rho_k}{\rho}=c_v T+\sum_k h_k Y_k\,, \tag{6}$$

$$\sigma_{xx}=\mu\left(\frac{\partial u_x}{\partial x_x}+\frac{\partial u_x}{\partial x_x}-\frac{2}{3}\delta_{xx}\frac{\partial u_x}{\partial x_x}\right)=\mu\frac{4}{3}\frac{\partial u_x}{\partial x_x} \tag{7}$$

Concentrations $Y_i=\rho_i/\rho$ of the species (H_2, H, O_2, O, H_2O, OH, HO_2, H_2O_2, M (N, N_2, NO, NO_2, Ar), are defined by the equations of chemical kinetics, which is solved with the aid of Gear method [11].

$$\frac{dY_i}{dt}=F_i(Y_1,Y_2,\ldots Y_N,T),\quad i=1,2,\ldots N\,. \tag{8}$$

The number of the reactions in Eq. (8) can be from 19 to 37 depending on the chemical scheme, where the right hand parts of Eq. (8) contain the rates of chemical reactions, which depend on temperature according to the Arrhenius law in a standard form [6]. Here we used standard notations: P, ρ, u, are pressure, mass density, and flow velocity, $E=\varepsilon+u^2/2$ is the total energy density, ε is the inner energy density, R_B - is the universal gas constant, m_i - the molar mass of i-species, $R_i=R_B/m_i$, n is the molar density, σ_{ij} is the viscous stress tensor, $c_v=\sum c_{vi}Y_i$ is the constant volume specific heat, c_{vi} - the constant volume specific heat of i-species, h_i - the enthalpy of formation of i-species, $\kappa(T)$ and $\mu(T)$ are the coefficients of thermal conductivity and viscosity, $D_i(T)$ - is the diffusion coefficients of i-species. The equations of state for the reactive mixture and for the combustion products are taken with the temperature dependence of the specific heats and enthalpies of each species borrowed from the JANAF tables and interpolated by the fifth-order polynomials [12]. The viscosity, thermal conductivity and diffusivity coefficients of the mixture are calculated from the gas kinetic theory using the Lennard-Jones potential [10]. The system of gas dynamics equations is solved using Lagrange-Euler method (Coarse-particle method) [13], which was modified and improved (see e.g. [14]).

The calculated induction periods and exothermal stage durations dependencies on initial temperature for stoichiometric hydrogen-oxygen and hydrogen-air mixture at 1atm, using chemical schemes [3-9], are shown in Figs.1 a) and b) together with available experimental data [15-23].

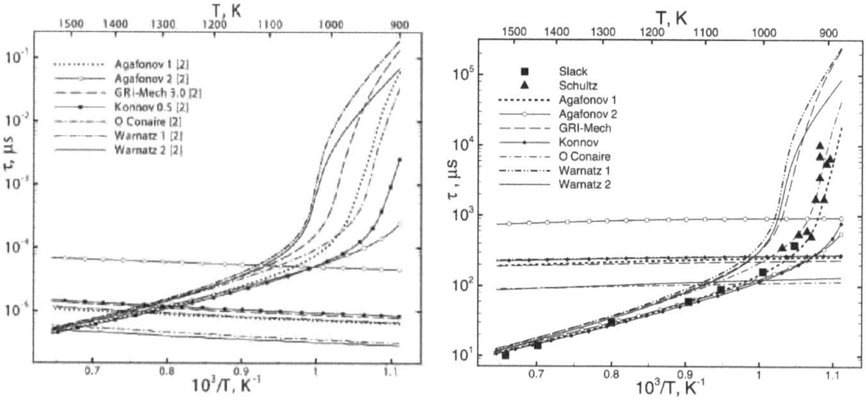

Fig. 1. (a) Induction periods and exothermal stage durations dependencies on initial temperature of stoichiometric H_2/O_2 at 1atm; (b) the same for H_2-air mixture at 1atm.

Almost linear curves intersecting the induction period dependencies represent an exothermal stage duration which determines the time scale of energy release inside the reaction zone. Despite the differences in the low temperature region, $T < 1100K$, all the kinetic schemes reproduce rather close values of the induction periods at normal conditions, within (10-15)% values for thermodynamic parameters and induction times. The differences in the induction time at low temperatures can be not essential if the endothermic induction time is larger than characteristic gas-dynamics time of the problem.

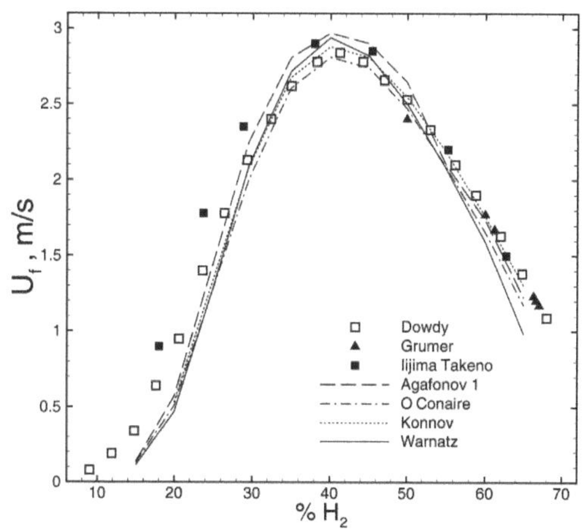

Fig. 2. Laminar flame speed calculated for the different fraction of H_2 in H_2-air mixture

The calculated for four kinetic schemes laminar flame speeds U_f at normal pressure and depending on the volume fraction of H_2 are presented in Fig. 2. It is seen that at normal pressure there is a satisfactory agreement with the experimental data. On the contrary, the differences in the induction time given by different schemes at low temperatures cause more essential difference for induction time calculated at elevated pressures shown in Fig. 3(a,b,c), where considerable difference from the experimental values for the induction period emerges in the low temperature region at pressure greater then 2.5atm, when three-body collisions become essential.

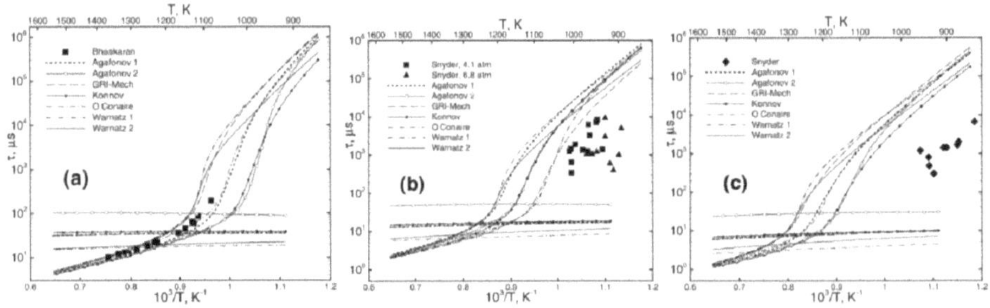

Fig. 3. Induction periods and exothermal stage durations dependencies on initial temperature of stoichiometric hydrogen-air mixture at (a): 2.5atm, (b): 5.0atm; (c): 8.8atm.

Only a little or uncertain and scarce experimental data, especially for hydrogen-oxygen are available for validation of the elementary reactions and the laminar flame speed at elevated pressures [20, 21]. Correspondingly, there are uncertainties in the chemical mechanism for low-temperatures and high pressures to be selected as a reliable one. The velocity-pressure dependence calculated for different kinetic schemes and experimental data are shown in Fig. 4. The calculations for some schemes are quite distinct from experimental data and the distinctions rise with the pressure increase. Nevertheless, experimental data and calculated values of the velocity-pressure dependence shown in Fig. 4 for hydrogen-oxygen can be approximated as $U_f \propto P^{(n-2)/2}$ for overall reaction order 2.74 (dotted line), which satisfactory well agrees with the usual theoretical concept [25]. On the contrary, in hydrogen-air mixture the overall reaction order changes with the pressure and the difference increases with the nitrogen dilution.

Thus, further improvement and the development of low-temperature kinetics (T<1000K) is particularly significant for the understanding of the ignition and self-ignition processes. Since only a scarce experimental data available for the high pressures range and at low temperatures, the problem of question is challenging. While speed of sound and therefore characteristic hydrodynamic time scales do not depend on pressure, the induction time, especially at the temperature range (1000÷1200)K is considerably sensitive to pressure. This and the different pressure dependencies given by different reduced schemes must be taken into account with a great care while modeling unsteady combustion processes.

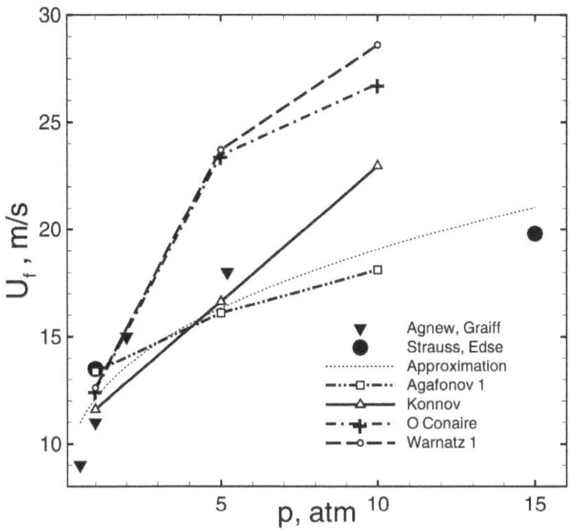

Fig. 4. Speed-pressure dependence of laminar flame for H_2/O_2

3 Combustion Regimes Initiated by Initial Temperature Gradient

In the most practical cases ignition arises from a small area of combustible mixture, which is locally heated by means of electric spark, hot wire, focused laser light and the like. Such local energy release results in formation of the initial nonuniform distribution of temperature and concentration of reagents which determines further evolution of the reaction wave depending on the mixture reactivity and the initial pressure. For the first time regimes of propagating chemical reaction wave ignited by the initial temperature gradient have been studied by Ya. B. Zeldovich, who introduced also the concept of *spontaneous reaction wave* [26]. The basic idea was that the reaction is ignited at the point of minimum induction time $\tau_{ind}(T(x))$ and correspondingly the maximum temperature and then spreads along the temperature gradient at neighboring locations where τ_{ind} is larger with velocity

$$U_{sp} = \left|(d\tau_{ind} / dx)\right|^{-1} = \left|(\partial\tau_{ind} / \partial T)^{-1} (\partial T / \partial x)^{-1}\right|. \tag{9}$$

For a one-step chemical reaction model the induction time, τ_{ind}, is defined as a time-scale of the maximum reaction rate. In the case of real chain branching chemistry this is the time scale for the stage of endothermic chain initiation and branching reactions. The value of U_{sp} depends only on the gradient steepness and does not depend on thermal conduction or sound speed. The Zeldovich's concept of the spontaneous reaction wave opened an avenue to study scenario of the reaction ignition and the evolution of spontaneous wave initiated by the initial non-uniformity in temperature or reactivity in different regimes of combustion wave and therefore of great fundamental and practical importance.

In the subsequent studies researchers have employed a one-step Arrhenius model, and their studies have been focused mainly on the regime of direct ignition of a detonation by the initial temperature gradient discovered by Zeldovich and co-authors [27]. All the same the models with a simplified one-step chemical kinetics may lead only to only poor accuracy if any, sometime giving even qualitatively incorrect interpretation. Subsequently it was shown using two-step and three-step models, which to some extent mimic the chain-branching kinetics with a simplified notional reaction scheme between a set of pseudo-species, that the one-step chemical model does not properly describe systems governed by chain-branching reaction, and the one-step model is not appropriate for simulating detonation initiation in systems governed by chain-branching reactions.

We consider regimes of chemical reaction wave in hydrogen-oxygen initiated be the initial temperature gradient depending on the steepness of the gradient and using a detailed chemical kinetic model [28, 29]. The initial conditions as quiescent and uniform mixture apart from a linear gradient in temperature (and hence density), with the left boundary at $x = 0$ being a solid reflecting wall, where $u(0,t) = 0$:

$$T(x,0) = T^* - (T^* - T_0)(x/L), \quad 0 \le x \le L, \tag{10}$$

$$P(x,0) = P_0, \; u(x,0) = 0. \tag{11}$$

The initial temperature gradient is characterized by the maximum temperature $T(0,0) = T^*$ at the top left edge, by the background mixture temperature $T(x > L, 0) = T_0$ outside the temperature gradient, and by the gradient steepness $(T^*-T_0)/L$. Here L is the "gradient scale", which characterizes the gradient steepness for the fixed value of (T^*-T_0) and can be viewed as the size of the area where the initial temperature gradient was created by the energy input or the like. The governing equations (1-8) are solved with the boundary conditions (10-11).

In what follows we consider the temperature gradients of different steepness at the initial pressure 1atm with $T^*=1500K$, $T_0=300K$. Fig. 5(a) and Fig. 6(a) show evolution of the reaction and pressure waves velocities for the case of the gentle, $L = 8cm$, and for the steeper, L=7cm, temperature gradients. The corresponding evolution of the temperature and the pressure profiles are shown in Fig. 5(b) and Fig. 6(b). The velocity of spontaneous wave initiated by the initial temperature gradient decreases as the wave propagates along the gradient. It reaches the minimum value at the point close to the cross-over temperature. The pressure waves generated during the exothermic stage of reaction can couple and evolve into a self-sustained detonation wave, or produce a flame and a decoupled shock depending on the gradient steepness. In the case of the gradient in Fig. 5(a) the spontaneous wave transforms into combustion wave and the pressure wave steepens into the shock wave. For a shallow gradient, such that the minimum speed of spontaneous wave is of the order of the sound speed $a_* = a(T^*)$ at the top of the gradient, the intensity of the shock wave formed ahead of the reaction wave is sufficient to accelerate the reaction in the flow formed behind the shock. The pressure peak formed at the reaction front grows at the expense of energy released in the reaction and the result of the positive feedback is a detonation wave shown in Fig. 6(a). On the contrary, for a steeper initial temperature gradient (L=7cm) the velocity of spontaneous wave in the minimum point, where the pressure wave overtakes the reaction wave, is not sufficient to sustain synchronous amplification of the pressure pulse in the flow behind the shock. As a result, the pressure wave runs ahead of the reaction wave, the velocity of the reaction wave decreases, and the resulting regime is deflagration, Fig.6(b).

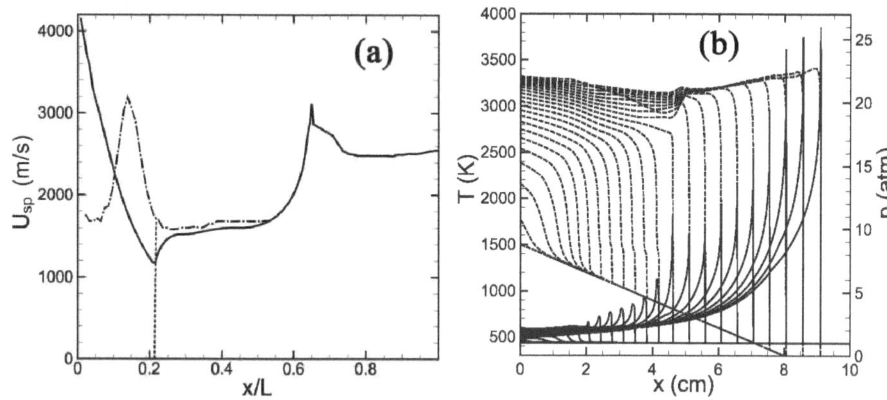

Fig. 5. (a): velocities of the spontaneous wave (solid lines) and pressure wave (dash-dotted lines) computed for the temperature gradient L=8cm, T*=1500K in H_2/O_2 mixture; P_0=1atm; (b): evolution of the temperature (dashed lines) and pressure (solid lines) profiles (intervals 2μs) during the detonation formation.

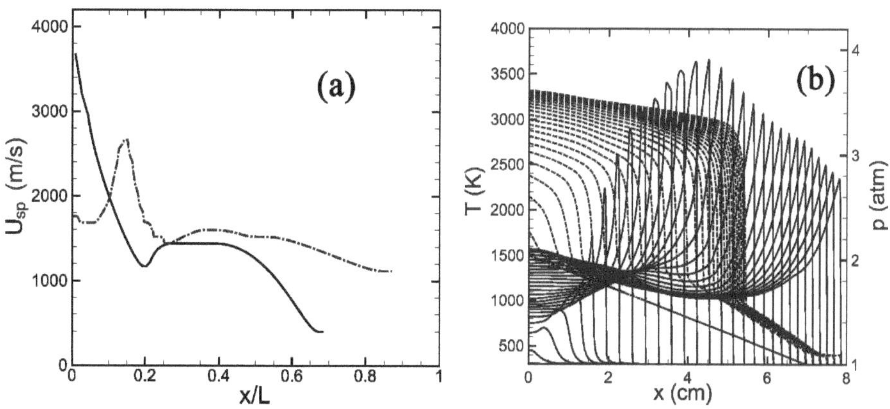

Fig. 6. (a): velocities of the spontaneous wave (solid lines) and pressure wave (dash-dotted lines) computed for the temperature gradient L=7cm, T*=1500K in H_2/O_2 mixture; P_0=1atm; (b): evolution of the temperature (dashed lines) and pressure (solid lines) profiles (intervals 2μs).

Possible regimes of the combustion wave inspired by the spontaneous wave initiated by the temperature gradient depending on the gradient steepness and were studied in [28, 29] using a detailed chemical scheme for H_2/O_2. The regimes depend on the gradient steepness and on the ratio between the speed of spontaneous wave at the point where the spontaneous wave velocity reaches minimum and the characteristic velocities of the problem: the normal laminar flame speed, U_f, the sound speeds $a_0 = a(T_0)$ and $a^* = a(T^*)$, the Neumann, a_N, and the Chapman-Jouguet, U_{CJ}, velocities. The obvious limiting case $\nabla T \to 0$, $U_{sp} \to \infty$ corresponds to the adiabatic explosion. Another limiting case of a very steep gradient igniting the deflagration is bounded from below by the minimum size of the hot region, for which the rate of heat removal from the "hot wall" is less than the normal flame velocity. Diagram in Fig.7 shows possible regimes depending on the gradient steepness and, correspondingly, on the ratio between the velocity of spontaneous wave at the minimum point and the characteristic velocities of the problem.

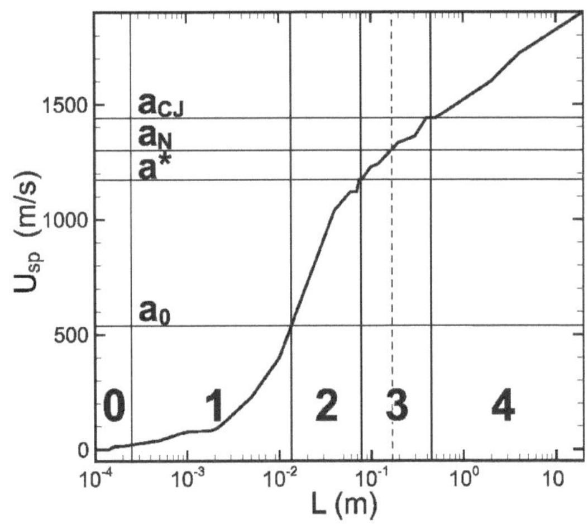

Fig. 7. Regimes of the reaction waves initiated by temperature gradients of different steepness in H_2/O_2

If $U_{sp} < U_f << a_s$ (domain 0), the rate of the heat transfer by thermal conduction is greater than the spontaneous wave velocity, and the resulting regime is a deflagration wave propagating due to thermal conduction with the normal flame velocity $U_f \approx 10 m/s$. For $U_f < U_{sp} < a_0$ the pressure wave overtakes the deflagration wave, and the fast deflagration wave propagates at nearly constant pressure (domain 1). If $a_0 < U_{sp} < a^*$, then pressure wave overtakes the reaction wave to form a weak shock wave that compresses and heats the gas further speeding up the deflagration wave (domain 2). If $a^* < \min\{U_{sp}\} < a_N < a_{CJ}$ the reaction wave accelerates behind the shock and the transition to detonation occurs due to the formation and amplification of the pressure peak at the front of the reaction wave (domain 3). If a_N<min{U_{sp}}<a_{CJ} then a quasi-stationary structure consisting of a shock wave and reaction zone is formed, which transforms into a detonation propagating down the temperature gradient. If a_{CJ}<min{U_{sp}}, then the intersection of the pressure wave and the spontaneous reaction wave creates a classical structure of a detonation wave with the leading shock wave initiating the reaction (domain 4).

The scale (steepness) of the temperature gradients capable to ignite one or another combustion regime for the detailed chemical model and a single-step model are considerably different. First, the reaction for a one-step model is exothermic for all temperatures, while chain branching reactions begins with neutral or endothermic induction stage and therefore, the gas-dynamics is effectively "switched-off" during the induction stage. On the contrary for a one-step model the spontaneous wave is considerably affected by the gas-dynamics from the very beginning, so that it is always the determining factor. The induction time for a single-step Arrhenius model is several orders of magnitude shorter than the real one, therefore velocity of spontaneous wave produced by the same temperature gradient is much smaller during the early stage for the chain-branching reaction. To match the induction time with experimental data for a one-step model some unrealistically high activation energy should be taken. As a result, the temperature gradient for initiating all possible combustion regimes is much steeper for a one-step model compared to that for a detailed chemical kinetics. Fig. 8(a) and Fig. 8(b) show scales of the temperature gradients calculated for the detailed chemical model for H_2/O_2 and for a one-step model.

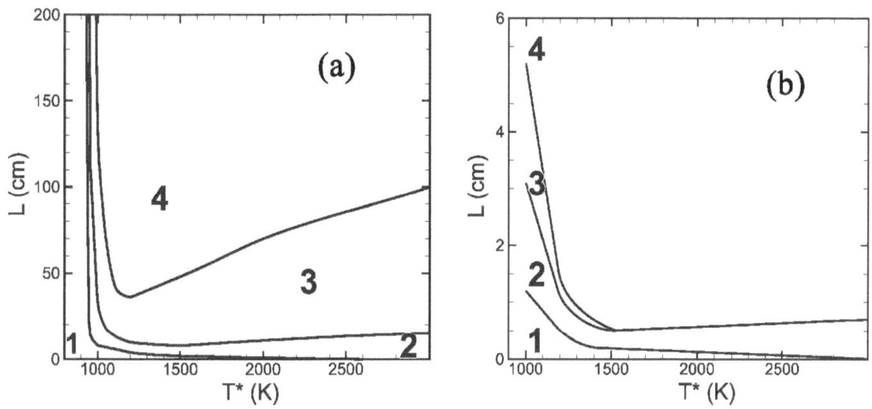

Fig. 8. Scales of the temperature gradient corresponding to the boundaries between regimes 1, 2, 3, 4: (a) for the detailed chemical model for H_2/O_2; (b)**:** for a one-step model, H_2/O_2

The induction time increases if a combustible mixture is diluted by neutral gas and for lower reactive mixtures. For example, in a stoichiometric hydrogen-air mixture, which can be viewed as H_2-O_2 diluted by nitrogen not involving in the chain-branching reaction, the induction time increases by 2-3 times in a wide temperature range compared to a stoichiometric hydrogen-oxygen mixture. At the same time the main features of the induction time dependence versus temperature and pressure remains nearly the same. So that the main difference between hydrogen-air compared to hydrogen-oxygen is the reduced velocity of spontaneous waves for the same temperature gradients. As a result, the temperature gradient required for realization various combustion regimes in the hydrogen-air mixture is significantly shallower than in the hydrogen-oxygen mixture. This is illustrated in figure 9, which shows the temperature gradient scales depending on T* calculated for the detailed chemical model for H_2-air mixture, and depicting the domains of the gradient steepness required for initiation of the regimes 1, 2, 3, 4 at initial pressure P_0=1atm.

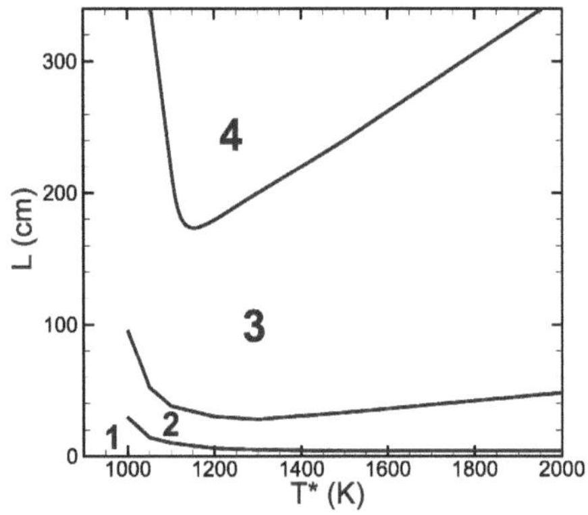

Fig. 9. Scales of the temperature gradient in H_2-air corresponding to the boundaries between regimes 1, 2, 3, 4 for a detailed chemical model

4 Combustion Regimes Initiated by Transient Energy Deposition

Transient thermal energy deposition into a reactive gas provides a source for ignition of either deflagration or detonation. Sufficiently fast and large energy addition can facilitate direct initiation of detonation. However the particular mechanisms of the direct initiation of detonation or deflagration can be different. In most practical cases ignition arises from a small volume of combustible mixture which is locally heated by energy input by means of an electric spark (e.g. a spark-plug in an engine combustor), hot wire, focused laser light and other related external sources. Such a transient energy addition process can generate a wide range of gas expansion processes depending on the amount and the rate of energy actually added and may result in the formation of the initial non-uniform distribution of temperature.

We consider classification of the propagation regimes of chemical reaction wave initiated by the transient energy deposition in gaseous mixtures using high resolution numerical simulations of reactive Navier-Stokes equations, including a multispecies transport model, the effects of viscosity, thermal conduction, molecular diffusion and a detailed chemical reaction mechanism for hydrogen-oxygen mixture [30]. Such a level of modeling allows clear understanding of the feedback between gasdynamics and chemistry which is the principal point when studying unsteady process of ignition and practically cannot be captured using a simplified gas-dynamical and chemical models. We assume uniform initial conditions and a transient external source of energy localized on the scale of the "hot spot" $0 \le x \le L$. Combustion regime initiated by the energy deposition depends on the interrelationship between the characteristic times of the problem: time of the hot spot transient heating Δt_Q; acoustic time $t_a = L / a(T)$, which defines the concomitant motion setup in the mixture, where $a(T)$ is the speed of sound; characteristic time of heat propagation, $t_T \sim L^2 / \chi$, where χ is the thermal conductivity coefficient; and the ignition time t_{ign}, which characterizes length of the endothermic chain initiation and branching reaction stage after or during the transient energy deposition, $t_{ign} = t_{ind}(T, P)$.

For a very fast energy deposition, such that $\Delta t_Q << t_a$, local heat addition occurs as a nearly constant volume process and the temperature elevation within the hot spot is accompanied by a concomitant pressure rise. Subsequent expansion of the hot spot driven by the large pressure gradient between the hot spot and the surrounding gas causes compression and shock waves propagating outside in the surrounding gas. If $\Delta t_Q >> t_a$ the acoustic waves have enough time for pressure equalization and the local heat addition occurs at nearly constant pressure. It should be noted that if the reaction has started, further heating does not influence the developed combustion regime. For a very short time of energy addition, $\Delta t_Q < t_{ign} << t_a$, the mixture in the hot spot can be heated to any temperature and the ignition itself will be determined by the induction time at that temperature and pressure. For longer energy deposition, the ignited regime depends on the size of the hot spot and correspondingly on the relation between Δt_Q and t_a.

Here we restrict ourselves by several examples demonstrating role of gasdynamics in the formation of different combustion regimes. We consider 1D model so that the governing equations are (1-8) with the initial conditions of constant pressure, the initial temperature, $T(t=0) = T_0$, and zero velocity of the unburned mixture. A solid reflecting wall are taken at x=0.

If $\Delta t_Q \sim 0.1\mu s < t_{ign} << t_a$, the time scales of the hot spot heating is shorter then acoustic time, and the heating causes almost uniformly fast elevation of pressure and temperature resulting in the volume explosion. The reaction starts after the end of the energy deposition. It should be noted that depending on the value of deposited energy the mixture of the hot spot achieves some temperature and pressure at which the reaction starts and the combustion regime then develops by the volumetric explosion at these conditions. If the power is large enough, subsequent events represent the decay of the initial discontinuity consisting of a compression wave propagating out of the hot spot, which steepens into the shock wave and the rarefaction wave propagating inside from the boundary of hot spot with the velocity equal to the local sound speed. The scenario is similar to a strong point explosion [31] and it results in the direct triggering of a detonation wave if a concomitant shock wave at the right boundary is strong enough (e.g. for L=1cm, $Q = 3.0 kJ/m^2$). In case of the rapid but relatively small energy deposition ($Q < 3.0 kJ/m^2$) the resulting regime is a fast deflagration wave with runaway shock wave.

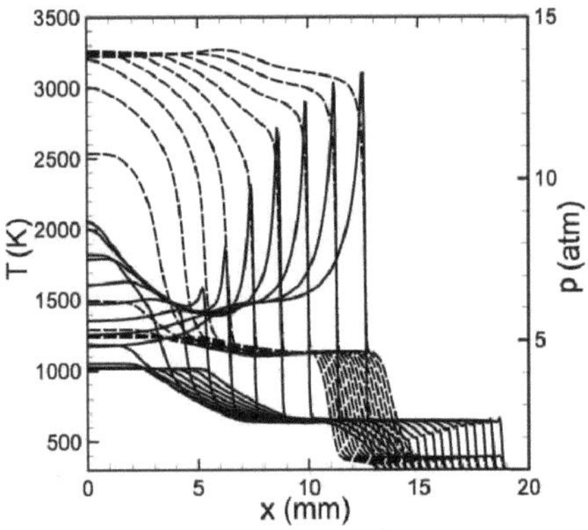

Fig. 10. Evolution of temperature and pressure profiles and a detonation formation in the energy release region ($L = 1cm$, $\Delta t_Q = 5\mu s$, $\Delta t = 0.5\mu s$).

A detonation can be also initiated through the Zeldovich gradient mechanism due to the shallow gradient formed by the rarefaction wave at the increased pressures in the hot spot region. Such scenario is shown in Fig.10. Here a relatively weak shock wave propagates out of the hot spot, and the rarefaction wave, which propagates to the left during the energy deposition, creates a shallow temperature and pressure gradients on the scale of about the hot spot size. Such temperature gradient cannot trigger detonation at the initial pressure $P_0 = 1atm$. However, since the pressure of the heated mixture increased during the heating up to about 4atm, this temperature gradient can produce a detonation through the Zeldovich gradient mechanism. The calculated temperature and pressure profiles, shown in Fig. 10, demonstrate the emergence of the spontaneous reaction wave, its coupling with the pressure wave leading to the detonation initiation.

The rapid energy deposition such that the heating time, Δt_Q, is less or comparable to the acoustic time scale of the volume, always results in the shock waves propagating away from the hot spot. A particular scenario of the resulting combustion regime depends on the size of the hot spot, though the basic physics appears to be similar to that described above. Different possible scenarios include either direct initiation of detonation through a constant volume explosion by a strong enough shock wave in the context of a thermal explosion, or through the Zeldovich gradient mechanism due to the shallow gradient formed by the rarefaction wave at the increased pressures in the hot spot region, or the shock waves propagating away from the hot spot producing ignition of the fast deflagration propagating behind the shock waves (regime 3 according to the classification of combustion regimes in previous section).

The scenario of thermal energy addition over a longer period of time is different. If the acoustic time is much less than the energy deposition time, $t_a \ll \Delta t_Q \le t_{ign}$, then there is enough time for pressure to be spatially homogenized by acoustic waves. In this case there are no strong compression waves emitted from the hot spot and the initiated combustion regimes depend essentially on the steepness of the temperature gradient, which is formed by the thermal wave and gas expansion in the vicinity of the hot spot. During the time of energy deposition the thermal wave propagates away from the hot spot at the distance $x_T \approx (\chi \Delta t_Q)^{1/2}$, and the expelled mass together with the thermal wave give rise to the temperature gradient in the surrounding mixture behind the boundary of the hot spot.

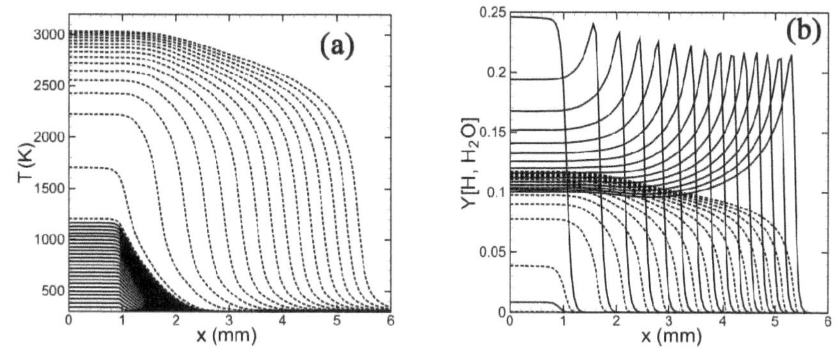

Fig. 11. a): Evolution of temperature profiles in the hot spot during energy deposition (solid lines) and during the deflagration wave formation (dashed lines); b) Evolution and formation of deflagration: H_2O concentration (dashed lines) and H-radical concentration (solid lines) profiles (L=1mm, Δt_Q=1000μs, P_0=1atm, Δt=5μs).

Fig. 11 shows the calculated temperature gradient formed at the end of the energy deposition. The thermal wave and the gas expansion are too slow to expand temperature and to form a shallow enough temperature gradient compatible with the detonation formation at atmospheric pressure. Long before the thermal wave moves away at a sufficiently long distance the temperature of the mixture rises to ignite the reaction, so that either a deflagration or a fast deflagration wave will be formed. However, since the coefficient of thermal conductivity does not depend on pressure, and the steepness of the temperature gradient for direct detonation initiation through the Zeldovich gradient mechanism decreases considerably with the increase in pressure, the temperature gradient created by the thermal wave can trigger detonation if the initial pressure is high enough (about 10atm).

5 Deflagration-to-Detonation Transition

It is known for a long time that a flame ignited near the closed end of a tube and propagating to the open end may spontaneously accelerate until a sudden abrupt increase of the flame velocity, which ends up with triggering a detonation. This phenomenon is known as the transition from a slow combustion (deflagration) to detonation or deflagration-to-detonation transition (DDT). Over the years deflagration-to-detonation transition has been one of the least understood process and was considered as one of the major challenges of combustion theory. Since the discovery of detonation and the first experimental study of the transition from deflagration to detonation by famous French scientists Mallard and Le Chatelier in 1883, significant efforts have been devoted to understanding the nature of the flame acceleration and the mechanism of DDT. A large number of experimental, theoretical and numerical studies had been taken in attempt to understand nature of the detonation formation because of its important applications ranging from industrial safety, nuclear power plants safety, potential application for micro-scale propulsion and power devices, and to deciphering of scenario of white dwarf thermonuclear explosion in Supernovae events. Yet many questions still remain unresolved or poorly understood. Significant insights have been obtained during the last decades by a long series of experimental, theoretical and numerical studies by different groups of authors.

Without a doubt one of the main difficulties for modeling and understanding of the mechanism of DDT is the necessity to use a detailed chemical reaction model. Most of the theoretical and numerical works have been made in attempt to understand DDT using a one-step kinetics model. Different answers have been given to the principle problem, including turbulent mixing, boundary layers, hydraulic resistance, etc. The widely popular conclusion obtained on the basis of simulations using a one-step kinetics model was that ignition centers (hot spots) are formed ahead of the flame front in the nearby unreacted material, that cause the development of detonation through the Zeldovich gradient mechanism ([32] and references within). However the experiments [33] and high resolution 2D [34-36] and 3D [37] simulations with detailed chemical reaction models have shown that temperature ahead of the flame front, in hot spots, is too low to ignite exothermic reaction on the time scales of the whole process duration. Moreover, the spatial scale of the temperature non-uniformity formed in the hot spots is at least by the order of magnitude smaller than that capable to ignite detonation via the Zeldovich gradient mechanism. Therefore the results of simulations, which are based on a one-step chemical kinetics, should be treated as an artificial as such models allow ignition at any low temperatures not consistent with practical combustible mixtures.

5.1 Flame Acceleration in Tube with No-Slip Walls: Mechanism of DDT

From the very beginning of DDT studies the very fact of the flame acceleration in tubes has been considered as an important factor that influences the DDT process. The flame acceleration and the transition to detonation have been studied using the one-dimensional and multi-dimensional analyses taking into account that the acceleration rate can be enhanced by external turbulence or intrinsic flame instabilities. The classical formulation of the problem in question and an experimental set-up is an initially planar laminar flame, which is ignited near the closed end and propagates to an open end of the tube. At the beginning, expansion of the combustion products causes the flow ahead of the flame front, which remains almost uniform in the bulk of the channel with the longitudinal flow velocity $u = (\Theta-1)U_f$, which drops to zero within the boundary layer. Here $\Theta = \rho_u / \rho_b$ is the density ratio of the unburned ρ_u and burned ρ_b gases, respectively and U_f is the normal flame velocity.

The upstream flow in a channel acquires a non-uniform velocity profile due to the non-slip boundary conditions at the wall. For a smooth non-slip walls the upstream flow remains almost uniform in the bulk of the channel with the longitudinal flow velocity, which drops to zero within a thin boundary layer. The velocity profile for a steady motion in a tube with non-slip walls is the Poiseuille flow with parabolic velocity profile. However, for a non-stationary flow a certain time is required for establishing the Poiseuille parabolic velocity profile. The time needed for the development of the Poiseuille parabolic velocity profile in the flow ahead of the flame can be estimated using the Blasius solution for the boundary layer (H. Blasius, L. Prandtl) and taking into account that the Poiseuille flow is established when thickness of the boundary layer became of the order of the channel width [34]. For a laminar boundary layer of thickness $\delta \approx 2.95\sqrt{L\nu / u} \approx L / \sqrt{Re}$, where L is the characteristic length of the flow, and Re is the Reynolds number, we obtain time, which is required for establishing the Poiseuille flow: $t_P \approx D^2 / 36\nu$. This time is much longer than duration of the whole process for typical experimental conditions. For example, $t_P \approx 2\,s$ for the channel of width 5cm, while duration of the whole process is about 1 ms.

Since the flame velocity at each point of the flame front is the sum of the normal flame velocity relative to the moving gas ahead and the velocity of the upstream flow ahead of the flame, the shape of the flame front reproduces according to Huygen's principle the velocity profile in the upstream flow. The flame skirt stretched backward within the boundary layer results in the increase of the flame front surface and, correspondingly, in the increase of the burning rate, which is proportional to the relative increase of the flame surface. With accuracy $\delta / D << 1$, the flame velocity is:

$$U_{fL} = \Theta U_f \cdot \exp\{\alpha U_f t / D\}, \tag{12}$$

where α is numerical factor of the order of unity. The accelerating flame acts as a piston producing compression waves in the flow ahead. The time and the coordinate where the compression wave steepens into a shock wave are determined by the condition that the Riemann solution for a compression wave becomes a multi-valued function. During the initial stage of the exponential increase of the flame velocity, the shock waves are formed at the coordinate $x = X_{sh}$ far ahead of the flame front (located at $x = X_f$) at the distance about $X_{sh} - X_f \approx (5 \div 7)D$ from the flame front [34, 35]. Then, thickness of the boundary layer can be estimated as $\delta_1 \sim (X_{sh} - X_f) / \sqrt{Re}$, where $Re = (\Theta - 1)U_f D / \nu \sim \Theta U_f D / \nu$ is the Reynolds number in the upstream flow. Theoretical estimate for the boundary layer thickness gives $\delta \approx (0.3 \div 0.4)\,mm$. From the numerical calculations [35, 36] the boundary layer thickness is about 0.4mm for D=10mm.

As the edge of the flame front is stretched along the wall within a boundary layer, a narrow fold is formed between the edge of the flame front skirt and the wall. As the fold becomes deeper, the angle at the fold's tip became smaller, some part of the flame front near the fold's tip, approaching the wall, is quenched due to the heat loss at the wall. Simple geometric consideration [34] shows that the corresponding time of the flame skirts quenching is $t_2 \approx D\delta / 2\Theta L_f U_f$. This reduces the total surface of the flame front area and thus decreases the rate of the flame acceleration. During the first stage, within the time interval t_2, the flame acceleration is constant with accuracy of the first order terms of series expansion in $L_f / D << 1$, which means that during the next stage, $t > t_2$, the flame velocity-time dependence can be approximated as

$$U_{fL} \simeq \Theta U_f \left[1 + \beta (t / \tau_f)^n \right], \; 0 < n < 1 \tag{13}$$

It can be shown [34] that for a piston moving with the velocity-time dependence (13) the Riemann solution for a simple travelling wave is multi-valued everywhere for any values of $0<n<1$. Therefore, during the second stage, $t > t_2$, the compression wave produced by the flame steepens into the shock directly on the surface of the flame front. The specific value of the exponent in (12) does not matter since we are interested only in the location where the shock wave is formed.

Contrary to a stationary flame, the flow with the accelerating flame is not isobaric. In the latter case pressure is growing at about the same rate as the flame velocity. From the time when the compression waves steepen into the shock close to the flame front, the unreacted mixture of considerably higher density compressed and heated in the shock starts entering the flame front and produces a narrow pressure peak on the scale of the flame width. While during the first stage the flame acceleration is related to the stretching of the flame front within a boundary layer, during the second stage subsequent acceleration of the flame is due to its coupling with the shock wave formed at the flame front. Consequently, there are two feedback mechanisms leading to the increase of the flame speed. One is driven by the increased temperature, and hence reactivity, of the mixture due to the shock, and the other by the increased density and hence amount of reacting fuel entering the flame front. A higher flame speed creates a higher gas velocity ahead of the flame. After the flame was accelerated up to the local sound speed the shocks and the pressure peaks do not run away from the reaction zone, where they were formed, and the pressure peak is localized directly in the reaction zone. Consequently, the amplitude of the pressure peaks at the flame front continues growing due to the combustion of larger amount of unburned mixture entering the reaction zone during this stage. Eventually, the pressure peak becomes strong enough to affect the reaction rate. The increase of pressure enhances reaction rate and the increased heat release in the reaction zone creates a positive feedback coupling between the pressure pulse and the heat released in the reaction. It results in more violent increase of the pressure peak, which finally steepens into the shock strong enough for formation of a detonation wave. Duration of the last stage of actual transition to detonation can be estimated as $L_f / a_s \approx 6\mu s$, which is consistent with the results of numerical simulations.

5.2 Numerical Simulation of Flame Acceleration and DDT

Direct numerical simulations of the flame propagating in channels with non-slip walls filled with the stoichiometric hydrogen-oxygen mixture at initial temperature T_0=298K and initial pressure P_0=1atm were performed for 2D channels of different widths from 1mm to 10mm [34-36] and for 3D rectangular channel with cress section 10⊗10mm [37] with the minimum computational cell size $\Delta = 0.005\,\text{mm}$, which ensure high resolution of the flame front, which width is 0.26mm, and convergence of the solutions. Qualitatively results of 2D simulations and 3D simulations are similar. The grid system in the 3D simulations was built out from 50 million up to 400 million grid points with the minimum computational cell size $\Delta \approx 0.021\,\text{mm}$, which resolves flame front using 12 computational cells. A laminar planar flame was initiated near the left closed end and propagates to the right open end of the channel. The computations solved the multidimensional, time-dependent, reactive Navier-Stokes equations including the effects of compressible gas convection, molecular diffusion, thermal conduction, viscosity and different detailed chemical kinetics schemes for the reactive species H_2, O_2, H, O, OH, H_2O, H_2O_2, and HO_2.

Figure 12 shows the overall picture of the 3D flow, the flame front evolution and the transition to detonation. Cross sections of the 3D images in the plane (x, y, z=5mm) and the streamlines are shown at the right part of Fig. 12. To analyze the influence of flame surface topology, the additional small perturbations were imposed in the beginning of the flame propagation. They cause formation of the flame brush, which shortly develops in many well pronounced bulged tips. The compression waves produced by the accelerating flame are seen as slightly corrugated surfaces ahead the flame front. They steepened into the shock waves far ahead of the flame not shown in the frames. It is seen that the flow ahead of the flame remains almost uniform in the bulk of the channel. The boundary layer is seen as lighter white strip on the bottom and back walls of the channel. It should be emphasized that the 2D images representing by the cross sections in the plane (x, y) do not resemble the experimental schlieren photographs, but rather similar to those obtained in 2D simulations. The shape of the flame propagating in the channel depends on the small perturbations imposed in the beginning of the process.

Shortly before the transition to detonation the shock waves are formed close to the leading bulged tip of the flame and practically "sits" at the flame tip (frame t=487μs). Then, after 8μs the flame turns in a detonation wave. The pre-detonation time estimated as time when the pressure reaches its maximum, and when the strong shock arises on the flame tip, overcomes the sonic barrier and forms a detonation behind its front, is 0.55 ms for the 3D channel and 1.35ms for the 2D channel. It should be noted also, that temperature ahead of the flame front does not exceed 500-600 K all the time till the moment of the actual transition to detonation, so that the rates of chemical reactions ahead of the flame are negligibly small, and temperature non-uniformities if any ahead of the flame can not produce a detonation through the Zeldovich gradient mechanism.

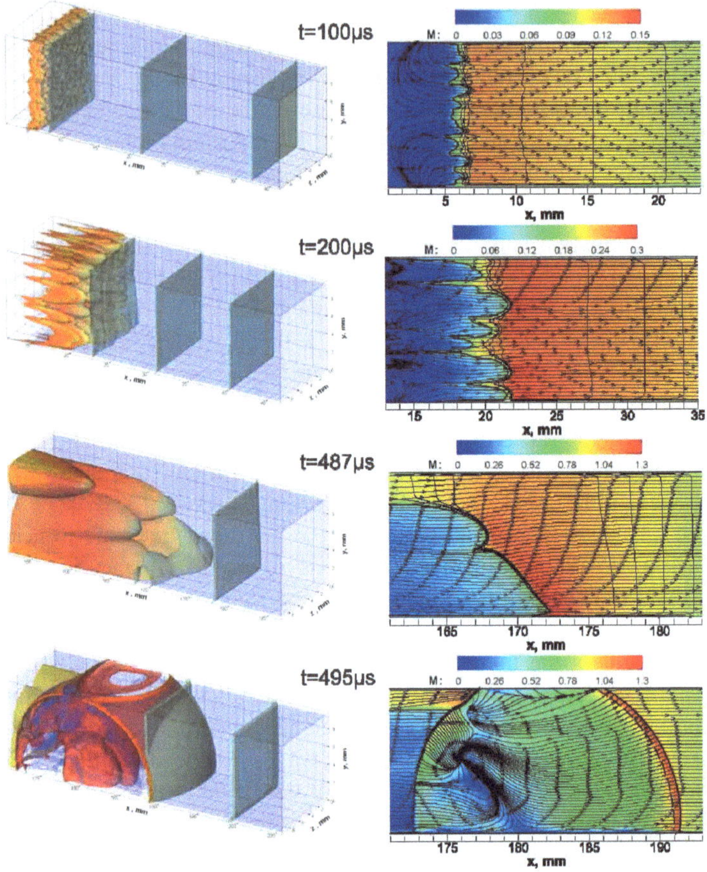

Fig. 12. The flame structure and compression (shock) waves (green surfaces) at sequential times 100, 200, 487 and 495μs (left) and the corresponding 2D cross sections of 3D images in the plane (x, y, z=5mm) at the right part.

The temporal evolution of the flame speed and the pressure peak at the leading flame tip throughout the process of the flame acceleration and DDT computed for 3D channel is shown in Fig. 13(a). The exponential increase of the flame velocities is shown by dashed lines for 2D and 3D channels in Fig. 13(b), while subsequent velocity time dependence can be approximated as $U_{fL} \approx \Theta U_{f0}\left[1+\beta(t/\tau_f)^n\right]$ with $0 < n < 1$. The computations confirmed the theoretical prediction: after the end of the exponential stage, a larger amount of the fresh fuel starts entering the flame front, which results in the development of a narrow pressure peak on the scale of the flame width, which are shown in Fig. 14(a, b) for 2D and 3D channels. Consequently, the amplitude of the pressure peaks at the flame front continues growing due to the combustion of larger amount of compressed mixture entering the reaction zone during this stage.

For both 2D and 3D cases the velocity-time dependence plots demonstrate the same feature of several distinctive stages of the flame acceleration: initial stage of flame expansion out from the ignition zone; the stage of exponential increase of the flame velocity; the stage when the rate of the flame acceleration decreases compared with the previous exponential stage; the final stage of sharp increase of the flame velocity and actual transition to detonation. For 3D case the run-up distance and duration of the flame propagating before the transition to detonation are about three times shorter compared to 2D channel of the same width. The pre-detonation time estimated as time when the pressure reaches its maximum, and when the strong shock arises on the flame tip, overcomes the sonic barrier and forms a detonation behind its front, is 0.55 ms for the 3D channel and 1.35ms for the 2D channel.

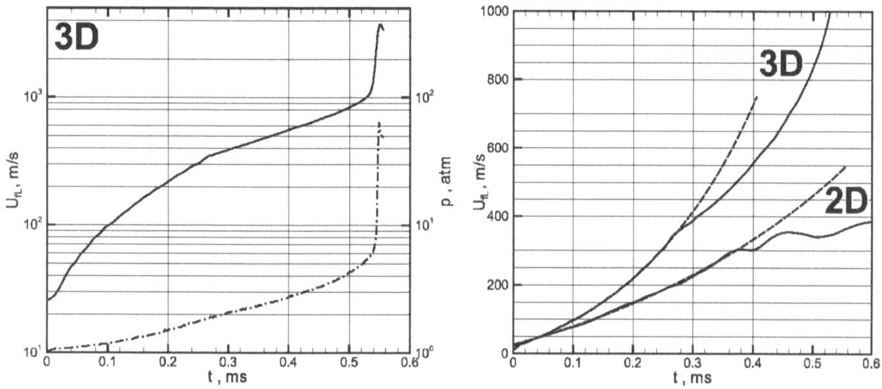

Fig. 13. a) Flame velocity (solid lines) and the pressure peak (dash-dotted lines) time dependences computed for 3D channel; b) The initial stage computed for 3D and 2D channels of width D=10mm (right).

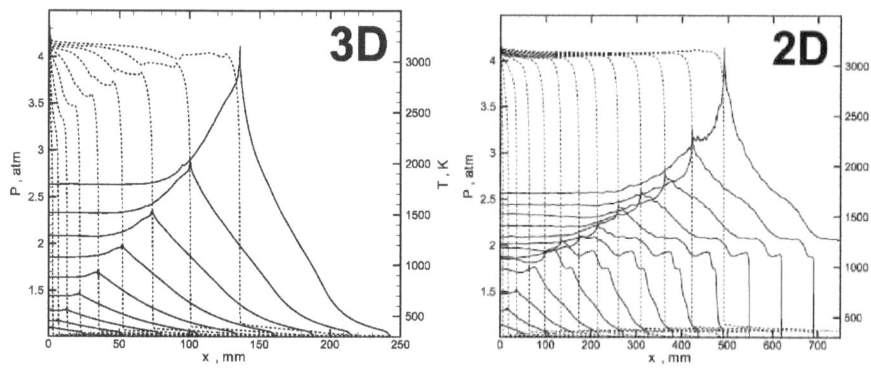

Fig. 14. Temperature (dashed lines) and pressure (solid lines) profiles corresponding to leading point of the flame front represent the flame structure and the pressure peak formation from t_0=50μs, Δt=50μs, t_f=500μs.

To capture the features of flame acceleration and the origins of detonation one needs accurately resolve both time and space scales, which are far beyond the limits of the nowadays temporary resolution of the cameras used in DDT experiments. In most cases one can recognize DDT using only two sequential snapshots. The first one shows the flow pattern in the vicinity of the flame front, and the second one shows already the detonation wave propagating at the background of the previous flow pattern. On the contrary, the available time and space resolution of the process using computer modeling is much higher and can bring more detailed information about the DDT origin. Besides, 3D computer simulations provide us with 3D images of the process not available with the schlieren technique, which in any case provides us only with 2D projection of the three dimensional flow, until holographic photos are not available for combustion experimental technique. As an example, Fig. 15(a) shows computed shadow photos for the last stage of the transition to detonation shown in Fig.12. The frames (a, b, c) corresponding to time instants 487μs, 492μs and 495μs show the detonation formation. The frames (d, e) represent the same phase of the process obtained experimentally [33] with the time resolution $\Delta t = 100\mu s$. The frame (d) corresponds to the frame (a), while the frame (e) shows already developed detonation wave as in the frame (c). The information about how the flame was transformed into the detonation is lost and cannot be obtained at the experimental shadowgraphs. On the other hand the qualitative similarity of the flow patterns confirms that the process in natural experiment evolves via the mechanism similar to the described above.

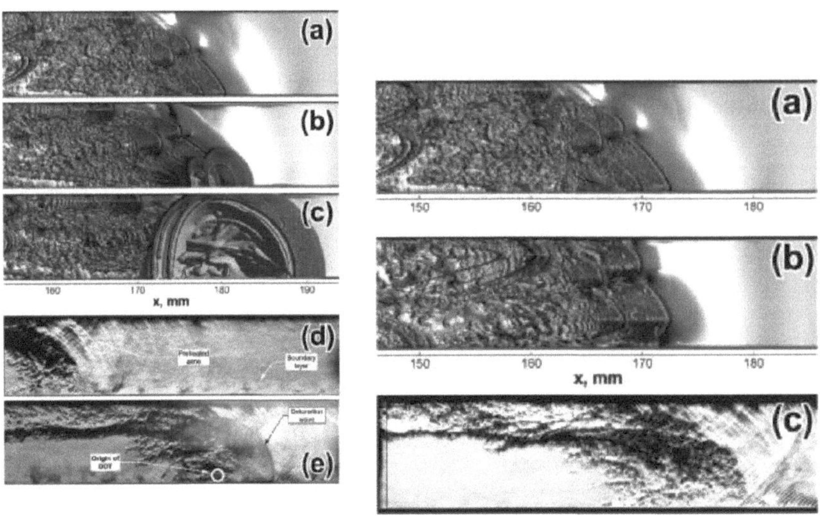

Fig. 15. a) Computational shadow photos for the last stage of DDT at: a) 487μs, b) 492μs, c) 495μs. Frames d) and e) are the experimental shadow photos with time resolution Δt=100μs [33]; b) (a) and ((b) are numerically obtained shadow photos in the XY and XZ cross sections at t=487μs, (c) – experimental photo [33].

Fig. 15b shows numerically obtained shadow photos using 3D simulations in the XY and XZ cross sections on the frames (a) and (b) correspondingly at 487μs, the frame (c) is an experimentally obtained shadow photo [33]. The shadowgraphs look like a turbulent one which is usually named "turbulent flame brush". However, as it can be observed from the 3D images of Fig. 12 either flame or flow are not turbulent until the detonation formation. The "turbulent" pattern here is a result of interference of light beams refracted on the different pieces of cellular flame surface. This issue shows us that the cellular laminar flame can be visualized and interpreted as a turbulent and it may cause misinterpretation of the studying phenomena.

6 Conclusion

This lecture aimed to communicate two messages regarding fundamentals of unsteady combustion processes. First, it should be noted that the understanding of combustion phenomena related to unsteady processes obtained on the basis of a one-step chemical model can be incorrect even qualitatively and therefore may require revision on the basis of a detailed chemical model. The second is the complex nature of chemical reaction mechanisms and its response to the various pressure-dependent chain mechanisms. Consequently efforts towards combustion modeling using detailed reaction mechanisms should be conducted with caution, especially in the case of unsteady processes, when wide variations in pressure, temperature, and composition are embedded in the phenomena under study. It is further noted that although most of the illustrations presented herein were based on hydrogen-oxygen, the same concepts carry over to hydrocarbon oxidation because their mechanisms are similarly affected by chain reactions, and because they must necessarily contain the hydrogen oxidation mechanism as a key component. Finally, it should be noted that the difference in induction time given by different chemical schemes for the low temperature region has little effect on the calculated flame speed at the normal or low pressures. However, at high pressures, when three body collisions become important, the induction time is shifted to lower temperatures and more reliable schemes are those which give closer to the experimental values of the induction time at low temperatures.

Acknowledgment. The computations were performed on resources provided by the Swedish National Infrastructure for Computing (SNIC) at the Center for Parallel Computers at the Royal Institute of Technology in Stockholm and the National Supercomputer Centers in Linkoping. The author is grateful to A. Kiverin, M. Ivanov, and I. Yakovenko for many fruitful discussions.

References

1. Maas, U., Warnatz, J.: Ignition Processes in Hydrogen-Oxygen Mixture. Combust. Flame 74, 53–69 (1988)
2. Sloane, T.M., Ronney, P.D.: A Comparison of Ignition Phenomena Modeled with Detailed and Simplified Kinetics. Combust. Sci. and Tech. 88, 1–13 (1993)

3. Konnov, A.: Remaining uncertainties in the kinetic mechanism of hydrogen combustion. Combust. Flame 152, 507–528 (2008)
4. Conaire, M.O., Curran, H.J., Simmie, J.M., Pitz, W.J., Westbrook, C.K.: A comprehensive modeling study of hydrogen oxidation. Int. Journal of Chemical Kinetics 36, 603–622 (2004)
5. Konnov, A.A.: Refinement of the Kinetic Mechanism of Hydrogen Combustion. Khimicheskaya Fizika 23, 5–18 (2004)
6. Warnatz, J., Maas, U., Dibble, R.W.: Combustion. Physical and chemical fundamentals, modeling and simulations, experiments, pollutant formation. Springer, Berlin (2006)
7. Kusharin, A.Y., Agafonov, G.L., Popov, O.E., Gelfand, B.E.: Detonability of H2/CO/C02/Air Mixtures. Combust. Sci. and Tech. 135, 1–6 (1998)
8. Kusharin, A.Y., Popov, O.E., Agafonov, G.L.: Burning velocities of oxygen-hydrogen mixtures with steam. Chem. Phys. Rep. 14, 584–594 (1995)
9. Smith, G.P., Golden, D.M., Frenklach, M., Moriarty, N.W., Eiteneer, B., Goldenberg, M., et al.: Gri-mech 3.0 (2002), `http://www.me.berkeley.edu/grimech`
10. Hirschfelder, J.O., Curtiss, C.F., Bird, R.B.: Molecular theory of gases and liquids. Wiley, New York (1964)
11. Hairer, E., Wanner, G.: Solving ordinary differential equations. Stiff and differential-algebraic problems. Springer, New York (1996)
12. McBride, B.J., Gordon, S., Reno, M.A.: Coefficients for calculating thermodynamic and transport properties of individual species. NASA Technical Memorandum 4513 (1993)
13. Belotserkovsky, O.M., Davydov Yu, M.: Coarse-particle method in hydrodynamics. Publ. Inc. Nauka, Mir, Moscow (1982) (in Russian)
14. Liberman, M.A., Ivanov, M.F., Valuev, D.M., Eriksson, L.-E.: Numerical modeling of knocking onset and hot spot formation by propagating flame in SI engines. Combust. Sci. and Tech. 178, 1613–1647 (2006)
15. McBride, B.J., Gordon, S., Reno, M.A.: NASA Technical Memorandum 4513 (1993)
16. Hermanns, R.T.E., Konnov, A.A., et al.: Energy & Fuels 21, 1977–1981 (2007)
17. Dowdy, D.R., Smith, D.B., et al.: Proc. Combust. Inst. 23, 325–332 (1991)
18. Grumer, J., Cook, E.B., Kubala, T.A.: Combust. Flame 3, 437–446 (1995)
19. Iijima, T., Takeno, T.: Combust. Flame 65, 35–43 (1986)
20. Snyder, A.D., Robertson, J., Zanders, D.L., Skinner, G.B.: Shock Tube Studies of Fuel-Air Ignition Characteristics, Technical Report AFAPL-TR-65-93 (1965)
21. Bhaskaran, K.A., Gupta, M.C., Just, T.H.: Combust. Flame 21, 45–48 (1973)
22. Slack, M., Grillo, A.: Investigation of Hydrogen-Air Ignition Sensitized by Nitric Oxide and by Nitrogen Dioxide, Grumman Research Department Report RE-537 (1977)
23. Schultz, E., Shepherd, J.: Validation of Detailed Reaction Mechanisms for Detonation Simulation, Report No. FM99-5, Graduate Aeronautical Laboratories (2000)
24. Zeldovich, Y.B., Barenblatt, G.I., Librovich, V.B., Makhviladze, G.M.: The Mathematical Theory of Combustion and Explosion. Plenum, New York (1985)
25. Zeldovich, Y.B.: Regime classification of an exothermic reaction with nonuniform initial conditions. Combust. Flame, 39, 211–226 (1980)
26. Zel'dovich, Y.B., Librovich, V.B., Makhviladze, G.M., Sivashinsky, G.I.: On the development of detonation in a non-uniformly preheated gas. Astronautica Acta 15, 313–321 (1970)
27. Liberman, M.A., Kiverin, A.D., Ivanov, M.F.: On Detonation Initiation by a Temperature Gradient for a Detailed Chemical Reaction Models. Phys. Letters, A 375, 1803–1808 (2011)

28. Liberman, M.A., Kiverin, A.D., Ivanov, M.F.: Regimes of chemical reaction waves initiated by nonuniform initial conditions for detailed chemical reaction models. Phys. Rev. E. 85, 056312/1-16 (2012)
29. Kiverin, A.D., Kassoy, D.R., Ivanov, M.F., Liberman, M.A.: Mechanisms of Ignition by Transient Energy Deposition: Regimes of Combustion Waves Propagation. Phys. Rev. E. 87, 033015/1-10 (2013)
30. Zeldovich Ya. B. and Raizer Yu. P., Physics of Shock waves and High-Temperature Hydrodynamic Phenonema. Academic Press, New-York (1966)
31. Oran, E.S., Gamezo, V.N.: Origins of the deflagration-to-detonation transition in gas-phase combustion. Combustion and Flame, 148, 4–47 (2007)
32. Kuznetsov, M., Liberman, M., Matsukov, I.: Experimental study of the preheat zone formation and deflagration-to-detonation transition. Combust. Sci. Technol. 182, 1628–1644 (2010)
33. Liberman, M.A., Ivanov, M.F., Kiverin, A.D., Kuznetsov, M.S., Chukalovsky, A.A., Rakhimova, T.V.: Deflagration-to-detonation transition in highly reactive combustible mixtures. Acta Astronautica 67, 688–701 (2010)
34. Ivanov, M.F., Kiverin, A.D., Liberman, M.A.: Flame Acceleration and Deflagration-to-Detonation Transition in Stoichiometric Hydrogen/Oxygen in Tubes of Different Diameters. Int. J. Hydrogen Energy 36, 7714–7728 (2011)
35. Ivanov, M.F., Kiverin, A.D., Liberman, M.A.: Hydrogen-oxygen flame acceleration and transition to detonation in channels with no-slip walls for a detailed chemical reaction model. Phys. Rev. E83, 56313–56316 (2011)
36. Ivanov, M.F., Kiverin, A.D., Yakovenko, I.S., Liberman, M.A.: Hydrogen-oxygen flame acceleration and deflagration-to-detonation transition in three-dimensional rectangular channels with no-slip walls. Int. J. Hydrogen Energy 38, 16427–16440 (2013)

Numerical Investigation of Reactive and Non-reactive Richtmyer-Meshkov Instabilities

Sergio Bengoechea, Lewin Stein, Julius Reiss, and Jörn Sesterhenn

Institut für Strömungsmechanik und Technische Akustik ISTA,
FG Numerische Fluiddynamik, Technische Universität Berlin,
Müller-Breslau-Straße 8, 10623 Berlin, Germany
{bengoech,stein,reiss}@tnt.tu-berlin.de, joern.sesterhenn@tu-berlin.de

Abstract. The Richtmyer-Meshkov instability (RMI) arises from the interaction of a shock wave with a density gradient in a fluid. The density gradient can be caused by temperature and/or species concentration. The RMI leads to strong mixing and in the case of reactive flows an existing flame will be accelerated. Such acceleration processes can lead to a transition to detonation. This work presents first results in the simulation of these instabilities driven by a shock induced acceleration. It examines the case of single-mode interface perturbations and compares its growth and behaviour for reactive and non-reactive flows. The chemistry is described by a multiple-species one-step Arrhenius based kinetics model. The skew-symmetric finite difference formulation of the Navier-Stokes equations is used to simulate the flow. Ideal gases of stoichiometric premixed methane-air, i.e. $CH_4+2O_2+7.52N_2$, are considered. The starting point is the validation of the physical model. Subsequently the role of the RMI in the transition to detonation is analysed by testing the influence of the shock Mach number on the onset of detonations.

Keywords: Richtmyer-Meshkov instability, Deflagration-to-detonation transition, Methane-air, One-step kinetics model.

1 Introduction

The interface between two fluids or fluid states is a highly ordered and unstable structure. Any infinitesimal perturbation on the interface will be amplified in case of an external force or acceleration. The initial perturbation undergoes a linear growth followed by a highly non-linear one before developing into a turbulent regime. Due to this growth, shear layers develop and multiple Kelvin-Helmholtz instabilities (KHI) appear as secondary instabilities. The non-linear effects start to dominate the evolution and finally a strong turbulent mixing results [10].

The density gradient, which initially drives the instability, can be produced by a temperature gradient and/or from different species mixture. Such a density gradient is inherent to reactive flows, where the reaction strongly heats the fluid and changes its composition.

R. King (ed.), *Active Flow and Combustion Control 2014*, Notes on Numerical Fluid Mechanics and Multidisciplinary Design 127, DOI: 10.1007/978-3-319-11967-0_21

The acceleration can be generated by sound or shock waves. This momentarily very strong acceleration triggers the RMI. Body forces like gravity induce a continuous acceleration on the density interface and are classified as Rayleig-Taylor instabilities (RTI). The strength of the acceleration scales with the strength of the shock waves, therefore for strong shocks, strong mixing is expected. The different speeds of sound between the different phases give rise to further shock reflection and refraction, which further influences the sensitive reaction process.

Depending on the initial conditions, the propagation of reactive zones can be fundamentally different. Roughly, two important states are distinguished: deflagration and detonation. In the isobaric combustion (deflagration) a reactive fluid is preheated in the preheating-zone by an existing flame through heat fluxes. Through the increase in temperature the reaction starts in a typically small reaction zone with a strong heat release. The completed reaction sets the temperature at the adiabatic flame temperature, which in turn supports the preheating-zone. The flame speed is small in laminar flames and might be accelerated by an increased mixing rate due to turbulent mixing. The flame acceleration results from the fact that more volume per unit time reaches the reaction temperature. The pressure gradient over the flame is small compared with the change in internal energy, the reaction is isobaric [15]. The detonation in contrast is a shock wave, which is followed by the reaction zone. The shock wave heats the fresh gases so that the reaction is initiated, and in turn the gas expansion due to combustion sustains the shock wave. The combustion process is close to isochoric [14].

The transition from deflagration to detonation (DDT) is a complex process. Usually the flame itself is not transformed into a detonation wave, but the flame acceleration, shock waves and local turbulent mixing set the conditions which are suitable to create a detonation [12].

The final aim of this research is to create a DDT in controlled and reproducible manner for the use in Pulse Detonation Engines (PDE) and the RMI is a generic configuration to create DDT [12]. The RMI can lead to a detonation either by a strong mixing process, which results into a flame acceleration, or by the influence of a shock wave. Strong shocks are needed to create a DDT in such a configuration, which are not expected in PDEs, nevertheless it will help to understand the process and to examine model assumptions.

1.1 The Richtmyer-Meshkov Instability

A shock wave crossing through a perturbed density interface generates a RMI. The initial perturbations on the interface will be enhanced by the refraction-reflection effects of the wave, forming a wide varity of fluid motions. The left side of Figure 1 shows a basic schematic of the initial equilibrium conditions, where the three initial states are marked with the roman numbers I, II, III. After the refraction of the shock wave a distorted shock wave is transmitted into fluid III, a rarefaction wave is reflected in fluid I and a contact discontinuity moving slower in the transmitted shock direction is formed. Note that this early post-shock state mimics the shock tube problem [6]. As a result a counterclockwise vorticity $\overrightarrow{\omega} = \overrightarrow{\nabla} \times \overrightarrow{v}$ on one side of the perturbation and a clockwise on the other are generated. This unstable vortex pair leads the interface to its subsequent deformation. The dominating misalignment of pressure and density gradients, i.e. flat shock wave and density perturbation, is responsible for the baroclinic term $\frac{\vec{\nabla}\rho\times\vec{\nabla}p}{\rho^2}$ of the vorticity equation for the underlying mechanism of vorticity generation [7, 10]. In the RMI case the alternating vorticity sign mechanism according to the perturbation shape interface is depicted on the right side of Figure 1.

For early times and low initial amplitudes, the growth of the perturbation on the interface is a rapid-linear compressible regime. The created curls first invert the concave-convex initial perturbation. For later times and as the amplitude increases the system develops into a highly non-linear and approximately incompressible regime. The crest and trough start to evolve asymetrically, creating a funnel. At a certain point, the smaller scales become important, causing the roll-up of the funnel, and the characteristic jellyfish-like shape of the RMI materializes [10].

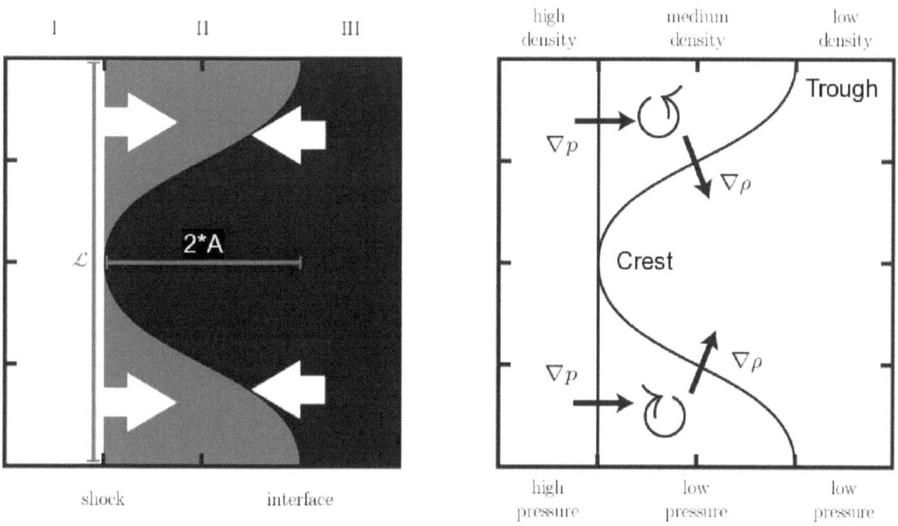

Fig. 1. Initial conditions (left) and mechanism (right) of a sinusoidally-perturbed RMI with amplitude A and wave number $k = 2\pi/\mathcal{L}$ [6]

Another physical viewpoint can be used to explain the instability evolution. The transmitted waves in the center-crest region are slightly diverging while the reflected waves tend to converge. For the lateral-trough region the reverse situation takes place. Therefore following the refraction of the incident shock wave, there exists in fluid II a positive pressure disturbance near the crest and a negative pressure disturbance near the trough, while in fluid III the opposite pressure disturbance sets in. This pressure distribution localized on the interface ensures the continuous increase of the density perturbation after the incident shock wave has passed [10].

The dynamic flows associated with a RMI are important in many physical enviroments. In astrophysics this kind of instability is related to supernova 1987A and to Type Ia explosions [9, 12] and it is also important in inertial confinement fusion (ICF) [9, 10]. In combustion systems the mixture relation between fuel and oxidizer can be improved with the use of a RMI [10]. Shock wave and flame interaction takes place in many combustion engines, either enhancing its performance or driving it to accidental explosions [11, 12], as engine knock [8, 12] or DDT [11].

This work is a preliminar study and investigates the application of a conservative, skew-symmetric finite difference scheme on the RMI. Further tests are therefore necessary.

This report is organized as follows: Section 2 outlines the physical model and parameters used in this study. The details of the numerical approach are described in Section 2.2. In Section 3 we describe the numerical results for the validation case of the steady laminar flame followed by the non-reactive and reactive RMI. The effects of the Mach number on the transition to detonation are included in this section as well. Finally, Section 4 presents the conclusions.

2 Physical Model

The compressible Navier-Stokes equations are used to describe the mass, momentum and energy conservation of the flow. This set of equations is complemented with N transport equations for the mass fraction of the N species, with $k = 1, ..., N$. Additional terms for the changes in mass fractions and internal energy need to be included in the transport and energy equations. These are respectively $\dot{\omega}_k$, the mass reaction rate for the k-th species and $\dot{\omega}_T$ the heat release due to combustion. The energy equation is not presented in conservative form but in skew-symmetric form. The advantage of this form is the correct treatment of the kinetic energy, the full scheme is conservative although the equation is not in divergence form [1].

$$\sqrt{\rho}\frac{\partial\sqrt{\rho}}{\partial t} + \frac{1}{2}\frac{\partial \rho u_i}{\partial x_i} = 0 \tag{1}$$

$$\sqrt{\rho}\frac{\partial\sqrt{\rho}u_i}{\partial t} + \frac{1}{2}[\frac{\partial \rho u_j u_i}{\partial x_j} + \rho u_j\frac{\partial u_i}{\partial x_j}] + \frac{\partial p}{\partial x_i} = \frac{\partial \tau_{ij}}{\partial x_j} \tag{2}$$

$$\frac{\partial \rho e}{\partial t} + \frac{\partial\left(u_i\rho(e+\frac{p}{\rho})\right)}{\partial x_i} - u_i\frac{\partial p}{\partial x_i} = \frac{\partial u_i\tau_{ij}}{\partial x_j} - u_j\frac{\partial \tau_{ji}}{\partial x_i} + \frac{\partial \Phi_i}{\partial x_j} + \dot{\omega}_T \tag{3}$$

$$\frac{\partial \rho Y_k}{\partial t} + \frac{\partial \rho u_i Y_k}{\partial x_i} = \frac{\partial}{\partial x_i}(\rho D_k\frac{\partial Y_k}{\partial x_i}) + \dot{\omega}_k, \tag{4}$$

where ρ is the density and u_i is the i-th velocity component. The internal energy of the ideal gas $e = p/(\rho(\gamma - 1))$ is used, with the further assumption of a constant adiabatic exponent γ. This neglects the mixture and the temperature dependence of the heat capacities. The first approximation is motivated by the dominance of nitrogen in air. The second is a more severe approximation, as the considered temperature range is large, however it is a frequent assumption in many theoretical procedures [11–13, 15]. This first study adopt them since these assumptions reduce the computational effort. In general the pressure p has to be determined numerically from the internal energy. Y_k is the mass fraction for the k-th species. The diffusion is described by Fick's law, with D_k as the mass diffusion coefficient for the k-th species. The friction and heat conduction are

$$\tau_{ij} = \mu(\frac{\partial u_i}{\partial x_j} + \frac{\partial u_j}{\partial x_i}) + \delta_{ij}(\mu_d - \frac{2}{3}\mu)\frac{\partial u_r}{\partial x_r} \tag{5}$$

$$\Phi_i = -\lambda\frac{\partial T}{\partial x_i} \tag{6}$$

μ represents the shear viscosity and μ_d the bulk viscosity, δ_{ij} is the Kronecker symbol.

To implement the viscosity temperature dependence in gases the shear viscosity μ is calculated with the Sutherland law [18]. The bulk viscosity is considered to be negligible. The Lewis number is equal to one in this model as in [11–13]. The mass diffusion coefficient D_k for the k-th species is calculated by

$$D_k = \frac{\mu}{\rho Pr Le} \tag{7}$$

with a constant Prandtl number $Pr = \frac{\mu C_p}{\lambda}$. The thermal conduction coefficient λ is calculated from the set of parameters Le, Pr, μ and the specific heat capacity at constant pressure $C_p = \frac{\gamma R_{spec}}{\gamma - 1}$. The specific gas constant is defined as $R_{spec} = \frac{R}{W}$, with the gas constant R and the molar mass of the gas mixture W.

The unusual form of the momentum equation and the non-standard choice of the variable $\sqrt{\rho}$ is the so called skew-symmetric form, [1–3]. It allows to have not only perfect conservation of mass, momentum and total energy (included chemical) in a high-order finite difference code but also to avoid numerical dissipation by the scheme. Dissipation is introduced only by the friction terms and by an explicitly added dissipation due to an adaptive high-order shock-capturing filter. The filter proposed in [5] is used for this purpose.

2.1 Reaction Mechanism

The global reaction

$$CH_4 + 2O_2 + 7.52N_2 \rightarrow CO_2 + 2H_2O + 7.52N_2 \tag{8}$$

is modelled by a one-step irreversible reaction. The change rates for the mass fractions are

$$\dot{\omega}_k = W_k \nu_k K_f \prod_{l=1}^{N} \left(\frac{\rho Y_l}{W_l}\right)^{\nu_l^{'}} \tag{9}$$

where W_k is the molar mass of species k, $\nu_k = \nu_k^{''} - \nu_k^{'}$ where $\nu_k^{''}$ and $\nu_k^{'}$ are the molar stoichiometric coefficients of species k. The reaction rate K_f is described by the Arrhenius law of the form

$$K_f = A_f T^{\beta} e^{-\frac{T_a}{T}}, [15] \tag{10}$$

The heat release by the reaction is

$$\dot{\omega}_T = -\sum_{k=1}^{N} \Delta h_{f,k}^{0} \dot{\omega}_k = -Q\dot{\omega}_{fuel} = -Q\dot{\omega}_{CH_4}, \tag{11}$$

where $\Delta h_{f,k}^{0}$ is the mass enthalpy of formation of species k at the reference temperature and Q the heat release per unit mass of fuel [15, 16]. The exponents $\nu_k^{'}$ of (9) correspond to the reaction orders [8]. These are assumed to be equal to one for all k species, since nonunity reaction orders add numerical stiffness to the computation [16]. The preexponential factor A_f and the activation temperature T_a are set to match the laminar flame speed and thickness of the methane-air combustion to those reported in [13, 15, 16]. The leading temperature T^{β} dependence is weak compared with the exponential. For this reason, $\beta = 0$ is considered in the following.

It is important to note that the chosen chemical reaction model needs to reproduce the chemical kinetics accurately enough while still allowing the computationally of the experiment.

2.2 Numerical Approach

Equations (1-4) are implemented as a finite differences code by discretizing all derivatives by explicit central differences of 6-th order. At the boundaries, discretisations with the Summation By Parts (SBP) property of [19] are used, see [1] for details. The implentation is in Fortran with MPI parallelisation by using a layer decomposition approach for the derivatives [4].

In x-direction, the inflow and outflow boundary conditions are set to non-reflecting. The reference state at the inflow is prescribed for every time step, while at the outflow it is taken from the next inner point. The boundaries in y- and z-direction are both periodic.

The time integration is performed with the 4-th order explicit Runge-Kutta method (RK4). For perfect conservation an implicit time integration needs to be used [2, 3]. The RK4 method is used in order to save computational time.

Three computations were executed, namely in $1-D$, $2-D$ and $3-D$ spaces. These correspond to the adiabatic premixed steady and unsteady flame validation, the effects of a higher Mach number on the transition to detonation and the reactive and non-reactive RMI, respectively. The details of the computations are listed in Table 1.

Table 1. Numerical parameters

	$1-D$	$2-D$	$3-D$	
x, y, z length	0.01	0.01, 0.01	0.03, 0.01, 0.01	m
n_x, n_y, n_z	1024	2048 x 2048	512 x 512 x 128	points
CFL-number	0.3	0.5	0.9	dimensionless
dt	$9.84 \cdot 10^{-10}$	$1.53 \cdot 10^{-9}$	$1.8 \cdot 10^{-8}$	s
Filter frequency	∞	1	15	time steps
Time integration method	Gauss 2-th	Runge-Kutta 4-th	Runge-Kutta 4-th	
Iteration tolerance	10^{-10}	-	-	

Note that for the $1-D$ case an implicit time method is used, which assures perfect conservation. The iteration tolerance in Table 1 corresponds to the terminal condition of the iterative solver, i.e. residual $\leq 10^{-10}$. This value was chosen based on the simulations reported in [3]. With this residual value the machine accuracy of the conserved quantities is reached, see [3]. On the other hand, in the $2-D$ and $3-D$ simulations no iterative solver is used, since an explicit time method is applied.

3 Numerical Results

This section presents the results of the three numerical investigations carried out in this study.

3.1 Laminar Premixed Flame

The laminar premixed flame is used as a basic test for the code. The reference steady solution is obtained by a shooting method and taken as initial condition for this case, see e.g. [13, 15] for a description. No filter was used.

Figure 2 shows the reference solution delivered by the shooting method and the result of the Navier-Stokes equations after 5.6 ms. The simulation and the steady state solution are in very good agreement. Note that the pressure presents a slightly deviation from the initial condition, due to the high sensitivity from the flame to pressure variations (isobaric flame). The adiabatic flame temperature is 2250 K, the laminar flame speed 0.4 m/s and the flame thickness $3 \cdot 10^{-4}$ m. These results coincide with those reported in [13, 15, 16]. On the other hand, Figure 3 depicts the laminar unsteady flame after 2.8 ms and 5.6 ms together with the initial steady state. In contrary to the steady flame simulation, in this case the coordinate system is fixed. The structure of the unsteady flame and its physical properties agree with the results obtained for the steady flame.

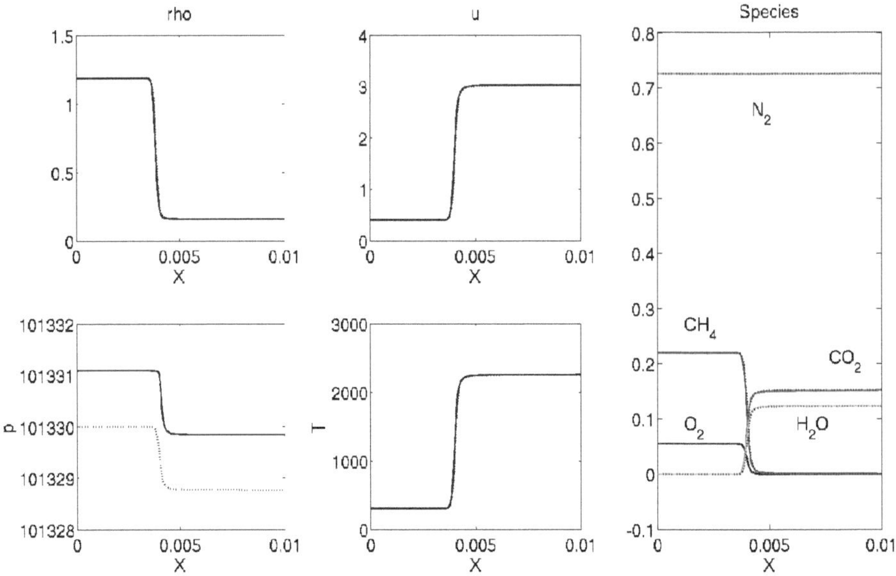

Fig. 2. Laminar steady flame results of the full Navier-Stokes equations after 5.6 ms (solid line). Initial steady state by the shooting method (dotted line).

Figure 4 depicts the change from temporal step n to $n+1$ of the mass, momentum and total energy in the form: $\int_x (\rho^{n+1} - \rho^n)dx$, $\int_x (\rho u^{n+1} - \rho u^n)dx$ and $\int_x (p^{n+1} - p^n)dx$ respectively. This figure proves the conservation of the scheme for the $1-D$ laminar steady and unsteady flames. The fluctuations till approximately $2 \cdot 10^6$ time steps are caused by the non-perfect initial solution.

A detailed list of the parameters applied in this approach is provided in Table 2, Appendix C.

3.2 Richtmyer-Meshkov Instability

The results delivered by the shooting method in the $1-D$ laminar flame are set as initial conditions in the density interface for the reactive RMI. For the non-reactive case the initial density discontinuity matches the reactive case. This was achieved by holding the pressure constant and increasing the temperature. A single-sinusoidal perturbation is applied at the density interface.

The incoming shock is characterized by the Rankine-Hugoniot condition, see [17], with a shock Mach number M_s with respect to state II equal to 1.5. The shock has velocity s in the positive x-direction and the density interface velocity is set to $-s + 0.45s$ into the negative x-direction. The coordinate system then moves with velocity $0.45s$ into the positive x-direction following the instability, in order to save computational domain.

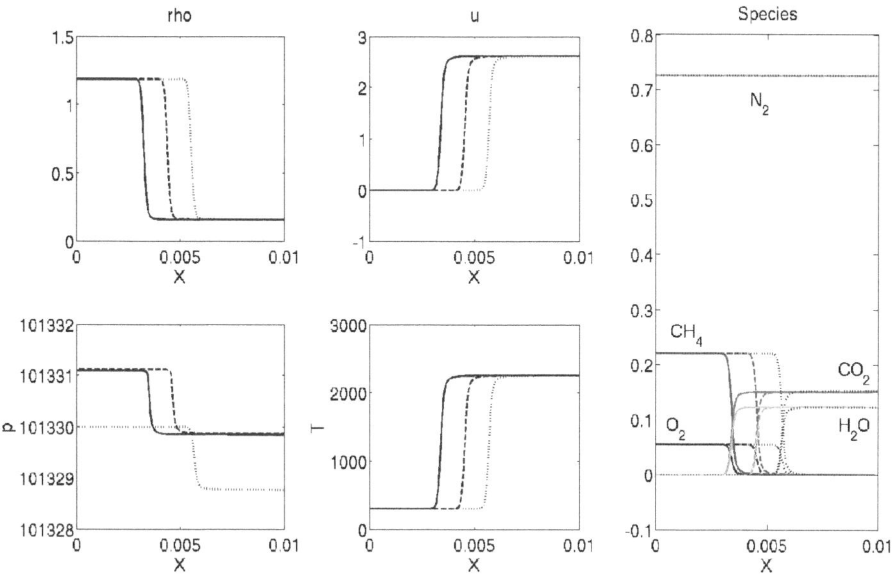

Fig. 3. Laminar unsteady flame results of the full Navier-Stokes equations after 2.8 *ms* (dashed line) and 5.6 *ms* (solid line). Initial steady state by the shooting method (dotted line).

The evolution of the RMI is represented in the Figures 7, 8, 9 and 10 in Appendix A. These figures show temperature, density, vorticity, heat release and fuel mass fraction Y_{CH_4} on the slice $z = 0.005\ m$. The time steps from 0 to 9000 with 1000 steps spacing are depicted.

The pressure-density misalignment triggers the vorticity mechanism, two big vortices arise and enhance the mixing process. Besides the two dominant and very stable vortices, no secondary instabilities are seen due to the low Reynolds number. The physical size of the system was chosen small to resolve the flame front accurately and the viscosity was adjusted intentionally to air viscosity since the flow is roughly composed by this material. This results in a diffusive regime and no small structures are observed.

The non-reactive and reactive RMI are analogous until the formation of the funnel. The non-reactive case presents a pure mixing process, whereas in the reactive case the fresh gas burns as it mixes. The latter case shows a more symmetric shape since its contours thin down. The combustion stabilizes the RMI while the flame surface and the heat release increases. The heat release was integrated over the whole domain and its relative change over time is depicted in Figure 5. In this figure one can observe a rise at the first part of the simulation, up to a point where it stays almost constant. It can be extracted that the temperature rise caused by the shock is not sufficient to reach the detonation threshold.

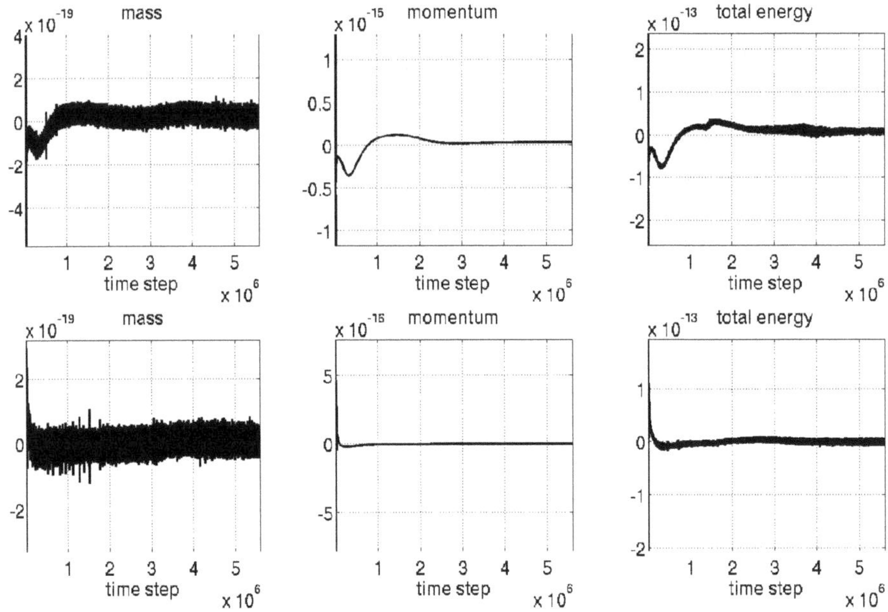

Fig. 4. Conservation magnitudes after 5.6 ms ($5.6 \cdot 10^6$ time steps). Upper row corresponds to the steady flame case, lower row to the unsteady flame case.

3.3 Effect of the Mach Number

This current section presents the results for the interaction of a higher shock Mach number $M_s = 2.75$ with the deflagration front described in Section 3.2. The study is restricted to short initial times and to local effects between shock and flame, thus the domain can be reduced. Making use of the periodicity, the initial perturbation is displaced π in y-direction.

In Figure 11, Appendix B, the temperature, pressure and heat release are plotted. In this figure one sees the outcome of secondary reflected waves due the interaction of the incoming shock with the flame front. The two reflected pressure waves converge at the center of the domain, where the pressure waves focus. The increase in the heat release is extremly high at the moment of the addition, as Figure 6 shows. This fact might indicate the initiation of a detonation, although the simulation could not be kept stable for further times due to the incapacity of the adaptive filter to handle the jump discontinuity. This point needs further investigation.

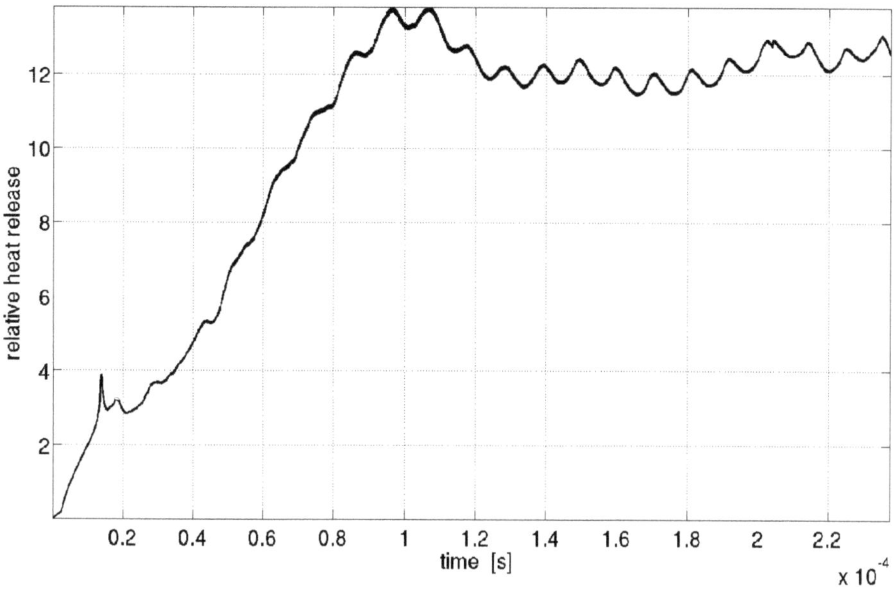

Fig. 5. Relative change of the heat release over time, $\frac{\partial \dot{\omega}_T}{\partial t}$

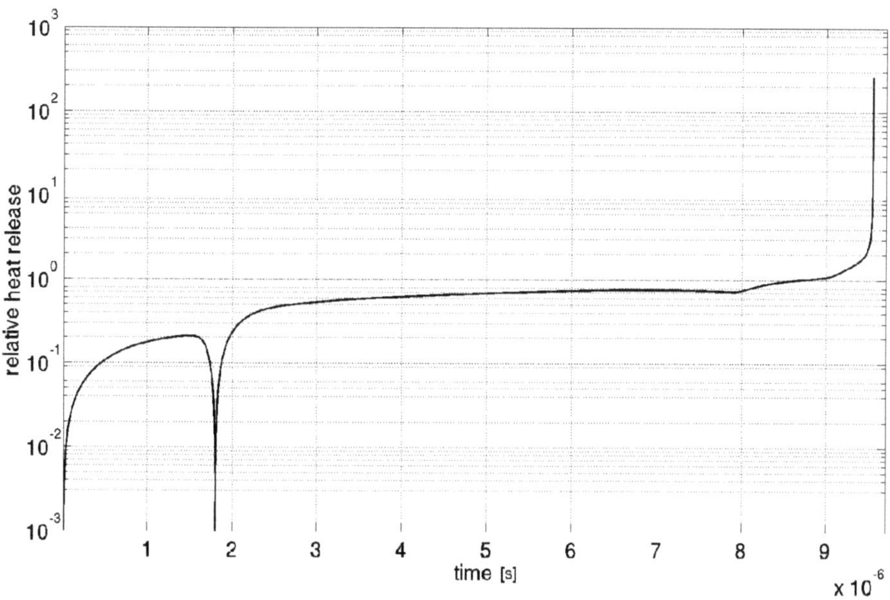

Fig. 6. Relative change of heat release over time $\frac{\partial \dot{\omega}_T}{\partial t}$

4 Conclusions

A detailed validation of the used model is important to create trustworthy simulations. In our case, the validation results are in very good agreement with the deflagration of methane-air reported in the literature. Qualitative results of the dynamics and the mixing process of the RMI are presented for reactive and non-reactive flows. The impact on the deflagration of the increase of the Mach number, which might potentially lead to a detonation was shown.

Acknowledgements. The authors gratefully acknowledge support by the Deutsche Forschungsgemeinschaft *DFG* as part of collaborative research center *SFB-1029* "Substantial efficiency increase in gas turbines through direct use of coupled unsteady combustion and flow dynamics". The computations were carried out at the *Leibniz-Rechenzentrum* under the ID-Project *pr83lu*.

References

1. Reiß, J., Sesterhenn, J.: A conservative, skew-symmetric Finite Difference Scheme for the compressible Navier-Stokes Equations. Accepted for Publication to Computers & Fluids (2014); see also (arXiv:1308.6672)
2. Brouwer, J., Reiß, J., Sesterhenn, J.: Fully conservative finite-difference schemes of arbitrary order for compressible flow. AIP Conference Proceedings 1479, 2290–2293 (2012)
3. Brouwer, J., Reiß, J., Sesterhenn, J.: Conservative time integrators of arbitrary order for skew-symmetric finite-difference discretizations of compressible flow. Computers & Fluids 100, 1–12 (2014)
4. Schulze, J.: Adjoint based jet-noise minimization. PhD thesis, Faculty V, Technische Universiät Berlin, Berlin (2011)
5. Bogey, C., de Cacqueray, N., Bailly, C.: A shock-capturing methodology based on adaptive spatial filtering for high-order non-linear computations. Journal of Computational Physics 228, 1447–1465 (2009)
6. Stein, L.: A skew-symmetric, conservative Finite Difference Scheme for the simulation of reactive Flow. Master's thesis, TU-Berlin (2013)
7. Nakai, S., Takabe, H.: Principles of inertial confinement fusion physics of implosion and the concept of inertial fusion energy. Reports on Progress in Physics 59, 1071–1131 (1996)
8. Warnatz, J., Maas, U., Dibble, R.W.: Combustion, 4th edn. Springer, Heidelberg (2006)
9. Nishihara, K., Wouchuk, J.G., Matsuoka, C., Ishizaki, R., Zhakhovsky, V.V.: Richtmyer Meshkov instability: theory of linear and nonlinear evolution. Philosophical Transactions of the Royal Society A 368, 1769–1807 (2010)
10. Brouillette, M.: The Richtmyer-Meshkov instability. Annual Review of Fluid Mechanics 34, 445–468 (2002)
11. Khokhlov, A.M., Oran, E.S., Chtchelkanova, A.Y., Wheeler, J.C.: Interaction of a shock with a sinusoidally perturbed flame. Combustion and Flame 117, 99–116 (1999)
12. Oran, E.S., Gamezo, V.N.: Origins of the deflagration-to-detonation transition in gas-phase combustion. Combustion and Flame 148, 4–47 (2007)

13. Kessler, D.A., Gamezo, V.S., Oran, E.S.: Simulations of flame acceleration and deflagration-to-detonation transitions in methaneair systems. Combustion and Flame 157, 2063–2077 (2010)
14. Lee, J.H.: The detonation phenomenon. Cambridge University Press (2008)
15. Poinsot, T., Veynante, D.: Theoretical and numerical combustion, 3rd edn. (2005)
16. Fernández-Tarrazo, E., Sánchez, A.L., Liñán, A., Williams, F.A.: A simple one-step chemistry model for partially premixed hydrocarbon combustion. Combustion and Flame 147, 32–38 (2006)
17. Bartlmä, F.: Gasdynamik der Verbrennung. Springer, Wien (1975)
18. White, F.M.: Viscous Fluid Flow, 2nd edn. McGraw-Hill, Inc. (1991)
19. Strand, B.: Summation by Parts for Finite Difference Apprximations for d/dx. Journal of Computational Physics 110, 47–67 (1994)
20. Baehr, H.D.: Thermodynamik. 7. Auflage (1989)
21. Chase, M.W.: NIST-JANAF Thermochemical Tables, 4th edn. (1998)

A Appendix

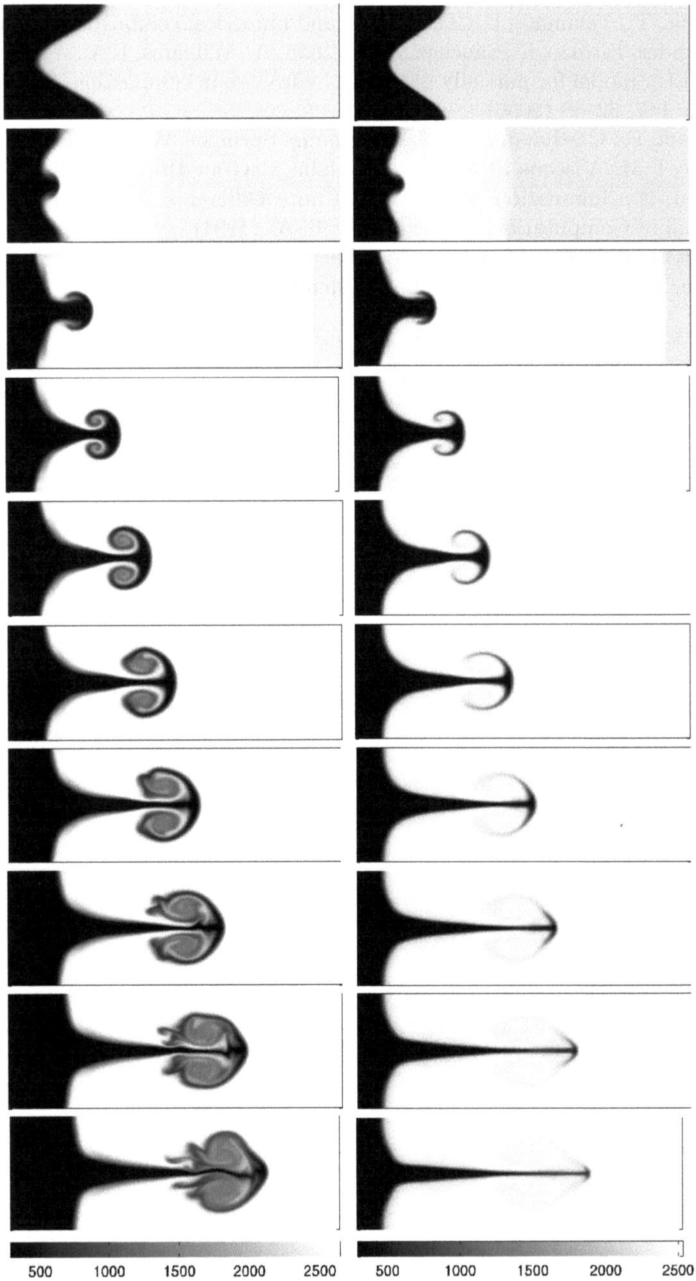

Fig. 7. Temperature for non-reactive (left) and reactive (right) case. Time steps 0, 1000, 2000, 3000, 4000, 5000, 6000, 7000, 8000, 9000 from top to bottom.

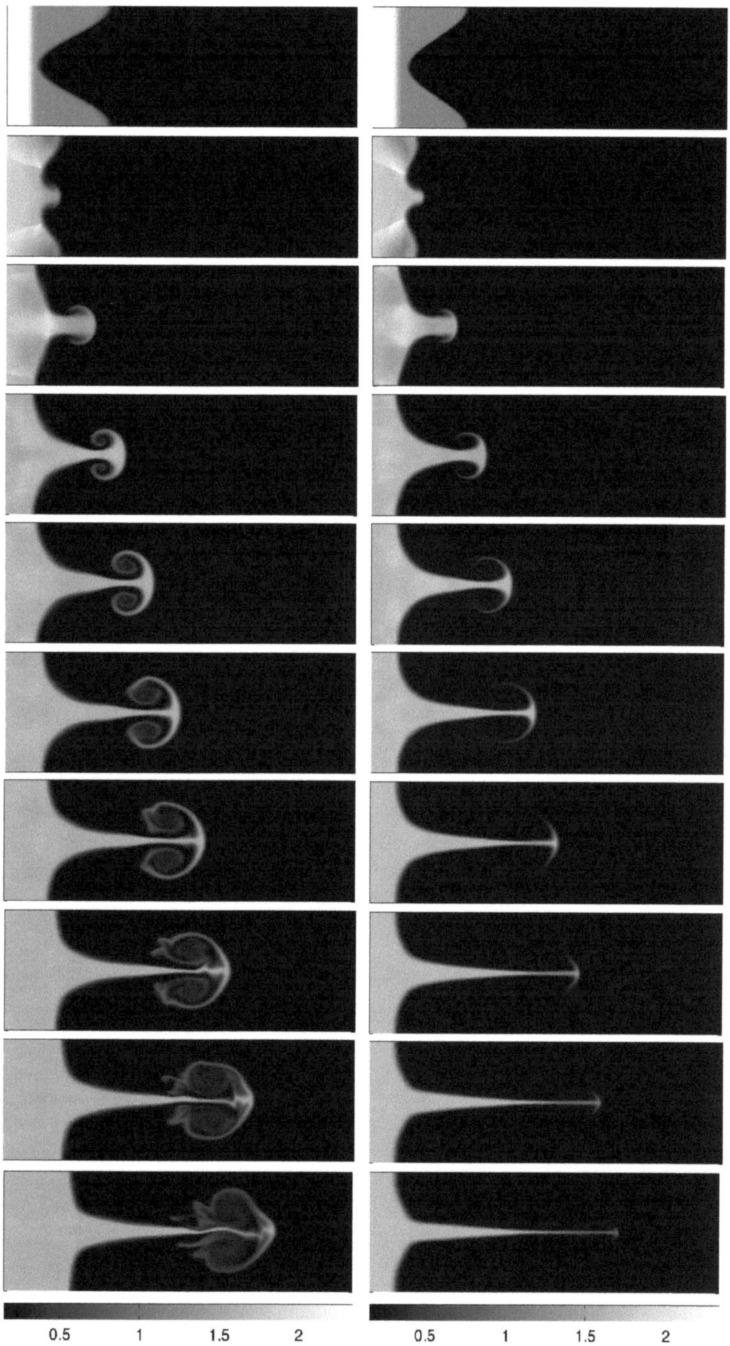

Fig. 8. Density for non-reactive (left) and reactive (right) case. Time steps 0, 1000, 2000, 3000, 4000, 5000, 6000, 7000, 8000, 9000 from top to bottom.

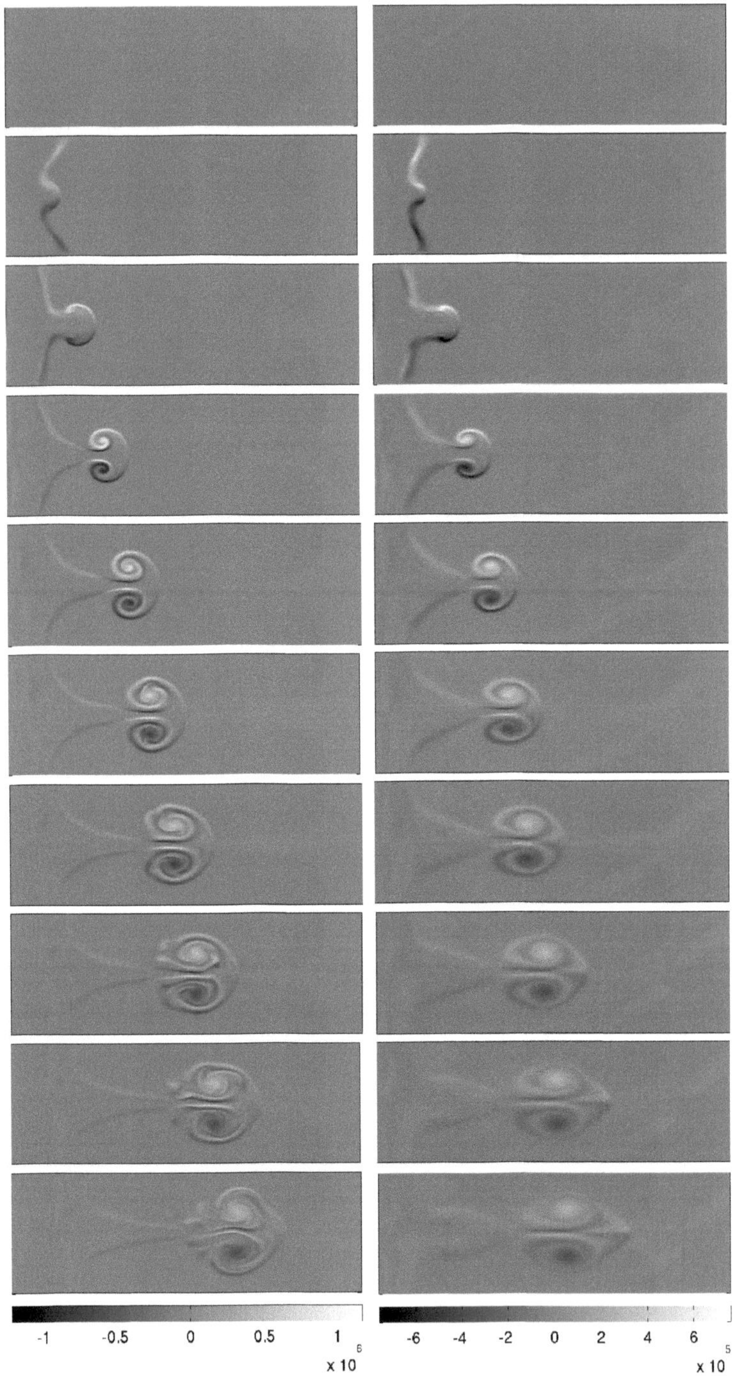

Fig. 9. Vorticity for non-reactive (left) and reactive (right) case. Time steps 0, 1000, 2000, 3000, 4000, 5000, 6000, 7000, 8000, 9000 from top to bottom.

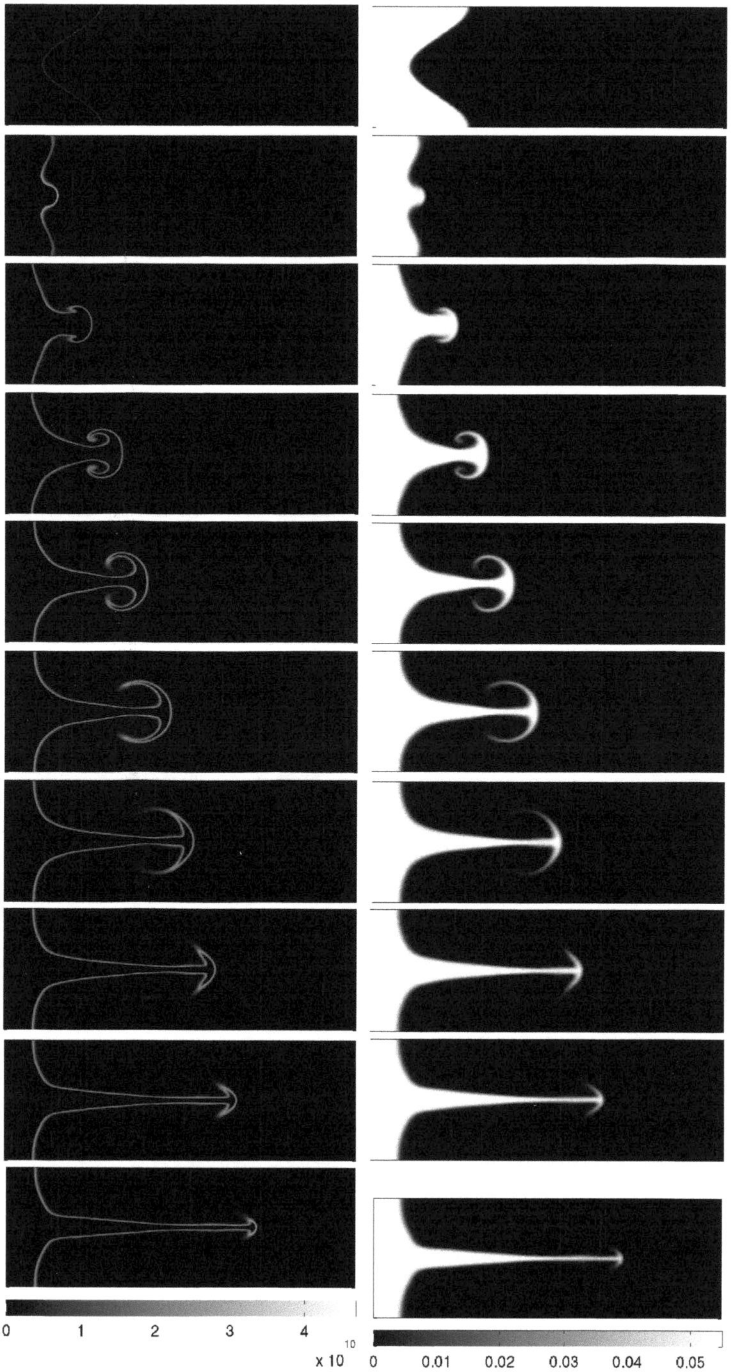

Fig. 10. Heat release (left) and fuel CH_4 (right) case. Time steps 0, 1000, 2000, 3000, 4000, 5000, 6000, 7000, 8000, 9000 from top to bottom.

B Appendix

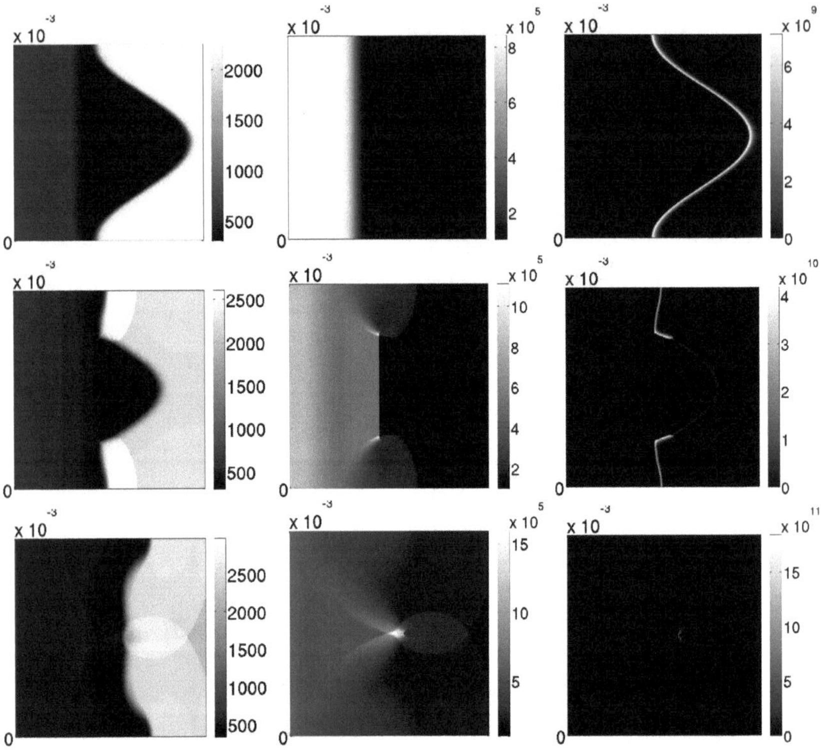

Fig. 11. Temperature (left), pressure (center) and heat release (right). Time steps 0, 3000, 6210 from top to bottom.

C Appendix

Table 2. Physical parameters for the laminar flame

Re	7.713107822005537	-	Reynolds number
Pr	0.71	-	Prandtl number
Le	1	-	Lewis number
γ	1.2552405	-	adiabatic index
M	0.001220384245504415	-	Mach number
T_0	298.15	$[K]$	Reference temperature
p_0	101330	$[Pa]$	Reference pressure
R	8.3144621	$[\frac{J}{molK}]$	Gas constant
W	0.0289647	$[\frac{Kg}{mol}]$	Air molar mass [20]
ρ_0	1.18396293233776	$[\frac{Kg}{m^3}]$	Reference density
C_p	1411.70021160307	$[\frac{J}{kgK}]$	Specific heat capacity
μ_0	$1.842 \cdot 10^{-5}$	$[Pa \cdot s]$	Reference viscosity in Sutherland's law
W_{CH_4}	0.0160425	$[kg/mol]$	CH_4 molar mass [21]
W_{O_2}	0.0319988	$[kg/mol]$	O_2 molar mass [21]
W_{N_2}	0.0280134	$[kg/mol]$	N_2 molar mass [21]
W_{CO_2}	0.0440095	$[kg/mol]$	CO_2 molar mass [21]
W_{H_2O}	0.0180153	$[kg/mol]$	H_2O molare mass [21]
A_f	1242379500	$[\frac{m^3}{mol \cdot s}]$	Prexponential factor
T_a	15900	$[K]$	Activation temperature [16]
Q	$50100 \cdot 10^3$	$[\frac{J}{Kg}]$	Mass heat release [15, 16]
ϕ	1	-	Equivalence ratio
s	4	-	Mass stoichiometric ratio [15]
$Y^{st}_{CH_4}$	0.055	-	CH_4 stoichiometric mass fraction [15]
$Y^{st}_{O_2}$	0.22	-	O_2 stoichiometric mass fraction
$Y^{st}_{N_2}$	0.725	-	N_2 stoichiometric mass fraction
$Y^{st}_{CO_2}$	0.152	-	CO_2 stoichiometric mass fraction
$Y^{st}_{H_2O}$	0.123	-	H_2O stoichiometric mass fraction

Part IV
Reduced Order Modeling

LQG-Balanced Truncation Low-Order Controller for Stabilization of Laminar Flows

Peter Benner and Jan Heiland

Max Planck Institute for Dynamics of Complex Technical Systems, Magdeburg, Germany
{benner,heiland}@mpi-magdeburg.mpg.de
http://www.mpi-magdeburg.mpg.de/csc

Abstract. Recent theoretical and simulation results have shown that Riccati based feedback can stabilize flows at moderate Reynolds numbers. We extend this established control setup by the method of *LQG-balanced truncation*. In view of practical implementation, we introduce a controller that bases only on outputs rather than on the full state of the system. Also, we provide a very low dimensional observer so that the control actuation can be computed in an online fashion.

Keywords: Navier-Stokes equation, model order reduction, stabilization, output feedback.

1 Introduction

The control of flows is of high interest in practical applications and a field of ongoing research [15,16]. We consider the particular aspect of stabilizing a flow so that it remains close to a desired state. This is of practical importance in systems where a flow regime is but one component and where stable quasi-stationary working conditions are required as, e.g., in pipelines [22] or in bubble column reactors [12].

We consider the generic control setup consisting of a plant, a controller that can act on the plant, and a measurement or observation unit that delivers information on the current state of the plant. In the realm of flow control, the plant may be modelled by semi-discrete Navier-Stokes equations for the evolution of the velocity $v(t)$ and the pressure $p(t)$ for time $t > 0$ in an incompressible flow:

$$M\dot{v} = -N(v)v - \frac{1}{Re}Lv + J^T p + f, \tag{1}$$

$$0 = Jv - g, \tag{2}$$

$$v(0) = \alpha. \tag{3}$$

We assume that the controller can act on the system by changing certain components of the inhomogeneity $f \leftarrow f + Bu$ in (1) and that the observations y are in a linear relation with v. The assumption that u appears in the right hand side of (1) is justified for so called *distributed control* setups that are of limited

R. King (ed.), *Active Flow and Combustion Control 2014*, Notes on Numerical Fluid Mechanics and Multidisciplinary Design 127, DOI: 10.1007/978-3-319-11967-0_22

practical relevance, see [10] for an application example. Recent analytical [21] and numerical results [3] show, however, that mathematical models with controls of *distributed* type can be used to determine feedback relations for boundary control.

The well-understood approach of *open-loop control* [11] does not apply well for stabilization since it cannot react to perturbations. One rather resorts to *closed-loop control* where the controller decides based on the current state information. Recent publications [3,5] have reported successful applications of static state-feedback boundary control for the stabilization of flows at moderate Reynolds numbers in simulations. The presented approach employs low-rank Newton-ADI iterations [4] to solve for Riccati-based feedback gains and, thus, despite its generality, it is feasible for high-dimensional systems.

The novelties we propose are of practical impact. In view of a generic control setup, where the full state is generally not available, we consider a *Kalman observer* that determines the feedback control from a few measurements. Combined with the *LQG-balanced truncation* model reduction technique, we design a reduced controller of very low dimension that can be evaluated in a negligible amount of time and, thus, is suitable for online feedback control.

To derive the low-order controller, we extend the method of *LQG-balanced truncation* [20] to the case of linearized Navier-Stokes equations. We will consider the linearization about a steady-state solution. This linearization is commonly used in the linear stability analysis of Navier-Stokes equations [13]. The DAE structure of the state equations (1)–(2) will be treated implicitly, as it was derived for the *Lyapunov equations* used in *balanced truncation* in [9] and as it was exploited for the solution of *Riccati equations* in [2,3]. Furthermore, since the controller acts only in the differential equation in (1), and since, though implicitly, we only consider the system that is projected onto the differential part, we do not need to account for an *improper* component of the system's transfer function as it was laid out in [7,23]. For the same reasons, the relation of our work to [19] is only marginal.

The paper is structured as follows. In Section 2, we introduce the method of *LQG-balanced truncation.* In view of implementation, we formulate the equations with the presence of a mass matrix. In Section 3, we extend the theory and the equations to the case of the linearized Navier-Stokes equations. To show applicability, we present a numerical example in Section 4. We conclude the paper by summarizing remarks on the potential impact and on the shortcomings of the presented results.

2 LQG-Balanced Truncation for Low-Order Controller Design

We start with a review of known results for reduced-order controller synthesis by *LQG-balanced truncation.* For further reference, we illustrate basic system theoretic concepts and the derivation of a balanced system, cf. [26], by an example problem.

Consider the linear time-invariant system

$$M\dot{v} = Av + Bu, \tag{4}$$
$$v(0) = 0, \tag{5}$$
$$y = Cv, \tag{6}$$

with matrices A, $M \in \mathbb{R}^{n_v,n_v}$, $B \in \mathbb{R}^{n_v,n_u}$, and $C \in \mathbb{R}^{n_y,n_v}$, a state $x(t) \in \mathbb{R}^{n_v}$, an input $u(t) \in \mathbb{R}^{n_u}$, and an output $y(t) \in \mathbb{R}^{n_y}$. The mass matrix M is assumed to be symmetric and positive definite.

We will call (4)–(6) a state space system that realizes a transfer function $G: \mathcal{U} \to \mathcal{Y}$, where $\mathcal{U}$ and $\mathcal{Y}$ denote the spaces of inputs and outputs, respectively. We will frequently write $G = (M, A, B, C, D)$.

By *LQG regulator* we denote the input u that minimizes the *LQG costs*, given as the expected value of

$$\lim_{t_f \to \infty} \frac{1}{2t_f} \int_{-t_f}^{t_f} y(t)^T y(t) + u(t)^T u(t) dt, \tag{7}$$

subject to (4)–(6) with inputs and outputs corrupted by additive Gaussian white noise with zero mean and the spectrum being the identity. We refer the reader to standard textbooks such as [17] for a thorough introduction on the *LQG control problem* or [20] for an overview directed to the *LQG-balanced truncation* setup.

We recall that a transfer function has infinitely many realizations, see, e.g., the textbook [26]. Particularly, for invertible transformation matrices $\mathcal{W} \in \mathbb{R}^{n_v,n_v}$ and $\mathcal{V} \in \mathbb{R}^{n_v,n_v}$ one has that

$$G = (M, A, B, C, D) \text{ equals } \tilde{G} = (\mathcal{W}^T M \mathcal{V}, \mathcal{W}^T A \mathcal{V}, \mathcal{W}^T B, C\mathcal{V}), \tag{8}$$

in the sense that G and $\tilde{G}$ map the same inputs u onto the same outputs y. Among the realizations, *minimal realizations* are those for which the state v has minimal dimension. For a *minimal realization*, one has the following characteristic properties:

Proposition 1 (cf. [20], Prop. 3.3). *Let* $G = (M, A, B, C)$ *be* minimal. *Then there exists a unique symmetric positive definite stabilizing solution* $X_c \in \mathbb{R}^{n_v,n_v}$ *to the* control algebraic Riccati equation (CARE)

$$A^T X_c M + M X_c A - M X_c B B^T X_c M + C^T C = 0 \tag{9}$$

and there exists a unique symmetric positive definite stabilizing solution $X_o \in \mathbb{R}^{n_v,n_v}$ *to the* filter algebraic Riccati equation (FARE)

$$A X_o M + M X_o A^T - M X_o C^T C X_o M + B B^T = 0. \tag{10}$$

Then the LQG regulator *is given via the control law*

$$u = -B^T M X_c \hat{x}, \tag{11}$$

where $\hat{x}$ *is the state of the* Kalman *observer*

$$M\dot{\hat{x}} = (A - X_o M C^T C - B B^T X_c M)\hat{x} + X_o M C^T y. \tag{12}$$

Equations (9)–(12) are derived from the standard case with $M = I$ through a scaling of (4) by M^{-1} and by redefining $X_c \leftarrow M^{-1}X_cM^{-1}$ and $X_o \leftarrow M^{-1}X_oM^{-1}$.

In the pioneering work [14] it was found that for minimal realizations the eigenvalues of the product of the solutions of the CARE and FARE are invariant under equivalence transformations of the system. Then, as in the case of *balanced truncation*, cf. [1], balanced realizations can be defined, interpreted, and truncated. We adapt the results and arguments given in [20] to the case with a mass matrix M.

Proposition 2 (cf. [20], Prop. 3.4). *Let* $G = (M, A, B, C)$ *be* minimal *and let* X_c *and* X_o *be the unique symmetric positive definite stabilizing solution to the CARE and FARE, respectively. Consider transformations as in (8) with invertible matrices* $\mathcal{W}$ *and* $\mathcal{V}$ *satisfying* $\mathcal{W}^T M \mathcal{V} = I$*. Then,*

- *the unique symmetric positive definite stabilizing solutions* $\tilde{X}_c$ *and* $\tilde{X}_o$ *to the CARE (9) and FARE (10) associated with the transformed system* $\tilde{G}$ *fulfill*

$$\tilde{X}_c = \mathcal{V}^T M X_c M \mathcal{V} \text{ and } \tilde{X}_o = \mathcal{W}^T M X_o M \mathcal{W} \tag{13}$$

- *and* $X_c M X_o$ *and* $\tilde{X}_c \tilde{X}_o$ *have the same eigenvalues.*

Remark 1. Because of $I = \mathcal{W}^T M \mathcal{V}$, the relations in (13) are the same as the commonly used relations

$$\tilde{X}_c = \mathcal{W}^{-1} X_c \mathcal{W}^{-T} \text{ and } \tilde{X}_o = \mathcal{W}^T M X_o M \mathcal{W}, \tag{14}$$

cf., e.g., [1, Ch. 4.3] for the balanced truncation case with $M = I$. In view of model reduction, for which we will apply non-square transformations $\mathcal{W}$ and $\mathcal{V}$, we will use the formulation (13).

From the positive definiteness of the matrices M, X_c, and X_o, it follows that the eigenvalues $\{\mu_i^2\}_{i=1}^{n_v}$ of $X_c M X_o$ are real and positive. Because of their invariance properties they are called *LQG-characteristic values* [20].

Furthermore, there exist transformation matrices $\mathcal{W}$, $\mathcal{V}$, with $\mathcal{W}^T M \mathcal{V} = I$ such that for the solutions of $\tilde{X}_c$ and $\tilde{X}_o$ of the CARE and FARE associated with the transformed $\tilde{G} = (I, \mathcal{W}^T A \mathcal{V}, \mathcal{W}^T B, C\mathcal{V})$, it holds that

$$\tilde{X}_c = \tilde{X}_o = S, \tag{15}$$

where S is the diagonal matrix of the *LQG-characteristic values* $\{\mu_i\}_{i=1}^{n_v}$ in descending order. In line with the notion for *balanced truncation*, this realization of G is called *LQG-balanced realization.*

If one has factored the matrices $X_c = UU^T$ and $X_o = LL^T$ of any minimal realization, then one can compute the transformations $\mathcal{W}$ and $\mathcal{V}$, that realize (15), as follows:

Algorithm 1

1. From the factors U and L compute the auxiliary matrix U^TML and a *Singular Value Decomposition*: $U^TML = ZSY^T$, where Z are Y orthonormal matrices containing the left and right singular vectors and S is the diagonal matrix of the singular values U^TML.
2. Set $\mathcal{W} := UZS^{-1/2}$ and $\mathcal{V} = LYS^{-1/2}$.

With $\mathcal{W}$ and $\mathcal{V}$ chosen by Algorithm 1 we directly confirm that

$$\mathcal{W}^TM\mathcal{V} = S^{-1/2}Z^TU^TMLYS^{-1/2} = S^{-1/2}Z^TZSY^TYS^{-1/2} = I \tag{16}$$

and with (15) that

$$\tilde{X}_c = \mathcal{V}^TMX_cM\mathcal{V} = S^{-1/2}Y^TL^TMUU^TMLYS^{-1/2} = S \tag{17}$$

and

$$\tilde{X}_o = \mathcal{W}^TMX_oM\mathcal{W} = S^{-1/2}Z^TU^TMLL^TMUZS^{-1/2} = S. \tag{18}$$

Thus, the singular values in S from Algorithm 1 are the *LQG-characteristic values*, cf. (15). Following [14], we interpret small values in S as associated with states in the LQG-balanced realization that are both difficult to control and to observe.

Consequently, if in Algorithm 1 we set

$$\mathcal{W} = \mathcal{W}_k := LY_kS_k^{-1/2} \text{ and } \mathcal{V} = \mathcal{V}_k = UZ_kS_k^{-1/2}, \tag{19}$$

where S_k is the diagonal matrix of the n_k largest singular values and where Z_k and Y_k^T contain the left and right singular vectors that correspond to the n_k largest singular values, we can define a truncated system as

$$\tilde{G}_k = (\mathcal{W}_k^TM\mathcal{V}_k, \mathcal{W}_k^TA\mathcal{V}_k, \mathcal{W}_k^TB, C\mathcal{V}_k) =: (I_k, A_k, B_k, C_k). \tag{20}$$

With similar arguments as in (17) and (18), one confirms that

$$\tilde{X}_{ck} := \mathcal{V}_k^TMX_cM\mathcal{V}_k = S_k \text{ and } \tilde{X}_{ok} = \mathcal{W}_k^TMX_oM\mathcal{W}_k = S_k. \tag{21}$$

Also, one can prove that $\tilde{X}_{ck}$ and $\tilde{X}_{ok}$ are the unique stabilizing solutions of the CARE and FARE for the truncated system $\tilde{G}_k$ [20, Rem. 3.7].

Finally, we define a truncation of the controller defined in (11) and (12) via

$$u = -B_k^T\tilde{X}_{ck}\hat{x}_k, \tag{22}$$

and the reduced observer

$$\dot{\hat{x}}_k = (A_k - \tilde{X}_{ok}C_k^TC_k - B_kB_k^T\tilde{X}_{ck})\hat{x}_k + \tilde{X}_{ok}C_k^Ty_k, \tag{23}$$

which is the *Kalman* observer of the system $\tilde{G}_k$, cf. [20].

Remark 2. The presented analysis requires that the considered realization is minimal. In theory this is not a restriction, since one can remove the unobservable and uncontrollable states from the system [26]. We will only require that the formally stated CARE (9) and FARE (10) uniquely define positive (semi-)definite solutions, which is sufficient to compute truncated balanced realizations as for the stable linear case [25]. Furthermore, we apply the *LQG-balanced truncation* (LQGBT) method to large-scale systems, where one rather computes low rank factors U_k and L_k that approximate $X_c \approx U_k U_k^T$ and $X_o \approx L_k L_k^T$. For such cases, Algorithm 1 readily provides truncating transformation matrices $\mathcal{W}_k$ and $\mathcal{V}_k$.

3 LQGBT for Stabilization of Flows

We consider semi-discrete Navier-Stokes equations modeling the evolution of the velocity v and the pressure p in an incompressible flow,

$$M\dot{v} = -N(v)v - \frac{1}{Re}Lv + J^T p + Bu + f \tag{24}$$

$$0 = Jv - g \tag{25}$$

$$v(0) = \alpha \tag{26}$$

$$y = Cv \tag{27}$$

starting from an initial state α.

Systems as (24)–(27) arise, e.g., in a finite element discretization of a flow problem with distributed control and observation. Here M is the symmetric positive mass matrix, L is the matrix associated with the discretized diffusion operator, and $N(v)$ accounts for the convection that is linear in v. The discrete divergence operator is given by the matrix J and the discrete gradient by J^T. The control and the observation are modeled with the matrices B and C. The boundary conditions are resolved in the definition of the discrete operators and appear in the source terms f and g.

We assume that J is of full rank, which possibly requires pinning the pressure p to a certain level, e.g. by setting one component to zero. Also we assume that f and g in (24) and (25) only contain stationary boundary conditions and, thus, are constant in time.

The presented theoretical results hold for any spatial discretization method that meet the assumptions on M and J. In particular, stable mixed finite element schemes, cf. [6], are a suitable choice.

Assume that α is a steady state solution to (24)–(26) and write the unsteady solution as $v = \alpha + v_\delta$ with a deviation v_δ that we assume to be small. Then, neglecting the quadratic term in v_δ, from Equations (24)–(27) we derive a linear model for the deviation in the output caused by the actuation Bu:

$$M\dot{v}_\delta = Av_\delta + J^T p + Bu \tag{28}$$
$$0 = Jv_\delta \tag{29}$$
$$v_\delta(0) = 0 \tag{30}$$
$$y_\delta = Cv_\delta, \tag{31}$$

where $Av_\delta := -N(v_\delta)\alpha - N(\alpha)v_\delta - \frac{1}{Re}Lv_\delta$.

Equations (28)–(30) are used in linear stability analysis of flows as they describe the propagation of small disturbances of a given flow [13]. We will design a controller that damps the output of System (33)–(31) to prevent small deviations in a solution to the flow equations from evolving.

Because of the constraint $Jv_\delta = 0$, the framework of Section 2 does not simply apply. Therefore, we reformulate System (28)–(31) in terms of the *underlying ODE* that describes the motion of the parts of v_δ that are not seen by the constraint. We point out that the following algebraic manipulations are purely for extending the *LQG-balanced truncation* framework to the projected equations. Once we have arrived at the CARE (9) and FARE (10), we give references to a solution approach that avoids the numerically unfeasible computation of projectors.

Recall that M is symmetric positive definite and that J is of full rank. Thus, we can define the projector

$$\mathcal{P} := I - M^{-1}J(J^T M^{-1} J)^{-1} J \tag{32}$$

that comes with the following properties, cf., e.g., [9]:

Proposition 3. *Consider System (28)–(31), let v_δ be a solution to it, and let $\mathcal{P}$ be defined as in (32). Then,*

(a) $v_\delta(t) = \mathcal{P}v_\delta(t)$,
(b) $\mathcal{P}^T J^T p(t) = 0$, for all time $t > 0$, and
(c) $\mathcal{P}^T M = M\mathcal{P}$.

Using the relations of Proposition 3, we can state that v_δ is a solution to (28)–(31) if, and only if, it solves

$$M\dot{v}_\delta = \mathcal{P}^T Av_\delta + \mathcal{P}^T Bu \tag{33}$$
$$v_\delta(0) = 0 \tag{34}$$
$$y_\delta = Cv_\delta. \tag{35}$$

For System (33)–(35), the CARE (9) and the FARE (10) can be written as

$$A^T \mathcal{P}^T X_c M + MX_c \mathcal{P}A - MX_c \mathcal{P}BB^T \mathcal{P}^T X_c M + C^T C = 0 \tag{36}$$

and

$$\mathcal{P}AX_oM + MX_oA^T\mathcal{P}^T - MX_oC^TCX_oM + \mathcal{P}BB^T\mathcal{P}^T = 0. \qquad (37)$$

Low-rank factors approximating the solutions to Equations (36) and (37) can be obtained efficiently by combining *low-rank Newton-ADI* iterations [4] with the ideas of [9] that realize the application of $\mathcal{P}$ in an implicit fashion. See, in particular, the references [2,3,5], where Equation (36) is solved to define a stabilizing feedback for flow problems.

Remark 3. By construction, cf. [9], the resulting transformation matrix $\mathcal{W}_k$ fulfills $\mathcal{W}_k = \mathcal{W}_k\mathcal{P}^T$. Thus, a reduced system to (28)–(31),

$$\tilde{G}_k = (I_k, A_k, B_k, C_k) := (\mathcal{W}_k^T M \mathcal{V}_k, \mathcal{W}_k^T A \mathcal{V}_k, \mathcal{W}_k^T B, C\mathcal{V}_k),$$

can be obtained without resorting to the reformulation (33)–(35).

4 Numerical Example

Motivated by theoretical results [21], in recent numerical investigations, *LQ regulators* for a linearization of the flow equations about a stationary solution were considered [2,3]. It was shown that the linear feedback law can be successfully applied in the nonlinear equations to keep the flow quasi-stationary.

To illustrate the applicability of *LQG-balanced truncation* based low-order controllers in the stabilization of unsteady flows, we consider the two-dimensional cylinder wake with distributed control and observation, as depicted in Figure 1. As the computational domain we consider the rectangle $[0, 2.2] \times [0, 0.41]$ with the cylinder of radius 0.05 centered at $(0.2, 0.2)$, The spatial coordinates we denote by x_1 and x_2. The results presented are for *Reynolds number* $Re = 133$, calculated with the peak inflow velocity and the cylinder diameter.

As boundary conditions for the velocities, we impose a parabolic inflow profile at the left boundary and *no-slip* conditions on the top and the bottom wall. At the outflow, we employ *do-nothing* conditions.

The implementation was done in Python [8]. For the spatial discretization we used the Python interface *dolfin* to *FeNiCS* [18].

Fig. 1. Setup of the 2D cylinder wake with control distributed in Ω_c and observation distributed in Ω_o

The spatial discretization was realized with a nonuniform triangulation of the computational domain and *Taylor-Hood* finite elements [24]. The considered discretization resulted in $n_v = 9356$ and $n_p = 1288$.

For $n_u \in \mathbb{N}$, we set the input space $\mathcal{U} := \mathcal{C}(0,T;U \times U)$, where U is spanned by n_u linear hat functions equally distributed on the unit interval $[0,1]$.

We define the domain of control to be $\Omega_c = [0.27, 0.32] \times [0.15, 0.25]$, cf. Figure 1, and the input operator $B: \mathcal{U} \to \mathcal{C}(0,T;\mathcal{C}(\Omega;\mathbb{R}^2))$ via

$$B_1 u(t;x_1,x_2) = \begin{cases} \begin{bmatrix} u_1(t;\theta(x_1)) \\ u_2(t,\theta(x_1)) \end{bmatrix}, & \text{if } (x_1,x_2) \in \Omega_c, \\ 0, & \text{elsewhere,} \end{cases} \tag{38}$$

with the affine linear function θ_c mapping $[0.27, 0.32]$ onto $[0,1]$.

For n_y, we define the output space $\mathcal{Y}$ similar to $\mathcal{U}$. As the domain of observation, we use $\Omega_o = [0.6, 0.7] \times [0.15, 0.25]$, cf. Figure 1, and for a $v \in \mathcal{C}(0,T;\mathcal{C}(\Omega;\mathbb{R}^2))$, we define the observation operator $C: v \to y \in \mathcal{Y}$ via

$$Cv(t)(\eta) = \begin{bmatrix} y_{x_1}(t;\eta) \\ y_{x_2}(t;\eta) \end{bmatrix} = \int_{0.6}^{0.7} \mathcal{P}_{\mathcal{Y}} \begin{bmatrix} v_1(t;x_1,\theta_o(\eta)) \\ v_2(t;x_1,\theta_o(\eta)) \end{bmatrix} dx_1, \tag{39}$$

where θ_o is an affine linear mapping adjusting $[0,1]$ to $[0.15, 0.25]$ and where the projector to the finite dimensional subspace $\mathcal{P}_{\mathcal{Y}}: [L^2(0,1)]^2 \to Y \times Y$ is chosen as the orthogonal L^2 projection.

By definition, C is defined such that it measures the velocity $v(t)$ in Ω_o averaged in x_1 direction. The x_2-dependence is approximated by n_y basis functions of Y. The input operator B maps the input into Ω_c such that it is constant in x_1 direction and space-varying in x_2.

In the semi-discrete setting v is typically assumed in a finite dimensional space of piecewise polynomial functions over Ω. By standard results, for continuous inputs, a solution v of the *ordinary differential equation* (33-34) and, thus, of (24-26) is continuous in time on its interval of existence. Thus, the choice of the input and output spaces and operators are justified.

In the presented examples we have chosen $n_u = n_y = 3$, meaning that the x_1 and x_2 components of both input and output signal are described by 3 nodal values each.

We set α to be the steady-state solution and consider the linearized system as in (28-31). Then we compute low-rank factors U_k and L_k that approximate the solutions to the projected CARE (36) and the projected FARE (37), $X_c \approx U_k U_k^T$ and $X_o \approx L_k L_k^T$, using the methodology described in [2]. As stabilizing initial guesses, we used solutions for lower *Reynolds numbers*. From the factors U_k and L_k, we compute $\mathcal{W}_k$ and $\mathcal{V}_k$ as defined in Equation (19) and the reduced system as $\tilde{G}_k = (I_k, A_k, B_k, C_k)$, cf. Remark 3.

For varying thresholds κ, we cut off all *LQG-characteristic values* $\mu_i < \kappa$, to get reduced systems of various size, see Table 1.

The reduced systems $\tilde{G}_k$ give good approximations to the linearized equations (28-31) both in in frequency (Figure 3) and time domain (Figure 2). Clearly, the reduced (linear) system cannot reproduce the nonlinear dynamics (Figure 2(e)).

Table 1. Dimension n_k of the reduced system for varying thresholds κ and the integrated deviation from the target output $d_k := \int_0^{12} \|y_k(t) - y_\alpha\|_2 dt$

κ	n_k	d_k
10^{-4}	42	0.01803696
10^{-3}	22	0.01803715
10^{-2}	13	0.01804118
10^{-1}	4	0.01818847
10^{-0}	2	3.16595044

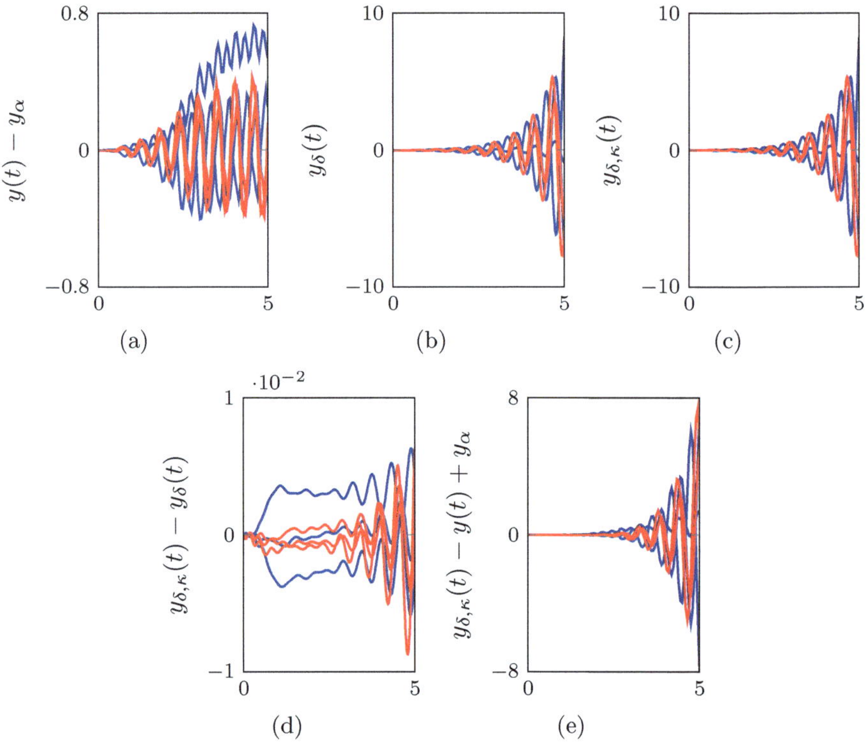

Fig. 2. Outputs and output differences of a step function in one input component for $t \in [0, 5]$ for (a) the nonlinear system (24-27), for (b) the linearized system (28-31) with $n_v = 9356$, and for (c) the reduced linearized system with $n_k = 13$. Here, y_α is the output of the linearization point. The (unstable) linearized system is well approximated (d), unlike the nonlinear system (e). The blue lines depict the x_1-components of the output signals and the red lines correspond to the x_2-components.

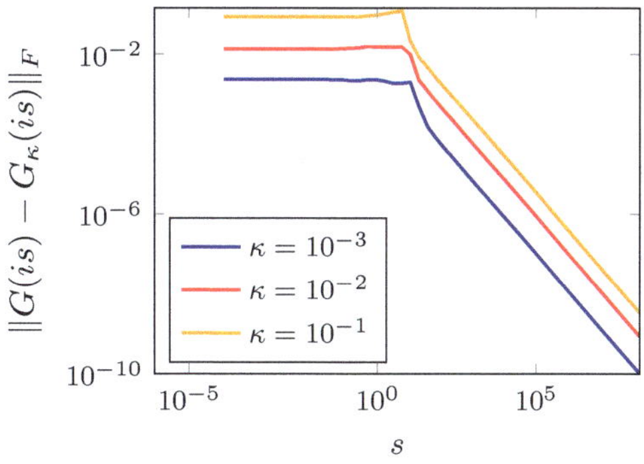

(a)

Fig. 3. The error in the frequency response for varying thresholds κ measured in the Frobenius norm with i denoting the imaginary unit and the transfer functions in frequency domain as defined, e.g., in [9].

However, the reduced controller, as defined via Equations (20), (22), and (23), successfully stabilizes the closed-loop system:

$$M\dot{v} = -N(v)v - \frac{1}{Re}Lv + J^T p - BB_k^T \tilde{X}_{ck}\hat{x}_k + f, \tag{40}$$
$$0 = Jv - g, \tag{41}$$
$$v(0) = \alpha, \tag{42}$$
$$y_k = Cv, \tag{43}$$
$$\dot{\hat{x}}_k = (A_k - \tilde{X}_{ok}C_k^T C_k - B_k B_k^T \tilde{X}_{ck})\hat{x}_k + \tilde{X}_{ok}C_k^T(y_k - y_\alpha), \tag{44}$$
$$\hat{x}_k(0) = 0, \tag{45}$$

where α is the steady-state solution and $y_\alpha := C\alpha$.

For numerical testing, we integrated the nonlinear closed-loop system starting from $v(0) = \alpha + \epsilon$, where $\epsilon := 10^{-3}\mathcal{P}\mathbf{e}$ and $\mathbf{e} \in \mathbb{R}^{n_v}$ is the vector of ones. The perturbation ϵ was introduced to trigger the instabilities. We considered the time interval $(0, T]$ discretized by a constant time step of length 0.005. As the numerical integration scheme for the state equations (40-41) we used the trapezoidal rule with an explicit treatment of the nonlinear part. The observer equation (44-45) was numerically integrated using the *implicit Euler* scheme. The control was lagged by one time-step, so that the current input was computed from the measurements of the previous states.

As can be seen from Figure 4(b), plotting the measurement signal $y(t)$ versus the time, the uncontrolled system is unstable and soon attains a state of periodic fluctuations. This is characteristic for the cylinder wake at moderate Reynolds numbers. If the loop is closed, then the system stays in a quasi stationary state (Figure 4(a)). Also, see Figure 5 for a plot of the deviation of the output from the starting value for different orders of reduction. We want to point out that the flow field was indeed stabilized, see Figure 6 for snapshots of the stabilized and the uncontrolled velocity fields.

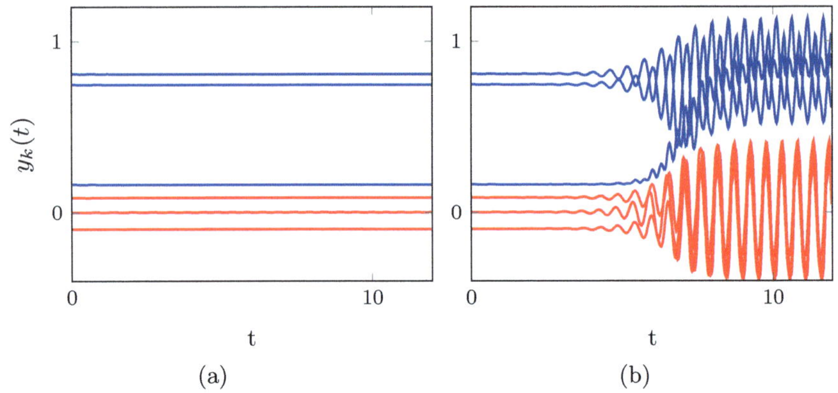

Fig. 4. The measured signal y_k over time $t \in [0, 12]$ of the closed loop system (a) with a reduced controller of dimension $n_k = 13$, compared to the response of the uncontrolled system (b). The blue lines depict the x_1-components of $y(t)$ and the red lines correspond to the x_2-components of $y(t)$.

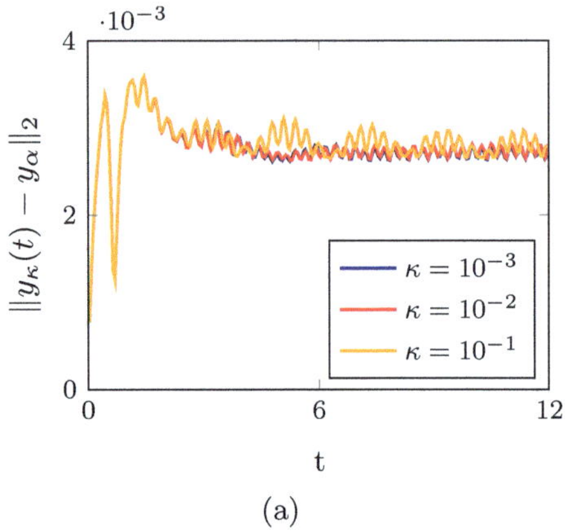

Fig. 5. The deviation of the measured output y_k from the steady-state output y_α for varying thresholds κ. For smaller κ the outputs become visually indistinguishable.

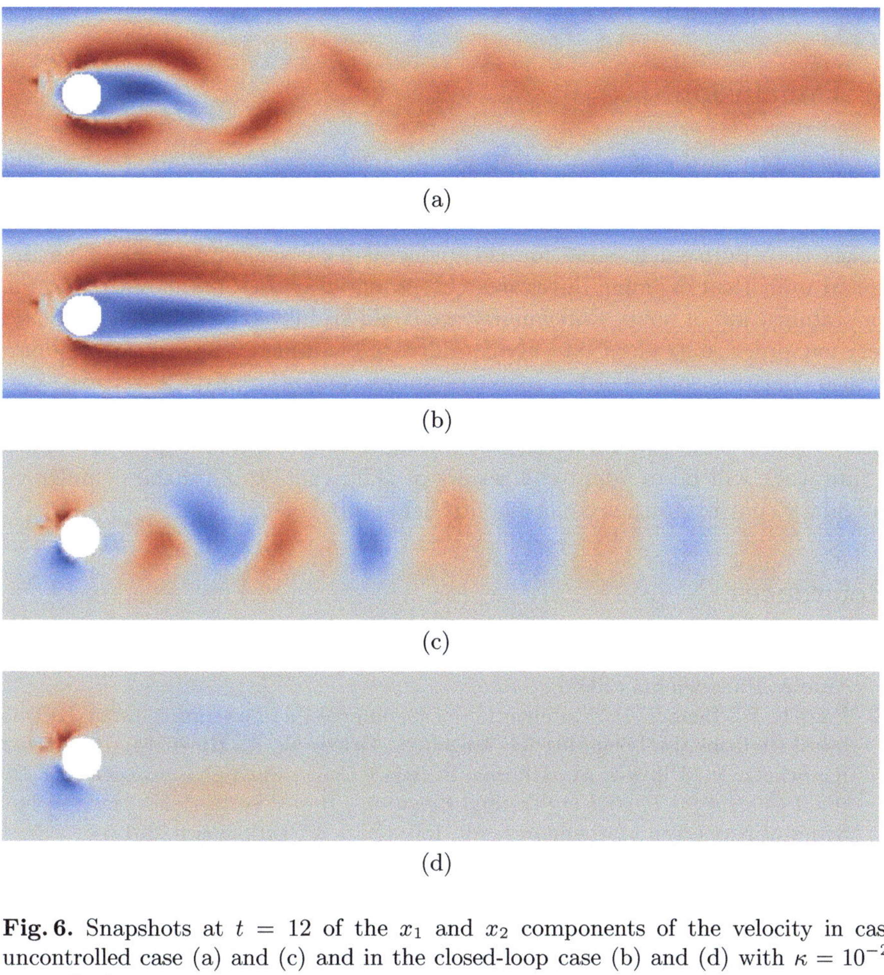

Fig. 6. Snapshots at $t = 12$ of the x_1 and x_2 components of the velocity in case uncontrolled case (a) and (c) and in the closed-loop case (b) and (d) with $\kappa = 10^{-2}$, respectively.

Remark 4. Note that the feedback control is defined only by the current output. Furthermore, the estimated state $\hat{x}_k(t)$ is obtained via the solution of an observer system of reduced dimension. In the presented example, a reduced system of order $n_k = 4$ has been enough for stabilization, cf. Table 1 and Figure 5. Recalling that the observer (44-45) is a linear time-invariant system, for a constant time-step length, all factors for the chosen numerical integration scheme can be precomputed. Thus, the effort for the update of $\hat{x}_k$ reduces to two small matrix-vector multiplications.

5 Conclusion

We have provided a generalization of *LQG-balanced truncation* to the setup of flow equations with distributed control. We have shown that recent results on projected low-rank Newton-ADI iterations can also be applied here, which makes the approach feasible for very general large-scale problems. This presented numerical example illustrates the potential of *LQG-balanced truncation* for stabilization of flows. Particularly, we could stabilize the cylinder wake by a very low-order controller that only considers a low-dimensional output. We have argued that this controller is capable of online-control in physical setups.

A shortcoming of the presented theory is the assumption that the control acts distributed in space, which is difficult to realize in a control setup. One task for future work will be to adapt the results of [3,5], that stabilize the cylinder via boundary control, to the reduced controller setup.

References

1. Antoulas, A.C.: Approximation of Large-Scale Dynamical Systems. SIAM Publications, Philadelphia (2005)
2. Bänsch, E., Benner, P.: Stabilization of incompressible flow problems by Riccati-based feedback. In: Leugering, G., Engell, S., Griewank, A., Hinze, M., Rannacher, R., Schulz, V., Ulbrich, M., Ulbrich, S. (eds.) *Constrained Optimization and Optimal Control for Partial Differential Equations*, Basel, Switzerland. International Series of Numerical Mathematics, vol. 160, pp. 5–20. Birkhäuser (2012)
3. Bänsch, E., Benner, P., Saak, J., Weichelt, H.K.: Riccati-based boundary feedback stabilization of incompressible Navier-Stokes flow. Preprint SPP1253-154, DFG-SPP1253 (2013)
4. Benner, P., Li, J.-R., Penzl, T.: Numerical solution of large Lyapunov equations, Riccati equations, and linear-quadratic control problems. Numer. Lin. Alg. Appl. 15(9), 755–777 (2008)
5. Benner, P., Saak, J., Stoll, M., Weichelt, H.: Efficient solution of large-scale saddle point systems arising in Riccati-based boundary feedback stabilization of incompressible Stokes flow. SIAM J. Sci. Comput. 170, S150–S170 (2013)

6. Girault, V., Raviart, P.-A.: Finite Element Methods for Navier-Stokes Equations. Theory and Algorithms. Springer, Berlin (1986)
7. Gugercin, S., Stykel, T., Wyatt, S.: Model reduction of descriptor systems by interpolatory projection methods. SIAM J. Sci. Comput. 35(5), B1010–B1033 (2013)
8. Heiland, J.: lqgbt-oseen – Python module for LQG-BT of linearized flow equations, v1.0 (2014), `https://github.com/highlando/lqgbt-oseen`
9. Heinkenschloss, M., Sorensen, D.C., Sun, K.: Balanced truncation model reduction for a class of descriptor systems with applications to the Oseen equations. SIAM J. Sci. Comput. 30(2), 1038–1063 (2008)
10. Hinze, M.: Control of weakly conductive fluids by near wall Lorentz forces. SFB609-Preprint 19-2004, Technische Universität Dresden (2004)
11. Hinze, M., Kunisch, K.: Second order methods for optimal control of time-dependent fluid flow. SIAM J. Cont. Optim. 40(3), 925–946 (2001)
12. Hooshyar, N., Hamersma, P.J., Mudde, R.F., van Ommen, J.R.: Intensified operation of slurry bubble columns using structured gas injection. Canadian J. Chem. Engrg. 88(4), 533–542 (2010)
13. Johnson, C., Rannacher, R., Boman, M.: Numerics and hydrodynamic stability: Toward error control in computational fluid dynamics. SIAM J. Numer. Anal. 32(4), 1058–1079 (1995)
14. Jonckheere, E.A., Silverman, L.M.: A new set of invariants for linear systems – application to reduced order compensator design. IEEE Trans. Automat. Control 28, 953–964 (1983)
15. King, R. (ed.): Active flow control. Papers contributed to the conference 'Active flow control 2006, September 27-29. Springer, Berlin (2006)
16. King, R. (ed.): Active Flow Control II: Papers Contributed to the Conference 'Active Flow Control II 2010', May 26-28. Notes on Numerical Fluid Mechanics and Multidisciplinary Design. Springer, Berlin (2010)
17. Locatelli, A.: Optimal Control: An Introduction. Birkhäuser, Basel (2001)
18. Logg, A., Ølgaard, K.B., Rognes, M.E., Wells, G.N.: FFC: the FEniCS form compiler. In: Automated Solution of Differential Equations by the Finite Element Method, pp. 227–238. Springer, Berlin (2012)
19. Möckel, J., Reis, T., Stykel, T.: Linear-quadratic gaussian balancing for model reduction of differential-algebraic systems. Internat. J. Control 84(10), 1627–1643 (2011)
20. Mustafa, D., Glover, K.: Controller design by $\mathcal{H}_\infty$-balanced truncation. IEEE Trans. Automat. Control 36(6), 668–682 (1991)
21. Raymond, J.-P.: Local boundary feedback stabilization of the Navier-Stokes equations. In: Control Systems: Theory, Numerics and Applications, Proceedings of Science. SISSA, Rome, March 30-April 1 (2005), `http://pos.sissa.it`
22. Storkaas, E., Skogestad, S., Alstad, V.: Stabilization of desired flow regimes in pipelines. In: AIChE Annual Meeting (2001)
23. Stykel, T.: Balanced truncation model reduction for semidiscretized Stokes equation. Linear Algebra Appl. 415(2-3), 262–289 (2006)
24. Taylor, C., Hood, P.: A numerical solution of the Navier-Stokes equations using the finite element technique. Internat. J. Comput. & Fluids 1(1), 73–100 (1973)
25. Tombs, M., Postlethwaite, I.: Truncated balanced realization of a stable non-minimal state-space system. Internat. J. Control 46(4), 1319–1330 (1987)
26. Zhou, K., Doyle, J., Glover, K.: Robust and Optimal Control. Prentice-Hall, Upper Saddle River (1996)

Model Reduction for DAEs with an Application to Flow Control

Jeffrey T. Borggaard and Serkan Gugercin

Department of Mathematics, Virginia Tech
Interdiciplinary Center for Applied Mathematics, Virginia Tech
Blacksburg, VA, 24061-0123, USA
{jborggaard,gugercin}@vt.edu

Abstract. Direct numerical simulation of dynamical systems is of fundamental importance in studying a wide range of complex physical phenomena. However, the ever-increasing need for accuracy leads to extremely large-scale dynamical systems whose simulations impose huge computational demands. Model reduction offers one remedy to this problem by producing simpler reduced models that are both easier to analyze and faster to simulate while accurately replicating the original behavior. Interpolatory model reduction methods have emerged as effective candidates for very large-scale problems due to their ability to produce high-fidelity (optimal in some cases) reduced models for linear and bilinear dynamical systems with modest computational cost. In this paper, we will briefly review the interpolation framework for model reduction and describe a well studied flow control problem that requires model reduction of a large scale system of differential algebraic equations. We show that interpolatory model reduction produces a feedback control strategy that matches the structure of much more expensive control design methodologies.

1 Introduction

Direct numerical simulation of dynamical systems has been one of the few available means for studying many complex systems of scientific and industrial value as dynamical systems are the basic framework for modeling, optimization, and control of these complex systems. Examples include chemically reacting flows, fluid dynamics, and signal propagation and interference in electric circuits. However, the ever increasing complexity and need for improved accuracy lead to the inclusion of greater detail in the model, and inevitably finer discretizations. Combined with the potential coupling to other complex systems, this results in extremely large-scale dynamical systems with millions of degrees of freedom to simulate. The simulations in these settings can be overwhelming; which is the main motivation for model reduction. The goal is to construct reduced models with significantly lower number of degrees of freedom that are easier to analyze and faster to simulate yet accurately approximate the important features in the underlying full-order large-scale simulations.

R. King (ed.), *Active Flow and Combustion Control 2014*, Notes on Numerical Fluid Mechanics and Multidisciplinary Design 127, DOI: 10.1007/978-3-319-11967-0_23

There is a tremendous amount of literature on model reduction. Here we only include a partial list of various applications settings where model reduction has been applied with great success: In fluid flow [25, 32, 38, 48, 49, 65] and design of feedback control systems [9, 11, 34, 42, 52, 53, 63], in optimization [4, 5, 8, 12, 26, 43, 75], in nonlinear inverse problems [21,24,26,30,45,71], in optimal design [3,46,47], in the analysis of structural mechanics [18, 33, 51, 64, 68], in circuit theory [10, 13, 14, 19, 23, 27, 28, 55, 62], and in structural mechanics, such as [18, 33, 51, 64, 68]. For a detailed discussion of several model reduction topics, see [6, 14, 15, 41, 54].

The active flow control application we consider in this paper is a well studied flow control problem of stabilizing the von Kármán vortex shedding behind a circular cylinder by controlling the rotational velocity of the cylinder. Upon linearizing the Navier-Stokes equations about a desired steady-state solution, the resulting large-scale linear systems of differential algebraic equations (DAE) is reduced by the interpolatory model reduction framework recently developed in [35]. We use this reduced model to design the feedback control strategy and compare these results to other feedback control laws found in the literature.

2 Description of the Flow Control Problem

Suppression of the vortex shedding behind a bluff body is a classical flow control problem with numerous applications ranging from minimizing drag to reducing the cross-stream lift-induced fatigue cycling. A number of experimental and computational studies have shown the effectiveness of cylinder surface suction and zero-mass actuation to completely suppress the von Kármán vortex street at modest Reynolds numbers that are slightly higher than the critical Reynolds number (bifurcation parameter) $Re_c \approx 47$, cf. [37, 58], based on the cylinder diameter and inflow velocity.

An alternate strategy capitalizes on the Magnus effect produced by cylinder rotation [60]. In a sequence of experiments in [69], the authors showed that a rotationally oscillating cylinder using carefully selected choices of frequency, amplitude, and phase angle could *effectively eliminate the wake* for moderate flows of $Re \approx 61$ and $Re \approx 110$. At higher values of the Reynolds number, it was not possible to eliminate the wake, but using good choices of frequency and amplitude made it possible to achieve nearly 20% drag reduction. Other experimental studies [29, 56, 70] suggest that matching the oscillation frequency to the vortex shedding frequency maximizes the impact on the flow (at this range of Re). This was confirmed numerically in [57].

A number of active feedback control approaches for the rotating/oscillating cylinder have appeared in the literature over the past fifteen years, including [1, 17, 20, 39, 59, 61, 66, 73, 74].

In the remainder of this section we describe our feedback control strategy based on linearizing the Navier-Stokes equations about a steady-state flow and controlling the discrepancy between the actual flow and the steady-state flow (cf. [22]), discretizing the associated linear state space model and then setting up the discrete flow control problem.

The fluid flow about a rotating circular cylinder can be described using the Navier-Stokes equations

$$\frac{\partial \mathbf{v}}{\partial t} + \mathbf{v} \cdot \nabla \mathbf{v} = -\nabla p + \frac{1}{Re} \nabla \cdot \tau(\mathbf{v}) + \mathcal{B}u,$$
$$\nabla \cdot \mathbf{v} = 0,$$

where $\mathbf{v}$ is the fluid velocity, p is the pressure, $\tau(\mathbf{v}) = \nabla \mathbf{v} + \nabla \mathbf{v}^T$ is the viscous stress tensor, and $\mathcal{B}u$ is the prescription of the Dirichlet boundary conditions on the cylinder surface with $u(t)$ representing the instantaneous tangential velocity component. Our strategy is to linearize these equations about a desired flow profile, then use this linearized model to regulate the discrepancy between the actual flow and the desired flow. For this study, we selected the steady-state solution at $Re = 60$ created from a uniform free-stream velocity profile. Although the steady-state solution $(\bar{\mathbf{v}}, \bar{p})$ exists at this low Reynolds number, it is an unstable equilibrium solution of the Navier-Stokes equations, solving

$$\bar{\mathbf{v}} \cdot \nabla \bar{\mathbf{v}} = -\nabla \bar{p} + \frac{1}{Re} \nabla \cdot \tau(\bar{\mathbf{v}}),$$
$$\nabla \cdot \bar{\mathbf{v}} = 0,$$

computed with uniform inflow velocity $\bar{\mathbf{v}} = (1, 0)$ and zero velocity on the cylinder surface $\bar{\mathbf{v}} = \bar{u}\hat{t} = (0, 0)$. For computational purposes, we consider the finite flow domain Ω consisting of the unit diameter cylinder centered at the origin embedded in a rectangular flow domain $(-7, 15) \times (-7, 7)$. For boundary conditions, we specify the uniform inflow velocity at the $\xi = -7$ boundary and stress-free outflow boundary conditions on the $\eta = -7$, $\xi = 7$, and $\eta = 7$ edges.

If we write $\mathbf{v} = \bar{\mathbf{v}} + \mathbf{v}'$ and $p = \bar{p} + p'$, then the flow fluctuations $(\mathbf{v}', p')$ satisfy the equations

$$\frac{\partial \mathbf{v}'}{\partial t} = -\mathbf{v}' \cdot \nabla \bar{\mathbf{v}} - \bar{\mathbf{v}} \cdot \nabla \mathbf{v}' - \nabla p' + \frac{1}{Re} \nabla \cdot \tau(\mathbf{v}') + \mathcal{B}u + \mathbf{F}(\mathbf{v}') \quad (1)$$
$$0 = \nabla \cdot \mathbf{v}', \quad (2)$$

where $\mathbf{F}(\mathbf{v}')$ satisfies $\|\mathbf{F}(\mathbf{v}')\| = O(\|\mathbf{v}'\|^2)$. The velocity fluctuation satisfies homogeneous boundary conditions at $\xi = -7$ and stress-free boundary conditions on the remaining exterior boundaries. Using $\mathcal{B}u(\cdot)$ to drive $\mathbf{v}' \to \mathbf{0}$ is equivalent to using $\mathcal{B}u(\cdot)$ to control the flow $(\mathbf{v}, p)$ to $(\bar{\mathbf{v}}, \bar{p})$. We seek to achieve this using linear control theory. Note that ignoring the nonlinear term $\mathbf{F}(\mathbf{v}')$ in (1) produces the Oseen equations and often arises when developing linear feedback flow control strategies for the Navier-Stokes equations, cf. [22].

At this point, we follow the standard strategy for calculating the linear feedback control laws for this problem (known as the *reduce-then-control* approach). We first develop a suitable discretization for equations (1)–(2) which results in a large system of DAEs and formulate the associated linear control problem for this approximate model. The solution to the resulting control problem is challenging and typically requires the use of suitable model reduction methods. The presentation of a new model reduction strategy for this class of problems will then be provided in Section 3.2.

2.1 Finite Element Discretization and the DAE Control Problem

We use a standard Taylor-Hood (P2-P1) finite element pair to find approximations to both $(\bar{\mathbf{v}}, \bar{p})$ and $(\mathbf{v}', p')$, cf. [36]. The nodal values of the fluctuating velocity components are denoted by $\mathbf{x}_1(t)$ while those for the pressure are denoted by $\mathbf{x}_2(t)$. We considered several choices for the controlled output variable y, but for the computations below, we define

$$y_{[2i-1,2i]}(t) = \frac{1}{|\Omega_i|} \int_{\Omega_i} \mathbf{v}'(t,\xi)\, d\xi\, d\eta, \qquad t > 0, \tag{3}$$

for six different patches downstream of the cylinder, located at $\Omega_1 = [1, 2.5] \times [0, 2]$, $\Omega_2 = [2.5, 4] \times [0, 2]$, $\Omega_3 = [4, 5.5] \times [0, 2]$, and three more reflected about the ξ axis. For each patch, we recover two components of the average fluctuating velocity. This is discretized as

$$\mathbf{y}(t) = \mathbf{C}_1\mathbf{x}_1(t) \qquad \text{(and generally } \mathbf{y}(t) = \mathbf{C}\mathbf{x} + \mathbf{D}\mathbf{u}\text{)},$$

and leads to $p = 12$ controlled output variables.

We now describe the flow control problem as well as the discretized version, the DAE control problem. The ultimate objective is to minimize the average fluctuation of the velocity from the smooth steady-state flow by optimally prescribing the rotational motion of the cylinder. For well-posedness, we place a penalty on activating the control. Thus, the control problem is

$$\min_u \int_0^\infty \left\{ \mathbf{y}^T(t)\mathbf{y}(t) + u^T(t) R u(t) \right\}\, dt,$$

where $R > 0$ is a preselected constant (taken as 10 in this study), and subject to the constraint that the flow satisfies (1)–(2) from some initial perturbed flow state.

Upon discretization, the problem becomes: Find a control $\mathbf{u}(\cdot)$ that solves

$$\min_{\mathbf{u}} \int_0^\infty \left\{ \mathbf{x}_1^T(t)\mathbf{C}_1^T\mathbf{C}_1\mathbf{x}_1(t) + \mathbf{u}^T(t)\mathbf{R}\mathbf{u}(t) \right\}\, dt, \tag{4}$$

subject to

$$\begin{bmatrix} \mathbf{E}_{11} & \mathbf{0} \\ \mathbf{0} & \mathbf{0} \end{bmatrix} \begin{bmatrix} \dot{\mathbf{x}}_1(t) \\ \dot{\mathbf{x}}_2(t) \end{bmatrix} = \begin{bmatrix} \mathbf{A}_{11} & \mathbf{A}_{21}^T \\ \mathbf{A}_{21} & \mathbf{0} \end{bmatrix} \begin{bmatrix} \mathbf{x}_1(t) \\ \mathbf{x}_2(t) \end{bmatrix} + \begin{bmatrix} \mathbf{B}_1 \\ \mathbf{0} \end{bmatrix} \mathbf{u}(t), \tag{5}$$

where $\mathbf{E}_{11} \in \mathbb{R}^{n_1 \times n_1}$ is the mass matrix for the velocity fluctuation variables and has full rank. The matrix $\mathbf{A}_{11} \in \mathbb{R}^{n_1 \times n_1}$, $\mathbf{A}_{21} \in \mathbb{R}^{n_2 \times n_1}$, and $\mathbf{B}_1 \in \mathbb{R}^{n_1 \times m}$ (note that we use the Dirichlet map, cf. [44], and $m = 1$ for this problem). Since we consider stress-free outflow conditions, $\mathbf{A}_{21}$ has full rank and $\mathbf{A}_{21}\mathbf{E}_{11}^{-1}\mathbf{A}_{21}^T$ is nonsingular. Additionally, since the tangential velocity control doesn't add mass to the domain, the term $\mathbf{B}_2$ does not appear above.

To capture the von Kármán vortex street, as well as to resolve the influence of cylinder rotations on the flow for this modest Reynolds number of 60, we use a mesh with about 5,400 elements. However, in most flow control problems, typical dimensions of n_1 and n_2 prohibit the straight-forward application of linear control methods to the problem above. Therefore, we investigate the use of interpolatory model reduction methods to create modest size problems from which we can develop suitable feedback control laws.

3 Interpolatory Projections

In this section, we describe the details of the interpolatory model reduction methodology we employ. We will explain the interpolation techniques for both the general DAE framework and the index-2 Oseen model arising in our application as explained in Section 2.

3.1 Interpolatory Model Reduction of DAEs

Consider the following system of differential algebraic equations (DAEs) given in the state-space form:

$$\begin{aligned} \mathbf{E}\,\dot{\mathbf{x}}(t) &= \mathbf{A}\mathbf{x}(t) + \mathbf{B}\mathbf{u}(t), \\ \mathbf{y}(t) &= \mathbf{C}\mathbf{x}(t) + \mathbf{D}\mathbf{u}(t), \end{aligned} \tag{6}$$

where $\mathbf{x}(t) \in \mathbb{R}^n$ represent the internal variables, $\mathbf{u}(t) \in \mathbb{R}^m$ are the inputs (excitation) and $\mathbf{y}(t) \in \mathbb{R}^p$ are the outputs (observation) of the underlying dynamical system. In (6), $\mathbf{E} \in \mathbb{R}^{n \times n}$ is a *singular* matrix, thus leading to a DAE system, $\mathbf{A} \in \mathbb{R}^{n \times n}$, $\mathbf{B} \in \mathbb{R}^{n \times m}$, $\mathbf{C} \in \mathbb{R}^{p \times n}$, and $\mathbf{D} \in \mathbb{R}^{p \times m}$. The model reduction framework for linear dynamical systems, especially for the interpolatory methods, is best understood in the frequency domain. Towards this goal, let $\widehat{\mathbf{u}}(s)$ and $\widehat{\mathbf{y}}(s)$ denote the Laplace transforms of $\mathbf{u}(t)$ and $\mathbf{y}(t)$, respectively, and take the Laplace transformation of (6) to obtain

$$\widehat{\mathbf{y}}(s) = \mathbf{G}(s)\widehat{\mathbf{u}}(s), \text{ where } \mathbf{G}(s) = \mathbf{C}(s\mathbf{E} - \mathbf{A})^{-1}\mathbf{B} + \mathbf{D}. \tag{7}$$

In (7), $\mathbf{G}(s)$ is called the transfer function of (6). We will denote both the underlying dynamical system and its transfer function by $\mathbf{G}$.

In this setting of model reduction, the goal is to construct a reduced model of the form

$$\begin{aligned} \widetilde{\mathbf{E}}\,\dot{\widetilde{\mathbf{x}}}(t) &= \widetilde{\mathbf{A}}\widetilde{\mathbf{x}}(t) + \widetilde{\mathbf{B}}\mathbf{u}(t), \\ \widetilde{\mathbf{y}}(t) &= \widetilde{\mathbf{C}}\widetilde{\mathbf{x}}(t) + \widetilde{\mathbf{D}}\mathbf{u}(t), \end{aligned} \tag{8}$$

where $\widetilde{\mathbf{E}}, \widetilde{\mathbf{A}} \in \mathbb{R}^{r \times r}$, $\widetilde{\mathbf{B}} \in \mathbb{R}^{r \times m}$, $\widetilde{\mathbf{C}} \in \mathbb{R}^{p \times r}$, and $\widetilde{\mathbf{D}} \in \mathbb{R}^{p \times m}$ with $r \ll n$ such that the reduced model output $\widetilde{\mathbf{y}}(t)$ approximates the original output $\mathbf{y}(t)$ for a wide range of input selections $\mathbf{u}(t)$ with bounded energy. As for the full model, we obtain the transfer function of the reduced model by taking the Laplace transform of (8):

$$\widetilde{\mathbf{G}}(s) = \widetilde{\mathbf{C}}(s\widetilde{\mathbf{E}} - \widetilde{\mathbf{A}})^{-1}\widetilde{\mathbf{B}} + \widetilde{\mathbf{D}}. \tag{9}$$

Thus, in the frequency domain, we can view the model reduction problem as a rational approximation problem in which we search for a reduced order rational function $\widetilde{\mathbf{G}}(s)$ to approximate the full order one $\mathbf{G}(s)$.

We will employ the commonly used Petrov-Galerkin projection framework to obtain the reduced model. We will construct two model reduction bases $\mathbf{V} \in \mathbb{R}^{n\times r}$ and $\mathbf{W} \in \mathbb{R}^{n\times r}$, approximate the full-order state $\mathbf{x}(t)$ by $\mathbf{V}\widetilde{\mathbf{x}}(t)$, and obtain the reduced-order model in (8) using

$$\widetilde{\mathbf{E}} = \mathbf{W}^T\mathbf{E}\mathbf{V}, \quad \widetilde{\mathbf{A}} = \mathbf{W}^T\mathbf{A}\mathbf{V}, \quad \widetilde{\mathbf{B}} = \mathbf{W}^T\mathbf{B}, \quad \text{and} \quad \widetilde{\mathbf{C}} = \mathbf{C}\mathbf{V}. \tag{10}$$

The feedthrough term $\widetilde{\mathbf{D}}$ will be chosen appropriately to enforce matching around $s = \infty$. For the case of ordinary differential equations (ODEs) where $\mathbf{E}$ is nonsingular, the generic choice is $\widetilde{\mathbf{D}} = \mathbf{D}$. However, for DAEs due to the eigenvalue of the matrix pencil $\lambda\mathbf{E} - \mathbf{A}$ at infinity, special care is needed in choosing $\widetilde{\mathbf{D}}$.

3.2 Model Reduction by Rational Tangential Interpolation

In model reduction by tangential interpolation, the goal is to construct a reduced transfer function $\widetilde{\mathbf{G}}(s)$ that interpolates $\mathbf{G}(s)$ at selected points in the complex plane along selected directions. The interpolation data consists of the interpolation points $\{\sigma_i\}_{i=1}^r \in \mathbb{C}$ together with the left tangential directions $\{\mathsf{c}_i\}_{i=1}^r \in \mathbb{C}^p$ and the right tangential directions $\{\mathsf{b}_i\}_{i=1}^r \in \mathbb{C}^m$. The usage of the terms "left" and "right" will be clarified once we define the interpolation problem: Given $\mathbf{G}(s)$ and the interpolation data, find a reduced model $\widetilde{\mathbf{G}}(s) = \widetilde{\mathbf{C}}(s\widetilde{\mathbf{E}} - \widetilde{\mathbf{A}})^{-1}\widetilde{\mathbf{B}} + \widetilde{\mathbf{D}}$ that satisfies, for $j = 1, \ldots, r$,

$$\begin{aligned} \mathsf{c}_i^T\mathbf{G}(\sigma_j) &= \mathsf{c}_i^T\widetilde{\mathbf{G}}(\sigma_j), \\ \mathbf{G}(\sigma_j)\mathsf{b}_j &= \widetilde{\mathbf{G}}(\sigma_j)\mathsf{b}_j, \qquad \text{and} \\ \mathsf{c}_i^T\mathbf{G}'(\sigma_j)\mathsf{b}_j &= \mathsf{c}_i^T\widetilde{\mathbf{G}}'(\sigma_j)\mathsf{b}_j. \end{aligned} \tag{11}$$

In other words, we require the reduced rational function $\widetilde{\mathbf{G}}(s)$ (the reduced model) to be a bitangential Hermite interpolant the original rational function $\mathbf{G}(s)$ (the full model). One might require interpolating higher-order derivatives of $\mathbf{G}(s)$ as well. Moreover, one might also choose different sets of interpolation points (i.e. the right and left interpolation points) along with the left and right tangential direction vectors. For brevity of the paper, we will consider only up to Hermite interpolation and choose one set of interpolation points. For the details of the general case, we refer the reader to [7,31,35].

The fundamental difference between model reduction of DAEs and ODEs is that due to the eigenvalue at infinity, the transfer function of a DAE system might contain a polynomial part. The reduced transfer function is required to exactly match the polynomial part of $\mathbf{G}(s)$; otherwise the error around $s = \infty$ can grow unbounded leading to unbounded model reduction error. Therefore, model reduction methods for DAEs aims to enforce matching of polynomial parts; see, e.g., [2,16,35,40,50,67] and the references therein.

Towards this goal, let $\mathbf{G}(s)$ be additively decomposed as

$$\mathbf{G}(s) = \mathbf{G}_{\text{sp}}(s) + \mathbf{P}(s), \tag{12}$$

where $\mathbf{G}_{\text{sp}}(s)$ is the strictly proper rational part, i.e., $\lim_{s\to\infty}\mathbf{G}(s) = 0$ and $\mathbf{P}(s)$ is the polynomial part of $\mathbf{G}(s)$. We will require that the reduced transfer function $\widetilde{\mathbf{G}}(s)$ have exactly the same polynomial part as $\mathbf{G}(s)$, i.e.,

$$\widetilde{\mathbf{G}}(s) = \widetilde{\mathbf{G}}_{\text{sp}}(s) + \widetilde{\mathbf{P}}(s), \quad \text{where} \quad \widetilde{\mathbf{P}}(s) = \mathbf{P}(s),$$

and $\widetilde{\mathbf{G}}_{\text{sp}}(s)$ is the strictly proper rational part. This will guarantee that the error transfer function does not contain a polynomial part and is simply given by $\mathbf{G}_{\text{err}}(s) = \mathbf{G}(s) - \widetilde{\mathbf{G}}(s) = \mathbf{G}_{\text{sp}}(s) - \widetilde{\mathbf{G}}_{\text{sp}}(s)$. For model reduction by tangential interpolation, [35, 72] showed how to construct the model reduction bases $\mathbf{V}$ and $\mathbf{W}$ so that the reduced-model of (10) satisfies the interpolation conditions (11) in addition to guaranteeing $\widetilde{\mathbf{P}}(s) = \mathbf{P}(s)$. As expected, the left and right deflating subspaces of the pencil $\lambda\mathbf{E} - \mathbf{A}$ corresponding to the finite and infinite eigenvalues will play a fundamental role in achieving this goal. The next result is a special case of Theorem 3.1 in [35] simplified to Hermite interpolation.

Theorem 1. *Given $\mathbf{G}(s) = \mathbf{C}(s\mathbf{E} - \mathbf{A})^{-1}\mathbf{B} + \mathbf{D}$, let $\mathbf{P}_l$ and $\mathbf{P}_r$ be the spectral projectors onto the left and right deflating subspaces of the pencil $\lambda\mathbf{E} - \mathbf{A}$ corresponding to the finite eigenvalues. Let the columns of $\mathbf{W}_\infty$ and $\mathbf{V}_\infty$ span the left and right deflating subspaces of $\lambda\mathbf{E} - \mathbf{A}$ corresponding to the eigenvalue at infinity. Let $\sigma_i \in \mathbb{C}$, for $i = 1, \dots, r$ be interpolation points such that $\sigma_i\mathbf{E} - \mathbf{A}$ and $\sigma_i\widetilde{\mathbf{E}} - \widetilde{\mathbf{A}}$ are nonsingular. Also let $\mathsf{b}_i \in \mathbb{C}^m$ and $\mathsf{c}_i \in \mathbb{C}^p$ be the corresponding tangential direction vectors for $i = 1, \dots, r$. Construct $\mathbf{V}_f$ and $\mathbf{W}_f$ such that*

$$(\sigma_i\mathbf{E} - \mathbf{A})^{-1}\mathbf{P}_l\mathbf{B}\mathsf{b}_i \in \operatorname{Im}(\mathbf{V}_f) \quad \textit{for} \quad i = 1, \dots, r, \tag{13}$$

$$\textit{and} \quad (\sigma_i\mathbf{E} - \mathbf{A})^{-T}\mathbf{P}_r^T\mathbf{C}^T\mathsf{c}_i \in \operatorname{Im}(\mathbf{W}_f) \quad \textit{for} \quad i = 1, \dots, r. \tag{14}$$

Then with the choice of $\mathbf{W} = [\,\mathbf{W}_f,\ \mathbf{W}_\infty\,]$, $\mathbf{V} = [\,\mathbf{V}_f,\ \mathbf{V}_\infty\,]$, and $\widetilde{\mathbf{D}} = \mathbf{D}$, the reduced-order model $\widetilde{\mathbf{G}}(s) = \widetilde{\mathbf{C}}(s\widetilde{\mathbf{E}} - \widetilde{\mathbf{A}})^{-1}\widetilde{\mathbf{B}} + \widetilde{\mathbf{D}}$ obtained via projection as in (10) satisfies the bitangential Hermite interpolation conditions (11) as well as $\widetilde{\mathbf{P}}(s) = \mathbf{P}(s)$.

Even though Theorem 1 resolves the tangential interpolation problem for DAEs, it comes with a numerical caveat that it explicitly uses the spectral projectors $\mathbf{P}_r$ and $\mathbf{P}_l$ in the model reduction step. For large-scale DAEs, computing $\mathbf{P}_r$ and $\mathbf{P}_l$ is, at best, very costly if not infeasible. Therefore, it is important to construct the model reduction bases without forming $\mathbf{P}_r$ and $\mathbf{P}_l$ explicitly. Fortunately, for the Stokes-type descriptor systems of index 2, [35, 72] recently showed how to apply interpolatory projections without forming $\mathbf{P}_r$ and $\mathbf{P}_l$ explicitly. This is what we consider next.

3.3 Interpolation Theorem for Stokes-type DAEs of Index 2

Recall the linearized DAE in (5), together with the output equation, appearing as the constraint for the optimal control problem:

$$\begin{bmatrix} \mathbf{E}_{11} & \mathbf{0} \\ \mathbf{0} & \mathbf{0} \end{bmatrix} \begin{bmatrix} \dot{\mathbf{x}}_1(t) \\ \dot{\mathbf{x}}_2(t) \end{bmatrix} = \begin{bmatrix} \mathbf{A}_{11} & \mathbf{A}_{21}^T \\ \mathbf{A}_{21} & \mathbf{0} \end{bmatrix} \begin{bmatrix} \mathbf{x}_1(t) \\ \mathbf{x}_2(t) \end{bmatrix} + \begin{bmatrix} \mathbf{B}_1 \\ \mathbf{B}_2 \end{bmatrix} \mathbf{u}(t), \tag{15}$$

$$\mathbf{y}(t) = \mathbf{C}_1\mathbf{x}_1(t) + \mathbf{C}_2\mathbf{x}_2(t) + \mathbf{D}\mathbf{u}(t), \tag{16}$$

where $\mathbf{E}_{11}$ is nonsingular, $\mathbf{B}_2 = \mathbf{0}$, $\mathbf{A}_{21}$ has full rank and $\mathbf{A}_{21}\mathbf{E}_{11}^{-1}\mathbf{A}_{21}^T$ is nonsingular. In this case, system (15) is of index 2. The next theorem from [35] will show how to construct a reduced model for (15) without requiring the deflating projectors. For details of the derivations, we refer the reader to [35]. Also, for balanced-truncation based model reduction of (15), see [40].

Theorem 2. *Given are the full-order DAE in (15), and the interpolation points* $\sigma_i \in \mathbb{C}$ *together with the tangential direction vectors* $\mathsf{b}_i \in \mathbb{C}^m$ *and* $\mathsf{c}_i \in \mathbb{C}^p$ $i = 1, \ldots, r$. *Let* $\mathbf{v}_i$ *and* $\mathbf{w}_i$ *solve*

$$\begin{bmatrix} \sigma_i\mathbf{E}_{11} - \mathbf{A}_{11} & \mathbf{A}_{21}^T \\ \mathbf{A}_{21} & \mathbf{0} \end{bmatrix} \begin{bmatrix} \mathbf{v}_i \\ \mathbf{z} \end{bmatrix} = \begin{bmatrix} \mathbf{B}_1\mathsf{b}_i \\ \mathbf{0} \end{bmatrix}, \tag{17}$$

and

$$\begin{bmatrix} \sigma_i\mathbf{E}_{11}^T - \mathbf{A}_{11}^T & \mathbf{A}_{21}^T \\ \mathbf{A}_{21} & \mathbf{0} \end{bmatrix} \begin{bmatrix} \mathbf{w}_i \\ \mathbf{q} \end{bmatrix} = \begin{bmatrix} \mathbf{C}^T\mathsf{c}_i \\ \mathbf{0} \end{bmatrix}. \tag{18}$$

for $i = 1, \ldots, r$. *Construct*

$$\mathbf{V} = [\mathbf{v}_1, \ldots, \mathbf{v}_r], \quad \textit{and} \quad \mathbf{W} = [\mathbf{w}_1, \ldots, \mathbf{w}_r]. \tag{19}$$

Define

$$\widetilde{\mathbf{D}} = \mathbf{D} - \mathbf{C}_2(\mathbf{A}_{21}\mathbf{E}_{11}^{-1}\mathbf{A}_{21}^T)^{-1}\mathbf{A}_{21}\mathbf{E}_{11}^{-1}\mathbf{B}_1.$$

Then the reduced model

$$\widetilde{\mathbf{G}}(s) = \mathbf{C}\mathbf{V}(s\mathbf{W}^T\mathbf{E}_{11}\mathbf{V} - \mathbf{W}^T\mathbf{A}_{11}\mathbf{V})^{-1}\mathbf{W}^T\mathbf{B}_1 + \widetilde{\mathbf{D}}. \tag{20}$$

satisfies the bitangential Hermite interpolation conditions (11) as well as $\widetilde{\mathbf{P}}(s) = \mathbf{P}(s)$.

Note that the expensive spectral projector computations are completely avoided; the only numerical cost is the need to solve $2r$ (sparse) linear systems arising in (17) and (18).

We also note that Theorem 2 can be directly extended to the case where the algebraic equation in (15) has the form $\mathbf{0} = \mathbf{A}_{21}\mathbf{x}_1(t) + \mathbf{B}_2\mathbf{u}(t)$ with $\mathbf{B}_2 \neq \mathbf{0}$. The numerical cost stays the same; see [35, 40] for details.

4 Numerical Results

We now apply the interpolatory model reduction algorithm described in Section 3.3 to the flow control problem described in Section 2.1. A relatively coarse mesh containing 5378 triangular elements was used to discretize flow solutions in the domain $\Omega = (-7, 15) \times (-7, 7) \backslash c$ where c is the cylinder centered at the origin with unit diameter. The steady-state flow corresponding to $Re = 60$ was computed on this mesh and the resulting $(\bar{\mathbf{v}}, \bar{p})$ was used to compute the discrete model for the flow fluctuations where $n_1 = 21,390$ and $n_2 = 2,777$. Plots of $\bar{\mathbf{v}}$ components appear in Fig. 1.

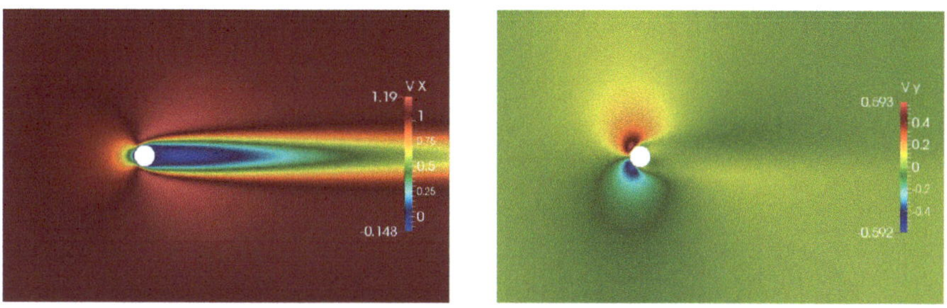

Fig. 1. Steady-state velocity contours (horizontal-left, vertical-right)

As mentioned in Section 3.3, the linearization around the steady-state solution at $Re = 60$ leads an unstable model; i.e the full-model DAE in (15) is unstable. There are two unstable poles are at $5.2480 \times 10^{-2} \pm \imath 7.6720 \times 10^{-1}$. We note that the model reduction framework we use does not require computing these unstable poles; we include them for comparison to the reduced model. The flexibility of the interpolatory model reduction is that even though the original model is unstable, Theorem 2 can still be applied as long as the interpolation points are not chosen as one of the poles. We have simply chosen 33 complex conjugate pairs (overall 66 points) on the imaginary axis as the interpolation points. The imaginary parts of the interpolation points varied from 10^{-3} to 10^{3}. Then, using Theorem 2, we have constructed our model reduction space $\mathbf{V} \in \mathbb{R}^{21390 \times 66}$. To preserve the symmetry in $\mathbf{E}_{11}$, we have simply set $\mathbf{W} = \mathbf{V}$. Thus, the reduced transfer function is a Lagrange interpolant in this case, not a Hermite interpolant. Due to the complex conjugate-pairs, construction of $\mathbf{V}$ required only 33 sparse linear solves. Then, using a short-SVD (a relatively minor computational task due to the small number of columns in $\mathbf{V}$), we have removed the linear dependent columns from $\mathbf{V}$, and reduced the dimension to $r = 60$; thus having a final reduced model of order $r = 60$.

An important requirement of the reduced model in this optimal control setting is that the reduced model should capture the unstable poles of the original model so that the controller design based on the reduced model can work effectively on the full-model. As for the full-order model, our reduced model has exactly two unstable poles. The unstable poles of $\mathbf{G}(s)$ and $\widetilde{\mathbf{G}}(s)$ are listed below:

$$\begin{aligned} \lambda_{\text{unstable}}(\mathbf{G}(s)) :& \quad 5.248019596820730 \times 10^{-2} \pm \imath\, 7.672028760928972 \times 10^{-1} \\ \lambda_{\text{unstable}}(\widetilde{\mathbf{G}}(s)) :& \quad 5.248030491505502 \times 10^{-2} \pm \imath\, 7.672029050490372 \times 10^{-1} \end{aligned}$$

As these numbers show, the unstable poles of $\mathbf{G}(s)$ are captured to a great accuracy as desired. To further illustrate the quality of the reduced model, in Fig. 2, we depict the singular values plots of frequency responses of $\mathbf{G}(s)$ and $\widetilde{\mathbf{G}}(s)$, i.e. $\|\mathbf{G}(\imath\omega)\|$ and $\|\widetilde{\mathbf{G}}(\imath\omega)\|$ vs $\omega \in \mathbb{R}$. As the figure shows, $\widetilde{\mathbf{G}}(s)$ replicates $\mathbf{G}(s)$ almost exactly.

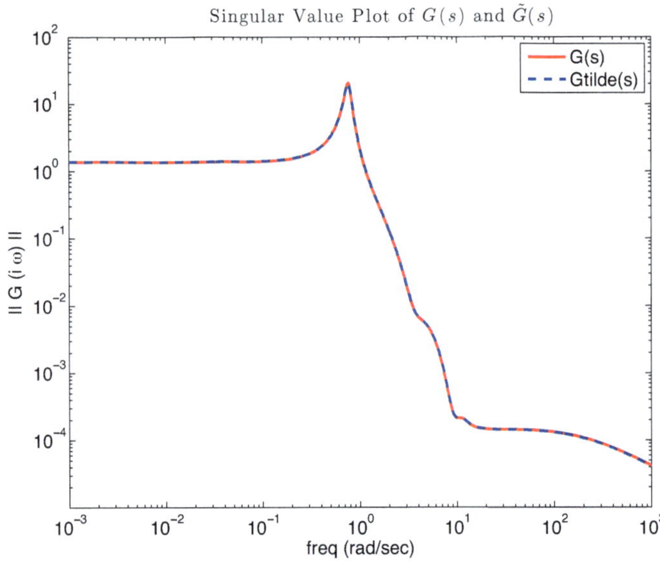

Fig. 2. The singular value plots of the full-model $\mathbf{G}(s)$ and the reduced model $\widetilde{\mathbf{G}}(s)$

We use the reduced-order model (20) to compute an approximate solution to the LQR problem (4). Therefore, since $\widetilde{\mathbf{D}} = \mathbf{0}$ in our example, we consider the solution to the problem

$$\min_{\mathbf{u}} \int_0^\infty \widetilde{\mathbf{x}}_1^T(t)\widetilde{\mathbf{C}}_1^T\widetilde{\mathbf{C}}_1\widetilde{\mathbf{x}}_1(t) + \mathbf{u}^T(t)\mathbf{R}\mathbf{u}(t)\; dt \tag{21}$$

subject to $\widetilde{\mathbf{x}}_1(\cdot)$ solving (8). The solution can be computed by solving the algebraic Riccati equation (using the `care` function in Matlab)

$$\widetilde{\mathbf{A}}_{11}^T\mathbf{P}\widetilde{\mathbf{E}}_{11} + \widetilde{\mathbf{E}}_{11}^T\mathbf{P}\widetilde{\mathbf{A}}_{11} - \widetilde{\mathbf{E}}_{11}^T\mathbf{P}\widetilde{\mathbf{B}}_1\mathbf{R}^{-1}\widetilde{\mathbf{B}}_1^T\mathbf{P}\widetilde{\mathbf{E}}_{11} + \widetilde{\mathbf{C}}_1^T\widetilde{\mathbf{C}}_1 = \mathbf{0}$$

for the positive definite, symmetric solution $\mathbf{P}$, then computing

$$\widetilde{\mathbf{K}} = \mathbf{R}^{-1}\widetilde{\mathbf{B}}_1^T \mathbf{P} \widetilde{\mathbf{E}}_{11}.$$

The solution to (21) is then $\mathbf{u} = -\widetilde{\mathbf{K}}\widetilde{\mathbf{x}}_1$. To find the representation of the control law in the original (full-order, discrete) variables, we can use

$$\mathbf{u} = -\underbrace{\widetilde{\mathbf{K}}\mathbf{V}^T}_{\mathbf{K}} \underbrace{\mathbf{V}\widetilde{\mathbf{x}}_1}_{\approx \mathbf{x}_1}.$$

Finally, we can consider the computation of $\mathbf{u}$ as an approximation of the infinite dimensional control problem where $\mathbf{v} = (v_\xi, v_\eta)$ for spatial variables (ξ, η)

$$u(t) = -\int_\Omega h_{v_\xi}(\xi,\eta)(v_\xi(\xi,\eta,t) - \bar{v}_\xi(\xi,\eta)) + h_{v_\eta}(\xi,\eta)(v_\eta(\xi,\eta,t) - \bar{v}_\eta(\xi,\eta))\, d\xi\, d\eta.$$

The finite element representations of the gains h_{v_ξ} and h_{v_η} corresponding, respectively, to the horizontal and vertical components of the velocity fluctuations are plotted in Fig. 3. These compare well to those calculated for the same

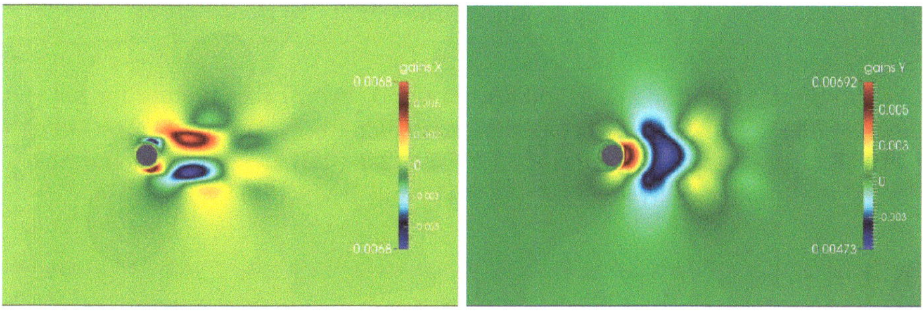

Fig. 3. Functional gains (horizontal h_{v_ξ}-left, vertical h_{v_η}-right)

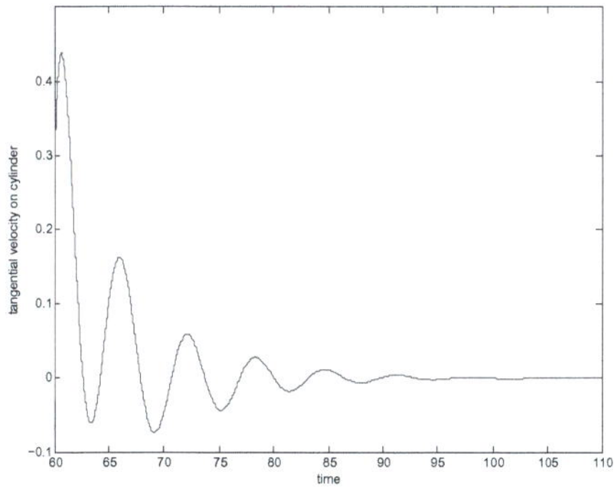

Fig. 4. Control Input: Tangential Cylinder Velocity

problem (with a different $\mathbf{C}$ operator and slightly higher Reynolds number of 100) appearing in Fig. 4 of [1]. Considering that the gains computed in this study were computed with dramatically less computational cost emphasizes the feasibility of this approach.

To verify that this control law stabilizes the flow, we simulated the von Kármán vortex street for 60 seconds, then applied the full-state feedback control. The control was able to nearly return the flow to the steady-state flow in 50 seconds of simulation. The control input is plotted in Fig. 4.

5 Conclusions and Future Work

We have shown that with modest cost, using interpolatory model reduction we can produce accurate reduced models for index-2 DAEs arising from flow control problems. Using this model reduction framework within a control setting led to qualitatively similar functional gains as computed using more expensive control algorithms. We will apply this approach to study more complicated settings including flow control problems with higher Reynolds numbers and finer discretization. We will also investigate the practical problem of building effective state estimators and reduced-order compensators, as well as the effect of different controlled output operators on the quality of the feedback control.

Acknowledgements. We would like to thank Prof. Peter Benner for his valuable comments. This work was supported in part by the Air Force Office of Scientific Research under contract FA9550-12-1-0173 and the National Science Foundation under contracts DMS-1016450 and DMS-1217156.

References

1. Akhtar, I., Borggaard, J., Stoyanov, M., Zietsman, L.: On commutation of reduction and control: Linear feedback control of a von Kármán street. In: Proc. 5th AIAA Flow Control Conference, Chicago, IL (2010), AIAA Paper 2010-4832
2. Alì, G., Banagaaya, N., Schilders, W., Tischendorf, C.: Index-aware model order reduction for linear index-2 DAEs with constant coefficients. SIAM Journal on Scientific Computing 35(3), A1487–A1510 (2013)
3. Amsallem, D., Farhat, C.: Interpolation method for the adaptation of reduced-order models to parameter changes and its application to aeroelasticity. AIAA Journal 46, 1803–1813 (2008)
4. Antil, H., Heinkenschloss, M., Hoppe, R.H.W.: Domain decomposition and balanced truncation model reduction for shape optimization of the Stokes system. Optimization Methods and Software 26(4-5), 643–669 (2011)
5. Antil, H., Heinkenschloss, M., Hoppe, R.H.W., Linsenmann, C., Wixforth, A.: Reduced order modeling based shape optimization of surface acoustic wave driven microfluidic biochips. Mathematics and Computers in Simulation 82(10), 1986–2003 (2012)
6. Antoulas, A.C.: Approximation of Large-Scale Dynamical Systems. SIAM Publications, Philadelphia (2005)

7. Antoulas, A.C., Beattie, C.A., Gugercin, S.: Interpolatory model reduction of large-scale dynamical systems. In: Mohammadpour, J., Grigoriadis, K. (eds.) Efficient Modeling and Control of Large-Scale Systems, pp. 2–58. Springer (2010)
8. Arian, E., Fahl, M., Sachs, E.: Trust-region proper orthogonal decomposition models by optimization methods. In: Proceedings of the 41st IEEE Conference on Decision and Control, Las Vegas, NV, pp. 3300–3305 (2002)
9. Atwell, J.A., King, B.B.: Proper orthogonal decomposition for reduced basis feedback controllers for parabolic equations. Mathematical and Computer Modelling 33, 1–19 (2001)
10. Bai, Z., Feldmann, P., Freund, R.W.: How to make theoretically passive reduced-order models passive in practice. In: Proceedings of the IEEE Custom Integrated Circuits Conference, pp. 207–210 (1998)
11. Banks, H.T., Beeler, S.C., Kepler, G.M., Tran, H.T.: Reduced order modeling and control of thin film growth in an HPCVD reactor. SIAM J. on Appl. Math. 62(4) (2002)
12. Baur, U., Benner, P., Beattie, C.A., Gugercin, S.: Interpolatory projection methods for parameterized model reduction. SIAM Journal on Scientific Computing 33, 2489–2518 (2011)
13. Baur, U., Benner, P., Greiner, A., Korvink, J.G., Lienemann, J., Moosmann, C.: Parameter preserving model order reduction for MEMS applications. Mathematical and Computer Modelling of Dynamical Systems 17(4), 297–317 (2011)
14. Benner, P., Hinze, M., ter Maten, E.J.W. (eds.): Model Reduction for Circuit Simulation. LNEE, vol. 74. Springer, Heidelberg (2011)
15. Benner, P., Mehrmann, V.L., Sorensen, D.C.: Dimension Reduction of Large-scale Systems: Proceedings of a Workshop Held in Oberwolfach, Germany, October 19-25, vol. 45. Springer (2005)
16. Benner, P., Sokolov, V.I.: Partial realization of descriptor systems. Systems & Control Letters 55(11), 929–938 (2006)
17. Bewley, T., Pralits, J., Luchini, P.: Minimal-energy control feedback for stabilization of bluff-body wakes based on unstable open-loop eigenvalues and left eigenvectors. In: Proceedings of the Fifth Conference on Bluff Body Wakes and Vortex-Induced Vibrations (BBVIV5), San Paolo, pp. 129–132 (2007)
18. Bindel, D.S., Bai, Z., Demmel, J.W.: Model reduction for RF MEMS simulation. In: Dongarra, J., Madsen, K., Waśniewski, J. (eds.) PARA 2004. LNCS, vol. 3732, pp. 286–295. Springer, Heidelberg (2006)
19. Bond, B.N., Daniel, L.: A piecewise-linear moment-matching approach to parameterized model-order reduction for highly nonlinear systems. IEEE Transactions on Computer-Aided Design of Integrated Circuits and Systems 26(12), 2116–2129 (2007)
20. Borggaard, J., Stoyanov, M.: A reduced order solver for Lyapunov equations with high rank matrices. In: Proc. 18th International Symposium on Mathematical Theory of Networks and Systems, Blacksburg, VA (2008), Paper 169
21. Bui-Thanh, T., Willcox, K., Ghattas, O.: Parametric reduced-order models for probabilistic analysis of unsteady aerodynamic applications. AIAA Journal 46(10), 2520–2529 (2008)
22. Burns, J.: Nonlinear distributed parameter control systems with non-normal linearizations: Applications and approximations. In: Smith, R.C., Demetriou, M.A. (eds.) Research Directions in Distributed Parameter Systems, pp. 17–53. SIAM (2003)

23. Daniel, L., Siong, O.C., Low, S.C., Lee, K.H., White, J.: A multiparameter moment matching model reduction approach for generating geometrically parameterized interconnect performance models. IEEE Transactions on Computer-Aided Design of Integrated Circuits and Systems 23(5), 678–693 (2004)
24. de Sturler, E., Gugercin, S., Kilmer, M.E., Chaturantabut, S., Beattie, C.A., O'Connell, M.: Nonlinear parametric inversion using interpolatory model reduction. arXiv preprint arXiv:1311.0922 (2013)
25. Deane, A.E., Kevrekidis, I.G., Karniadakis, G.E., Orszag, S.A.: Low-dimensional models for complex geometry flows: Application to grooved channels and circular cylinders. Physics of Fluids A: Fluid Dynamics 3, 2337 (1991)
26. Druskin, V., Simoncini, V., Zaslavsky, M.: Solution of the time-domain inverse resistivity problem in the model reduction framework part I. one-dimensional problem with SISO data. SIAM Journal on Scientific Computing 35(3), A1621–A1640 (2013)
27. Feldmann, P., Freund, R.W.: Efficient linear circuit analysis by Padé approximation via the Lanczos process. IEEE Transactions on Computer-Aided Design of Integrated Circuits and Systems 14, 639–649 (1995)
28. Feng, L.: Parameter independent model order reduction. Mathematics and Computers in Simulation 68(3), 221–234 (2005)
29. Filler, J.R., Marston, P.L., Mih, W.C.: Response of the shear layers separating from a circular cylinder to small-amplitude rotational oscillations. Journal of Fluid Mechanics 231, 481–499 (1991)
30. Galbally, D., Fidkowski, K., Willcox, K., Ghattas, O.: Nonlinear model reduction for uncertainty quantification in large-scale inverse problems. International Journal for Numerical Methods in Engineering 81(12), 1581–1608 (2010)
31. Gallivan, K., Vandendorpe, A., Van Dooren, P.: Model reduction of MIMO systems via tangential interpolation. SIAM J. Matrix Anal. Appl. 26(2), 328–349 (2005)
32. Gatski, T.B., Glauser, M.N.: Proper orthogonal decomposition based turbulence modeling. In: Instability, Transition, and Turbulence, pp. 498–510. Springer (1992)
33. Gawronski, W., Juang, J.: Model reduction for flexible structures. Control and Dynamic Systems 36, 143–222 (1990)
34. Gugercin, S., Antoulas, A.C., Beattie, C.A., Gildin, E.: Krylov-based controller reduction for large-scale systems. In: Proceedings of the 43rd IEEE Conference on Decision and Control (December 2004)
35. Gugercin, S., Stykel, T., Wyatt, S.: Model reduction of descriptor systems by interpolatory projection methods. SIAM Journal on Scientific Computing 35(5), B1010–B1033 (2013)
36. Gunzburger, M.: Finite Element Methods for Viscous Incompressible Flows. Academic Press (1989)
37. Gunzburger, M.D., Lee, H.C.: Feedback control of Karman vortex shedding. Transactions of the ASME 63, 828–835 (1996)
38. Hay, A., Borggaard, J.T., Pelletier, D.: Local improvements to reduced-order models using sensitivity analysis of the proper orthogonal decomposition. Journal of Fluid Mechanics 629, 41–72 (2009)
39. He, J.W., Glowinski, R., Metcalfe, R., Nordlander, A., Periaux, J.: Active control and drag optimization for flow past a circular cylinder: I. Oscillatory cylinder rotation. Journal of Computational Physics 163, 83–117 (2000)
40. Heinkenschloss, M., Sorensen, D.C., Sun, K.: Balanced truncation model reduction for a class of descriptor systems with application to the Oseen equations. SIAM Journal on Scientific Computing 30(2), 1038–1063 (2008)

41. Holmes, P., Lumley, J.L., Berkooz, G.: Turbulence, Coherent Structures, Dynamical Systems and Symmetry. Cambridge University Press, Cambridge (1996)
42. King, B.B.: Nonuniform grids for reduced basis design of low order feedback controllers for nonlinear continuous time systems. Math. Models & Meth. in Appl. Sci. 8, 1223–1241 (1998)
43. Kunisch, K., Volkwein, S.: Proper orthogonal decomposition for optimality systems. ESAIM: Mathematical Modelling and Numerical Analysis 42(1), 1–23 (2008)
44. Lasiecka, I.: Unified theory of abstract parabolic boundary problems–a semigroup approach. Applied Mathematics and Optimization 6, 287–333 (1980)
45. Lieberman, C., Willcox, K., Ghattas, O.: Parameter and state model reduction for large-scale statistical inverse problems. SIAM Journal on Scientific Computing 32(5), 2523–2542 (2010)
46. Lieu, T., Farhat, C.: Adaptation of aeroelastic reduced-order models and application to an F-16 configuration. AIAA Journal 45(6), 1244–1257 (2007)
47. Lieu, T., Farhat, C., Lesoinne, M.: Reduced-order fluid/structure modeling of a complete aircraft configuration. Computer Methods in Applied Mechanics and Engineering 195, 5730–5742 (2006)
48. Ly, H.V., Tran, H.T.: Modeling and control of physical processes using proper orthogonal decomposition. Journal of Mathematical and Computer Modeling 33, 223–236 (2001)
49. Mathelin, L., Le Maitre, O.: Robust control of uncertain cylinder wake flows based on robust reduced order models. Computers and Fluids 38(6), 1168–1182 (2009)
50. Mehrmann, V., Stykel, T.: Balanced truncation model reduction for large-scale systems in descripter form. In: Benner, P., Mehrmann, V., Sorensen, D.C. (eds.) MFCS 1976. Lecture Notes in Computational Science and Engineering, vol. 45, pp. 83–115. Springer, Heidelberg (1976)
51. Meyer, D.G., Srinivasan, S.: Balancing and model reduction for second-order form linear systems. IEEE Transactions on Automatic Control 41(11), 1632–1644 (1996)
52. Noack, B.R., Afanasiev, K., Morzyński, M., Tadmor, G., Thiele, F.: A hierarchy of low-dimensional models for the transient and post-transient cylinder wake. J. Fluid Mech. 497, 335–363 (2003)
53. Noack, B.R., Tadmor, G., Morzyński, M.: Actuation models and dissipative control in empirical Galerkin models of fluid flows. In: The 2004 American Control Conference, Boston, MA, U.S.A., June 30-July 2, pp. 0001–0006 (2004), Paper FrP15.6
54. Obinata, G., Anderson, B.D.O.: Model reduction for control system design. Springer, London (2001)
55. Odabasioglu, A., Celik, M., Pileggi, L.T.: PRIMA: passive reduced-order interconnect macromodeling algorithm. In: Proceedings of the 1997 IEEE/ACM International Conference on Computer-Aided Design, pp. 58–65 (1997)
56. Olinger, D.J., Sreenivasan, K.R.: Nonlinear dynamics of the wake of an oscillating cylinder. Physical Review Letters 60(9), 797–800 (1988)
57. Ou, Y.-R.: Control of oscillatory forces on a circular cylinder by rotation. Technical report, ICASE, ICASE Report No. 91–67 (1991)
58. Park, D.S., Ladd, D.M., Hendricks, E.W.: Feedback control of von Kármán vortex shedding behind a circular cylinder at low Reynolds numbers. Physics of Fluids 6(7), 2390–2405 (1994)
59. Pralits, J.O., Brandt, L., Giannetti, F.: Instability and sensitivity of the flow around a rotating circular cylinder. Journal of Fluid Mechanics 650, 513–536 (2010)
60. Prandtl, L.: The Magnus effect and windpowered ships. Naturwissenschaften 13, 93–108 (1925)

61. Protas, B.: Linear feedback stabilization of laminar vortex shedding based on a point vortex model. Physics of Fluids 16, 4473–4483 (2004)
62. Reis, T., Stykel, T.: Positive real and bounded real balancing for model reduction of descriptor systems. International Journal of Control 83(1), 74–88 (2010)
63. Rowley, C.W., Colonius, T., Murray, R.M.: POD based models of self-sustained oscillations in the flow past an open cavity. In: Proceedings of the 6th AIAA/CEAS Aeroacoustics Conference (2000), AIAA Paper Number 2000-1969
64. Rudnyi, E.B., Korvink, J.G.: Review: Automatic model reduction for transient simulation of MEMS-based devices. Sensors Update 11(1), 3–33 (2002)
65. Sirovich, L.: Turbulence and the dynamics of coherent structures. Part 1: Coherent structures. Quarterly of Applied Mathematics 45(3), 561–571 (1987)
66. Stoyanov, M.: Reduced Order Methods for Large Scale Riccati Equations. PhD thesis, Virginia Tech, Blacksburg, VA (May 2009)
67. Stykel, T.: Gramian-based model reduction for descriptor systems. Mathematics of Control, Signals, and Systems 16(4), 297–319 (2004)
68. Su, T.-J., Craig Jr., R.R.: Model reduction and control of flexible structures using Krylov vectors. Journal of Guidance, Control, and Dynamics 14(2), 260–267 (1991)
69. Taneda, S.: Visual observations of the flow past a circular cylinder performing a rotatory oscillation. Journal of the Physical Society of Japan 45(3), 1038–1043 (1978)
70. Tokumaru, P.T., Dimotakis, P.E.: Rotary oscillation control of a cylinder wake. Journal of Fluid Mechanics 224, 77–90 (1991)
71. Wang, J., Zabaras, N.: Using Bayesian statistics in the estimation of heat source in radiation. International Journal of Heat and Mass Transfer 48, 15–29 (2005)
72. Wyatt, S.: Issues in Interpolatory Model Reduction: Inexact Solves, Second-order Systems and DAEs. PhD thesis, Virginia Polytechnic Institute and State University (2012)
73. Xiao, M., Lin, Y., Myatt, J.H., Camphouse, R.C., Banda, S.S.: Rotation implementation of a circular cylinder in incompressible flow via staggered grid approach. Journal of Applied Mathematics & Computing 22(1-2), 67–82 (2006)
74. Xiao, M., Novy, M., Myatt, J., Banda, S.: A rotary control of the circular cylinder wake: An analytic approach. In: Proc. 1st AIAA Flow Control Conference (2002), AIAA Paper 2002-3075
75. Yue, Y., Meerbergen, K.: Accelerating optimization of parametric linear systems by model order reduction. SIAM Journal on Optimization 23(2), 1344–1370 (2013)

Model Reduction of Reactive Processes

Mathias Lemke[1], Agnieszka Międlar[2], Julius Reiss[1],
Volker Mehrmann[2], and Jörn Sesterhenn[1]

[1] Fachgebiet Numerische Fluiddynamik, Technische Universität Berlin, MB1,
Müller-Breslau-Straße 8, 10623 Berlin, Germany
{mlemke,reiss,sesterhenn}@tnt.tu-berlin.de
[2] Fachgebiet Numerische Mathematik, Technische Universität Berlin,
MA 4-5, Straße des 17. Juni 136, 10623 Berlin, Germany
{miedlar,mehrmann}@math.tu-berlin.de

Abstract. Interpolation based model reduction is applied to a reactive process. A zero-dimensional, perfectly stirred, constant pressure reactor with complex chemistry, modeled by the GRI3.0 scheme, is considered. The aim of this work is to analyze how interpolatory model reduction performs for combustion processes, where the solution is very sensitive to the choice of input parameters. In this study the initial temperature is chosen as varying parameter.

Keywords: reactive flow, interpolatory model reduction, reduced order model, singular value decomposition.

1 Introduction

Model reduction (MR) has developed to a vital tool in the area of fluid mechanics. It allows to determine, from calculated or measured data, a model describing quantities of interest with lower computational cost. Reduced models can be used to speed up a design process or to perform real time calculations feeding a control mechanism.

MR techniques very often build on modes, which are used to approximate a desired solution. The cost and the quality of such a reduced order model (ROM) depend on the number of modes, their properties and on the family of solutions which is to be reduced. Typically smooth solutions, which are slowly varying or periodic functions of the input parameters can be described by a few modes. However, discontinuities or steep gradients are common in systems of practical interest and pose a challenge. For example, these can be shocks in compressible and fast temperature changes in reactive flows.

We will study an interpolation based technique [3], where the reduced order model is created from modes spanning the solution space. Here the modes include the time dependence revealing the system dynamics. It makes no use of a specified structure of the original system model, which makes it a suitable candidate for the reduction of flows with discontinuities or steep gradients.

The main idea of the method is to enrich an interpolation based reduced order model (ROM) with a parameter-dependent confidence measure. A set of system

R. King (ed.), *Active Flow and Combustion Control 2014*, Notes on Numerical Fluid Mechanics and Multidisciplinary Design 127, DOI: 10.1007/978-3-319-11967-0_24

states is obtained for a number of parameter values s_i, $i = 1, \ldots, S$, for example, by solving the governing partial differential equations (PDE) and combined to a so-called training matrix. The singular value decomposition (SVD) of this matrix is performed to extract important features of the parameter dependent model. Each column of the matrix contains the full spatial-temporal solution for a given parameter, which results in a matrix of very large dimension. Fortunately, the generation of the training matrix and the computation of the SVD are parallelizable and thus feasible in practical applications. The left singular vectors depend on space and time, whereas the right singular vectors reveal the parameter dependence. In [3] the authors noticed a direct relation between the higher order modes and the right singular vectors. Namely, the latter become oscillatory in the range of parameters for which the PDE solution is very sensitive even for small parameter perturbations. This allows to select several smooth right singular vectors and use them within the interpolation based MR. Moreover, the quality of the ROM, its sensitivity, can be easily estimated with a provided error estimator.

We have investigated how well this approach performs near discontinuities present in a model problem of a homogeneous (0D) reactor with a methane air mixture. If the temperature is sufficiently high, autoignition occurs, and the temperature rapidly increases after a certain time, until the combustible is consumed. The ignition time [1] strongly depends on the initial temperature T_0. We found that the method introduced in [3] does not handle this problem satisfactorily and we propose a modification, which takes additional physical information into account. We illustrate that the modified approach gives good results.

2 Governing Equations of the Model Problem

For the simulation of a zero-dimensional (0D), perfectly stirred, constant pressure reactor the following set of equations can be used:

$$\frac{\partial \varrho Y_k}{\partial t} = \dot{\omega}_k \tag{1a}$$

$$\varrho C_p \frac{\partial T}{\partial t} = \dot{\omega}'_T, \tag{1b}$$

where ϱ denotes the density, Y_k the mass fraction of species $k = 1, \ldots, K$, T the temperature and $C_p = C_p(T)$ the isobaric specific heat capacity. The nomenclature closely follows [18]. The heat release $\dot{\omega}'_T$ is given by

$$\dot{\omega}'_T = -\sum_{k=1}^{K} h_k \dot{\omega}_k = -\sum_{k=1}^{K} h_{s,k} \dot{\omega}_k - \sum_{k=1}^{K} \Delta h^0_{f,k} \dot{\omega}_k, \tag{2}$$

with the sensible enthalpies $h_{s,k}$ and mass formation enthalpies $\Delta h^o_{f,k}$ at a certain reference temperature T_0. The rate $\dot{\omega}_k$ is defined as

$$\dot{\omega}_k = \sum_{j=1}^{K} \dot{\omega}_{kj}, \tag{3}$$

with the reaction rates $\dot{\omega}_{kj}$

$$\dot{\omega}_{kj} = \mathcal{Q}_j W_k \nu_{kj}, \qquad \text{for all reactions} \qquad j = 1, \ldots, J. \tag{4}$$

Here, W_k denotes the molecular weight and $\nu_{kj} = \nu''_{kj} - \nu'_{kj}$ the molar stoichiometric coefficients of species k in reaction j.

The progress rates $\mathcal{Q}_j$ are defined by:

$$\mathcal{Q}_j = K_{\text{forward},j} \prod_{k=1}^{K} [X_k]^{\nu'_{kj}} - K_{\text{reverse},j} \prod_{k=1}^{K} [X_k]^{\nu''_{kj}}, \tag{5}$$

with $[X_k]$ being the molar concentration of species k. The set of equations is closed with the law of ideal gas:

$$p = \varrho \frac{R}{W} T, \tag{6}$$

with R being the ideal gas constant and W the mean molecular weight.

$$\frac{1}{W} = \sum_{k=1}^{K} \frac{Y_k}{W_k}. \tag{7}$$

The temperature dependence of the specific heat capacity is captured by use of corresponding NASA polynomials [2,15].

2.1 Arrhenius Approach

The rate coefficients $K_{\text{forward},j}$ and $K_{\text{reverse},j}$ are modeled by an Arrhenius approach [17,18,23]:

$$K_{\text{forward},j} = A_{fj} T^{\beta_j} \exp\left(\frac{-E_{\text{a},j}}{RT}\right), \tag{8}$$

where A_{fj} corresponds to the so-called pre-exponential or frequency factor and $E_{\text{a},j}$ to the activation energy. β_j reflects an additional temperature dependence. The reverse rate is given by:

$$K_{\text{reverse},j} = \frac{K_{\text{forward},j}}{C \exp\left(\frac{\Delta G_j^0}{RT}\right)}, \tag{9}$$

following from the equilibrium with ΔG_j^0 as Gibbs free energy [23]. The term C is defined as:

$$C = \left(\frac{p_{\text{atm}}}{RT}\right)^{\sum_{k=1}^{N} \nu_{kj}} \tag{10}$$

and acts as a unit correction, with the atmospheric pressure p_{atm}.

2.2 Reaction Mechanism

The coefficients A_f, β and E_a are obtained from model reaction schemes, which provide a reduced description of the full reaction kinetics. Except for certain reactions, the coefficients are determined experimentally and often optimized toward a certain application, e.g. pure oxygen [13] or natural gas flames [19].

Within this work we will use the well known GRI3.0 [7] reaction scheme. It is designed to model natural gas combustion under atmospheric conditions, which enables it to cover a large variety of applications. The GRI3.0 scheme includes 325 reactions and $K = 53$ species consisting of 5 elements, i.e., O, H, C, N, Ar. It contains hydrocarbons as well as nitrogen oxides.

We will now follow the notation of [5]. The GRI3.0 reaction mechanism involves three-body, Lindemann and Troe falloff reaction types. For the three-body reaction $A + B + M \leftrightarrows AB + M$ the progress rates are weighted by:

$$[M] = \sum_{k=1}^{K} \epsilon_k [X_k], \tag{11}$$

with reaction scheme-dependent efficiencies ϵ_k and possible collision partners M. The Lindemann form of the falloff reactions [14] is implemented by

$$K_{\mathrm{forward},j} = \frac{K_{0,j}\,[M]}{1 + \frac{K_{0,j}[M]}{K_{\infty,j}}}, \tag{12}$$

where $K_{0,j}$ defines the low- and $K_{\infty,j}$ the high pressure limit. The corresponding kinetic parameters are given within the reaction scheme. The approach can be generalized by defining a reduced pressure p_red

$$p_{\mathrm{red},j} = \frac{K_{0,j}\,[M]}{K_{\infty,j}}, \tag{13}$$

and

$$K_{\mathrm{forward},j} = K_{\infty,j}\left(\frac{p_{\mathrm{red},j}}{1 + p_{\mathrm{red},j}}\right) F_{\mathrm{falloff}}(T, p_{\mathrm{red},j}), \tag{14}$$

with the so-called *falloff function* F_falloff. For Troe type falloff reactions [4] a four parameter function is used. Again, all the values are given within the reaction scheme.

2.3 Numerical Validation

We have performed all our simulations with a serial MATLAB (8.1.0.604) implementation of the equations (1a) – (1b). Since these equations are stiff, the integrated solver `ode15s` is used. Also other integration methods can be applied. However, additional attention has to be paid to the size of time steps. If the step size is too large the reaction of the mixture maybe missed in the simulation.

In order to validate our implementation, a comparison with the open-source chemical kinetics software CANTERA [5] is carried out. Two cases are analyzed:

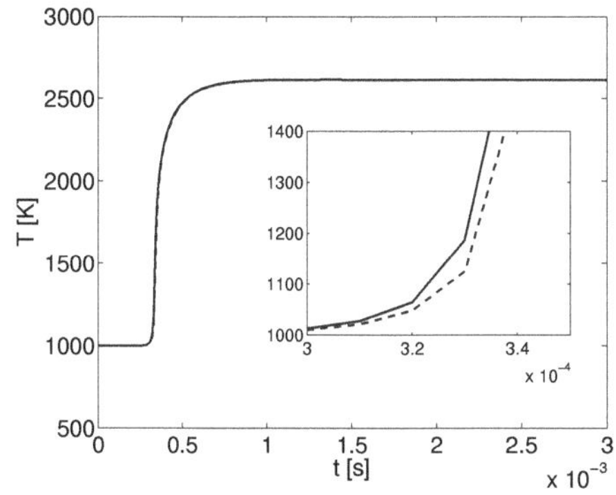

Fig. 1. Case (a): Temperature history of a hydrogen combustion obtained with CANTERA (solid) and our implementation (dashed).

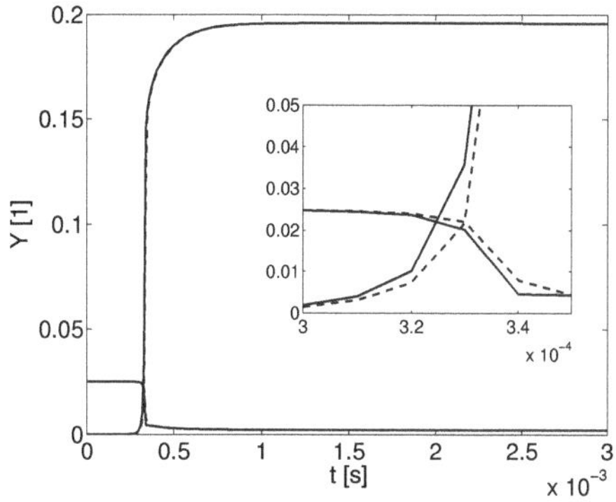

Fig. 2. Case (a): Mass fractions for H_2 and H_2O obtained with CANTERA (solid) and our implementation (dashed).

the combustion of (a) hydrogen and (b) methane. For (a) the initial mixture is given by $Y_{H_2} = 0.025$, $Y_{O_2} = 0.2$ and $Y_{N_2} = 0.775$. For (b) $Y_{CH_4} = 0.05$, $Y_{O_2} = 0.2$ and $Y_{N_2} = 0.75$ is chosen. In both cases the initial temperature is $T_0 = 1000$ K.

As we may observe in Figs. 1 and 2 and Figs. 3 and 4, both codes match well, albeit using two different time stepping schemes. The reaction process over time is well described. Only a small delay in temperature increase may be observed. It should be noted that, without any limitations, also CANTERA can be directly used for below model reduction process. However, since a goal of our future work is to introduce an adjoint based model reduction approach, we implemented our own solver. It has been shown in [11,12], that incorporating adjoint equations offers several further options within the framework of reactive compressible flows.

3 Model Reduction

Reduced order models are often determined by a decomposition of a full system into modes. These modes can be derived from different principles such as the proper orthogonal decomposition (POD) [9,10,22], balanced truncation [16,8] or dynamic mode decomposition (DMD) [20,21]. Since the model reduction techniques, dedicated only to the spatial modes, are computationally still expensive (the reduced system dynamics have to be solved), they are inappropriate for reducing huge systems, e.g. direct numerical simulations of reactive flows containing shocks.

Therefore we focus on interpolation based methods building on modes of the full spatial-temporal space, where the system dynamics are captured by the interpolation. We investigate the combustion in a homogeneous (0D) reactor described by (1a)–(1b). By this, we restrict ourselves to study only the time

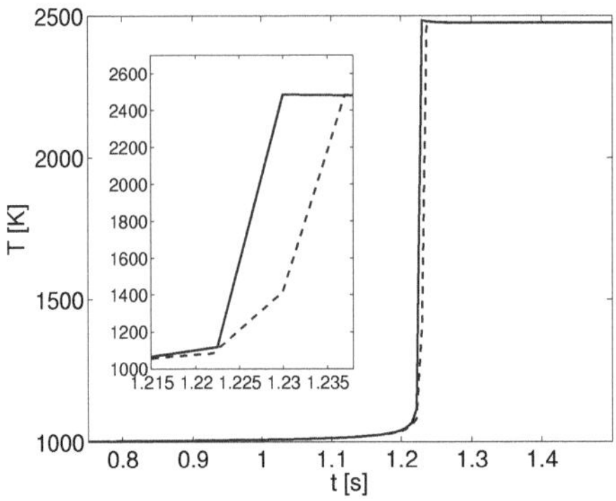

Fig. 3. Case (b): Temperature history of a methane combustion obtained with CANTERA (solid) and our implementation (dashed)

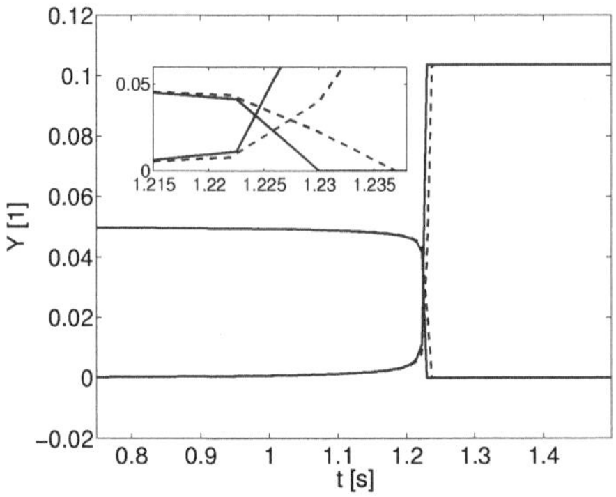

Fig. 4. Case (b): Mass fractions for CH_4 and H_2O obtained with CANTERA (solid) and our implementation (dashed)

evolution. No transport effects are considered. However, a main feature of reactive flows is captured: the sudden inset of the reaction resulting in as steep increase of the temperature. The corresponding time strongly depends on the initial temperature T_0 and therefore, we use this initial temperature as a free parameter in our model reduction.

The task of the MR is to predict the behavior of the system for an arbitrarily chosen initial temperature. Our main goal is to investigate how well the proposed method deals with the suddenly and strongly varying behavior in temperature and mass fractions. This example partly mimics difficulties of more complex processes, like deflagration-to-detonation transition (DDT). The crucial point is, how well the reaction process can be predicted from a few training runs at given (but different) parameters.

In order to set up an interpolatory ROM which can be used to investigate the considered system in dependence of the initial temperature T_0, we simulated our model problem with different values T_0, see Fig. 5.

We applied the method proposed in [3] to observe how a global interpolatory approach can handle the nearly discontinuous behavior. The method includes a strategy to improve the interpolation by neglecting higher modes in a (heuristic) controlled manner and provides a heuristic error estimator. Since the underlying SVD can be also calculated for very large systems, it can be applied to our future target: full complex simulations with billions of points. However, in spite of the initial hopes, the results are not satisfactory, as described in the following Sect. 3.1. As a result, we suggest a modification of the method by enriching it with additional physical information, here provided by interpolated values of the ignition time. This strongly improves the original model reduction approach. The extended framework and the results are presented in Sect. 3.2.

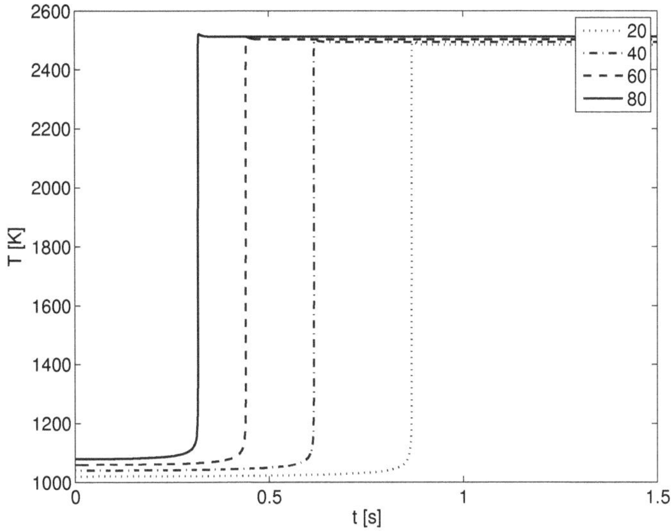

Fig. 5. Temperature T for homogeneous (0D) reactor model for initial temperature $T_0 = 1000K + \Delta T_0$ with $\Delta T_0 = 20, 40, 60, 80$ K

3.1 Basic MR Method: MRa

Following [3] let us define the so-called training matrix F containing the solutions $f(t, (Y_k, T), T_0^s)$ (training runs) of our model problem for S different values of the initial temperature T_0

$$F = \left[f\left([t^n, k], T_0^{(1)} \right), \ldots, f\left([t^n, k], T_0^{(s)} \right) \ldots, f\left([t^n, k], T_0^{(S)} \right) \right]. \tag{15}$$

Therein f corresponds to a column vector of the length $(K+1) \cdot N$ containing the spatial-temporal solution for the initial temperature $T_0^{(s)}$. N corresponds to the number of used time steps. $K+1$ is the number of degrees of freedom for every time step t^n, i.e. the number of species and the temperature. Time and species index form a combined index. This super-index is marked by brackets $[\cdot, \cdot]$.

By computing the skinny (thin) SVD [6, Sect. 2.4.3] we get the following decomposition of F:

$$F = U \Sigma V^T, \qquad \text{with } \Sigma = \begin{bmatrix} \sigma_1 & & \\ & \ddots & \\ & & \sigma_S \end{bmatrix} \tag{16}$$

and

$$U = \left[u_1([t^n, k]), \ldots, u_S([t^n, k])\right] \in \mathbb{C}^{(K+1)\cdot N \times S}, \tag{17}$$

$$V = \begin{bmatrix} v_1\left(T_0^{(1)}\right) & \ldots & v_S\left(T_0^{(1)}\right) \\ \vdots & \ddots & \vdots \\ v_1\left(T_0^{(S)}\right) & \ldots & v_S\left(T_0^{(S)}\right) \end{bmatrix} \in \mathbb{C}^{S \times S} \tag{18}$$

with orthonormal columns. The training runs are decomposed into the modes U with the weights Σ and V. The training matrix can be approximated by the modes corresponding to the largest singular values.

To interpolate a solution for a different arbitrary initial temperature T_0 the weights V instead of the solution can be interpolated. Here, a linear interpolation is used, but also higher interpolation schemes are also possible. Selecting the R largest singular values yields

$$f(t^n, k, T_0) \approx \widetilde{f}(t^n, k, T_0) = \sum_{r=1}^{R} \sigma_r u_r(t^n, k) \widetilde{v}_r(T_0),$$

with

$$\widetilde{v}_r(T_0) = v_r\left(T_0^{(s)}\right) + \frac{v_r\left(T_0^{(s+1)}\right) - v_r\left(T_0^{(s)}\right)}{T_0^{(s+1)} - T_0^{(s)}} \left(T_0 - T_0^{(s)}\right) \tag{19}$$

for $T_0^{(s)} \leq T_0 \leq T_0^{(s+1)}$.

Since the full solution for all time steps is obtained by interpolation, no differential equations have to be solved in addition to capture the system dynamics.

In order to assess the ROM, we compare it with the solution of the homogeneous reactor implementation, introduced in Sect. 2.3, for an initial temperature of $T_0 = 1050$K. We use our ROM with the training set $T_0 = 1020$, 1040, 1060, 1080K, see Fig. 5. As in the validation case (b), the initial mass fraction distribution is given by $Y_{CH_4} = 0.05$, $Y_{O_2} = 0.2$ and $Y_{N_2} = 0.75$. We notice that the obtained ROM *does not* describe the combustion process correctly. Several jumps in temperature, which correspond directly to ignition times of the training set, are visible, see Fig. 6. As for the temperature, also the mass fractions are not correct, see Fig. 7.

As suggested by the authors of [3], reducing the number of right singular vectors should improve the results due to an oscillatory behavior of $v(T_0)$. For the presented examples $R = 3$ is used. Despite finding the expected oscillatory behavior of our modes, reducing their number worsened the results. Consequently the method does not work as initially hoped.

This may be caused by selecting too few training data sets. However, the determination of these is in general very expensive and should be avoided especially for high-dimensional problems. Since the algorithm is unable to handle the occurring strong variations correctly, it does not qualify for our purposes. Fortunately, a slight modification of the method resulted in a much better approach.

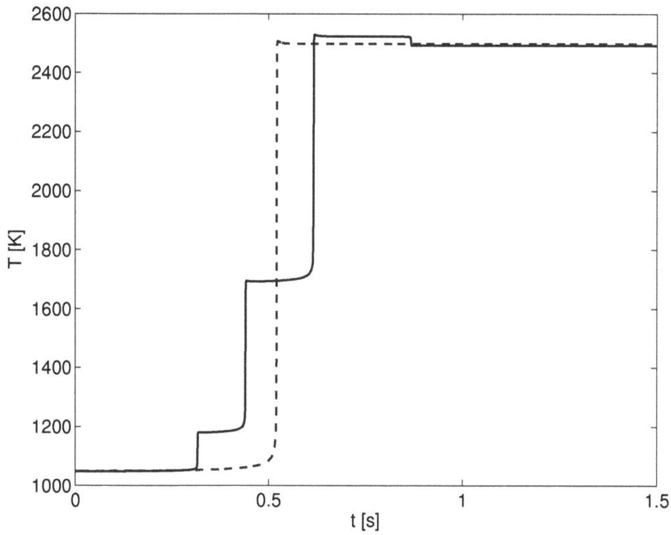

Fig. 6. Temperature T for homogeneous (0D) reactor solution with $T_0 = 1050$ K obtained by the full simulation (dashed) and the ROM(MRa) (solid)

3.2 Modified MR Method: MRb

The main idea to improve the ROM discussed in the previous Sect. is to incorporate additional physical information, such that the discontinuities or sharp gradients, or at least their effects, will be removed within the MR framework.

One may find that the discontinuities in the mentioned case are defined by the strong temperature increase. Since the physical process is independent of the initial time it is possible to change the time frame. Therefore, the values of the solutions $f(t, (Y_k, T), T_0^s)$ are time-shifted, such that their maximum temperature gradients match in time. No interpolation is involved during the shifting procedure, thus the shift accuracy is limited to the resolution of the time grid. Outside the simulated time interval a constant extrapolation is used. To determine the time shift we use an interpolation by cubic polynomials with respect to the varying parameter T_0, see Fig. 8. For larger ranges of T_0 the shift may be approximated by a logarithmic approach. The shifted values are now used as an input training set for the procedure described in the previous Sect. The number of the temperature dependent modes is chosen again as $R = 3$. Now the ignition event is represented properly, see Fig. 9 and Fig. 10. Moreover, even small species fractions, like NO are well described.

So far the presented method can be viewed as an interpolation of a PDE solution using training data in an orthonormal basis obtained by means of a singular value decomposition. A simple interpolation of the solution directly from the (shifted) training data will provide a similar result. But, also the storage requirement will be similar, if the number of modes is not much smaller than the number of training runs, as every column of $u_r(k, t^n) = U([k, t^n], r)$ is as big as a full training solution.

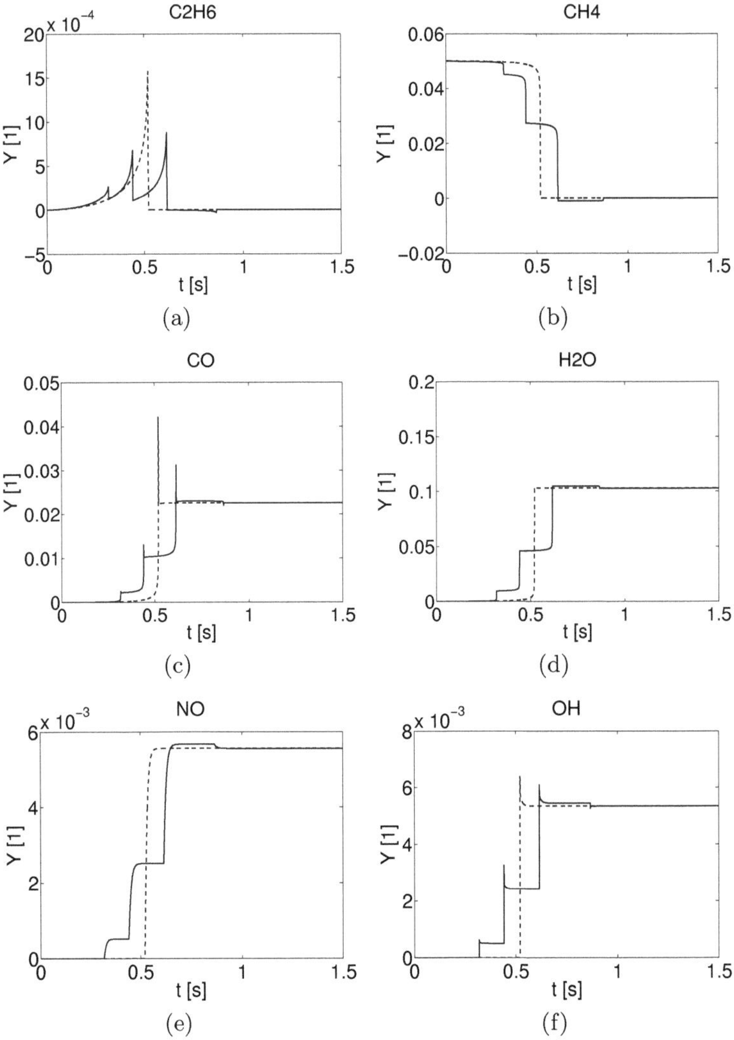

Fig. 7. Mass fractions Y for $C_2H_6, CH_4, CO, H_2O, NO, OH$ obtained by the full simulation (dashed) and the ROM(MRa) (solid)

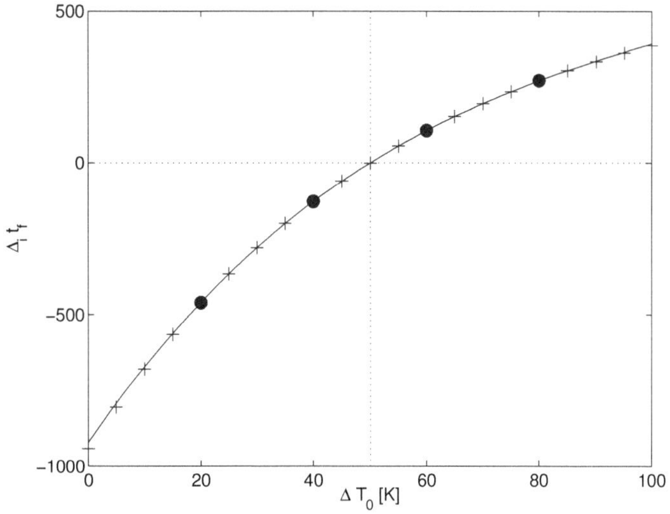

Fig. 8. Shown is the ignition time shift $\Delta_i t_f$ (time index) with respect to the initial temperature deviation ΔT_0. The filled circles correspond to the training set $T_0 = 1020, 1040, 1060, 1080$K and the line to a cubic function based on these values. The dotted line represents the target initial temperature $T_0 = 1050K$. The crosses correspond to benchmark simulations. The shifts are well approximated by a cubic polynomial in the considered range.

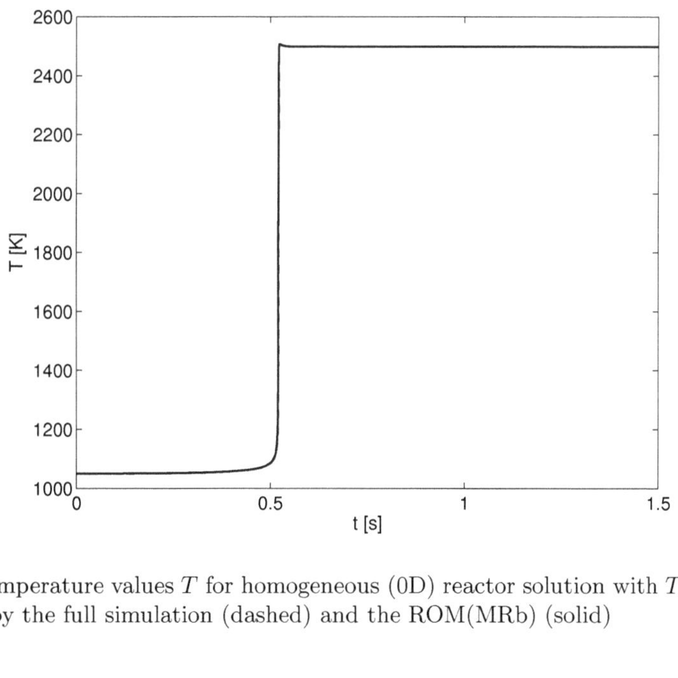

Fig. 9. Temperature values T for homogeneous (0D) reactor solution with $T_0 = 1050$K obtained by the full simulation (dashed) and the ROM(MRb) (solid)

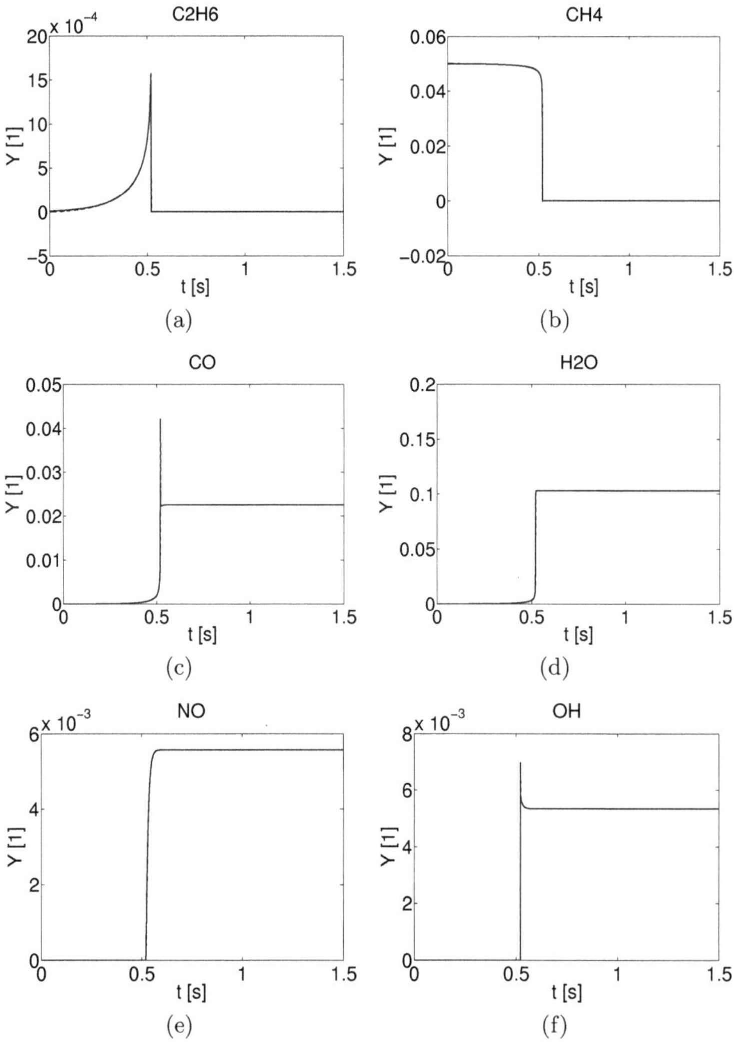

Fig. 10. Mass fractions Y for $C_2H_6, CH_4, CO, H_2O, NO, OH$ obtained by the full simulation (dashed) and the ROM(MRb) (solid)

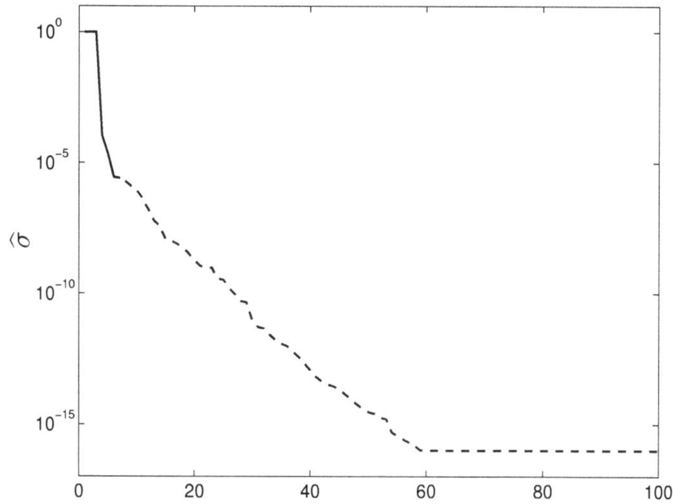

Fig. 11. First 100 singular values $\widehat{\sigma}_q$ defined in (21). For the reconstruction of the modes U the modes $\widehat{u}_1, \ldots, \widehat{u}_7$ are used. The dashed part of the curve corresponds to the neglected singular values.

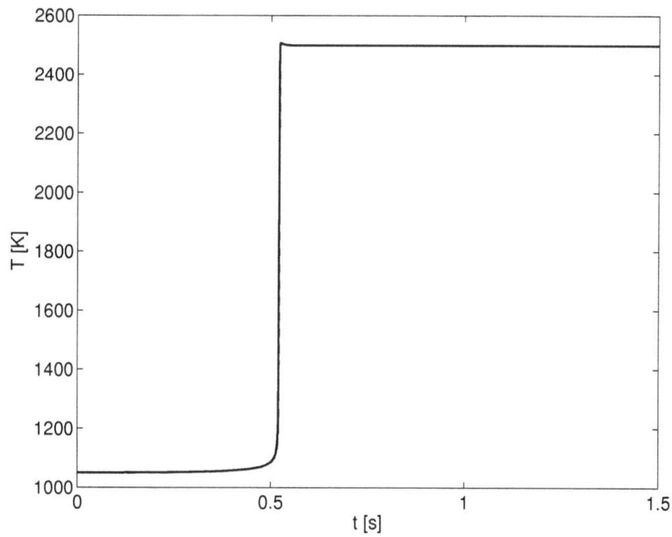

Fig. 12. Temperature values T for homogeneous (0D) reactor solution with $T_0 = 1050\text{K}$ obtained by the full simulation (dashed) and the ROM(MRb + additional decomposition of U) (solid)

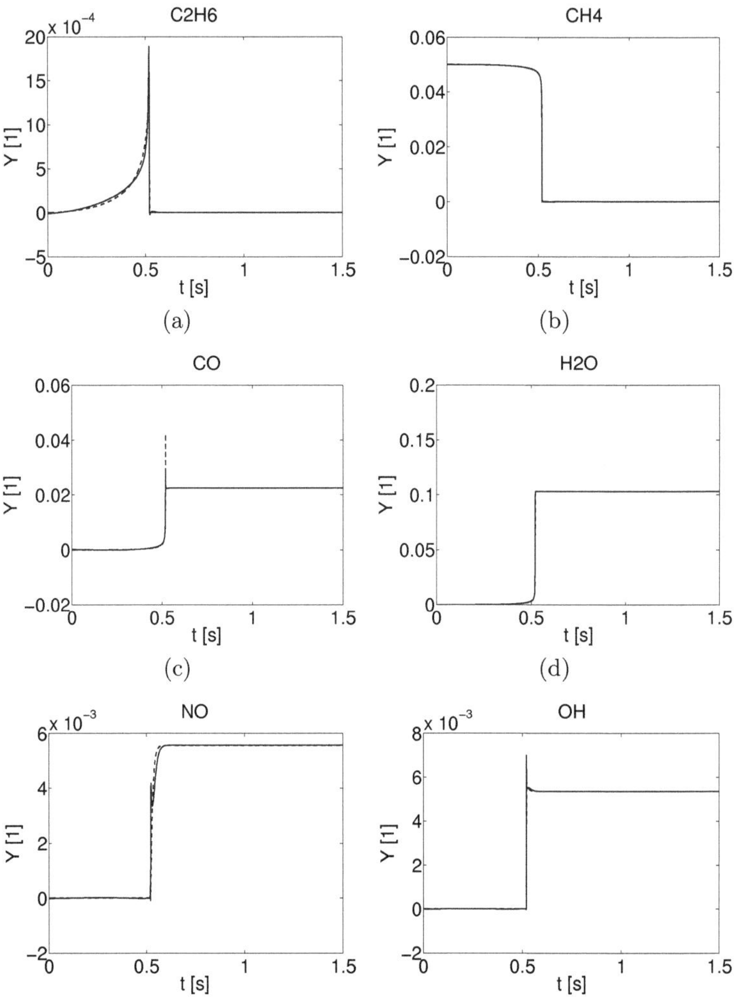

Fig. 13. Mass fractions Y for $C_2H_6, CH_4, CO, H_2O, NO, OH$ obtained by the full simulation (dashed) and the ROM(MRb + additional decomposition of U) (solid)

The advantage of the proposed method is the possibility to reduce its storage requirements. The basic idea is an additional SVD decomposition of the obtained modes U given by $U = \widehat{U}\widehat{\Sigma}\widehat{V}$. Technically the first R modes are reshaped, such that the first index contains only the time

$U(t^n, [k, r]) =$

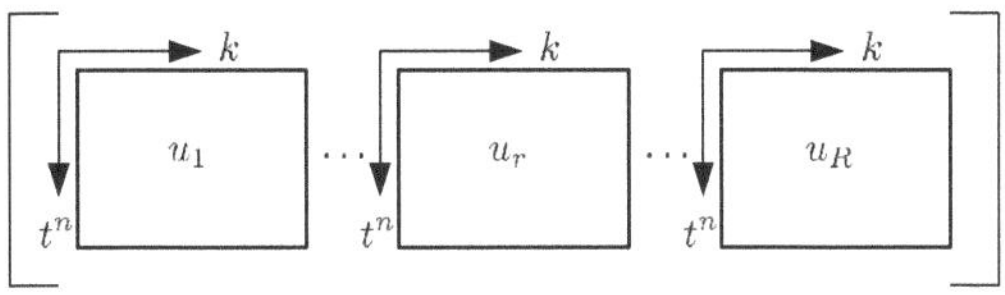

, (20)

and

$$U(t^n, [k, r]) = \sum_{q=1}^{(K+1)R} \widehat{u}_q(t^n)\widehat{\sigma}_q\widehat{v}_q^r(k)^T \tag{21}$$

holds. The distribution of the singular values $\widehat{\sigma}_q$ is shown in Fig. 11.

If only the first $Q \ll (K+1)\,R$ modes corresponding to the largest singular values are taken into account, the full reduction can be summarized as

$$\boxed{f(t^n, k, T_0) \approx \widetilde{f}(t^n, k, T_0) = \sum_{r=1}^{R}\sum_{q=1}^{Q} \sigma_r \widehat{u}_q(t^n)\widehat{\sigma}_q\widehat{v}_q^r(k)\widetilde{v}_r(T_0).} \tag{22}$$

The resulting formula (22) is cheap to evaluate, no dynamics has to be simulated and the storage requirements are strongly reduced compared with a simple interpolation approach.

We found, that the interpolation data set can be reduced, while maintain a suitable interpolation. For the considered case with $R = 3$ the decomposition of U with $Q = 7$ lead to a reduction of the required data to about 14% of *one* training set. The results are shown in Figs. 12 and 13.

4 Conclusions

In this paper, we have extended a well-recognized interpolation based model reduction technique by incorporating additional physical information within the interpolation procedure. Preliminary numerical investigations on a homogeneous (0D) reactor with detailed chemistry allowed to construct a proper ROM. The generalization to more complex systems is non-trivial and subject of current research. Our next aim is to investigate the quality of the method when applied to more complex systems including transport and involving more than one discontinuity.

Acknowledgements. The authors gratefully acknowledge support by the Deutsche Forschungsgemeinschaft (DFG) as part of collaborative research center SFB 1029 "Substantial efficiency increase in gas turbines through direct use of coupled unsteady combustion and flow dynamics".

References

1. Annamalai, K., Puri, I.K.: Combustion Science and Engineering. CRC Press, Taylor & Francis Group, LLC (2007)
2. Burcat, A., McBride, B.: 1994 Ideal Gas Thermodynamic Data for Combustion and Air- Pollution Use. Technion Report TAE 697 (December 1993)
3. Constantine, P.G., Gleich, D.F., Hou, Y., Templeton, J.: Model Reduction with MapReduce-enabled Tall-and-Skinny Singular Value Decomposition. ArXiv e-prints, arXiv:1306.4690 (math.NA) (2013)
4. Gilbert, R.G., Luther, K., Troe, J.: Theory of Thermal Unimolecular Reactions in the Fall-off Range. II. Weak Collision Rate Constants. Berichte der Bunsengesellschaft für physikalische Chemie 87, 169 (1983)
5. Goodwin, D.: Cantera: An object-oriented software toolkit for chemical kinetics, thermodynamics, and transport processes. Caltech, Pasadena (2014), http://code.google.com/p/cantera
6. Golub, G.H., Van Loan, C.F.: Matrix Computations, 4th edn. Johns Hopkins University Press, Baltimore (2013)
7. Smith, G.P., Golden, D.M., Frenklach, M., Moriarty, N.W., Eiteneer, B., Goldenberg, M., Bowman, C.T., Hanson, R.K., Song, S., Gardiner Jr., W.C., Lissianski, V.V., Qin, Z. (2014), http://www.me.berkeley.edu/gri_mech/
8. Gugercin, S., Antoulas, A.C.: A survey of model reduction by balanced truncation and some new results. Internat. J. Control 77(8), 748–766 (2004)
9. Kunisch, K., Volkwein, S.: Control of the Burgers Equation by a Reduced-Order Approach Using Proper Orthogonal Decomposition. Journal of Optimization Theory and Applications 102, 345–371 (1999)
10. Kunisch, K., Volkwein, S.: Galerkin proper orthogonal decomposition methods for parabolic problems. Numerische Mathematik 90, 117–148 (2001)
11. Lemke, M., Reiss, J., Sesterhenn, J.: Adjoint-based analysis of thermoacoustic coupling. In: AIP Conference Proceedings, vol. 1558, pp. 2163–2166 (2013)
12. Lemke, M., Reiss, J., Sesterhenn, J.: Adjoint based optimisation of reactive compressible flows. Combustion and Flame (2014), http://dx.doi.org/10.1016/j.combustflame.2014.03.020
13. Liberman, M.A., Ivanov, M.F., Kiverin, A.D., Kuznetsov, M.S., Chukalovsky, A.A., Rakhimova, T.V.: Deflagration-to-detonation transition in highly reactive combustible mixtures. Acta Astronautica 67, 688–701 (2010)
14. Lindemann, A.: Trans. Faraday Soc. 17, 598 (1922)
15. McBride, B.J., Gordon, S., Reno, M.A.: Coefficients for Calculating Thermodynamic and Transport Properties of Individual Species. NASA Report TM-4513 (October 1993)
16. Mullis, C.T., Roberts, R.A.: Synthesis of minimum roundoff noise fixed point digital filters. IEEE Transactions on Circuits and Systems, CAS-23, 551–562 (1976)
17. Oran, E.S., Boris, J.P.: Numerical simulation of reactive flow. Cambridge University Press (2005)
18. Poinsot, T., Veynante, D.: Theoretical and Numerical Combustion, 3rd edn. (2012)
19. Chemical-Kinetic Mechanisms for Combustion Applications. San Diego Mechanism web page, Mechanical and Aerospace Engineering Combustion Research). University of California at San Diego, http://combustion.ucsd.edu

20. Schmid, P.J., Sesterhenn, J.L.: Decomposition. Mode Decomposition of numerical and experimental data. Bull. Amer. Phys. Soc., San Antonio/TX (2008)
21. Schmid, P.J.: Dynamic mode decomposition of numerical and experimental data. J. Fluid Mech., 256, 5–28 (2010)
22. Volkwein, S.: Model reduction using proper orthogonal decomposition. Lecture notes, Univeristy Konstanz (2011), `http://www.math.uni-konstanz.de/numerik/personen/volkwein/teaching/scripts.php`
23. Warnatz, J., Maas, U., Dibble, R.W.: Verbrennung, vol. 2. Springer, Berlin (1997)

Author Index

Abel, Dirk 181
Albin, Thivaharan 181
An, Xuanhong 55

Bengoechea, Sergio 343
Ben-Harav, Amos 71
Benner, Peter 365
Berndt, Phillip 281
Bobusch, Bernhard C. 281
Borggaard, Jeffrey T. 381
Burnazzi, Marco 101

Cai, Liming 299
Carter, Campbell 247
Chen, Kevin K. 115
Chughtai, Saulat S. 131

da Franca, Aline Aguiar 181

Eshel, Ben 71

Gray, Joshua A.T. 265
Greenblatt, David 71
Grimaud, Lou 55
Gugercin, Serkan 381

Haw, Willie 247
Heiland, Jan 365

Jozefik, Zoltan 197

Kerstein, Alan R. 197
King, Rudibert 39, 229
Klein, Rupert 281
Kruse, Stephan 181

Lautman, Ronen 71
Lehrheuer, Bastian 163
Lemke, Mathias 397
Liberman, Michael A. 317

Mack, Marion 3
Mehrmann, Volker 397
Międlar, Agnieszka 397
Moeck, Jonas P. 265
Morcinkowski, Bastian 163
Müller-Vahl, Hanns 71

Niehuis, Reinhard 3
Nijs, Martin 163
Nitsche, Wolfgang 23, 39

Oevermann, Michael 213
Ombrello, Timothy 247

Paschereit, Christian O. 213, 265, 281
Peltzer, Inken 23
Pischinger, Stefan 163
Pitsch, Heinz 181, 299

Radespiel, Rolf 101
Reiss, Julius 343, 397
Rowley, Clarence W. 115

Schmidt, Heiko 197
Schrödinger, Christina 213
Sesterhenn, Jörn 147, 343, 397
Staats, Marcel 23, 39
Stein, Lewin 343
Steinberg, Simon J. 39

Tam, Chung-Jen 247

Varea, Emilien 181

Werner, Herbert 131
Wild, Jochen 87
Wilke, Robert 147
Williams, David R. 55
Wolff, Sascha 229

MIX
Papier aus verantwortungsvollen Quellen
Paper from responsible sources
FSC® C105338

If you have any concerns about our products,
you can contact us on
ProductSafety@springernature.com

In case Publisher is established outside the EU,
the EU authorized representative is:
Springer Nature Customer Service Center GmbH
Europaplatz 3, 69115 Heidelberg, Germany

Printed by Libri Plureos GmbH
in Hamburg, Germany